本书由广州再生医学与健康广东省实验室资助

HUMAN GENOME EDITING
Science, Ethics, and Governance

人类基因组编辑
科学·伦理·管理
（中英对照）

"人类基因编辑：科学、医学和伦理决策"委员会
美国科学院研究理事会　编

裴端卿 等 译

科学出版社
北 京

图字：01-2019-2325号

内 容 简 介

基因组编辑是一个强有力的新工具，可以精确地改变生物体的遗传物质。最近的科学进展使基因组编辑比以往任何时候都更有效率、更精确和更灵活。这些进展促使全球范围内人们对通过基因组编辑改善人类健康的方式产生了兴趣。这些技术的发展和应用引起许多决策者和利益相关方的关切，如是否有适当的制度来管理这些技术，以及公众如何、何时参与这些决策。本书从人类基因组编辑和管理的总体原则、基因编辑技术的基础研究应用、体细胞基因组编辑、可遗传性基因组编辑、基因强化、公众参与等方面进行了阐述。

本书可供从事基因组编辑相关研究的科研人员及企业人员参考使用。

This is a translation of *Human Genome Editing: Science, Ethics, and Governance*, National Academy of Sciences; National Academy of Medicine; National Academies of Sciences, Engineering, and Medicine; Committee on Human Gene Editing: Scientific, Medical, and Ethical Considerations ©2017 National Academy of Sciences. First Published in English by National Academies Press. All rights reserved.

图书在版编目（CIP）数据

人类基因组编辑：科学·伦理·管理：汉英对照 / 美国"人类基因编辑：科学、医学和伦理决策"委员会，美国科学院研究理事会编；裴端卿等译. —北京：科学出版社，2019.6

书名原文：Human Genome Editing: Science, Ethics, and Governance

ISBN 978-7-03-061360-8

Ⅰ. ①人⋯ Ⅱ. ①美⋯ ②美⋯ ③裴⋯ Ⅲ. ①人类基因－基因组－研究－汉、英 Ⅳ. ①Q987

中国版本图书馆CIP数据核字（2019）第103655号

责任编辑：王 静 罗 静 刘 晶 / 责任校对：郑金红
责任印制：张 伟 / 封面设计：刘新新

科学出版社出版
北京东黄城根北街16号
邮政编码：100717
http://www.sciencep.com

北京虎彩文化传播有限公司 印刷
科学出版社发行 各地新华书店经销

*

2019年6月第 一 版　开本：889×1194 1/16
2019年9月第二次印刷　印张：19 1/2
字数：560 000

定价：180.00元

（如有印装质量问题，我社负责调换）

《新生物学丛书》专家委员会

主　任：蒲慕明

副主任：吴家睿

专家委员会成员（按姓氏汉语拼音排序）

昌增益	陈洛南	陈晔光	邓兴旺	高　福
韩忠朝	贺福初	黄大昉	蒋华良	金　力
康　乐	李家洋	林其谁	马克平	孟安明
裴　钢	饶　毅	饶子和	施一公	舒红兵
王　琛	王梅祥	王小宁	吴仲义	徐安龙
许智宏	薛红卫	詹启敏	张先恩	赵国屏
赵立平	钟　扬	周　琪	周忠和	朱　祯

译校者名单
List of Translators

（按姓氏笔画排序）

王 飞	王 波	王付卉	王学聪	石 曦	匡俊企	邢 琦
朱洁滢	朱艳玲	朱菲艳	刘 鹤	孙 昊	孙 薇	苏整会
李 婷	李长朋	李宇航	李林鹏	杨 肖	吴 芳	吴琳琳
岑晓彤	何江平	何松蔚	余致君	张 田	张一心	张炳文
张祖明	张梦丹	张燕琪	陈金龙	陈聪玲	林立龙	周纯华
郑 辉	单永礼	赵 圆	钟肖芬	侯红明	姚 姣	徐慧娟
高铭蔚	郭 婧	郭 琳	郭宜平	黄 可	曹 凡	龚举成
商必志	梁丽宁	梁泽川	梁锦川	喻 佩	裴端卿	樊琛语

《新生物学丛书》丛书序

当前,一场新的生物学革命正在展开。为此,美国国家科学院研究理事会于 2009 年发布了一份战略研究报告,提出一个"新生物学"(New Biology)时代即将来临。这个"新生物学",一方面是生物学内部各种分支学科的重组与融合,另一方面是化学、物理、信息科学、材料科学等众多非生命学科与生物学的紧密交叉与整合。

在这样一个全球生命科学发展变革的时代,我国的生命科学研究也正在高速发展,并进入了一个充满机遇和挑战的黄金期。在这个时期,将会产生许多具有影响力、推动力的科研成果。因此,有必要通过系统性集成和出版相关主题的国内外优秀图书,为后人留下一笔宝贵的"新生物学"时代精神财富。

科学出版社联合国内一批有志于推进生命科学发展的专家与学者,联合打造了一个 21 世纪中国生命科学的传播平台——《新生物学丛书》。希望通过这套丛书的出版,记录生命科学的进步,传递对生物技术发展的梦想。

《新生物学丛书》下设三个子系列:科学风向标,着重收集科学发展战略和态势分析报告,为科学管理者和科研人员展示科学的最新动向;科学百家园,重点收录国内外专家与学者的科研专著,为专业工作者提供新思想和新方法;科学新视窗,主要发表高级科普著作,为不同领域的研究人员和科学爱好者普及生命科学的前沿知识。

如果说科学出版社是一个"支点",这套丛书就像一根"杠杆",那么读者就能够借助这根"杠杆"成为撬动"地球"的人。编委会相信,不同类型的读者都能够从这套丛书中得到新的知识信息,获得思考与启迪。

《新生物学丛书》专家委员会
主 任:蒲慕明
副主任:吴家睿
2012 年 3 月

美国国家科学院、工程院和医学院

美国国家科学院是在 1863 年根据时任美国总统林肯签署的一项国会法案成立的，这是一个私营的非政府机构，旨在就科学和技术问题向美国政府提供建议。成员由对研究做出突出贡献的同行选出，由 Marcia K. McNutt 博士担任院长。

美国国家工程院是在 1964 年根据美国国家科学院的章程成立的，旨在将工程实践引入国家建设。成员由对工程做出特别贡献的同行选举产生，由 C. D. Mote Jr. 博士担任院长。

美国国家医学院（原医学研究院）是在 1970 年根据美国国家科学院的章程成立的，旨在就医疗和健康问题为美国政府提供建议。成员由为医学和健康做出杰出贡献的同行选举产生，由 Victor J. Dzau 博士担任院长。

这三个研究院合称为美国国家科学院、工程院和医学院，为美国政府提供独立、客观的分析和建议，并开展其他活动来解决复杂的问题，为公共政策决策提供信息。美国国家科学院还鼓励教育和研究，确认对知识的杰出贡献，并增进公众对科学、工程和医学的了解。

有关美国国家科学院、工程院和医学院的更多信息，请访问 www.national-academies.org。

The National Academies of
SCIENCES • ENGINEERING • MEDICINE

The **National Academy of Sciences** was established in 1863 by an Act of Congress, signed by President Lincoln, as a private, nongovernmental institution to advise the nation on issues related to science and technology. Members are elected by their peers for outstanding contributions to research. Dr. Marcia K. McNutt is president.

The **National Academy of Engineering** was established in 1964 under the charter of the National Academy of Sciences to bring the practices of engineering to advising the nation. Members are elected by their peers for extraordinary contributions to engineering. Dr. C. D. Mote, Jr., is president.

The **National Academy of Medicine** (formerly the Institute of Medicine) was established in 1970 under the charter of the National Academy of Sciences to advise the nation on medical and health issues. Members are elected by their peers for distinguished contributions to medicine and health. Dr. Victor J. Dzau is president.

The three Academies work together as the **National Academies of Sciences, Engineering, and Medicine** to provide independent, objective analysis and advice to the nation and conduct other activities to solve complex problems and inform public policy decisions. The National Academies also encourage education and research, recognize outstanding contributions to knowledge, and increase public understanding in matters of science, engineering, and medicine.

Learn more about the National Academies of Sciences, Engineering, and Medicine at **www.national-academies.org**.

美国国家科学院、工程院和医学院

　　本报告记录了专家创作委员会基于证据的共识，包括根据委员会和委员会审议收集的信息得出的发现、结论和建议。报告由同行评审，并由美国国家科学院、工程院和医学院批准。

　　会议记录了在研讨会、专题讨论会或其他会议上的陈述和讨论。会议记录中的陈述和意见是与会者的陈述和意见，没有得到其他与会者、计划委员会，或者国家科学院、工程院和医学院的认可。

　　有关美国国家科学院其他产品和活动的信息，请访问 nationalacademies.org/whatwedo。

The National Academies of
SCIENCES • ENGINEERING • MEDICINE

Reports document the evidence-based consensus of an authoring committee of experts. Reports typically include findings, conclusions, and recommendations based on information gathered by the committee and committee deliberations. Reports are peer reviewed and are approved by the National Academies of Sciences, Engineering, and Medicine.

Proceedings chronicle the presentations and discussions at a workshop, symposium, or other convening event. The statements and opinions contained in proceedings are those of the participants and have not been endorsed by other participants, the planning committee, or the National Academies of Sciences, Engineering, and Medicine.

For information about other products and activities of the National Academies, please visit nationalacademies.org/whatwedo.

"人类基因编辑：科学、医学和伦理决策"委员会
COMMITTEE ON HUMAN GENE EDITING:
SCIENTIFIC, MEDICAL, AND ETHICAL CONSIDERATIONS

R. ALTA CHARO (*Co-Chair*), Sheldon B. Lubar Distinguished Chair and Warren P. Knowles Professor of Law & Bioethics, University of Wisconsin–Madison

RICHARD O. HYNES (*Co-Chair*), Investigator, Howard Hughes Medical Institute, Daniel K. Ludwig Professor for Cancer Research, Massachusetts Institute of Technology

DAVID W. BEIER, Managing Director, Bay City Capital

ELLEN WRIGHT CLAYTON, Craig Weaver Professor of Pediatrics, Professor of Law, Vanderbilt University

BARRY S. COLLER, David Rockefeller Professor of Medicine, Physician in Chief, and Head, Allen and Frances Adler Laboratory of Blood and Vascular Biology, Rockefeller University

JOHN H. EVANS, Professor, University of California, San Diego

JUAN CARLOS IZPISUA BELMONTE, Professor, Gene Expression Laboratory, Salk Institute for Biological Studies

RUDOLF JAENISCH, Professor of Biology, Massachusetts Institute of Technology

JEFFREY KAHN, Andreas C. Dracopoulos Director, Johns Hopkins Berman Institute of Bioethics, Johns Hopkins University

EPHRAT LEVY-LAHAD, Director, Fuld Family Department of Medical Genetics, Shaare Zedek Medical Center; Faculty of Medicine, Hebrew University of Jerusalem

ROBIN LOVELL-BADGE, Senior Group Leader, Laboratory of Stem Cell Biology and Developmental Genetics, The Francis Crick Institute

GARY MARCHANT, Regents' Professor of Law, Arizona State University **JENNIFER MERCHANT**, University Professor, Université de Paris II (Panthéon-Assas)

LUIGI NALDINI, Professor of Cell and Tissue Biology and of Gene and Cell Therapy, San Raffaele University, and Director of the San Raffaele Telethon Institute for Gene Therapy

DUANQING PEI, Professor and Director General of Guangzhou Institutes of Biomedicine and Health, Chinese Academy of Sciences

MATTHEW PORTEUS, Associate Professor of Pediatrics, Stanford School of Medicine **JANET ROSSANT**, Senior Scientist and Chief of Research Emeritus, Hospital for Sick Children,

University of Toronto

DIETRAM A. SCHEUFELE, John E. Ross Professor in Science Communication and Vilas Distinguished Achievement Professor, University of Wisconsin–Madison

ISMAIL SERAGELDIN, Founding Director, Bibliotheca Alexandrina **SHARON TERRY**, President & CEO, Genetic Alliance

JONATHAN WEISSMAN, Professor, Department of Cellular and Molecular Pharmacology, University of California, San Francisco

KEITH R. YAMAMOTO, Vice Chancellor for Science Policy and Strategy, University of California, San Francisco

研究人员(*Study Staff*)

KATHERINE W. BOWMAN, Study Director

MONICA L. GONZALEZ, Associate Program Officer

JOANNA R. ROBERTS, Senior Program Assistant

ANDREW M. POPE, Director, Board on Health Sciences Policy

FRANCES E. SHARPLES, Director, Board on Life Sciences

顾问(*Consultants*)

RONA BRIERE, Editor

HELAINE RESNICK, Editor

题　　词

　　Ralph Cicerone 博士（1943—2016）于 2015 年担任美国国家科学院院长，当时他与美国国家医学院院长合作宣布了一项关于人类基因组编辑的倡议，该倡议包括科学、伦理和管理。他指出，美国国家科学院、工程院和医学院将努力引导，为许多新兴和有争议的遗传学和细胞生物学领域制定负责任的综合政策，如人类胚胎干细胞研究、人类克隆和"功能获得"研究。最值得注意的是它包含了 1975 年阿西洛马尔会议之前的关键事件。但 Cicerone 博士在接受《自然》杂志采访时说，阿西洛马尔时代和今天有着重要的区别，因为 1975 年很少有研究人员从事重组 DNA 的研究，而现代的基因组编辑技术更易于使用，并且能被广泛使用，这使他得出结论，这种情况需要一种"比阿西洛马尔更国际化的方案"。

　　Cicerone 博士言行一致。他与来自中国和英国的科学院和医学院合作，发起了一项与国际首脑会议有关的倡议。由此，承诺今后将举行首脑会议，并成立一个研究委员会，成员来自加拿大、中国、埃及、法国、德国、以色列、意大利、西班牙、英国和美国，或在这些国家工作。本报告是该委员会工作的成就，也是对这位伟大的美国国家科学院院长的致敬。

Dedication

Dr. Ralph Cicerone (1943-2016) was President of the National Academy of Sciences in 2015 when, in partnership with the President of the National Academy of Medicine, he announced a human genome-editing initiative that would encompass science, ethics, and regulation. He noted that the National Academies of Sciences, Engineering, and Medicine have led the effort to develop responsible, comprehensive policies for many emerging and controversial areas of genetics and cell biology, such as human embryonic stem cell research, human cloning, and "gain-of-function" research. Most notable was its involvement in key events leading up to the 1975 Asilomar conference. But there are important differences between the Asilomar era and today, Dr. Cicerone said in an interview with *Nature*, because few researchers were pursuing recombinant DNA research in 1975. Modern genome-editing techniques are easy to use and widely accessible, leading him to conclude that the situation requires an approach that is "really more international than Asilomar ever had to be."

Dr. Cicerone was as good as his word. In collaboration with science and medicine academies from China and the United Kingdom, an initiative was launched with an international summit. From this came a commitment to future summits, and the formation of a study committee, with members hailing from or working in Canada, China, Egypt, France, Germany, Israel, Italy, Spain, the United Kingdom, and the United States. This report is the culmination of the work by that committee and is dedicated to this great leader of the National Academy of Sciences.

致　　谢

本报告由选取具有不同观点和技术专长的个人对草稿进行审查。本次独立审查的目的是提供坦诚和批评性的评论,以帮助本机构尽可能完善已公布的报告,并确保报告符合机构标准和对研究费用的相关制度标准。为保护审议程序的完整性,审查意见和稿件草稿保密。我们要感谢以下各人对本报告的审核：

Eli Adashi,布朗大学

George Annas,波士顿大学

Dana Carroll,犹他大学

Michael Dahlstrom,爱荷华州立大学

Hank Greely,斯坦福大学

J. Benjamin Hurlbut,亚利桑那州立大学

Maria Jasin,纪念斯隆-凯特林癌症中心

James Lawford-Davies,亨普森律师事务所,英国

Andrew Maynard,亚利桑那州立大学

Krishanu Saha,威斯康星大学

Fyodor Urnov,阿尔蒂乌斯研究所

Keith Wailoo,普林斯顿大学

尽管上述审查人员提供了许多建设性意见和建议,但并不要求他们认可这些结论或建议,他们也没有看到报告发布前的最终版本。该报告的审查由 Harvey Fineberg(穆尔基金会)和 Jonathan Moreno(宾夕法尼亚大学)负责。他们负责确保按照机构程序对本报告进行独立审查,并仔细考虑所有审查意见。本报告的最终内容完全由编写委员会和机构负责。

Reviewers

This report was reviewed in draft form by individuals chosen for their diverse perspectives and technical expertise. The purpose of this independent review is to provide candid and critical comments that will assist the institution in making its published report as sound as possible and to ensure that the report meets institutional standards for objectivity, evidence, and responsiveness to the study charge. The review comments and draft manuscript remain confidential to protect the integrity of the deliberative process. We wish to thank the following individuals for their review of this report:

Eli Adashi, Brown University
George Annas, Boston University
Dana Carroll, University of Utah
Michael Dahlstrom, Iowa State University
Hank Greely, Stanford University
J. Benjamin Hurlbut, Arizona State University
Maria Jasin, Memorial Sloan Kettering Cancer Center
James Lawford-Davies, Hempsons Law Firm, United Kingdom
Andrew Maynard, Arizona State University
Krishanu Saha, University of Wisconsin-Madison
Fyodor Urnov, Altius Insitute
Keith Wailoo, Princeton University

Although the reviewers listed above provided many constructive comments and suggestions, they were not asked to endorse the conclusions or recommendations, nor did they see the final draft of the report before its release. The review of this report was overseen by **Harvey Fineberg** (Moore Foundation) and **Jonathan Moreno** (University of Pennsylvania). They were responsible for making certain that an independent examination of this report was carried out in accordance with institutional procedures and that all review comments were carefully considered. Responsibility for the final content of this report rests entirely with the authoring committee and the institution.

序　言

　　基因组编辑是一套比以往更准确、更灵活地改变 DNA 的方法，曾被《自然 - 方法》杂志誉为 2011 年年度方法，CRISPR/CAS9 基因组编辑系统被《科学》杂志评为 2015 年年度突破。由于它可以用于洞察基本的生物过程，以及为人类健康可能带来的进步，这项技术在全球范围内引起了人们的兴趣。但是随着研究的进步，出现了许多问题，包括在避免不必要的影响的同时达到预期效果的技术方面，以及一系列的用途，这些用途不仅可以治愈病人，而且还可以预防我们自己和后代的疾病，甚至改变与健康需求无关的特征。现在是时候考虑这些问题了。使用编辑过的人体细胞进行的临床试验已经在进行中，并且已经有了更多的预期。为了有助于应用基因组编辑以广泛促进人类福祉，有必要审查其提出的科学、伦理和社会问题，并评估管理系统确保其负责任的发展和使用。这样做还需要阐明一些更崇高的原则，这些原则应该是这些系统的基础。

　　这些并非易事，但我们非常感谢加入我们的委员会成员。他们愿意并且深思熟虑地将他们不同的观点带到我们的讨论中，我们感谢他们对这项研究的承诺，以及他们在过去一年中投入了如此多的时间和精力。和他们一起工作是一种乐趣和荣幸。本报告还得到了与会发言者的许多介绍和讨论，这些发言者的贡献提供了丰富的信息和见解。我们感谢他们与我们分享他们的研究和观点。最后，我们代表委员会感谢美国国家科学院、工程院和医学院的工作人员，他们在整个研究过程中与我们一起工作，他们的想法和支持对于项目的成功至关重要；同时感谢该研究的发起人，他们对该研究的潜力有着广阔的视野。

R. Alta Charo 和 Richard O. Hynes

"人类基因编辑：科学、医学和伦理决策"委员会的共同主持人

Preface

Genome editing—a suite of methods for creating changes in DNA more accurately and flexibly than previous approaches—was hailed as the 2011 Method of the Year by *Nature Methods* , and the CRISPR/Cas9 system of genome editing was named the 2015 Breakthrough of the Year by *Science*. The technology has excited interest across the globe because of the insights it may offer into fundamental biological processes and the advances it may bring to human health. But with these advances come many questions, about the technical aspects of achieving desired results while avoiding unwanted effects, and about a range of uses that may include not only healing the sick, but also preventing disease in this and future generations, or even altering traits unrelated to health needs. Now is the time to consider these questions. Clinical trials using edited human somatic cells are already underway, and more are anticipated. To help direct the use of genome editing toward broadly promoting human well being, it is important to examine the scientific, ethical, and social issues it raises, and assess the capacity of governance systems to ensure the technologies' responsible development and use. Doing so also entails articulating the larger principles that should underlie such systems.

These were not easy tasks, but we are profoundly grateful to the committee members who joined us in tackling our charge. They willingly and thoughtfully brought their diverse perspectives to bear on our discussions, and we thank them for their commitment to this study and for devoting so much of their time and energy over the past year. It has been a pleasure and a privilege to work with them. The report was also informed by many presentations and discussions with speakers whose contributions provided a wealth of information and insight. We thank them for sharing their research and viewpoints with us. Finally, on behalf of the committee, we would like to thank the staff of the National Academies of Sciences, Engineering, and Medicine for working alongside us throughout the study—their ideas and support were crucial to bringing the project to fruition—and thank the sponsors of the study, who had an expansive vision for its potential.

R. Alta Charo and Richard O. Hynes, *Co-Chairs*
Committee on Human Gene Editing: Scientific, Medical, and Ethical Considerations

目 录

总论 ·· 1
 基因组编辑应用和政策问题概述 ··· 2
 人类基因组编辑的应用 ··· 4
 监管人类基因组编辑的原则 ··· 10
 建议 ·· 12

1 概述 ·· 14
 研究现况 ··· 16
 研究背景 ··· 19
 研究方法 ··· 25
 报告概要 ··· 25

2 人类基因组编辑监管的总体原则 ·· 27
 人类基因组编辑的监管原则 ··· 27
 美国基因治疗技术的监管 ··· 32
 其他国家的监管细节 ··· 53
 总结和建议 ··· 53

3 基因组编辑技术的基础研究应用 ·· 57
 基因组编辑基本方法 ··· 57
 基因组编辑技术的飞速发展 ··· 63
 人类细胞和组织基础实验研究进展 ··· 65
 哺乳动物繁殖与发育基础实验研究进展 ·· 66
 基础研究中的伦理和监管问题 ·· 75
 总结和建议 ··· 76

4 体细胞基因组编辑 ··· 78
 研究背景 ··· 79
 基因编辑较传统基因治疗和早期方案的优势 ··· 81
 基于核酸酶的同源与非同源重组编辑修复技术 ······································· 85
 人类体细胞基因编辑技术的潜在应用 ·· 86
 基因编辑的设计和应用涉及的科技考量 ··· 88
 体细胞基因编辑产生的伦理和监管问题 ··· 97
 总结和建议 ··· 103

5 可遗传性基因组编辑 · · · · · · 105
- 潜在应用方向和替代方法 · · · · · · 107
- 相关的科学和技术问题 · · · · · · 110
- 以纠正引发疾病性状为目的的种系编辑带来的伦理监管问题 · · · · · · 112
- 规则制定 · · · · · · 123
- 总结和建议 · · · · · · 126

6 基因强化 · · · · · · 130
- 人类遗传多样性及"正常"和"自然"的定义 · · · · · · 131
- 了解公众对基因强化的态度 · · · · · · 132
- 如何区分基因治疗和基因强化 · · · · · · 137
- 种系（遗传）基因组的编辑和强化 · · · · · · 147
- 总结和建议 · · · · · · 154

7 公众参与 · · · · · · 155
- 公众参与：广义概念 · · · · · · 157
- 美国的实践 · · · · · · 160
- 国际上的实践 · · · · · · 164
- 公众参与性研讨活动的经验 · · · · · · 166
- 更进一步 · · · · · · 167
- 总结和建议 · · · · · · 168

8 原则和建议概要 · · · · · · 171
- 人类基因组编辑监管原则 · · · · · · 171
- 美国监督人类基因组编辑的现有机制 · · · · · · 173

参考文献 · · · · · · 183

附录 A 基因组编辑的基础科学 · · · · · · 199
附录 B 国际研究监管法规 · · · · · · 236
附录 C 资料来源及方法 · · · · · · 246
附录 D 组委会成员简介 · · · · · · 251
附录 E 术语表 · · · · · · 264

Contents

Summary 1
    OVERVIEW OF GENOME-EDITING APPLICATIONS AND POLICY ISSUES 2
    APPLICATIONS OF HUMAN GENOME EDITING 4
    PRINCIPLES TO GUIDE THE GOVERNANCE OF HUMAN GENOME EDITING 10
    RECOMMENDATIONS 12

1 Introduction 14
    STUDY CONTEXT 16
    BACKGROUND 19
    STUDY APPROACH 25
    ORGANIZATION OF THE REPORT 26

2 Oversight of Human Genome Editing and Overarching Principles for Governance 27
    PRINCIPLES FOR GOVERNANCE OF HUMAN GENOME EDITING 27
    REGULATION OF GENE THERAPY IN THE UNITED STATES 32
    GOVERNANCE IN OTHER NATIONS 55
    CONCLUSIONS AND RECOMMENDATION 56

3 Basic Research Using Genome Editing 57
    THE BASIC TOOLS OF GENOME EDITING 57
    RAPID ADVANCES IN GENOME-EDITING TECHNOLOGY 63
    BASIC LABORATORY RESEARCH TO ADVANCE UNDERSTANDING OF HUMAN
        CELLS AND TISSUES 65
    BASIC LABORATORY RESARCH TO ADVANCE UNDERSTANDING OF MAMMALIAN
        REPRODUCTION AND DEVELOPMENT 66
    ETHICAL AND REGULATORY ISSUES IN BASIC RESEARCH 76
    CONCLUSIONS AND RECOMMENDATION 77

4 Somatic Genome Editing 78
    BACKGROUND 79
    ADVANTAGES OF GENOME EDITING OVER TRADITIONAL GENE THERAPY AND
        EARLIER APPROACHES 81
    HOMOLOGOUS AND NONHOMOLOGOUS REPAIR METHODS USED FOR
        NUCLEASE-BASED GENOME EDITING 85

  POTENTIAL HUMAN APPLICATIONS OF SOMATIC CELL GENOME EDITING ··············86
  SCIENTIFIC AND TECHNICAL CONSIDERATIONS ASSOCIATED WITH THE DESIGN AND
    APPLICATION OF GENOME-EDITING STRATEGIES ···88
  ETHICAL AND REGULATORY ISSUES POSED BY SOMATIC CELL GENOME EDITING ········97
  CONCLUSIONS AND RECOMMENDATIONS ··103

5 Heritable Genome Editing ···105
  POTENTIAL APPLICATIONS AND ALTERNATIVES ··107
  SCIENTIFIC AND TECHNICAL ISSUES ···110
  ETHICS AND REGULATION OF EDITING THE GERMLINE TO CORRECT
    DISEASE-CAUSING TRAITS ···112
  REGULATION ··123
  CONCLUSIONS AND RECOMMENDATION ··127

6 Enhancement ··130
  HUMAN GENETIC VARIATION AND DEFINING "NORMAL" AND "NATURAL" ············131
  UNDERSTANDING PUBLIC ATTITUDES TOWARD ENHANCEMENT ·····························132
  DRAWING LINES: THERAPY VERSUS ENHANCEMENT···137
  GERMLINE (HERITABLE) GENOME EDITING AND ENHANCEMENT ····························147
  CONCLUSIONS AND RECOMMENDATIONS ··154

7 Public Engagement ··155
  PUBLIC ENGAGEMENT: BROAD CONCEPTS··157
  U.S. PRACTICES ··160
  INTERNATIONAL PRACTICES ···164
  LESSONS LEARNED FROM PUBLIC ENGAGEMENT ACTIVITIES ··································166
  MOVING FORWARD ···168
  CONCLUSIONS AND RECOMMENDATIONS ··169

8 Summary of Principles and Recommendations ···171
  OVERARCHING PRINCIPLES FOR GOVERNANCE OF HUMAN GENOME EDITING ······171
  EXISTING U.S. OVERSIGHT MECHANISMS FOR HUMAN GENOME EDITING ·············173

References ···183

Appendix A The Basic Science of Genome Editing ···199
Appendix B International Research Oversight and Regulations ··236
Appendix C Data Sources and Methods ···246
Appendix D Committee Member Biographies ··251
Appendix E Glossary ···264

总 论[1]
Summary[1]

基因组编辑[2]是一种对"生物的整套遗传物质——基因组"进行精确地插入、敲除和改变的新的强大工具。巨大核酸酶(meganucleases)、锌指核酸酶(zinc finger nucleases, ZNFs)、转录激活因子样效应核酸酶(transcription activator-like effector nucleases, TALENs)及CRISPR/Cas9(clustered regularly interspaced short palindromic repeat)等技术手段的快速发展和应用,使基因组编辑较之以前更加精准、高效、灵活和廉价。这些进步也促使人们尝试将基因组编辑技术应用于基础研究,以及疾病的治疗和预防当中。其中,在疾病治疗和预防方面,基因编辑技术的应用范围可从体细胞到生殖系细胞,即从成体病变器官的功能恢复到预防未来后代的遗传病发生均具有可行性。

与其他医学进步一样,每一种新技术的应用随之而来的都有它特定的利益、风险、规则、伦理和社会影响等问题。基因组编辑同样存在类似的重要问题,例如,如何平衡潜在利益与意外伤害风险;如何管理这些技术的应用;如何将社会价值融入临床和政策考量;如何尊重固有的民族文化差异,尤其是当这些民族文化差异可以决定

Genome editing[2] is a powerful new tool for making precise additions, deletions, and alterations to the genome—an organism's complete set of genetic material. The development of new approaches—involving the use of meganucleases; zinc finger nucleases (ZFNs); transcription activator-like effector nucleases (TALENs); and, most recently, the CRISPR/Cas9 system—has made editing of the genome much more precise, efficient, flexible, and less expensive relative to previous strategies. With these advances has come an explosion of interest in the possible applications of genome editing, both in conducting fundamental research and potentially in promoting human health through the treatment or prevention of disease and disability. The latter possibilities range from restoring normal function in diseased organs by editing somatic cells to preventing genetic diseases in future children and their descendants by editing the human germline.

As with other medical advances, each such application comes with its own set of benefits, risks, regulatory frameworks, ethical issues, and societal implications. Important questions raised with respect to genome editing include how to balance potential benefits against the risk of unintended harms; how to govern the use of these technologies; how to incorporate societal values into salient clinical and policy considerations; and how to respect the inevitable differences, rooted in national cultures, that will shape perspectives on whether and how

---

[1] 该总论不包含参考文献,其中涉及的参考文献将在随后报告的相应章节被引用。
[2] 术语"基因组编辑"在整篇报告中指通过添加、置换或移除特定DNA碱基来改变基因组序列的过程。用这一术语替代"基因编辑"是因为它更精确,因为编辑可以靶向到基因本身之外的序列,如基因表达调控区。

[1] This summary does not include references. Citations for the discussion presented in the summary appear in the subsequent report chapters.
[2] The term "genome editing" is used throughout this report to refer to the processes by which the genome sequence is changed by adding, replacing, or removing DNA base pairs. This term is used in lieu of "gene editing" because it is more accurate, as the editing could be targeted to sequences that are not part of genes themselves, such as areas that regulate gene expression.

人们对是否应该以及如何运用这些技术的看法的时候。

美国国家科学院（National Academy of Sciences）和美国国家医学科学院（National Academy of Medicine）在认识到人类基因组编辑技术的前景和问题并存后，召集成立了人类基因组编辑委员会（Committee on Human Gene Editing，下面简称"委员会"），并就本文所述的研究展开讨论。虽然基因组编辑技术在农业和非人类动物上也有潜在应用，委员会的工作重心还是集中讨论和研究其在人类中的应用。委员会的工作内容主要包括：评估目前基因组编辑技术中的科学要素，可能的临床应用，潜在风险和预期收益，是否可以建立衡量非预期效应的标准，现有的管理体系是否可以做到合理监管，用什么样的总体原则来监管人类基因组编辑。

基因组编辑应用和政策问题概述

近年来已经开展了一些将基于蛋白质对特异DNA序列识别能力的基因组编辑技术（如巨大核酸酶、ZNFs和TALENs等技术）应用于人类基因治疗的临床试验。此外，近年来还出现了基于RNA对特异DNA序列识别能力的基因组编辑技术，CRISPR/Cas9技术就是其中之一。CRISPR（clustered regularly interspaced short palindromic repeats）是最初在细菌中发现的短DNA重复序列。这些序列的发现催生了CRISPR/Cas9这一利用短RNA序列及Cas9（CRISPR associated protein 9，RNA定向核酸酶）或者类似核酸酶编辑特定DNA片段的新技术体系。相比以前的技术，CRISPR/Cas9基因组编辑技术在改变基因组方面具有明显的优势，已成为应用基因组编辑技术促进人类健康相关研究中的焦点。像巨大核酸酶、ZFNs和TALENs等基因组编辑技术一样，CRISPR/Cas9基因组编辑技术也是基于诱导特定位点DNA双链断裂和细胞内DNA修复机制对基因组进行精确编辑。但是，相比于其他类似技术，CRISPR/Cas9更加方便，成本也更低。

新的基因组编辑技术可以对基因组进行高效率和相对准确的改变，这些特性正在激发人们开发出一些安全、有效的治疗方法。除了整个基

to use these technologies.

Recognizing both the promise and concerns related to human genome editing, the National Academy of Sciences and the National Academy of Medicine convened the Committee on Human Gene Editing: Scientific, Medical, and Ethical Considerations to carry out the study that is documented in this report. While genome editing has potential applications in agriculture and non-human animals, this committee's task was focused on human applications. The charge to the committee included elements pertaining to the state of the science in genome editing, possible clinical applications of these technologies, potential risks and benefits, whether standards can be established for quantifying unintended effects, whether current regulatory frameworks provide adequate oversight, and what overarching principles should guide the regulation of genome editing in humans.

OVERVIEW OF GENOME-EDITING APPLICATIONS AND POLICY ISSUES

Genome-editing methods based on protein recognition of specific DNA sequences, such as those involving the use of meganucleases, ZFNs, and TALENs, are already being tested in several clinical trials for application in human gene therapy, and recent years have seen the development of a system based on RNA recognition of such DNA sequences. CRISPR (which stands for clustered regularly interspaced short palindromic repeats) refers to short, repeated segments of DNA originally discovered in bacteria. These segments provided the foundation for the development of a system that combines short RNA sequences paired with Cas9 (CRISPR associated protein 9, an RNA-directed nuclease), or with similar nucleases, and can readily be programmed to edit specific segments of DNA. The CRISPR/Cas9 genome-editing system offers several advantages over previous strategies for making changes to the genome and has been at the center of much discussion concerning how genome editing could be applied to promote human health. Like the use of meganucleases, ZFNs, and TALENs, CRISPR/Cas9 genome-editing technology exploits the ability to create double-stranded breaks in DNA and the cells own DNA repair mechanisms to make precise changes to the genome. CRISPR/Cas9, however, can be engineered more easily and cheaply than these other methods to generate intended edits in the genome.

The fact that these new genome-editing technologies can be used to make precise changes in the genome at a high frequency and with considerable accuracy is driving intense interest in research to develop safe and effective therapies that use these approaches and that offer options beyond simply replacing an entire gene. It is now possible

因的替换，现有的基因组编辑技术已经可以实现单个核苷酸的插入或删除、针对特定基因或基因元件的干扰、DNA单链的断裂、核苷酸的修饰，或者通过改变表观遗传学修饰调控基因的表达。在生物医学领域，基因组编辑技术有三大应用方向：基础研究、体细胞干预和生殖系细胞干预。

基础研究主要包括细胞、分子、生化、遗传和免疫学方面的机制，也包括影响生殖和疾病发生的调控机制，以及治疗应答的相关机制研究。根据美国联邦法律的相关规定，在不泄漏捐献者个人信息的前提下使用人类来源的细胞或组织的研究工作不属于人体受试者研究。虽然涉及人类细胞的基础研究大都使用体细胞（非生殖细胞），如皮肤细胞、肝细胞、肺细胞和心脏细胞等，但是也有一些基础研究使用的是生殖系（germline）细胞，包括人类早期胚胎、卵子、精子，以及可以产生卵子和精子的细胞。后面这类涉及生殖系细胞的研究尽管不涉及妊娠及可以遗传的改变，但是仍需建立关于如何收集和使用此类细胞，以及它们的使用目的的伦理和监管规则。

与基础研究不同，基因组编辑技术的临床研究涉及对人体受试者的干预。在美国及其他大多数管理体系健全的国家，潜在的临床应用必须先经过一个受监督的研究阶段，然后才能广泛地应用于患者。正如已有的用于疾病预防和治疗的基因疗法一样，靶向体细胞的基因组编辑技术在临床应用中只会影响患者自身，而不会影响其后代。与此相反，涉及生殖系干预的临床应用不仅影响其子女，而且可能会影响子女的后代。

围绕基因治疗和人类生殖医学的大量伦理、法律和社会问题为审议基因组编辑相关议题提供了参照。在认真执行和适当监督的情况下，基因疗法已经得到许多利益相关群体的支持。然而由于CRISPR/Cas9这样的技术将基因组编辑变得如此高效和精准，它使得以前只在理论上有可行性的一些应用成为可能。通过生殖系干预预防遗传性疾病就是这样一个例子。其他的潜在应用包括通过基因组编辑"增强"而不仅限于恢复或者保护健康。

基因组编辑的研究正处于从基础研究过渡到

to insert or delete single nucleotides, interrupt a gene or genetic element, make a single-stranded break in DNA, modify a nucleotide, or make epigenetic changes to gene expression. In the realm of biomedicine, genome editing could be used for three broad purposes: for basic research, for somatic interventions, and for germline interventions.

Basic research can focus on cellular, molecular, biochemical, genetic, or immunological mechanisms, including those that affect reproduction and the development and progression of disease, as well as responses to treatment. Such research can involve work on human cells or tissues, but unless it has the incidental effect of revealing information about an identifiable, living individual, it does not involve human subjects as defined by federal regulation in the United States. Most basic research on human cells uses somatic cells—nonreproductive cell types such as skin, liver, lung, and heart cells—although some basic research uses germline (i.e., reproductive) cells, including early-stage human embryos, eggs, sperm, and the cells that give rise to eggs and sperm. These latter cases entail ethical and regulatory considerations regarding how the cells are collected and the purposes for which they are used, even though the research involves no pregnancy and no transmission of changes to another generation.

Unlike basic research, clinical research involves interventions with human subjects. In the United States and most other countries with robust regulatory systems, proposed clinical applications must undergo a supervised research phase before becoming generally available to patients. Clinical applications of genome editing that target somatic cells affect only the patient, and are akin to existing efforts to use gene therapy for disease treatment and prevention; they do not affect offspring. By contrast, germline interventions would be aimed at altering a genome in a way that would affect not only the resulting child but potentially some of the child's descendants as well.

A number of the ethical, legal, and social questions surrounding gene therapy and human reproductive medicine provide a backdrop for consideration of key issues related to genome editing. When conducted carefully and with proper oversight, gene therapy research has enjoyed support from many stakeholder groups. But because such technologies as CRISPR/Cas9 have made genome editing so efficient and precise, they have opened up possible applications that have until now been viewed as largely theoretical. Germline editing to prevent genetically inherited disease is one example. Potential applications of editing for "enhancement"—for changes that go beyond mere restoration or protection of health—are another.

Because genome editing is only beginning to transition from basic research to clinical research applications, now is the time to evaluate the full range of its possible uses

临床研究应用的早期阶段，目前是评估其在人体中所有可能的用途，以及考虑如何推进和管理这些科学发展的适当时机。科学的快速发展让科学家、企业、健康倡导组织和从这些进步中受益的患者产生了极大的热情，同时也引起人们的担忧。如前所述，政策制定者和其他有关各方提出了他们的担忧，例如，这些技术的监管是否到位，以及基因组编辑最终的实际应用是否能真实反映社会价值观。

公众的投入和参与是促进科学和医学进步的重要因素，对于那些可遗传的、涉及生殖系细胞的基因组编辑，以及用于疾病治疗和预防之外其他目的的基因组编辑来说尤其如此。政策制定者和利益相关者的有效参与可以提高相关研究的透明度，赋予其合法性并改善相关政策的制定。目前有多种渠道可以让公众参与到这些讨论中来，包括公共信息宣传、正式征求公众意见和将公众舆论纳入政策等。

人类基因组编辑的应用

基因组编辑已经在实验室的基础研究中广泛应用，涉及体细胞（即非生殖系）的临床应用正处于早期试验阶段，该技术将来可能被用于涉及生殖系细胞的临床应用，该类应用将产生可遗传的基因组改变。

实验室基础科学研究

人类细胞和组织的基因组编辑的基础研究对促进生物医学的发展十分关键。利用基因组编辑技术研究体细胞，可以提升人们对疾病发生发展的相关分子机制的了解，从而为患者制订更好的治疗方案。涉及生殖系细胞的基因组编辑研究则可以帮助人们了解人体发育及生殖过程，从而促进再生医学和生育治疗等领域的发展进步。

基因组编辑在基础研究中所涉及的伦理问题与其他使用人类细胞或组织的基础研究一样，已经被严格的管理体系所规范。尽管人们对当前管理体系的局限性仍有争议，尤其是在如何合理使用配子、胚胎和胎儿组织等问题上还存在持久的争论，但是从这些条例的持久性可以看出，它们能够很好地监管基础科学研究。另一方面，即使在现有法律容许

in humans and consider how to advance and govern these scientific developments. The speed at which the science is developing has generated considerable enthusiasm among scientists, industry, health-related advocacy organizations, and patient populations that perceive benefit from these advances. It is also raising concerns, such as those cited earlier, among policy makers and other interested parties to voice concerns about whether appropriate systems are in place to govern the technologies, and whether societal values will be reflected in how genome editing is eventually applied in practice.

Public input and engagement are important elements of many scientific and medical advances. This is particularly true with respect to genome editing for potential applications that would be heritable—those involving germline cells—as well as those focused on goals other than disease treatment and prevention. Meaningful engagement with decision makers and stakeholders promotes transparency, confers legitimacy and improves policy making. There are many ways to engage the public in these debates, ranging from public information campaigns to formal calls for public comment and incorporation of public opinion into policy.

APPLICATIONS OF HUMAN GENOME EDITING

Genome editing is already being widely used for basic science research in laboratories; is in the early stages of development of clinical applications that involve somatic (i.e., nonreproductive) cells; and in the future might be usable for clinical applications involving reproductive cells, which would produce heritable changes.

Basic Science Laboratory Research

Basic laboratory research involving genome editing of human cells and tissues is critical to advancing biomedical science. Genome-editing research using somatic cells can advance understanding of molecular processes that control disease development and progression, potentially facilitating the ability to develop better interventions for affected people. Laboratory research involving genome editing of germline cells can help in understanding human development and fertility, thereby supporting advances in such areas as regenerative medicine and fertility treatment.

The ethical issues associated with basic science research involving genome editing are the same as those that arise with any basic research involving human cells or tissues, and these issues are already addressed by extensive regulatory infrastructures. There are, of course, enduring debates about limitations of the current system, particularly with respect to how it addresses the use of gametes, embryos, and fetal tissue, but the regulations are considered adequate for oversight of basic

的范畴内，涉及人类配子和胚胎的研究仍然需要遵守一些特别的条例。这些规定也适用于人类配子和胚胎中的基因组编辑研究。总之，在现有的伦理规范和管理制度下，在地方、州和联邦各个层面都已经可以有效管理实验室中的人类基因组编辑相关基础研究。

体细胞基因组编辑在疾病与残疾的治疗和预防中的临床应用

最近获批的一个临床试验是运用基因组编辑技术改造体（非生殖系）细胞从而达到治疗或预防疾病目的的一个典型案例。这一临床试验的对象是常规的化疗或放疗已经失效的晚期癌症患者。在此项研究中，通过基因组编辑技术改造患者的免疫细胞，达到靶向清除肿瘤细胞的目的。

体细胞是指身体组织中除了精子和卵子及其前体细胞之外的细胞，因此体细胞基因组编辑的效应将只限于受试个体而不会遗传给后代。通过改变体细胞的遗传物质进行疾病治疗（即"基因疗法"）并不是一个新的概念，基因组编辑技术在体细胞中的应用与基因疗法类似。由于基因疗法已经在相应的伦理规范和管理体系下被监管了相当长一段时间，从中得到的经验将为建立类似的针对体细胞的基因组编辑的法规和监督机制提供指导。

体细胞基因组编辑疗法在临床实践中具有多种应用途径。一种方式是患者体内取出相关细胞，如血细胞或骨髓细胞，对它们进行基因组编辑，再将这些细胞输回到同一患者体内。由于编辑是离体（*ex vivo*）进行的，在将细胞输回患者体内之前可以进行基因组编辑是否成功的验证。此外，也可以通过将基因组编辑载体注射到血液或目标器官中实现在体（*in vivo*）基因组编辑，但该技术仍然存在一些技术方面的挑战，如导入体内的基因组编辑工具在目的细胞中找到靶标基因的效率不够高，因此对患者的康复没什么益处。该疗法还可能带来一些副作用，如对生殖系细胞的意外伤害。在这种情况下，对生殖系细胞进行筛选检查就很有必要。尽管如此，针对B型血友病和I型黏多糖贮积症的体内编辑疗法已进入临床试验。

运用体细胞基因组编辑治疗或预防疾病和残疾涉及的科学、技术、伦理和监管问题只与患者个

science research, as evidenced by their longevity. Special considerations may come into play for research involving human gametes and embryos in jurisdictions where such research is permitted; in those cases, the current regulations governing such work will apply to genome-editing research as well. Overall, then, basic laboratory research in human genome editing is already manageable under existing ethical norms and regulatory frameworks at the local, state, and federal levels.

Clinical Uses of Somatic Cell Editing for Treatment and Prevention of Disease and Disability

An example of the application of genome editing to alter somatic (nonreproductive) cells for purposes of treating or preventing disease is a recently authorized clinical trial involving patients whose advanced cancer has failed to respond to such conventional treatments as chemotherapy and radiation. In this study, genome editing is being used to program patients' immune cells to target the cancer.

Somatic cells are all those present in the tissues of the body except for sperm and egg cells and their precursors. This means that the effects of genome editing of somatic cells are limited to treated individuals and are not inherited by their offspring. The idea of making genetic changes to somatic cells—referred to as "gene therapy"—is not new and genome editing for somatic applications would be similar. Gene therapy has been governed by ethical norms and subject to regulatory oversight for some time, and this experience offers guidance for establishing similar norms and oversight mechanisms for genome editing of somatic cells.

Somatic genome-editing therapies could be used in clinical practice in a number of ways. Some applications could involve removing relevant cells—such as blood or bone marrow cells— from a person's body, making specific genetic changes, and then returning the cells to that same individual. Because the edited cells would be outside the body (*ex vivo*), the success of the editing could be verified before the cells were replaced in the patient. Somatic genome editing also could be performed directly in the body (*in vivo*) by injecting a genome-editing tool into the bloodstream or target organ. Technical challenges remain, however, to the effective delivery of *in vivo* genome editing. Gene-editing tools introduced into the body might not find their target gene within the intended cell type efficiently. The result could be little or no health benefit to the patient, or even unintended harm, such as inadvertent effects on germline cells, for which screening would be necessary. Despite these challenges, however, clinical trials of *in vivo* editing strategies are already under way for hemophilia B and mucopolysaccharidosis I.

The primary scientific and technical, ethical, and regulatory issues associated with the use of somatic gene therapies to treat or prevent disease or disability concern

体相关。基因组编辑涉及的科学和技术问题，如尚未完全建立起来的检测和评价脱靶的标准，可以通过不断提高效率和准确性来解决。而基因组编辑涉及的伦理和管理问题，则可以在充分平衡患者的预期风险和潜在收益后融合到现有管理体制中去。

总体来说，委员会的结论是，针对人类临床研究、基因转移研究和现有的体细胞疗法建立起来的道德规范和管理制度，同样适用于以治疗或预防疾病和残疾为目的的体细胞基因组编辑。然而，由于脱靶效应因技术平台、细胞类型、靶标基因等因素而异，目前对体细胞基因组编辑的效率或特异性，以及可接受的脱靶率难以建立统一的标准。由于上述原因，体细胞基因组编辑实施途径具有多样性，因此管理者在衡量预期风险和收益时需要考虑基因组编辑系统的技术背景和临床应用目的。

生殖系编辑与可遗传的改变

虽然在动物中已经实现了个体生殖系细胞的基因组编辑，但该技术在人体内的实施还在安全性及后果预测等方面面临重大技术挑战。尽管如此，因为成千上万的人类遗传病是由单个基因突变所造成的[3]，生殖系编辑技术还是引起了人们极大的兴趣。对携带疾病相关突变的个体进行生殖系基因组编辑，可以让他们在不传递这些变异风险的情况下，拥有自己血缘关系上的孩子。虽然在可以预见的未来，生殖系基因组编辑技术的使用频率不太可能高到对这些疾病的流行病学发病率产生重要的影响，但是可以为一些家庭提供避免遗传疾病最有效或者最可以接受的选择，如避免现有的检测技术（如产前或胚胎植入前遗传诊断）不够有效，或者是诊断后需要舍弃患病胎儿，或接受选择性流产等较难接受的选择。

然而与此同时，生殖系基因组编辑存在很大争议，因为它产生的遗传变化会传递给后代，因此这项技术需要越过一个很多人认为不可逾越的界限。生殖系基因组编辑产生可遗传的变异将这一争议从个人层面转移到显然更为复杂的技术、社会和宗教层面，例如，对自然进行这样的干预是否

only the individual. The scientific and technical issues of genome editing, such as the as-yet incompletely developed standards for measuring and evaluating off-target events, can be resolved through ongoing improvements in efficiency and accuracy, while the ethical and regulatory issues would be taken into account as part of existing regulatory frameworks that involve assessing the balance of anticipated risks and benefits to a patient.

Overall, the committee concluded that the ethical norms and regulatory regimes developed for human clinical research, gene transfer research, and existing somatic cell therapy are appropriate for the management of new somatic genome-editing applications aimed at treating or preventing disease and disability. However, off-target effects will vary with the platform technology, cell type, target gene, and other factors. As a result, no single standard for somatic genome-editing efficiency or specificity—and no single acceptable off-target rate—can be defined at this time. For this reason, and because, as noted above, somatic genome editing can be carried out in a number of different ways, regulators will need to consider the technical context of the genome-editing system as well as the proposed clinical application in weighing anticipated risks and benefits.

Germline Editing and Heritable Changes

Although editing of an individual's germline (reproductive) cells has been achieved in animals, there are major technical challenges to be addressed in developing this technology for safe and predictable use in humans. Nonetheless, the technology is of interest because thousands of inherited diseases are caused by mutations in single genes.[3] Thus, editing the germline cells of individuals who carry these mutations could allow them to have genetically related children without the risk of passing on these conditions. Germline genome editing is unlikely to be used often enough in the foreseeable future to have a significant effect on the prevalence of these diseases but could provide some families with their best or most acceptable option for averting disease transmission, either because existing technologies, such as prenatal or preimplantation genetic diagnosis, will not work in some cases or because the existing technologies involve discarding affected embryos or using selective abortion following prenatal diagnosis.

At the same time, however, germline editing is highly contentious precisely because the resulting genetic changes would be inherited by the next generation, and the technology therefore would cross a line many have viewed as ethically inviolable. The possibility of making heritable changes through the use of germline genome

---

[3] OMIM, https://www.omim.org (accessed January 5, 2017); Genetic Alliance, http://www.diseaseinfosearch.org (accessed January 5, 2017).

合适，以及人们对先天残疾儿童的接受程度是否会受到影响。制定这一领域的政策需要仔细权衡文化习俗、儿童的生理和心理健康、家长的自主权，以及管理体系预防不规范使用或滥用该技术的能力。

鉴于所涉及的技术和社会问题，委员会的结论是：只有在慎重衡量其风险和收益是否符合现有的临床试验批准标准之后，才允许进行生殖系基因组编辑的临床研究和试验。即使这样，相关研究和试验仍需要有充分的理由并接受严格的监管。进行此类研究必须要小心和慎重，并且十分需要公众的广泛参与。

在美国，由于食品药品监督管理局（Food and Drug Administration，FDA）被禁止利用联邦经费去资助"意图引入可遗传的基因修饰而改造或修饰人类胚胎"的相关研究，因此当局不接受生殖系基因组编辑相关的研究计划[4]。其他一些国家目前则完全禁止生殖系基因组编辑试验。如果美国解除对这类试验的禁令，或者在法律不禁止此类试验的国家进行这些试验，那么有必要做到以下几点：只在最有需要的情况下才进行此类试验，用完善的监管体系保护受试者及其后代，制定防止其向不必要或是未知领域蔓延的保护措施。需要特别指出的是，只有在符合以下监管标准的前提下才允许进行生殖系基因组编辑：

- 缺乏其他可行的替代方案；
- 仅用于对严重疾病及生理缺陷的预防；
- 仅用于编辑那些被明确证明会导致这些疾病及生理缺陷的基因；
- 仅用于将这些基因组编辑为人群中最普遍存在的、与正常健康状态相关的、无副作用的亚型；
- 具备可靠的关于该实施方案的风险和潜在利益的临床前/临床试验数据；
- 对基因组编辑的临床试验对象的健康和安全的影响进行持续、严格的监管；
- 在尊重个人自主性的基础上，建立长期的、多代人的预后随访方案；

editing moves the conversation away from individual-level concerns and toward significantly more complex technical, social, and religious concerns regarding the appropriateness of this degree of intervention in nature and the potential effects of such changes on acceptance of children born with disabilities. Policy in this area will require a careful balancing of cultural norms, the physical and emotional well-being of children, parental autonomy, and the ability of regulatory systems to prevent inappropriate or abusive applications.

In light of the technical and social concerns involved, the committee concluded that germline genome-editing research trials might be permitted, but only following much more research aimed at meeting existing risk/benefit standards for authorizing clinical trials and even then, only for compelling reasons and under strict oversight. It would be essential for this research to be approached with caution, and for it to proceed with broad public input.

In the United States, authorities currently are unable to consider proposals for this research because of an ongoing prohibition on the U.S. Food and Drug Administration's (FDA's) use of federal funds to review "research in which a human embryo is intentionally created or modified to include a heritable genetic modification."[4] In a number of other countries, germline genome-editing trials would be prohibited entirely. If U.S. restrictions on such trials were allowed to expire or if countries without legal prohibitions were to proceed with them, it would be essential to limit these trials only to the most compelling circumstances, to subject them to a comprehensive oversight framework that would protect the research subjects and their descendants, and to institute safeguards against inappropriate expansion into uses that are less compelling or well understood. In particular, clinical trials using heritable germline editing should be permitted only if done within a regulatory framework that includes the following criteria and structures:

- absence of reasonable alternatives;
- restriction to preventing a serious disease or condition;
- restriction to editing genes that have been convincingly demonstrated to cause or to strongly predispose to the disease or condition;
- restriction to converting such genes to versions that are prevalent in the population and are known to be associated with ordinary health with little or no evidence of adverse effects;
- availability of credible pre-clinical and/or clinical data on risks and potential health benefits of the procedures;
- ongoing, rigorous oversight during clinical trials of the effects of the procedure on the health and safety of the research participants;
- comprehensive plans for long-term, multigener-ational

---

[4] *Consolidated Appropriations Act of 2016*, Public Law 114-113 (adopted December 18, 2015).

- 尊重患者隐私，同时能最大程度的透明化；
- 对个体健康及社会层面的利弊的反复评估，且包括公众的广泛参与；
- 切实的监管机制，确保本技术不被用于除预防严重疾病及生理缺陷以外的其他用途。

即使对于支持上述建议的人而言，他们的出发点也可能是不同的。对于那些认为该技术带来的利益是令人信服的人来说，以上标准可以在"妥善关注（due care）"及"科学诚信（responsible science）"的框架下促进社会福利。对于那些认为这项技术带来的利益还不足以弥补其引发的社会问题的人来说，他们可能会认同这些标准一旦被切实贯彻，他们担忧的事情就不会发生。值得注意的是，以上标准中像"其他可行的替代方法"及"严重疾病及生理缺陷"这样的术语是模糊的。不同团体基于自身不同的历史、文化、社会背景，结合公众意见及其自身话语权会对这些术语给出不同的解释。比如，医生和患者会根据个体自身的情况来决定是否将生殖系基因组编辑作为一个选项。不过，对于"严重疾病及生理缺陷"这个术语，美国FDA曾给出了具体的定义[5]。而对于那些完全反对生殖系基因组编辑的人来说，他们可能会认为以上的研究标准非常严格，一旦实施，将阻止所有涉及生殖系基因组编辑的临床试验。

将基因组编辑用于"能力强化"

目前对于基因组编辑技术的讨论主要集中在疾病和残疾的治疗与预防中。然而公众也担心这项技术被用于其他方面，例如，用于增强一些超越正常健康范畴的身体特征和能力。理论上，这种以增强能力为目的的基因组编辑既包括针对体细胞也包括针对生殖系细胞的基因组编辑。这种技术的应用会导致诸如公平、社会准则、个人自主性及政府职能等方面的问题。

首先，对于"能力强化"应该给予准确的定义，需要考量各个利益相关者对于"正常范围"理

- follow-up while still respecting personal autonomy;
- maximum transparency consistent with patient privacy;
- continued reassessment of both health and societal benefits and risks, with broad on-going participation and input by the public; and
- reliable oversight mechanisms to prevent extension to uses other than preventing a serious disease or condition.

Even those who will support this recommendation are unlikely to arrive at it by the same reasoning. For those who find the benefits sufficiently compelling, the above criteria represent a commitment to promoting well-being within a framework of due care and responsible science. Those not completely persuaded that the benefits outweigh the social concerns may nonetheless conclude that these criteria, if properly implemented, are strict enough to prevent the harms they fear. It is important to note that such concepts as "reasonable alternatives" and "serious disease or condition" embedded in these criteria are necessarily vague. Different societies will interpret these concepts in the context of their diverse historical, cultural, and social characteristics, taking into account input from their publics and their relevant regulatory authorities. Likewise, physicians and patients will interpret them in light of the specifics of individual cases for which germline genome editing may be considered as a possible option. Starting points for defining some of these concepts exist, such as the definition of "serious disease or condition" used by the U.S. FDA.[5] Finally, those opposed to germline editing may even conclude that, properly implemented, the above criteria are so strict that they would have the effect of preventing all clinical trials involving germline genome editing.

Use of Genome Editing for "Enhancement"

Although much of the current discussion around genome editing focuses on how these technologies can be used to treat or prevent disease and disability, some aspects of the public debate concern other purposes, such as the possibility of enhancing traits and capacities beyond levels considered typical of adequate health. In theory, genome editing for such enhancement purposes could involve both somatic and germline cells. Such uses of the technologies raise questions of fairness, social norms, personal autonomy, and the role of government.

To begin, it is necessary to define what is meant by "enhancement." Formulating this definition requires

---

[5] FDA对"严重疾病及生理缺陷"（serious disease or condition）的定义是：症状会严重影响日常生活的疾病。虽然通常不考虑短期症状及个体特异性症状，但是症状不需要是不可逆的，持久性和反复性的症状也可纳入。一种疾病是否严重取决于临床诊断的综合考量，涉及存活率、患者的日常生活，以及治疗情况下疾病恶化的可能（21 CFR 312.300(b)(1)）。

[5] While not drafted with the above criteria in mind, the FDA definition of "serious disease or condition" is "a disease or condition associated with morbidity that has substantial impact on day-to-day functioning. Short-lived and self-limiting morbidity will usually not be sufficient, but the morbidity need not be irreversible if it is persistent or recurrent. Whether a disease or condition is serious is a matter of clinical judgment, based on its impact on such factors as survival, day-to-day functioning, or the likelihood that the disease, if left untreated, will progress from a less severe condition to a more serious one" (21 CFR 312.300(b)(1)).

解的差异。例如，应用基因组编辑技术降低胆固醇水平偏高患者的胆固醇可以被认为是预防心脏疾病；但是应用基因组编辑技术降低正常的胆固醇水平则很难界定是否属于"能力强化"。另外，此类技术与目前使用的他汀类药物（statins）控制胆固醇技术有本质区别吗？又如，运用基因组编辑技术对肌营养不良患者增肌可以被认为是疾病治疗；但是如果将其用于没有疾病表征和能力衰竭的人身上，使他们在"正常范围"之内变得更加强壮，则可能被认为是能力强化。如果更进一步将基因组编辑技术用于把人的肌肉力量提升到（甚至超越于）人类的极限水平，则毫无疑问属于能力强化。

无论定义如何，将基因组编辑技术应用于能力强化都会引发社会不适感，如加剧对社会不公的恐惧、造成社会压力迫使人们不得不选择这一编辑自身基因组的技术。考虑到个体主观因素的强大作用，评价能力强化的利害关系是非常困难的，正因为如此，对于将基因组编辑用于能力强化的科研及市场应用的决策需要引入公开讨论（public discussion）机制。当政府针对这些应用制定政策时，其实际及预期的社会影响力也需要通过公开讨论来探索。委员会认为，对于基因组编辑在非疾病治疗和预防目的方面的运用，目前不应该提上日程；而日后任何有关是否或如何推进临床试验等的决定，都必须先经过公开的讨论。

公众参与

公众参与永远都是监管新技术的一个重要部分。如上所述，对于体细胞编辑来说，可能会被用于治疗预防疾病和残疾以外的其他用途（如能力强化）。在评估是否批准其临床试验之前，进行公开透明的、广泛的政策讨论是必需的。而对于可遗传的生殖系基因组编辑，社会力量的广泛参与，以及对健康与社会利弊的反复评估对于最终批准临床试验来说也是至关重要的。

目前，在美国的监管体系中已经引入了一些与公众对话和咨询的机制，其中包含了针对基因治疗的部分，这部分也可以将基因组编辑囊括进来。在一些情况下，管理条例和文件必须经过长期的公众评价和各机构回应之后才会被发布。各个州及联邦政府中由技术和社会科学领域的专家组成的生物伦

总　论　9

a careful examination of how various stakeholders conceptualize "normal." For example, using genome editing to lower the cholesterol level of someone with abnormally high cholesterol might be considered prevention of heart disease, but using it to lower cholesterol that is in the desirable range is less easily characterized, and would either intervention differ from the current use of statins? Likewise, using genome editing to improve musculature for patients with muscular dystrophy would be considered a restorative treatment, whereas doing so for individuals with no known pathology and average capabilities just to make them stronger but still within the "normal" range might be considered enhancement. And using the technology to increase someone's muscle strength to the extreme end of human capacity (or beyond) would almost certainly be considered enhancement.

Regardless of the specific definition, there is some indication of public discomfort with using genome editing for what is deemed to be enhancement, whether for fear of exacerbating social inequities or of creating social pressure for people to use technologies they would not otherwise choose. Precisely because of the difficulty of evaluating the benefit of an enhancement to an individual given the large role of subjective factors, public discussion is needed to inform the regulatory risk/benefit analyses that underlie decisions to permit research or approve marketing. Public discussion also is needed to explore social impacts, both real and anticipated, as governance policy for such applications is developed. The committee recommends that genome editing for purposes other than treatment or prevention of disease and disability should not proceed at this time, and that it is essential for these public discussions to precede any decisions about whether or how to pursue clinical trials of such applications.

Public Engagement

Public engagement is always an important part of regulation and oversight for new technologies. As noted above, for somatic genome editing, it is essential that transparent and inclusive public policy debates precede any consideration of whether to authorize clinical trials for indications that go beyond treatment or prevention of disease or disability (e.g., for enhancement). With respect to heritable germline editing, broad participation and input by the public and ongoing reassessment of both health and societal benefits and risks are particularly critical conditions for approval of clinical trials.

At present, a number of mechanisms for public communication and consultation are built into the U.S. regulatory system, including some designed specifically for gene therapy, whose purview would include human genome editing. In some cases, regulatory rules and guidance documents are issued only after extensive public comment and agency response. Discussion is fostered

理委员会将举办面向社会公开的讨论会。此外，美国国立卫生研究院（National Institutes of Health）的重组 DNA 咨询委员会（Recombinant DNA Advisory Committee）也针对基因治疗提供了与公众对话的渠道，其中包括对具体操作程序的评估，以及对监管部门提出建议。其他国家，如法国和英国，分别通过正式轮询或听证的机制保证接收到来自社会的不同意见。

监管人类基因组编辑的原则

基因组编辑委员会的职责之一就是建立一套各国政府都能参考使用的人类基因组编辑监管原则。这套原则被详细地列在延伸内容 S-1 中。委员会建议任何国家在建立其针对基因组编辑的管理条例和程序及职能的时候，都可以引入这些原则。

by the various state and federal bioethics commissions, which typically bring together technical experts and social scientists in meetings that are open to the public. And the National Institutes of Health's Recombinant DNA Advisory Committee offers a venue for general public discussion of gene therapy, for review of specific protocols, and for transmission of advice to regulators. Other countries, such as France and the United Kingdom, have mechanisms that involve formal polling or hearings to ensure that diverse viewpoints are heard.

PRINCIPLES TO GUIDE THE GOVERNANCE OF HUMAN GENOME EDITING

One of the charges to the committee was to identify principles that many countries might be able to use to govern human genome editing. The principles identified by the committee are detailed in Box S-1. The committee recommends that any nation considering governance of human genome editing can incorporate these principles and the responsibilities that flow therefrom into its regulatory structures and processes.

延伸内容 S-1

监管人类基因组编辑的原则
Principles for the Governance of Human Genome Editing

1. 福利提升原则：指在促进利益的同时防止对当事人产生有害影响，即生物伦理论著中常见的仁慈（beneficence）和不邪恶（nonmaleficence）原则。

由此原则产生的职责包括：①促进有关基因组编辑技术在提升个体健康和福祉方面的应用，如用于治疗和预防疾病，对于不确定性较高的前期试验最大程度降低受试者的风险；②在所有基因组编辑技术应用中确保利弊平衡。

2. 透明性原则：该原则要求公开及信息分享，即对利益相关者提供容易获得的、可理解的公开信息。

由此产生的职责包括：①致力于最大限度地、及时地公开信息；②在制定基因组编辑，以及其他新的、具有争议的技术的决策过程中，切实可行地

1. Promoting well-being: *The principle of promoting well-being supports providing benefit and preventing harm to those affected, often referred to in the bioethics literature as the principles of beneficence and nonmaleficence.*

Responsibilities that flow from adherence to this principle include (1) pursuing applications of human genome editing that promote the health and well-being of individuals, such as treating or preventing disease, while minimizing risk to individuals in early applications with a high degree of uncertainty; and (2) ensuring a reasonable balance of risk and benefit for any application of human genome editing.

2. Transparency: *The principle of transparency requires openness and sharing of information in ways that are accessible and understandable to stakeholders.*

Responsibilities that flow from adherence to this principle include (1) a commitment to disclosure of information to the fullest extent possible and in a timely manner, and (2) meaningful public input into the policy-making process related to human genome editing, as well as other novel and disruptive technologies.

纳入公众意见。

3. 妥善关注原则：指对于参与试验及医疗救治的病人，需向其提供细心妥善的关注，所有举措都需要有充足的依据。

由此产生的职责包括：整个程序要在适当监管下谨慎地展开，并根据未来技术和文化观点的改变而进行不断地重新评估。

4. 科学诚信原则：指从研发到临床的过程中，始终坚持最高的科研标准，符合专业及国际的规范。

由此产生的职责包括：①高质量的实验设计和数据分析；②对实验方案和结果进行适当的评估；③信息透明；④对错误或者误导性的数据或分析进行校正。

5. 尊重个人原则：该原则要求尊重所有个体的尊严，认同个人意愿的中心地位并尊重个人的决定。无论遗传背景如何，所有人在道义上是平等的。

由此产生的职责包括：①致力于保证所有个体的价值平等；②尊重并提倡个人的自主决定；③致力于防止过去曾发生过的"优生学"的重蹈覆辙；④致力于对残疾人的去污名化。

6. 公平原则：指要求相似的案例需要得到相似的处理，做到风险和利益的均等分配（分配正义，distributive justice）。

由此产生的职责包括：①对科研负担和利益的均等分配；②广泛及平等地享有人类基因组编辑临床应用所产生的利益。

7. 跨国合作原则：指在尊重文化差异的基础上，推动研究和监管方面的跨国合作。

由此产生的职责包括：①尊重各国的政策差异；②在管理规范及流程方面尽量协调一致；③不同科研团体和监管机构之间的跨国合作与数据共享。

3. Due care: *The principle of due care for patients enrolled in research studies or receiving clinical care requires proceeding carefully and deliberately, and only when supported by sufficient and robust evidence.*

Responsibilities that flow from adherence to this principle include proceeding cautiously and incrementally, under appropriate supervision and in ways that allow for frequent reassessment in light of future advances and cultural opinions.

4. Responsible science: *The principle of responsible science underpins adherence to the highest standards of research, from bench to bedside, in accordance with international and professional norms.*

Responsibilities that flow from adherence to this principle include a commitment to (1) high-quality experimental design and analysis, (2) appropriate review and evaluation of protocols and resulting data, (3) transparency, and (4) correction of false or misleading data or analysis.

5. Respect for persons: *The principle of respect for persons requires recognition of the personal dignity of all individuals, acknowledgment of the centrality of personal choice, and respect for individual decisions. All people have equal moral value, regardless of their genetic qualities.*

Responsibilities that flow from adherence to this principle include (1) a commitment to the equal value of all individuals, (2) respect for and promotion of individual decision making, (3) a commitment to preventing recurrence of the abusive forms of eugenics practiced in the past, and (4) a commitment to destigmatizing disability.

6. Fairness: *The principle of fairness requires that like cases be treated alike, and that risks and benefits be equitably distributed (distributive justice).*

Responsibilities that flow from adherence to this principle include (1) equitable distribution of the burdens and benefits of research and (2) broad and equitable access to the benefits of resulting clinical applications of human genome editing.

7. Transnational cooperation: *The principle of transnational cooperation supports a commitment to collaborative approaches to research and governance while respecting different cultural contexts.*

Responsibilities that flow from adherence to this principle include (1) respect for differing national policies, (2) coordination of regulatory standards and procedures whenever possible, and (3) transnational collaboration and data sharing among different scientific communities and responsible regulatory authorities.

建 议

基于上述考量，委员会针对基因组编辑包括体细胞及生殖系的基因组编辑的基础研究和临床试验提出了一系列建议。对这些建议的摘要总结见延伸内容 S-2。

RECOMMENDATIONS

In light of the considerations detailed above, the committee made a series of recommendations targeted to basic research and to clinical applications, both somatic and germline. A summary of the key messages in these recommendations is found in Box S-2.

延伸内容 S-2

监管和运用人类基因组编辑：建议摘要

Oversight and Use of Human Gene Editing: Summary of Recommendations

科研及临床应用的总原则
考虑并应用监管人类基因组编辑的总原则 (2.1)
提升福利原则
透明性原则
妥善关注原则
科学诚信原则
尊重个人原则
公平原则
跨国界合作原则

实验室基础研究
用现有的规章制度来监管实验室中的人类基因组编辑的研究 (3.1)

体细胞基因组编辑
利用现行的针对基因治疗的规章制度来监管体细胞基因组编辑的实验室研究和应用 (4.1)
对用于治疗疾病和残疾的临床试验及应用目前应当加以限制 (4.2)
如果需要应用，需在考虑利弊的基础上对其安全性和功效进行评估 (4.3)
在扩大应用之前需要有广泛的公众参与 (4.4)

生殖系（可遗传）基因组编辑
在有充分理由支持下，为了治疗或预防严重疾病及生理缺陷，允许进行生殖系基因组编辑临床试验，但必须要有严格的监管系统将其限制在特定的规范内 (5.1)

Global Principles for Research and Clinical Use
Consider and apply the global principles in governance of human genome editing (2.1)
Promoting well-being
Transparency
Due care
Responsible science
Respect for persons
Fairness
Transnational cooperation

Basic Laboratory Research
Use existing regulatory processes to oversee human genome editing laboratory research (3.1)

Somatic Genome Editing
Use existing regulatory processes for human gene therapy to oversee somatic human genome editing research and uses (4.1)
Limit clinical trials or therapies to treatment and prevention of disease or disability at this time (4.2)
Evaluate safety and efficacy in the context of risks and benefits of intended use (4.3) Require broad public input prior to extending uses (4.4)

Germline (Heritable) Genome Editing
Permit clinical research trials only for compelling purposes of treating or preventing serious disease or disabilities, and only if there is a stringent oversight system able to limit uses to specified criteria (5.1)

能力强化

在目前情况下,不允许将基因组编辑应用于除治疗和预防疾病及残疾以外的其他用途(6.1)

对于将体细胞基因组编辑用于治疗或预防疾病和残疾之外的其他用途,需要鼓励公开讨论和政策辩论(6.2)

公众参与

如需将基因组编辑技术扩展到除治疗或预防疾病和残疾以外的其他用途,则在此之前需要广泛的公众参与(7.1)

在可遗传的生殖系基因组编辑的临床试验之前,需要有不断地重新评估和公众参与(7.2)

对于"能力强化",在政策制定的程序中需要有公众参与(7.3)

在资助基因组编辑技术时,考虑纳入加强公众参与的政策研究(7.4)

对人类基因组编辑技术在伦理、立法及社会影响方面进行长期评估(7.5)

Enhancement

Do not proceed at this time with human genome editing for purposes other than treatment or prevention of disease and disability *(6.1)*

Encourage public discussion and policy debate with respect to somatic human genome editing for uses other than treatment or prevention of disease and disability *(6.2)*

Public Engagement

Public input should precede any clinical trials for an extension of human genome editing beyond disease treatment and prevention *(7.1)*

Ongoing reassessment and public participation should precede any clinical trials of heritable germline editing *(7.2)*

Incorporate public participation into the human genome editing policy process about "enhancement" *(7.3)*

When funding genome editing research, consider including research on strategies to improve public engagement *(7.4)* and for long-term assessment of ethical, legal and social implications of human genome editing *(7.5)*

1
概 述
Introduction

基因组编辑⁶是一种对"生物的整套遗传物质——基因组"进行精确地插入、敲除和改变的新型强大工具。巨大核酸酶（meganucleases）、锌指核酸酶（zinc finger nucleases，ZNFs）、转录激活因子样效应核酸酶（transcription activator-like effector nucleases，TALENs）及 CRISPR/Cas9（clustered regularly interspaced short palindromic repeat）等技术手段的快速发展和应用使基因组编辑较之以前更加精准、高效、灵活和廉价。这些进步也促使人们尝试将基因组编辑技术应用于基础研究，以及疾病的治疗和预防中。在疾病治疗和预防方面，基因编辑技术的应用范围包括体细胞及生殖系细胞，从恢复成体病变器官的功能到预防未来后代遗传病的发生。

与其他医学进步一样，每一种新技术的应用也会带来特定的利益、风险、规则、伦理和社会影响等问题。基因组编辑相关的重要问题有：如何平衡潜在利益与意外伤害风险；如何管理这些技术的应用；如何将社会价值融入临床和政治考量；如何尊重固有的民族文化差异，尤其是当这些民族文化差异可以决定人们对是否应该应用、如何运用这些技术的看法的时候。

美国国家科学院（National Academy of Sciences，NAS）和美国国家医学科学院（National Academy

Genome editing⁶ is a powerful new tool for making precise additions, deletions, and alterations to the genome—an organism's complete set of genetic material. The development of new approaches—involving the use of meganucleases; zinc finger nucleases (ZFNs); transcription activator-like effector nucleases (TALENs); and, most recently, the CRISPR/Cas9 system—has made editing of the genome much more precise, efficient, flexible, and less expensive relative to previous strategies. With these advances has come an explosion of interest in the possible applications of genome editing, both in conducting fundamental research and potentially in promoting human health through the treatment or prevention of disease and disability. The latter possibilities range from editing somatic cells to restore normal function in diseased organs to editing the human germline to prevent genetic diseases in future children and their descendants.

As with other medical advances, each application comes with its own set of benefits, risks, regulatory questions, ethical issues, and societal implications. Important questions raised with respect to genome editing include how to balance potential benefits against the risk of unintended harms; how to govern the use of these technologies; how to incorporate societal values into salient clinical and policy considerations; and how to respect the inevitable differences, rooted in national cultures, that will shape perspectives on whether and how to use these technologies.

---

⁶ 术语"基因组编辑"在这一整篇报告中指通过添加、置换或移除特定DNA碱基来改变基因组序列的过程。用这一术语替代"基因编辑"是因为它更精确，因为编辑可以靶向到基因本身之外的序列，如基因表达调控区。

⁶ The term "genome editing" is used throughout this report to refer to the processes by which the genome sequence is changed by adding, replacing, or removing DNA base pairs. This term is used in lieu of "gene editing" because it is more accurate, as the editing could be targeted to sequences that are not part of genes themselves, such as areas that regulate gene expression.

of Medicine，NAS)⁷在认识到人类基因组编辑技术的前景和问题并存后，召集成立了人类基因组编辑委员会（Committee on Human Gene Editing：Scientific，Medical，and Ethical Considerations，下面简称"委员会"），并就本文所述的研究展开讨论。虽然基因组编辑技术在农业和非人类动物上也有潜在应用⁸，本委员会的工作重心（延伸内容1-1）还是集中讨论其在人类中的应用⁹。委员会的工作内容主要包括：评估目前基因组编辑技术中的科学要素，可能的临床应用，潜在风险和预期收益，是否可以建立衡量非预期效应的标准，现有的管理体系是否可以做到合理监管，用什么样的总体原则来监管人类基因组编辑。

Recognizing both the promise and concerns related to human genome editing, the National Academy of Sciences (NAS) and the National Academy of Medicine (NAM)⁷ convened the Committee on Human Gene Editing: Scientific, Medical, and Ethical Considerations to carry out the study that is documented in this report. While genome editing has potential applications for use in agriculture and nonhuman animals,⁸ this committee's task (see Box 1-1) was focused on human applications.⁹ The charge to the committee included elements pertaining to the state of the science in genome editing, possible clinical applications of these technologies, potential risks and benefits, whether standards can be established for quantifying unintended effects, whether current regulatory frameworks provide adequate oversight, and what overarching principles should guide the regulation of genome editing in humans.

延伸内容 1-1

任 务 声 明

Statement of Task

这个报告将评价人类基因组编辑技术在生物医药研究和医学应用方面的科学依据及其在临床、伦理、法律和社会各方面的影响。包括人类生殖系编辑在内的人类基因组编辑应用相关问题如下：

1. 人类基因组编辑的科学现状，以及未来研究的发展方向和面临的挑战是什么？

The study will examine the scientific underpinnings as well as the clinical, ethical, legal, and social implications of the use of human genome editing technologies in biomedical research and medicine. It will address the following issues related to human gene editing, including editing of the human germline:

1. What is the current state of the science of human gene editing, as well as possible future directions and challenges to further advances in this research?

---

⁷ NAS 和 NAM 在本文以国家科学院（National Academies）统一代称，在讨论到其他国家的学术机构时用美国国家科学院（U.S. National Academies）代称。2016 年以前，NAM 也被称为医学研究所（IOM）。

⁸ 2017 年 1 月，FDA 发布修订后指导意见用以规范对有意改变植物和非人动物基因组 DNA 的行为管理框架。这些行为就包括通过基因组编辑改变基因组 DNA 在内。这一指导意见不影响对作为人类应用的人类药物、设备和生物制剂的管理框架。详见 FDA "Regulation of Intentionally Altered Genomic DNA in Animals—Draft Guidance"（January 2017）http://www.fda.gov/downloads/AnimalVeterinary/GuidanceComplianceEnforcement/GuidanceforIndustry/ucm113903.pdf （accessed January 30，2017） and Genome Editing in New Plant Varieties Used for Foods；意见征求稿见 https://www.regulations.gov/document?D=FDA-2016-N-4389-0001 （accessed January 30，2017）。

⁹ 联邦部门和机构的监管与生物技术应用的调控框架可见于"现代化生物技术产品的监管体系：生物技术的调控框架最终版本2017版"（January 4，2017）和"国家战略现代化生物技术产品的监管体系"（September 2016）。https://obamawhitehouse.archives.gov/blog/2017/01/04/increasing-transparency-coordination-andpredictability-biotechnology-regulatory（January 30，2017）.）。

---

⁷ The NAS and the NAM are referred to throughout this report simply as the National Academies, or the U.S. National Academies when discussed in relation to the academies of other nations. Until 2016, the NAM was known as the Institute of Medicine (IOM).

⁸ In January 2017, the FDA issued revised draft guidances addressing the regulatory pathway for intentionally altered genomic DNA in plants and non-human animals. This would include DNA intentionally altered through genomic editing. The guidances do not affect the regulatory pathway for human applications that are regulated as human drugs, devices and biologics. See FDA "Regulation of Intentionally Altered Genomic DNA in Animals—Draft Guidance" (January 2017) at http://www.fda.gov/downloads/AnimalVeterinary/GuidanceComplianceEnforcement/ GuidanceforIndustry/ucm113903.pdf (accessed January 30, 2017) and Genome Editing in New Plant Varieties Used for Foods; Request for Comments at https://www.regulations.gov/document?D=FDA-2016-N-4389-0001 (accessed January 30, 2017).

⁹ The regulatory roles of the federal departments and agencies, and the overall framework for regulation of applications of biotechnology, are outlined in "Modernizing the Regulatory System for Biotechnology Products: Final Version of the 2017 Update to the Coordinated Framework for the Regulation of Biotechnology" (January 4, 2017) and "National Strategy for Modernizing the Regulatory System for Biotechnology Products" (September 2016). https://obamawhitehouse.archives.gov/blog/2017/01/04/increasing-transparency-coordination-and-predictability-biotechnology-regulatory (accessed January 30, 2017).

2. 在人类疾病的治疗方面，该技术最有希望的临床应用是什么？对于这些应用，目前有没有其他替代方案？
3. 人类基因组编辑的功效和风险是什么，何种技术能够降低人类基因组编辑的风险及提高特异性和有效性？技术进展能否产生新的临床应用及减少安全隐患？
4. 是否或能否建立明确的科学标准来量化基因组改变过程中的脱靶效应？如果可以，如何将其应用于人类疾病治疗？
5. 现行的人体研究相关的伦理和法律标准对包括生殖系编辑在内的人类基因组编辑是否已经足够了？
6. 什么样的原则和框架能为人体细胞和生殖系编辑提供合理的监督？如何决定哪种人类基因组编辑技术的应用是可以实施的？应建立什么保障措施合理指导基因组编辑的研究和应用？
7. 提供国际上处理这类问题的案例。协调政策的前景如何？从不同地区各自的监管方式中我们能学到什么？

委员会将讨论这些问题，并提供一个包含结论和建议的报告。任何一个希望对人类基因组编辑建立指南的国家，都可以采用本报告提供的框架和原则，或在此基础上作相应的修改。这份报告还包括一份着重针对美国的建议。

2. What are the potential clinical applications that may hold promise for the treatment of human diseases? What alternative approaches exist?
3. What is known about the efficacy and risks of gene editing in humans, and what research might increase the specificity and efficacy of human gene editing while reducing risks? Will further advances in gene editing introduce additional potential clinical applications while reducing concerns about patient safety?
4. Can or should explicit scientific standards be established for quantifying off-target genome alterations and, if so, how should such standards be applied for use in the treatment of human diseases?
5. Do current ethical and legal standards for human subjects research adequately address human gene editing, including germline editing? What are the ethical, legal, and social implications of the use of current and projected gene-editing technologies in humans?
6. What principles or frameworks might provide appropriate oversight for somatic and germline editing in humans? How might they help determine whether, and which applications of, gene editing in humans should or should not go forward? What safeguards should be in place to ensure proper conduct of gene-editing research and use of gene-editing techniques?
7. Provide examples of how these issues are being addressed in the international context. What are the prospects for harmonizing policies? What can be learned from the approaches being applied in different jurisdictions?

The committee will address these questions and prepare a report that contains its findings and recommendations. The report will provide a framework based on fundamental, underlying principles that may be adapted and adopted by any nation that is considering the development of guidelines. The report will also include a focus on advice for the United States.

研究现况

NAS 和 NAM 就人类基因组编辑的倡议

由于基因组编辑广阔的应用前景，以及面临的相关监管和伦理道德问题，NAS 和 NAM 提出倡议，建议深入探讨此类问题并就如何解决此类问题展开美国与国际之间的对话。人类基因组编辑倡议的第一项活动是"人类基因组编辑国际峰会"的召开——与中国科学院（Chinese Academy of Sciences）和英国皇家学院（U.K. Royal Society）的

STUDY CONTEXT

The NAS and the NAM Human Gene-Editing Initiative

In light of the promise of genome editing and the associated regulatory and ethical issues, the NAS and the NAM established an initiative to explore these issues in greater depth and facilitate U.S. and international dialogue on how to address them. The first activity of this Human Gene-Editing Initiative was the convening of the International Summit on Human Gene Editing: A Global Discussion jointly with the Chinese Academy of Sciences and The Royal Society of the United Kingdom.

全球共同讨论。这个为期三天的会议讨论了基因组编辑技术方面的进展、这些技术对患者的潜在应用价值,以及使用这些技术可能会引起的伦理和社会问题。委员会就这次会议发表了一份总结声明(NASEM,2016d)。委员会主席 David Baltimore 指出"我们希望我们的这次讨论能成为有意义的持续性国际对话的基础"(NASAM,2016d)。三个国家支持并拥护声明倡导的继续基因组编辑研究,对可遗传的编辑审慎处理,并就这一主题进行持续的公开讨论[10]。这一次峰会以及 NAS 和 NAM 在相关主题的其他研究共同为本文提供了重要意见(延伸内容 1-2)。

This 3-day event addressed a number of scientific advances in the development of modern genome-editing tools, potential medical uses of these tools in human patients, and ethical and social issues their uses might pose. The organizing committee released a statement that summarized its conclusions from the meeting (NASEM, 2016d). Panel chair David Baltimore also noted "we hope that our discussion here will serve as a foundation for a meaningful and ongoing global dialogue" (NASEM, 2016d). All three nations embraced the statement's call for continued research on gene editing, further deliberation with regard to heritable changes, and a continued public discourse on the topic.[10] The summit provided important input to the present study, as did other studies by the NAS and the NAM on related topics (see Box 1-2).

延伸内容 1-2

NAS 和 NAM 的相关研究
Related Studies of the NAS and the NAM

非人类基因组编辑

因为如 CRISPR/Cas9 这样的基因组编辑方法比较简单,在实验室可以用多种方法得到遗传信息发生变化的微生物、非人体细胞或人体细胞。目前针对人类基因组编辑技术的相关研究,是美国国家科学院针对基因组编辑技术的一系列研究的一部分。美国国家科学院的研究还包括以下方面:

(1)转基因作物:经验和前景。这个报告通报了利用转基因技术研发的食物农作物在安全性、环境、监管和其他方面的问题。转基因作物可以通过多种方法获得,而新的基因组编辑技术是其中一种(NASEM,2016c)。

(2)转基因系统面临的问题:科学先进性;方向指向不确定性;研究与公共价值一致性。这个报告集中于 CRISPR/Cas9 技术使遗传改变在没有选择优势的情况下获得传播这一现状。这种情况最常见于农作物的病虫害防治中。可遗传的基因组变

Genome Editing in Nonhuman Organisms

Because genome-editing methods such as the CRISPR/Cas9 system are simply tools, they can be applied in myriad ways to achieve genetic changes in cells in the laboratory, in microbes, in nonhuman organisms, or in human subjects. The present study, which focuses on the use of genome editing in humans, is part of a broader examination by the U.S. National Academies of the implications of genome editing across a number of applications that also includes projects addressing the following:

- *Genetically Engineered Crops: Experiences and Prospects*—This report addresses safety, environmental, regulatory, and other aspects of food crops developed through the use of genetic engineering technology. Such crops can be produced using a number of methods, and new genome-editing tools are among them (NASEM, 2016c).
- *Gene Drives on the Horizon: Advancing Science, Navigating Uncertainty, and Aligning Research with Public Values*—This report focuses on a specific application enabled by CRISPR/Cas9 technology that allows genetic changes to spread in a population in the absence of selective advantage.

---

[10] 美国国家科学院院长 Ralph J. Cicerone、美国国家药学院院长 Victor J. Dzau、中国科学院院长白春礼、英国皇家学院主席 Venki Ramakrishna 的联合声明 http://www8.nationalacademies.org/onpinews/newsitem.aspx?RecordID=12032015b(2017,1,24)。

[10] Statement by Ralph J. Cicerone, President, U.S. National Academy of Sciences; Victor J. Dzau, President, U.S. National Academy of Medicine; Chunli Bai, President, Chinese Academy of Sciences; Venki Ramakrishnan, President, The Royal Society; http://www8.nationalacademies.org/onpinews/newsitem.aspx?RecordID=12032015b (accessed January 24, 2017).

化在生态系统中的传播增加了应用基因组编辑技术在科学上的复杂性，以及在伦理和监管上的挑战（NASEM，2016b）。

（3）利用基因组编辑技术修饰动物基因组并应用于研究的研讨会：出于科学和伦理考量（动物实验室研究所圆桌会议［Institute for Animal Laboratory Research，ILAR，Roundtable］）。基因组编辑工具可以用于产生新的实验动物模型从而更好地为疾病研究服务，也可以使牲畜产生所需要的特征。研讨会探讨了动物使用基因组编辑的相关伦理和监管[a]。

（4）未来生物技术产品和生物技术系统监管能力的提升。利用生物技术获得的产品数量在大幅度增加，而这些产品的监管框架仍是几十年前建立的。这一研究将分析美国现有的监管系统对通过基因组编辑等一系列技术获得的产品进行监管的能力和专业性。本研究不涉及人类药物和医药设备的开发和监管[b]。

临床研究与应用的其他内容

美国国家科学院最近的一些报告不仅是针对基因组编辑，而且与基因组编辑技术的临床应用密切相关。

（1）临床基因转化协议的监管和总结：评估DNA重组咨询委员会的作用——这一报告总结了DNA重组咨询委员会（Recombinant DNA Advisory Committee，RAC）的作用，并建议其更加明智地利用其咨询权审查具体协议，主要针对基因疗法的新应用，或者为公众讨论基因疗法提供途径和渠道（IOM，2014）。

（2）线粒体置换技术：考虑伦理、社会、道德问题——分析替换配子或胚胎中的线粒体DNA所带来的特殊机遇和潜在问题（NASEM，2016e）。

（3）胚胎干细胞研究指南（IOM，2005；NRC and IOM，2007，2008，2010）——为新出现的可以

---

[a] ILAR圆桌会议——用于研究动物基因组编辑修饰的科学和伦理考量。http://nas-sites.org/ilar-roundtable/roundtable-activities/gene-editing-to-modify-animalgenomes-for-research (2016,10,21)。
[b] 未来生物技术产品和加强生物技术监管体系的机遇。http://nas-sites.org/biotech (2016,10,21)。

This technology is not applicable to every species and is most commonly proposed for uses such as insect vector control. The ability of heritable genetic changes to spread through an ecosystem raises its own complex set of scientific, ethical, and governance challenges (NASEM, 2016b).

- *Workshop on Gene Editing to Modify Animal Genomes for Research: Scientific and Ethical Considerations (Institute for Animal Laboratory Research [ILAR] Roundtable)*—Genome-editing tools can be used to produce laboratory animal models enabling better study of diseases, as well as to produce livestock with desired traits. The workshop explored animal uses of genome editing, along with associated ethical and regulatory considerations.[a]
- *Future Biotechnology Products and Opportunities to Enhance Capabilities of the Biotechnology Regulatory System*—As the types of products that can be created through biotechnology expand, these products are evaluated in regulatory frameworks initially created decades ago. This ongoing study is examining capabilities and expertise that may be needed by U.S. regulatory systems to assess and regulate future products that could be created through a number of technologies, including genome editing. The study is not addressing the development and regulation of human drugs or medical devices.[b]

Other Studies on Clinical Research and Applications

Not specific to genome editing but pertinent to discussions about all of its clinical applications are several other recent National Academies reports.

- *Oversight and Review of Clinical Gene Transfer Protocols: Assessing the Role of the Recombinant DNA Advisory Committee* was a report on the role of the U.S. Recombinant DNA Advisory Committee (RAC), recommending that it move toward more judicious use of its advisory power to review specific protocols, and that it focus primarily on novel applications of gene therapy or on providing a venue for broad public debate about the therapy (IOM2014).
- *Mitochondrial Replacement Techniques: Ethical, Social, and Policy Considerations:* presents an analysis of the special opportunities and concerns associated with making changes in mitochondrial DNA in gametes or embryos (NASEM, 2016e).

---

[a] *ILAR roundtable—gene editing to modify animal genomes for research scientific and ethical considerations*. http://nas-sites.org/ilar-roundtable/roundtable-activities/gene-editing-to-modify-animal-genomes-for-research (accessed October 21, 2016).
[b] *Future biotechnology products and opportunities to enhance capabilities of the biotechnology regulatory system*. http://nas-sites.org/biotech (accessed October 21, 2016).

带来公共利益和争议的技术提供自我监管途径的路线图。

公众参与科学传播的研究

（1）科学有效的交流：有研究发现，人们很少仅靠科学信息做决定。他们也会同时考虑自身的目标与价值，仅依赖知识不足以达到有效交流的目的（NASEM，2016a）。

（2）公众参与环境评估与决策（NRC，2008）描述了公众参与如何提高公共决策的质量和合法性，增强各方之间的信任与理解。

（3）了解风险：在民主社会的风险决策——风险评估必须能够响应要解决的问题和受影响的各方利益（NRC，1996）。

与美国国家科学院的人类基因组编辑倡议相一致，这些研究代表了试图解决基因组编辑使用所带来的科学、伦理及管理问题的一系列努力。

委员会秉承基因组编辑国际峰会的倡议和精神，开展一项长达一年之久的更为深入的研究计划。如任务声明（延伸内容1-1）中所述，委员会严格审查了人类基因组编辑方面的科研现状、潜在的应用，以及如何管理这一强有力的新技术所带来的伦理问题。委员会学习讨论并综合了其他学术讨论和专家意见达成一致后形成了该份报告。我们期待着由英国皇家学会和中国科学院举办的其他相关活动，包括2017年在中国举办的另一个国际峰会。

研 究 背 景

美国及国际政策的讨论

最早是由那些从事开发基因组编辑工具和推进其临床应用的相关学术团体提出要对此技术的影响进行细致的论证。在2015年，由包括CRISPR/Cas9技术的开发人员在内的一群研究者和伦理学家在美国加利福尼亚州的纳帕举行了峰会，会后对

- *Guidelines for Embryonic Stem Cell Research* (IOM, 2005; NRC and IOM, 2007, 2008, 2010): outline the regulatory landscape and provide a roadmap for professional self-regulation for emerging technologies that generate considerable public interest and controversy.

Studies on Public Engagement and Science Communication

- *Communicating science effectively: A research agenda* finds that people rarely make decisions based only on scientific information; they also consider their own goals and values, and a focus on knowledge alone is not enough to achieve communication goals (NASEM 2016a).
- *Public participation in environmental assessment and decision making* (NRC 2008) describes how public participation can improve the quality and legitimacy of policy decisions, and enhance trust and understanding among all parties.
- *Understanding risk: Informing decisions in a democratic society* describes how risk characterization must be responsive to the problem to be solved and to the interests of the parties affected (NRC 1996).

In concert with the National Academies' Human Gene-Editing Initiative, these studies represent a series of efforts exploring scientific, ethical, and governance issues raised by potential uses of genome-editing.

This committee was convened to continue the dialogue initiated by the International Summit and to undertake a year-long, in-depth consensus study. As specified in its statement of task (see Box 1-1), the committee examined the state of the science in human genome editing, its potential applications, and the ethical issues that need to be considered in deciding how to govern the use of these powerful new tools. This report is the product of that study and, as with all other Academies consensus studies, underwent peer review by an independent panel of experts. Additional activities of the Chinese Academy of Sciences and The Royal Society of the United Kingdom are anticipated, including another international summit to take place in China in 2017.

BACKGROUND

U.S. and International Policy Discussions

Among the earliest calls for a detailed examination of the implications of genome-editing technologies were those made by members of the scientific community

进行人类基因组编辑研究提出了相应的要求，并对其可接受的应用提出了指导建议（Baltimore et al., 2015）。在同一年，一些发表在学术期刊和大众杂志上的文章及评论呼吁重视由CRISPR/Cas9等类似的基因组编辑技术所引发的学术和伦理挑战（Editing Humanity, 2015; Bosley et al., 2015; Lanphier et al., 2015; Maxmen, 2015; Specter, 2015）。

随后，专业机构、国际组织和美国国家科学院及医学院通过发表基因组编辑适当用途，特别是可遗传基因修饰的潜在可能，进一步优化了对这一问题的阐述。这些组织团体包括：英国医学科学院（U.K. Academy of Medical Sciences）及其合作伙伴，欧洲科学与新技术伦理委员会（European Group on Ethics in Science and New Technologies）（欧盟委员会主席的谏言团队），国际干细胞研究学会（International Society for Stem Cell Research, AMS et al., 2015; Council of Europe, 2015; EGE, 2016; Friedmann et al., 2015; Hinxton Group, 2015; ISSCR, 2015）。联合国教科文组织（United Nations Educational, Scientific, and Cultural Organization, UNESCO）在2015年亦颁布了更新后的针对基因组编辑技术使用的指导意见。包括法国国家医学会（Académie Nationale de Médecine, France）、法国国家健康与医学研究院（Institut Nationale de la Santé et de la Récherche Médicale, France）、德国柏林-勃兰登堡科学与人文学院（Berlin-Brandenburg Academy of Sciences and Humanities; BBAW, 2015）、德国国家科学院与工程科学院（National Academy of Sciences Leopoldina in partnership with the Deutsche Akademie der Technikwissenshaften）、德国联邦科学与人文学院（Union of German Academies of Sciences and Humanities）、德国研究基金会（German Research Foundation; Leopoldina, 2015）、欧洲医学院校联盟（Federation of European Academies of Medicine）、英国医学科学院（UK Academy of Medical Sciences; FEAM and UKAMS, 2016）、荷兰皇家艺术与科学研究院（Royal Netherlands Academy of Arts and Sciences; KNAW, 2016）、纳菲尔德生物伦理理事会（Nuffield Council on Bioethics; Nuffield Council, 2016b）等在内的其他团体组织（延伸内容1-3）也启动了相应的措施，对基因组编辑的应用做出了更为细致的规范。

engaged in developing these tools and advancing their clinical applications. In 2015 a group of investigators and ethicists, including CRISPR/Cas9 developers, met in Napa, California, and subsequently published a request for the community to explore the nature of human genome editing and provide guidance on its acceptable uses (Baltimore et al., 2015). That same year, a number of articles and commentaries appearing in scientific journals and the popular press called attention to scientific and ethical challenges that would be posed by CRISPR/Cas9 and similar genetic tools (Bosley et al., 2015; Editing Humanity, 2015; Lanphier et al., 2015; Maxmen, 2015; Specter, 2015).

Professional bodies, international organizations, and national academies of sciences and medicine further raised the profile of genome editing by issuing statements on its appropriate uses, particularly in reference to the potential for creating heritable genetic modifications. Among others, they included the U.K. Academy of Medical Sciences and a number of collaborative partners; the European Group on Ethics in Science and New Technologies, an advisory body to the president of the European Commission; the Council of Europe; and the International Society for Stem Cell Research (AMS et al., 2015; Council of Europe, 2015; EGE, 2016; Friedmann et al., 2015; Hinxton Group, 2015; ISSCR, 2015). The United Nations Educational, Scientific and Cultural Organization (UNESCO) (2015) issued updated guidance to reflect genome-editing advances. Others launched activities to examine the implications of genome editing in greater detail, including the Académie Nationale de Médecine (France) (ANM, 2016); Institut Nationale de la Santé et de la Récherche Médicale (France) (INSERM; Hirsch et al., 2017); Berlin-Brandenburg Academy of Sciences and Humanities (BBAW, 2015); National Academy of Sciences Leopoldina in partnership with the Deutsche Akademie der Technikwissenshaften (National Academy of Science and Engineering: "acatech"); Union of German Academies of Sciences and Humanities, and German Research Foundation (Leopoldina 2015); Federation of European Academies of Medicine; UK Academy of Medical Sciences (FEAM and UKAMS, 2016); Royal Netherlands Academy of Arts and Sciences (KNAW, 2016); Nuffield Council on Bioethics (Nuffield Council, 2016b), and others (see Box 1-3).

延伸内容 1-3

全世界关于此次研究学习和公众讨论的节选
Excerpts from Selected Calls Around the World for Continued Study and Public Discussion

中国、英国和美国

这是人类历史上一个重要的时刻，我们有责任让大众了解这项技术，特别是那些可能会影响到下一代的应用（NASEM，2016d）。

法国

"我们的建议包括建立一个由不同学科专家组成的欧洲委员会来评价 CRISPR/Cas9 技术使用的范围、效率及安全性，并重新评估关于在生殖细胞系上的所有遗传修饰的禁令"（Hirsch et al.，2017）。

"［我们建议］针对生殖细胞和胚胎细胞的基因组编辑对未出生的孩子及后代的基因组可能的影响开展多学科讨论，并将此作为所有医疗技术应用大辩论的一部分"（ANM，2016）。

德国

"一个客观的讨论是很重要的，借此清晰透明地将有关技术研究和发展现状传达给所有的利益相关者，并确保任何被采纳的决定都具有健全的科学依据"（Leopoldina，2015）。

荷兰

"公开辩论能够给患者、技术提供者及社会大众提供讨论存在争议问题的机会，以不断发展的科技洞察力去评估生殖细胞应用中潜在的风险、优势和条件，并且促进更好的实践和进一步制定规范"（KNAW，2016）。

英国

"必要的是全球的利益相关者能够积极地、更早地参与到这场辩论之中，其中的利益相关者包括但不限于生物医学家和社会学家、伦理学家、医疗保健方面的专家、科学研究的资助者、管理者、受影响的患者及其家属，还应有广泛的社会大众"（AMS et al.，2015）。

China, the U.K. and the U.S.

This is an important moment in human history and we have a responsibility to provide all sections of society with an informed basis for making decisions about this technology, especially for uses that would affect generations to come (NASEM, 2016d).

France

"Our recommendations include setting up a European committee of experts from different disciplines to assess the scope, efficacy and safety of CRISPR–Cas9, and reviewing the ban on all genetic modifications to the germline" (Hirsch et al., 2017).

"[We recommend the] establishment of multidisciplinary discussions on the questions posed by the techniques for the germline and embryonic genome editing...considered as part of a wider debate on all the medical technologies...with potential effects on the genome of unborn children and, possibly, that of subsequent generations" (ANM, 2016).

Germany

"It is important to have an objective debate that informs all stakeholders in a clear and transparent manner about the status of research and development into the techniques, and to ensure that any decisions taken are based on sound scientific evidence" (Leopoldina, 2015).

The Netherlands

"Public debate would give patients, care providers and society an opportunity to discuss controversial issues, to assess the risks, advantages and conditions of potential germline applications based on growing scientific insight, and to develop good practices and further regulation" (KNAW, 2016).

UK

"Active early engagement with a wide range of global stakeholders will therefore be needed, which should include, but not be limited to, biomedical and social scientists, ethicists, healthcare professionals, research funders, regulators, affected patients and their families, and the wider public" (AMS et al., 2015).

技术

新的或者是改进后的技术使得研究新的问题和产生新的解决方案成为可能，从而促进科学进步。在健康与医疗领域，科学家和临床医生一直想运用分子生物学的技术来帮助他们了解胚胎发育、生理学、免疫和神经系统等基础的生物学知识，进而治疗或预防疾病。目前，多种疾病发生机制研究已取得很大进展，包括镰状红细胞贫血症、肌肉营养不良、囊性纤维化、耳聋、矮小症、失明等。这些疾病的发生发展都有遗传因素参与。其中一些疾病是简单的单基因改变的结果，但更多的是更为复杂的、由遗传和环境及其他因素相互作用的结果。针对这些疾病的理解目前还停留在比较初级的阶段。除此之外，遗传序列本身也仅仅描绘了生物图谱的一部分。因此，如何以及何时调控基因的表达，包括表观基因组[11]的作用，还需要人类去积极地探索。受控基因表达和表观修饰的改变影响组织的发育与分化，并在癌症和胚胎发育等领域产生临床影响。

应用改变DNA序列的工具来帮助研究者理解或提升相应基因的功能并不是新鲜的概念。最近几年，我们已经有了一系列能够更简便、更易控制、更准确地在细胞中改变DNA序列的基因组编辑工具。这些基因组编辑工具可利用外源的核酸酶在DNA上的特定位点进行剪切，再利用细胞内部的程序来修补缺损DNA，从而使得遗传密码子增加、修改或删除。基因和基因组编辑技术在实验室应用并进一步适应其他科学挑战的速度充分反映了其在科学和临床领域中的强大影响。

最早的基因组编辑技术依赖蛋白质与特定基因序列的结合，主要包括重组酶（也称之为巨大核酸酶）、锌指核酸酶（ZFNs）、转录激活因子样效应物核酸酶（TALENs）。然而，最近出现的RNA介导识别特定序列的技术简化了基因组编辑的过程。最早在2012~2013年就有相关的研究项目描述了源于自然细菌抵抗病毒机制的CRISPR/Cas9酶改变包括人在内的DNA的作用机制（Cho et al., 2013; Cong et al., 2013; Jinek et al., 2012, 2013; Mali

---

[11] 术语"表观基因组"指影响到基因表达的，发生在与基因结合的DNA、蛋白质和RNA上的一系列化学修饰。

The Technologies

New or improved tools facilitate scientific progress by making it possible to investigate new kinds of questions and to generate new solutions. In the area of health and medicine, scientists and clinicians have long sought to apply the techniques of molecular biology to understand basic biology—including embryonic development, physiology, and the immune and nervous systems—and to treat or prevent disease. Much progress has been made in elucidating the role of genetics in diseases, ranging from sickle-cell anemia, muscular dystrophy, and cystic fibrosis, to such conditions as deafness, short stature, and blindness. The development of many such diseases and conditions has a genetic component. Some result from straightforward single-gene changes, but most involve a complex interplay of genetic, environmental, and other factors that remain only imperfectly understood. Furthermore, genetic sequences themselves paint only part of the biological picture. Regulation of how and when genes are turned on and off, including the role of the epigenome,[11] continues to be actively explored. Controlled gene expression and epigenetic alterations influence how tissues develop and differentiate and have clinical ramifications in such areas as cancer and embryonic development.

Tools that enable investigators to alter DNA sequences in order to understand or improve their function are not new. Recent years, however, have seen the development of a suite of genome-editing tools that allow for easier, better controlled, and more accurate changes to DNA inside cells. These tools are based on exogenous enzymes that cut DNA at specific locations, combined with endogenous processes that repair the broken DNA, thereby enabling letters of the genetic code to be added, modified, or deleted. The speed with which this technology has been adopted in research laboratories and further adapted to tackle additional scientific challenges is a reflection of how powerful a technique the editing of genes and genomes will be for the scientific and clinical communities.

The earliest applications of nuclease-based genome-editing methods employed targeted recognition of specific DNA sequences by proteins: homing nucleases (also known as meganucleases), ZFNs, and TALENs. However, the recent development of RNA- based targeting has greatly simplified the process of genome editing. The first publications on the subject, in 2012-2013, explained how the CRISPR/Cas9 system, derived from a natural bacterial defense mechanism against infecting viruses,

---

[11] The term "epigenome" refers to a set of chemical modifications to the DNA of the genome and to proteins and RNAs that bind to DNA in the chromosomes to affect whether and how genes are expressed.

et al., 2013)。这是基因组编辑领域内的一大进步。这一技术迅速被全世界的科学家采用，并加快了包括在实验室中利用该技术改造细胞来研究特定基因的功能、利用实验室的干细胞或实验动物开发用于研究人类疾病的模型、创造提高食物产量的植物或动物、开发用于人类的新型医疗手段等进程。基因组编辑技术正迅速地成为实验室和生物技术公司中不可或缺的核心技术，并已开始向临床领域转化（Cyranoski，2016；Reardon，2016；Urnov et al.，2010)。

问题

个体层面

与其他种类的医疗干预手段一样，基因组编辑技术能否应用于治疗患者取决于对安全性和有效性的理解，以及预期效益与不良反应之间的合理平衡。基因组编辑技术治疗主要对可影响特定靶标功能的DNA序列进行修改，且极力避免改变其他DNA序列。后一种情况的发生称之为脱靶效应。大部分脱靶效应并不导致明显的后遗症，但是少量的脱靶效应可能导致一些损伤性后遗症。这些后遗症的好坏则取决于这些遗传改变所处的位点及其作用。总之，基因组编辑技术在研究和开发新的治疗手段的过程中都有相同的问题：哪些情况和疾病适合使用这类技术，怎样识别和评价脱靶效应及其潜在的副作用，哪些类型的患者更适合这些研究。如本报告所描述的一样，仅针对受试者本身（不包括其后代）的基因组编辑技术的监管体系在美国和其他许多国家已经出现，且这些监管制度还可以进一步改善。

社会层面

基因组编辑技术的社会影响力因其具体应用不同而异。对于那些不具有遗传特性或其效应仅局限于某个个人的基因组编辑技术来说，它的社会影响与传统的医疗手段并没有太大的差异。不同的是，可能被后代所继承的基因组改变会引发很多问题，包括：基因组编辑的长期效果及其可预测性程度，人类刻意地改变他们的遗传背景是否适当（Frankel and Chapman，2000；Juengst，1991；Parens，1995）。除此之外，确定基因组编辑技术广泛的应

can be harnessed to make controlled genetic changes in any DNA, including that of human cells (Cho et al., 2013; Cong et al., 2013; Jinek et al., 2012, 2013; Mali et al., 2013). This was a game-changing advance. These methods have rapidly been adopted by scientists worldwide and have greatly accelerated fundamental research that has included altering cells in the laboratory to study the functions of particular genes, developing models for studies of human diseases using stem cells or laboratory animals, creating modified plants and animals to improve food production, and developing therapeutic uses in humans. Genome editing has rapidly become an invaluable core technology in research laboratories and biotechnology companies, and is already moving into clinical trials (e.g., Cyranoski, 2016; Reardon, 2016; Urnov et al., 2010).

The Issues

Individual-Level Concerns

As with other types of medical interventions, whether genome editing can be used in patients will depend largely on understanding the safety and efficacy of the treatment and evaluating whether the anticipated benefits are reasonable with respect to the risk of adverse effects. Treatments based on genome editing are intended to make controlled modifications to specific portions of the DNA that affect the functions of their target(s) while avoiding changes to other portions whose alteration is not desired. The latter alterations, referred to as off-target events, could have consequences, many unnoticeable but others damaging, depending on their location and their effects. In general, human genome editing raises questions common to the process of researching and developing new treatments: which conditions or diseases are most suitable to address with these technologies, how to identify and evaluate off-target events and other potential side effects, and which patients are most appropriate for studies. As described in this report, regulatory systems for addressing the individual-level concerns associated with genome editing already exist in the United States and many other countries, but can be improved.

Societal-Level Concerns

The use of genome editing also has significant social dimensions that vary depending on the proposed application. The use of a genome-editing treatment whose effects are nonheritable and are restricted to an individual patient may not differ greatly from the use of a traditional drug or medical device. By contrast, making changes that may be inherited by future generations raises questions about the extent to which the long-term effects of proposed edits can be predicted and whether it

用范围对传统疾病和残疾构成的常规理念来说是另一个巨大的挑战。社会上对于基因组编辑技术致力于增强人类能力的担忧尤为严重。此类应用亦会引发关于怎样定义和促进社会公平的担忧（President's Council on Bioethics，2003）。此外，与其他的遗传技术一样，基因组编辑技术的应用引起了人们对历史上的强制和滥用的优生计划的关注，这些计划是基于错误的科学认知，并服务于歧视的政治目的（Wailoo et al.，2012）。

安全性和有效性

尽管对于基因组编辑技术本身的讨论并非刚刚开始，但是过去的人类细胞遗传修饰费时、困难且昂贵，在专业的医疗机构之外是很难实现的。目前的基因组编辑技术，特别是 CRISPR/Cas9 体系扩大了可能的使用者范围。基因组编辑技术的快速发展和应用也缩短了对决定或开发哪些适当的管理机构的讨论时间。随着这些技术的安全性和有效性的不断提高，最关键的问题从科学家和临床医生是否可以运用基因组编辑技术来对 DNA 进行改变，逐渐转变成了他们是否应该这么做。生物黑客与 DIY 社区已经有关于对非人物种的 DIY 编辑和对基因组编辑工具使用的讨论（Brown，2016；Ledford，2015）。围绕可接受的基因组编辑技术应用于人类的棘手问题将取决于更多的科学认知，并且可能越来越多地涉及超越个人层面的风险和利益的权衡因素。

人类基因组编辑涉及的科学与伦理问题其实就是如何管理其应用，以便于适当应用，同时避免滥用。因各国的文化、政治和法律背景的差异，确定技术使用的限制和实施这些限制所需的监管机制将有所不同。但是，是否以及如何更好地推动人类基因组编辑向前发展，需要公开讨论和投入决策及跨国科学合作。科学家和其他利益相关者参与这样的活动有足够多的先例。同时本报告将充分参考一系列国际公约和宣言中提出的要点，如奥维耶多公约（Oviedo Convention，1997）、国际人类基因数据宣言（the International Declaration on Human Genetic Data，2003）、生物伦理和人权宣言（the Universal Declaration on Bioethics and Human Rights，Andorno，2005；UNESCO，2004a，2005）。

is appropriate for humans to purposely alter any aspect of their genetic future (Frankel and Chapman, 2000; Juengst, 1991; Parens, 1995). In addition, identifying the increased range of applications made possible by genome editing may be yet another challenge to conventional conceptions of what constitutes a disease or disability. Societal-level concerns are particularly acute with respect to genome-editing interventions aimed at enhancing human capabilities. Such applications also raise questions about how to define and promote fairness and equity (President's Council on Bioethics, 2003). Moreover, as with other genetic technologies, such genome-editing applications may raise concerns about coercive and abusive eugenics programs of the past, which were based on faulty science and served discriminatory political goals (Wailoo et al., 2012).

Looking Beyond Safety and Efficacy

Although the nature of the debate surrounding genome editing is not new, the tools available in the past for making genetic modifications in human cells were time-consuming, difficult, and expensive, and were unlikely to be used outside of specialized medical applications. Recent genome-editing technologies, particularly the CRISPR/Cas9 system, have greatly expanded the landscape of potential applications and potential users. Their rapid development and adoption also have shortened the timeline for discussion of what appropriate governance structures need to be identified or developed. As the safety and efficacy of these technologies continue to improve, the critical question will become not whether scientists and clinicians can use genome editing to make a certain change, but whether they should. There is already discussion of do-it-yourself (DIY) editing and the use of genome editing tools by the biohacker and DIY biology communities, albeit in non-human organisms (Brown, 2016; Ledford, 2015). Thorny issues around acceptable uses of the technology in humans will depend on more than scientific considerations, and may increasingly involve weighing factors beyond individual-level risks and benefits (NRC, 1996).

Layered on the scientific and ethical issues associated with human genome editing is the question of how to govern its application so as to facilitate its appropriate use and avoid its misuse. Determining the limits of the technologies' uses and the regulatory mechanisms needed to enforce these limits will vary according to each nation's cultural, political, and legal context. But whether and how best to move human genome editing forward has implications for transnational scientific cooperation that require ongoing public discussion and input into policy making. There is ample precedent for scientists and other stakeholders to engage in just such activities, and this

研 究 方 法

为了完成上述复杂的任务（延伸内容 1-1），委员会召集了基础和临床研究的专家、人类基因疗法的研发专家，以及美国和国际的法律专家。其中包括生物学家、伦理学家和社会科学家，并纳入了潜在受影响的患者和社会利益相关者的意见。因为人类基因组编辑所带来的伦理和社会问题是超越国界的，因而委员会不仅包括美国的成员，也包括来自加拿大、中国、埃及、法国、德国、以色列、意大利、西班牙和英国的公民，或目前在这些国家工作的人们。委员会成员的简要履历见附录 D。

这项研究不仅受到先前所述国际峰会（International Summit，在委员会的第一次会议前召开）的支持，而且得到重要文献综述、其他会议以及积极与委员会分享研究和观点的演讲者的协助。有关这项研究过程的详细信息在附录 C 中给出。

在评价新的基因组编辑工具的影响时，委员会还审查了其他人类医学技术方面的科学进展、伦理争议和监管结构，如人类辅助生殖技术、干细胞疗法、基因转移和线粒体替换技术。这些发展与基因组编辑相互交融，因为干细胞编辑对治疗和预防疾病在临床上有应用潜力，而生殖技术可与未来能应用的任何可遗传的基因组编辑结合使用。随着这些技术的快速发展，必须发展法律和监管框架、道德行为规范，以提供适当的人类使用和监督指导（Health Canada，2016；HFEA，2014；IOM，2005；NASEM，2016e；NRC and IOM，2007，2008；Nuffield Council，2016a；Präg and Mills，2015；Qiao and Feng，2014）。这里引用的报告为委员会评估人类基因组编辑工具的使用提供了基础，并在随后的相关章节中引用。

报 告 概 要

该报告首先回顾了委员会在管理人类基因组编辑中所体现的一整套国际规范（第 2 章），随后概述了美国关于基因组编辑的研究与临床应用的法规，并与其他国家监管体系作出比较。

report is intended to build on points raised by a number of international conventions and declarations, such as the Oviedo Convention (1997), the International Declaration on Human Genetic Data (2003), and the Universal Declaration on Bioethics and Human Rights (2005) (Andorno, 2005; UNESCO, 2004a, 2005).

STUDY APPROACH

To address its complex task (see Box 1-1), the committee included members with expertise in basic and clinical research, in the development of human genetic therapies, and in U.S. and international legal and regulatory frameworks. It included biologists, bioethicists, and social scientists, and incorporated perspectives from potentially affected patient and stakeholder communities. Because the ethical and social issues posed by human genome editing transcend national boundaries, the committee included not only U.S. members but also those who are citizens of or are currently working in Canada, China, Egypt, France, Germany, Israel, Italy, Spain, and the United Kingdom. Brief biographies of the committee members are found in Appendix D.

This study was informed not only by the International Summit described earlier, which immediately preceded the committee's first meeting, but also by review of the salient literature, additional meetings, and speakers who generously shared their knowledge with the committee. Further information on the process by which the committee conducted this study is provided in Appendix C.

In evaluating the implications of new genome-editing tools, the committee also reviewed scientific progress, ethical debates, and regulatory structures related to the use in humans of medical developments such as assisted reproductive technologies, stem cell therapies, gene transfer, and mitochondrial replacement techniques. These developments interface with those of genome editing since editing of stem cells has potential clinical applications for treating or preventing disease, and reproductive technologies would have to be used in combination with genome editing for any heritable application of the latter technologies. As these other technologies have advanced, legal and regulatory frameworks and ethical norms of conduct have been developed to provide guidance on their appropriate human uses and oversight (Health Canada, 2016; HFEA, 2014; IOM, 2005; NASEM, 2016e; NRC and IOM, 2007, 2008; Nuffield Council, 2016a; Präg and Mills, 2015; Qiao and Feng, 2014). The reports cited here helped provide a basis for the committee's assessment of the use of genome-editing tools in humans and are referenced in subsequent chapters where relevant.

基于这一原则和规范，第 3 章至第 6 章专述了人类基因组编辑技术在四个特定应用方面相关的科学问题、监管环境及伦理问题。体细胞的实验室研究，以及人类生殖细胞、配子和早期胚胎的非遗传性实验室研究放在第 3 章。第 4 章探讨了基因组编辑对体细胞干预的应用，侧重于包括胎儿在内的治疗应用。第 5 章讨论基因组编辑技术在生殖系细胞中的潜在应用和人类患者的临床治疗应用。第 6 章考虑了人类基因组编辑在增强人类功能方向的潜在用途，而非治疗或预防疾病和残疾。

随后的第 7 章从应用类别分析转向讨论未来公共投入在美国和其他国家对基因组编辑技术的管理作用。该章探索公众参与不同类别的基因组编辑应用，并探索这样的公众参与模式的潜在优势和局限性。

最后，第 8 章回到整体性的原则，以及在人类基因组编辑的大背景下这些原则所承担的责任上。这一章根据这些基本概念总结这篇报告的结论和建议。

ORGANIZATION OF THE REPORT

The report begins by reviewing international norms that are embodied in the set of overarching principles adopted by the committee for governance of human genome editing (Chapter 2). The chapter continues with an overview of the U.S. regulation of research and clinical application of genome editing, drawing comparisons where appropriate to other national systems of oversight.

With this grounding in principles and regulation, Chapters 3-6 delve into human genome-editing technology and the scientific issues, regulatory context, and ethical implications of four specific applications. Laboratory research conducted in somatic cells and non-heritable laboratory research in human germ cells, gametes, or early-stage embryos is covered in Chapter 3. Chapter 4 examines the uses of genome editing for somatic interventions focused on therapy, including fetal therapy. Chapter 5 addresses the use of genome-editing technology in germline cells for potential research and clinical therapeutic applications in human patients. Chapter 6 considers the potential use of human genome editing to enhance human functions rather than to treat or prevent disease or disability.

The subsequent chapter (Chapter 7) turns from analysis of these categories of application to the role of public input in determining how genome-editing technology should be governed in the future, both in the United States and in other countries. The chapter considers public engagement for different categories of genome-editing applications and explores strengths and limitations of potential models for undertaking such public engagement.

Finally, Chapter 8 returns to the set of overarching principles and the responsibilities that flow from them in the context of human genome editing. The chapter pulls together the report's conclusions and recommendations in light of these fundamental concepts.

2
人类基因组编辑监管的总体原则[12]
Oversight of Human Genome Editing and Overarching Principles for Governance[12]

人类基因组编辑的监管是在基因治疗监管的总体框架里。该框架嵌入在更大范围内的国际公约和规范里，这些公约和规范旨在保护人权，以及更确切地说是保护涉及人类项目与临床治疗的研究。从这些国际文书中，人们能够总结出在美国及世界范围内普遍适用的基因组编辑总体治理原则，并反映到世界各国所采用的具体法律法规中。

本章从描述委员会所采用的针对人类基因组编辑中的总体原则开始。这些原则来源于被广泛认可的国际法律法规，也会给这些国际法律法规反馈相关的结论和建议。然后，本章总结了美国对基因转移研究和治疗的监管，并简要回顾了其他国家使用的替代方法（其中一些在附录B进行了深入探讨）。第3章到第6章主要讨论美国监管体系处理由基因组编辑引发的特定技术与伦理问题的适用性。

人类基因组编辑的监管原则

Louis Pasteur 曾说："科学无国界，因为知识是全人类的遗产"。尽管科学是全球性的，但科学却是在各种政治制度和文化规范中进行的。找出超

Oversight of human genome editing fits within the overarching framework of oversight of gene therapy. That framework is embedded within the larger context of international conventions and norms for protection of human rights and, more specifically, for research involving human subjects and clinical care. From these international instruments, one can derive principles for governance of genome editing that have general application within the United States and across the globe, and are reflected in the specific statutory and regulatory rules that are adopted by various nations.

This chapter begins by describing the overarching principles for human genome editing adopted by the committee for this study, which are informed by those international instruments and national rules, and which in turn inform the conclusions and recommendations presented in this report. It then provides an in-depth look at U.S. governance of gene transfer research and therapy, and a brief review of alternate approaches used in other countries (some of which are explored in greater depth in Appendix B). Conclusions regarding the adequacy of U.S. oversight systems to deal with the specific technical and ethical issues raised by genome editing appear in Chapters 3 through 6.

PRINCIPLES FOR GOVERNANCE OF HUMAN GENOME EDITING

Louis Pasteur once said: "La science n'a pas de patrie, parce que le savoir est le patrimoine de l'humanité"

---

[12] 本章的部分内容摘自或者修改自美国医学研究所（IOM，2014）及国家工程医学科学院（NASEM，2016e）的报告。

[12] Portions of this chapter were adapted and updated from Institute of Medicine (IOM, 2014) and the National Academies of Sciences, Engineering, and Medicine (NASEM, 2016e).

越这些差异与分歧并且容纳文化多样性的原则是非常重要的，也是非常困难的。在整体的伦理原则上达成共识并以此为基础提出相关的执行建议无疑是困难的，因为没有一个伦理理论能够被哲学家和神学家同时接受，或者是因为很难找到一种方法从这些理论中推导出原理来。功利主义者可能对评估整体利弊的必要性表示同意，但也许不会同意是否去评价某条规范或者某条特定法案的后果。道义论者不仅将努力争取得到一个对基本行为规范的守护性列表，同时也将面临严格遵守规章制度可能会导致的直观上不可接受甚至具有破坏性的结果。其他理论也面对着类似的困境。

生物伦理学作为应用伦理学的一种形式，具有所有这些复杂性。我们可以从高级理论中推理出所有的原则和某种行为，也可以从特定情况演绎推理出一般原则，即反理论。关于我们是否应当制定出这样的高级理论或反理论，在很长一段时间都受到争议。当生命伦理纳入公共政策，而不是个人临床伦理分析之时，更广泛的如公民多元化的社会、民主理论、责任分配和福利等都应囊括在被关注范围内[13]。

这些理论是建立在功利结果论、道义论或美德伦理基础之上的，无论推论是否起始于理论，随着时间的推移都会出现。"反推平衡"这个概念包含了归纳和演绎推理，也结合了理论和以案例为基础的辩论。因为这是在不管个人精神或宗教取向的情况下，出于公众对推论理解的需要（Arras，2016），这个概念是可以被大众接受的。它有助于形成对全球具有影响力的声明和指导性文件。

第二次世界大战后短暂施行的《联合国人权宣言》（The Universal Declaration of Human Rights, UN, 1948）成为许多具体声明、公约及协议的基本文件。在文件的序言中指出，"对于人类大家庭所有成员而言，对其固有尊严、平等和不可剥夺的权利的认可是自由、正义与世界和平的基础……"。同时它的第一条款也指出："每个人在尊严及权利面前都是生来平等和自由的"。其他一些国际文件都应该建立在这个核心原则基础之上。例如，《儿童权利公约》（The Convention on Rights of the Child）要求为儿童最佳的发展提供条件，如

(Science has no homeland, because knowledge is the heritage of humanity). But while science is global, it proceeds within a variety of political systems and cultural norms. It is important to identify principles that can transcend these differences and divisions, while accommodating cultural diversity. This is no easy task. Achieving consensus around overarching ethical principles to undergird specific recommendations for action can be difficult, whether because no one theory of ethics has been accepted by philosophers and theologians or because no one algorithm for deriving principles from those theories has been found. Utilitarians may agree on the need to evaluate overall beneficial consequences, but may disagree on whether to evaluate the consequences of a rule or of a specific act. Deontologists not only will struggle to derive a defensible list of fundamental rules of behavior, but also will be confronted with specific cases in which adherence leads to results that are intuitively unacceptable or even destructive. Other theories suffer from similar complications.

Bioethics, as a form of applied ethics, has suffered from all these complexities. It has also been dogged by long-standing debates about whether the best approach is high theory, from which all principles and specific actions flow, or anti-theory, in which deductive reasoning from specific cases leads to generalizable principles. And when bioethics is incorporated into public policy making, as opposed to individual clinical ethics analyses, it is necessary to incorporate a wider range of concerns about multicultural civil society, theories of democracy, and just distribution of burdens and benefits.[13]

Regardless of whether reasoning begins with theories grounded in utilitarian consequentialism or deontology or virtue ethics, there has emerged over time what some deem "reflective equilibrium." This concept encompasses the use of both inductive and deductive reasoning, incorporating both theory and case-based casuistry, and accepting the need for reasoning that is understandable to the public, regardless of individual spiritual or religious orientation (Arras, 2016). It has helped shape influential statements and guidance documents across the globe.

The Universal Declaration of Human Rights (UN, 1948), adopted shortly after World War II, became the foundational document for many of the more particularized declarations, conventions, and treaties that followed. In its preamble, it states that "recognition of the inherent dignity and of the equal and inalienable rights of all members of the human family is the foundation of freedom, justice and peace in the world," and its very first provision reads, "All human beings are born free and equal in dignity and rights." Other international documents

---

[13] 有关进一步的讨论可参考 Arras (2016)。

[13] Further discussion of these issues can be found in Arras (2016).

医疗保健和卫生设施（The Universal Declaration of Human Rights，1990）。《残疾人权利公约》（the Convention on the Rights of Persons with Disabilities）强调"尊重残疾人内在的尊严"、"残疾人作为人类多样性之一，也是出于人道，我们应尊重他们的差异并接受残疾人"，以及"尊重残疾儿童的发展能力和尊重残疾儿童的权利来保护他们这个特殊人群"（UN，2006）。对于每一个国家而言，不是每个公约都对其具有全部或部分的法律约束力，但即使没有纳入国家法规或应用于国家法庭案件，这些公约背后的原则也已经成为全球规范的重要元素和指向。

其他国际活动也在更紧密地关注生物医学研究。国际医学组织理事会（The Council for International Organizations of Medical Sciences，CIOMS）是一个国际性的、非政府、非营利性组织。国际医学组织理事会和世界卫生组织（the World Health Organization，WHO）、联合国教科文组织（the United Nations Educational，Scientific and Cultural Organization，UNESCO）共同成立于1949年，其成员来自全球近50个机构，包含专业协会、国家科学院、研究委员会。除此之外，基于一些指导文件，对健康研究[14]领域也发行了国际准则，如世界医学协会的《赫尔辛基宣言》（the World Medical Association's Declaration of Helsinki，WMA，2013）和联合国教科文组织（2005）的《世界生物伦理与人权宣言》。2016版本的指南（van Delden and van der Graaf，2016）强调"研究需要有科学意义和社会价值，对于在资源匮乏地区的健康研究可以提供特殊指南，对于研究中涉及的弱势群体应该详细规定其范围，并且详细规定在什么条件下生物样品和与健康有关的数据可以用于研究"（CIOMS，2012）。基因组编辑相关政策问题详见指南1，它强调了需要具有保护和促进健康的知识，还强调了它和指南2、3、4的关系，这几个指南主要阐述了对于个人和团体之间、风险和收益之间的平衡与分配的公平性（包括在资源丰富和资源匮乏的国家人口分布情况）。此外，指南7和大众密切相关，不仅是良好政策的确立和政策合法化所需要的，也将有利于将科研成果转化为临床

build on this core principle. The Convention on Rights of the Child, for example, calls for providing conditions for optimal development, such as health care and sanitation (UNICEF, 1990). And the Convention on the Rights of Persons with Disabilities emphasizes "respect for inherent dignity," "respect for difference and acceptance of persons with disabilities as part of human diversity and humanity," and "respect for the evolving capacities of children with disabilities and respect for the right of children with disabilities to preserve their identities" (UN, 2006). Not every convention is legally binding in whole or part on every country, but even where not incorporated into domestic statutes or applied in domestic court cases, the principles underlying these conventions have become important elements of global norms and aspirations.

Other international activities are focused more closely on biomedical research. The Council for International Organizations of Medical Sciences (CIOMS) is an international, nongovernmental, nonprofit entity established in 1949 jointly by the World Health Organization (WHO) and the United Nations Educational, Scientific and Cultural Organization (UNESCO), whose members include nearly 50 organizations—professional societies, national academies, research councils—from across the globe. Among other things, it issues international guidelines for health research[14] based on such guidance documents as the World Medical Association's Declaration of Helsinki (WMA, 2013) and UNESCO's (2005) Universal Declaration on Bioethics and Human Rights. The 2016 version of the guidelines (van Delden and van der Graaf, 2016) stresses "the need for research having scientific and social value, by providing special guidelines for health-related research in low-resource settings, by detailing the provisions for involving vulnerable groups in research and for describing under what conditions biological samples and health-related data can be used for research" (CIOMS, 2012). Of particular relevance to genome-editing policy questions are Guideline 1, emphasizing the need to generate knowledge to protect and promote health, and its relationship to Guidelines 2, 3, and 4, which focus on fairness in the balance and distribution of risks and benefits to individuals and groups (including distribution among populations of high- and low-resource countries). Also of particular relevance is Guideline 7 on public engagement, needed not only to develop and legitimize good policy but also to help translate research into clinical benefit.

In the United States, the landmark 1979 *Belmont Report* of the National Commission for the Protection of Human Subjects in Biomedical and Behavioral research (HHS, 1979) focused on avoiding infliction of harm,

---

[14] See http://www.cioms.ch (accessed January 5, 2017).

成果。

1979年，美国国家委员会为保护在生物医学和行为研究中的人类受试者（HHS，1979）发布了具有里程碑意义的《贝尔蒙特报告》（Belmont Report），该报告旨在避免对受试者造成伤害，并对捐献者承担应有的责任，坚守公正的原则。多年来，研究伦理学的这些主体被不断诠释、扩充、深化和应用，并被纳入到美国的人类参与者管理研究系统中（21 CFR Part 50 和 45 CFR Part 46）。在实践中，对个人和社会而言，这带来了风险与利益之间的合理平衡和公平共享。这些原则特别注重尊重个人自主性，一般都要求知情并要求自愿参与，防止因无行为能力或环境而易受伤害者受到胁迫或虐待。

因为基因组编辑科学及应用都将超越国界，下文将详述在这些国际和国家标准的基础上管理人类基因组的核心原则。其中的一些原则与生物医学研究和护理相关，而另一些则在当前新兴的技术背景下非常重要，但都是管理人类基因组编辑的基础。

在这种情况下，委员会主要以以下几点作为原则：将保护和促进个人的健康和人类更好地生存作为关注的重点；关注新技术不断发展的情况；尊重个人权利；防止产生有害的社会影响；公平分配责任和利益。社会和法律文化的差异必然会导致不同国家用不同的国内政策来管理每个国家内基因组编辑的应用。尽管如此，一些原则仍然可以跨越国界共享。因此，尽管本文提供的总体原则是主要针对美国政府的，但是这些原则和它们应尽的责任在全球都是通用的。这些原则见延伸内容2-1。

accepting a duty of beneficence, and maintaining a commitment to justice. These pillars of research ethics have been interpreted, expanded, deepened, and applied over the years and incorporated into the U.S. system for governing research with human participants (21 CFR Part 50 and 45 CFR Part 46). In practice, they have resulted in a focus on ensuring a reasonable balance between risk and hoped-for benefits, to the individual and to society, and on ensuring that both risks and benefits are equitably shared. These principles also have come to incorporate particular attention to the need for respect of individual autonomy, in the form of generally requiring informed and voluntary participation, and the need to provide special protection against coercion or abuse of those who are vulnerable because of incapacity or circumstances.

Because both the science and the applications of genome editing will transcend national boundaries, the core principles for governance of human genome detailed below build on the foundations of these international and national norms. Some of these principles are generally relevant to biomedical research and care, while others are of particular importance in the context of an emerging technology, but all are foundational for the governance of human genome editing.

In this context, the committee focused on principles that are aimed at protecting and promoting the health and well-being of individuals; approaching novel technologies with careful attention to constantly evolving information; respecting individual rights; guarding against unwanted societal effects; and equitably distributing information, burdens, and benefits. Differences in social and legal culture inevitably will lead to different domestic policies governing specific applications of genome editing. Nonetheless, some principles can be shared across national borders. Thus, while the overarching principles presented here are aimed primarily at the U.S. government, they and the responsibilities that underlie them are universal in nature. The principles are listed in Box 2-1 and elaborated below.

延伸内容 2-1

人类基因编辑研究和临床应用的总体性原则

Overarching Principles for Research on and Clinical Applications of Human Gene Editing

基因组编辑对深入理解生物学，预防、改善或消除人类疾病都有着极大的意义。在这样的背景下，需要有责任感和从道德上以适当的方法来进行研究和临床应用。以下的一般原则是这些方法的基础。

Genome editing holds great promise for deepening understanding of biology and for preventing, ameliorating, or eliminating many human diseases and conditions. Along with this promise comes the need for responsible and ethically apwpropriate approaches to research and clinical use. The following general principles are essential foundations for those approaches:

1) 福利提升原则
2) 透明性原则
3) 妥善关注原则
4) 科学诚信原则
5) 尊重个人原则
6) 公平原则
7) 跨国界合作原则

1. Promoting well-being
2. Transparency
3. Due care
4. Responsible science
5. Respect for persons
6. Fairness
7. Transnational cooperation

1 福利提升原则：指在促进利益的同时防止对当事人产生有害影响，这些利益及有害影响符合生物伦理论著中有关仁慈（beneficence）和不邪恶（nonmaleficence）原则。

遵守这一原则的责任包括：①促进个人健康和幸福的人类基因组编辑的应用，如治疗或预防疾病，同时使具有高度不确定性的针对个体的早期应用中的风险最小化；②确保任何人类基因组编辑应用的风险和利益合理的平衡。

2 透明原则：指对利益相关者提供容易获得的、可理解的公开信息。

针对这一原则的职责包括：①致力于最大限度地、及时地公开信息；②在制定基因组编辑及其他新的、具有争议的技术政策的过程中，切实可行地纳入公众意见。

3 妥善关注原则：指对于参与研发过程及医疗救治的患者，需向其提供细心妥善的关注，所有举动都需要在证据充足的情况下进行。

针对这一原则的职责包括：在整个程序中对患者谨慎、持续地关注，监管机制需要提供定期的、面向未来和文化层次上的反复评估。

4 科学诚信原则：指从研发到临床的过程中，始终由最高标准的科研支撑，并且与职业的、国际的规范接轨。

针对这一原则的职责包括：①高质量的实验设计和数据分析；②对实验程序和数据的审查；③信息透明；④对错误或者误导性的数据或分析进行校正。

5 尊重个人原则：指要求考虑不同个体的尊严，尊重并把个人选择放在中心地位。无论个体的遗传背景如何，所有人都享有同等的道德价值。

1. Promoting well-being: *The principle of promoting well-being supports providing benefit and preventing harm to those affected, often referred to in the bioethics literature as the principles of beneficence and nonmaleficence.*

Responsibilities that flow from adherence to this principle include (1) pursuing applications of human genome editing that promote the health and well-being of individuals, such as treating or preventing disease, while minimizing risk to individuals in early applications with a high degree of uncertainty; and (2) ensuring a reasonable balance of risk and benefit for any application of human genome editing.

2. Transparency: *The principle of transparency requires openness and sharing of information in ways that are accessible and understandable to stakeholders.*

Responsibilities that flow from adherence to this principle include (1) a commitment to disclosure of information to the fullest extent possible and in a timely manner, and (2) meaningful public input into the policy-making process related to human genome editing, as well as other novel and disruptive technologies.

3. Due care: *The principle of due care for patients enrolled in research studies or receiving clinical care requires proceeding carefully and deliberately, and only when supported by sufficient and robust evidence.*

Responsibilities that flow from adherence to this principle include proceeding cautiously and incrementally, under appropriate supervision and in ways that allow for frequent reassessment in light of future advances and cultural opinions.

4. Responsible science: *The principle of responsible science underpins adherence to the highest standards of research, from bench to bedside, in accordance with international and professional norms.*

Responsibilities that flow from adherence to this principle include a commitment to (1) high-quality experimental design and analysis, (2) appropriate review and evaluation of protocols and resulting data, (3) transparency, and (4) correction of false or misleading data or analysis.

5. Respect for persons: *The principle of respect*

针对这一原则的职责包括：①致力于保证所有个体的平等对待；②尊重并提倡个人的自主决定；③致力于防止重蹈历史上曾发生过的"优生学"的覆辙；④致力于对残疾人的去污名化。

6 公平原则：指要求相似的案例需要得到相似的处理，做到风险和利益的均等分配（分配正义，distributive justice）。

针对这一原则的职责包括：①对科研负担和利益的均等分配；②对人类基因组编辑的临床转化成果广泛、公平的享有。

7 跨国界合作原则：指在尊重文化差异的基础上，需推动科研和监管方面的跨国界合作。

针对这一原则的职责包括：①尊重各国的政策差异；②如有可能，应该统一管理标准和程序；③跨国界合作还包括不同科研团体和监管机构之间的合作与数据共享。

在美国的管理框架下，这些原则坚持：对有自主能力的人需要保证自愿和充分的知情同意；对缺乏自主能力的人有特殊保护；有效地平衡危害和利益；努力减少任何时候发生危险的可能性；公平地选择研究参与者。

美国基因治疗技术的监管

在美国，体细胞和生殖系细胞的基因组编辑被纳入到转基因研究的相关框架中，一旦实验成功，则被纳入到基因治疗的框架内，其中涉及人类细胞和组织早期的实验室研究到临床前测试、临床实验，以及批准引入治疗和批准后的监督。在国家层次上，管理在各种情况下都可以是强制性的，例如，当一项工作被提交到美国食品药品监督管理局（FDA）审批的时候，它只能对那些使用联邦基金的人强制执行。监管也可以根据专业化的标准进行自我调控。除了国家标准，有时每个州对于如胚胎研究的特定项目也有特定的规则，或者州经费应用于如胚胎干细胞研究等有附加限制。因此，不像某些司法管辖区如英国，与胚胎相关的工作只受到单一司法框架的管理，美国对于工作阶段及经费来源有不同的规定，这些规定相互重叠、相互作用，最终提供了一个相对公平、全面的管理。

一般来说，实验室的工作由研究单位的生物

for persons requires recognition of the personal dignity of all individuals, acknowledgment of the centrality of personal choice, and respect for individual decisions. All people have equal moral value, regardless of their genetic qualities.

Responsibilities that flow from adherence to this principle include (1) a commitment to the equal value of all individuals, (2) respect for and promotion of individual decision making, (3) a commitment to preventing recurrence of the abusive forms of eugenics practiced in the past, and (4) a commitment to destigmatizing disability.

6. Fairness: *The principle of fairness requires that like cases be treated alike, and that risks and benefits be equitably distributed (distributive justice).*

Responsibilities that flow from adherence to this principle include (1) equitable distribution of the burdens and benefits of research and (2) broad and equitable access to the benefits of resulting clinical applications of human genome editing.

7. Transnational cooperation: *The principle of transnational cooperation supports a commitment to collaborative approaches to research and governance while respecting different cultural contexts.*

Responsibilities that flow from adherence to this principle include (1) respect for differing national policies, (2) coordination of regulatory standards and procedures whenever possible, and (3) transnational collaboration and data sharing among different scientific communities and responsible regulatory authorities.

In U.S. regulation, these principles underlie the insistence on voluntary, informed consent from competent persons; special protections for those lacking competence; a reasonable balance between the risks of harm and potential benefits; attention to minimizing risks whenever possible; and equitable selection of research participants.

REGULATION OF GENE THERAPY IN THE UNITED STATES

Both somatic and germline human genome editing would be regulated in the United States within the framework for gene transfer research and, once approved, for gene therapy, which applies to work with human tissues and cells from the early stages of laboratory research through preclinical testing, human clinical trials, approval for introduction into medical therapy and postapproval surveillance. At the national level, regulation may be mandatory in all cases—for example, when the work is to be submitted to the U.S. Food and Drug Administration (FDA) for approval—or it can be mandatory only for those who are using federal funds. Oversight also can proceed according to voluntary self-

安全委员会（institutional biosafety committees，IBCs）进行安全管理，很多情况下也受到联邦根据《临床实验室改进修正案》（Clinical Laboratory Improvement Amendments，CLIA）的质量监管[15]。在一些情况下，使用活体来源细胞的实验工作也要受到伦理审查委员会（institutional review boards，IRBs）的监管。伦理审查委员会的工作是保护供者的个人信息免受泄漏及保证合适的知情同意。除非细胞供者可以被认出，否则使用人胚胎的实验工作不用受伦理审查委员会监管。但是这项工作或许会被自愿的个体组织监管，例如，根据国家科学院建议建立的胚胎干细胞监管委员会（embryonic stem cell research oversight committee，ESCRO），或者根据国际干细胞研究学会（International Society for Stem Cell Research；ISSCR，2016a）建议建立的胚胎研究监管委员会（embryo research oversight committee，EMRO）。临床前的动物实验由动物关爱和利用协会根据《动物福利条例》调控和监管。临床试验要经过美国国立卫生研究院（NIH）的DNA重组建议委员会（Recombinant DNA Advisory Committee，RAC）讨论，但是不需要获得IRB、FDA的许可。

FDA认为人类基因组编辑技术是基因治疗工具。FDA在现有的生物产品（包括基因治疗产品）框架条例下管理人类基因组编辑。FDA也负责批准一系列的基因治疗方式，但是至今还没有任何一种获批进入市场。一旦获得批准，它不仅要受到FDA现有条例的监管，若有必要的话，FDA还可以进一步限制它们的用途。FDA的监管需要符合生物类监管条例，以及很多情况下要进行药品类监管条例的复审。

一旦基因治疗被引入到临床，FDA不仅会继续开展生物安全方面的监管，同时也会着手对是否是标注的用途开展正式研究来重新审视生物安全和治疗效率。上市后产品的使用可能会超出治疗批准的适应证。对一个已经获批的生物制品开展在正式标注之外用途的研究一般不被视为标签外用法（"off-label"）需要FDA的监管。但是在研究之外，标签外用法是完全合法的，在药物这方面也已经成为内科医生们的一种常见做法，

regulation pursuant to professional guidelines. In addition to national rules, individual states have at times issued rules on specific topics, such as embryo research, or attached restrictions to the use of state funds, such as for embryonic stem cell work. As a result, unlike some jurisdictions, such as the United Kingdom, in which work with embryos generally falls under a single statutory framework or regulatory body, the United States has individual rules related to stage of work and source of funding that overlap and interact in a manner that, in the end, provides fairly comprehensive coverage.

In general, laboratory work is subject to local oversight by institutional biosafety committees (IBCs) whose focus is on safety, and in many cases to federal oversight for quality assurance under the Clinical Laboratory Improvement Amendments as well.[15] In some cases, laboratory work using cells from identifiable living donors also is subject to review by institutional review boards (IRBs), whose focus is on protecting donors from the effects of being identified, and on ensuring appropriate informed consent. Laboratory work using human embryos does not fall within IRB jurisdiction unless the progenitor-donors are identifiable, but this work may be overseen by voluntary oversight bodies, such as embryonic stem cell research oversight committees (ESCROs) created pursuant to NAS/IOM recommendations or the embryo research oversight committees (EMROs) recently created pursuant to recommendations of the International Society for Stem Cell Research (ISSCR, 2016a). Preclinical animal work is subject to regulation and oversight by institutional animal care and use committees pursuant to the Animal Welfare Act. Clinical trials may be the subject of discussion and advisory protocol review by the National Institutes of Health (NIH) Recombinant DNA Advisory Committee (RAC), but will nonetheless require approval by an IRB and permission from the FDA.

Human genome editing technologies are considered to be gene therapies with regard to FDA oversight, and the agency regulates human genome editing under the existing framework for biological products, which includes gene therapy products. The FDA has authorized a number of gene therapy trials but has not yet approved a gene therapy for market. If one is approved, it will still be subject to the FDA's ongoing monitoring and, if necessary, restrictions on their use. This FDA oversight entails review under rules governing biologics and, in many cases, under rules governing drugs.

Once gene therapies are introduced into clinical care, not only will the FDA maintain surveillance to detect safety concerns, but formal studies of the labeled uses also may be conducted to take a fresh look at the safety

---

[15] See https://www.cms.gov/Regulations-and-Guidance/Legislation/CLIA/index.html?redirect=/clia (accessed January 5, 2017).

一旦基因组编辑被批准，基因组编辑的生物产品也是有标签外用法的。除了专业学科的建议，内科医生也会利用他们的经验和信息来源。他们在州一级由其许可和纪律机构管理，被保险公司对于患者在新的发明方面的保险责任范围所限制，被在医疗事故中需要承担的侵权责任赔偿所束缚。

表 2-1 展示了一个由基因组编辑所创造的医用产品的发展过程中的主要监管步骤。这些独立的步骤和注意事项会在这章的后面内容中进行详细的阐述。

实验室基础研究的监管

对包括体细胞、生殖细胞、胚胎及胎儿在内的人类细胞和组织的研究监管专注解决几个关键问题。就大多数细胞和组织而言，首要的问题是提供者能否接受任何形式的现金或实物支付。目前在生殖细胞研究领域存在一个非常敏感的话题，争论的

and efficacy of the therapy. Postmarket use may also encompass uses that go beyond the indications for which a therapy was approved. Formal studies of an approved biologic for a use other than specified in the labeling would generally not be considered off-label use and would require FDA oversight. But outside of a study, "off label" use in clinical care is entirely legal, and has become a common practice among physicians with respect to drugs, and might be available for a gene transfer product using genome editing once it is approved. Physicians use their own expertise and sources of information, as well as the advice of professional societies. They are regulated at the state level by their licensing and disciplinary bodies, may be limited by availability of patients' insurance coverage for novel interventions, and are constrained by the prospect of tort liability for medical malpractice should they be deemed negligent or reckless.

Table 2-1 provides a summary of the major steps in the anticipated regulatory pathway for the development of a new medical product created using genome editing. The individual steps and considerations listed in this table are discussed in greater detail in the remainder of the chapter.

表 2-1 美国对于基因组编辑创造的医用产品的管理途径的概括

步骤	主要监管单位（美国体系）	注意事项
细胞和组织（非胚胎）的实验室研究，包括人诱导多能性干细胞（induced pluripotent stem cell，iPSC）	①生物安全委员会 ②伦理审查委员会 ③受到 NIH 基金资助者必须遵守 NIH 对于人干细胞的准则（iPSC 细胞系的某些用途是禁止的） ④受到 NIH 基金资助者必须遵守 NIH 对于人干细胞的准则（只可以使用获得批准的人胚胎干细胞系，而胚胎干细胞的某些用途是禁止的）	①实验室工作人员安全 ②组织提供者的安全、隐私和权利，充分的知情同意
人胚胎和胚胎多能干细胞的实验室研究	①研究单位的胚胎干细胞研究监管或者胚胎研究监管（自愿但是广泛存在） ②利用联邦经费开展涉及胚胎的形成、销毁及一定程度上的损伤的研究是被禁止的 ③其他州级法律	与应用人胚胎和胚胎多能干细胞的研究相关的伦理问题（联邦和州的层次）
临床前动物研究	①美国农业部 ②人道关爱和实验动物利用的公共健康服务政策 ③研究单位动物关爱和使用委员会	人道主义关爱，试验设计，疼痛最小化
临床试验（新药研发申请，investigational new drug，IND）	①伦理审查委员会 ②研究单位生物安全委员会 ③DNA 重组咨询委员会（Recombinant DNA Advisory Committee，美国国立卫生研究院，NIH）（仅咨询） ④FDA，组织和先进疗法办公室，生物制品评估和研究中心（Center for Biologics Evaluation and Research，CEBR）	①预期危险和潜在收益的平衡 ②合适试验方案和知情同意
新医学产品的申请（生物制品执照申请）	FDA，CEBR	评估安全性和有效性数据
批准的医学产品（进入市场后的评估）	FDA，CEBR	患者长期的安全性

TABLE 2-1 Summary of U.S. Regulatory Pathway for a Medical Product Created Using Genome Editing

Step	Primary Regulatory Authorities (U.S. System)	Examples of Considerations
Laboratory research in cells and tissues (nonembryonic), including human induced pluripotent stem cells (iPSCs)	• Institutional biosafety committee • Institutional review board (certain uses of human tissue) • NIH-funded researchers must comply with NIH Guidelines for Human Stem Cell Research (certain uses of iPSC lines are prohibited) • NIH-funded researchers must comply with NIH Guidelines for Human Stem Cell Research (only hESC lines approved by NIH may be used; certain uses of hESC lines are prohibited)	• Laboratory worker safety • Tissue donor safety, privacy, and rights (human cells and tissue). Adequacy of consent process
Laboratory research in human embryonic stem cells or embryos	• Institutional embryonic stem cell research oversight or embryo research oversight committees (voluntary but widespread) • Prohibition on use of federal funds for research in which human embryos are created for research purposes, destroyed, or subject to a certain level of risk of harm • Additional state laws as applicable	• Special ethical concerns and regulations (federal and state) associated with research using human embryo and hESC lines
Preclinical animal studies	• U.S. Department of Agriculture • Public Health Service Policy on Humane Care and Use of Laboratory Animals • Institutional animal care and use committee	• Humane care, study design, and pain minimization
Clinical trials (Investigational New Drug [IND] application)	• Institutional review board • Institutional biosafety committee • Recombinant DNA Advisory Committee (National Institutes of Health) (advisory) • U.S. Food and Drug Administration (FDA), Office of Tissues and Advanced Therapies, Center for Biologics Evaluation and Research (CBER)	• Balance of anticipated risks and benefits to human subjects • Appropriate protocol design and informed consent
New medical product application (Biologic Licensing Application)	• FDA CBER	• Evaluation of safety and efficacy data
Licensed medical product (postmarket measures)	• FDA CBER	• Long-term patient safety

焦点与其说是在是否可以使用配子进行研究的伦理问题，不如说是配子获得的方式及是否涉及不当利诱的伦理问题。在存在涉及面更加广泛的人类生殖和受孕产物的相关管理法规的国家中，与胚胎和胎儿组织相关的管理条例也会受到这些法规的规范。对于所有的组织而言，关注点应该在于组织的获取是否应该得到提供者的许可和它的利用是否会给提供者的隐私造成危害。如果组织是从尸体上获得而不是从活体捐赠者上获得的，这些规则也可以随之改变。

在美国，用作研究的人类组织的捐献方式是多种多样的。法律主要是从以下几个方面来管理捐献：这些人类组织是否是合适的临床手术遗留物，最重要的是为了研究专门通过新的干预获取的，以及由此获得的组织标本是否附有易于确定供者身份的信息。当组织就是为一个特定研究通过物理干预方法获取时（如抽血），提供者的主体是人，IRB 就会监管志愿者的招募过程、收集的步骤及获

Oversight of Laboratory-Based Research

Rules governing research with human cells and tissues, including somatic cells, gametes, embryos, and fetal tissue tend to focus on several key issues. For most cells and tissues, an initial question is whether the donor can receive any kind of payment, in cash or kind. This has been a particularly sensitive issue with respect to gametes used in research, with debate being focused less on the ethics of research using gametes and more on the ethics of how they are obtained and whether it involves anything that resembles undue inducement. For embryos and fetal tissue, rules are influenced by broader legal regimes governing human reproduction and products of conception, to the extent such regimes exist in a given country. And for all tissues, attention is given to whether the tissue is obtained with required permissions from the donor and whether its use poses any risk to the donor's privacy. These rules can change, of course, when tissue is obtained from cadavers rather than live donors.

In the United States, human tissue is donated for research in various ways. Rules governing that donation depend on several factors, the most important of which are

得知情同意的信息等内容[16]。然而，就像下面讨论的那样，仅仅同意使用已经切除的组织并不能确定供体是人，除非这个组织有使供体身份易于确认的信息。

一旦组织被获取，它就可以被用作实验室研究，受到IBC制定的关于重组DNA研究条例的监控。无论是否对组织进行基因组编辑，这种管理模式都是相同的。

重组DNA研究和研究机构的生物安全委员会

很多对仅在实验室内开展的涉及人类组织和细胞的研究（不涉及使用非人灵长类或人类的临床前研究）的规定和要求都将关注的重点放在保证实验环境内的工作人员的安全上。而对于受《涉重组或合成核酸分子研究指导方针》（NIH指导方针）监管的研究，则需要研究单位生物安全委员会（IBC）的评估和批准。NIH指导方针适用于所有受NIH资助或主导的研究，事实上虽然很多机构没有被这样要求，它们仍然遵守NIH指导方针。在研究单位的层次上，IBC评估几乎所有的使用重组或合成核酸分子的研究。IBC保证研究的实施过程符合NIH指导方针，并评估研究对人类健康和环境的潜在风险。生物安全评价通过评估研究进行合理的物理和生物学控制，并保证从事研究的工作人员受到充分的训练以安全地从事该工作。

研究机构的生物安全委员会需要至少5名重组或合成核酸分子技术领域的专家，其中至少2人独立于开展该研究的机构之外。在条件允许并不损害机构自身隐私及利益的情况下，NIH指导方针鼓励科研机构向公众开放生物安全委员会。NIH指导方针还要求机构安全委员会在公众要求时公开提供会议备忘。

人类组织的使用及伦理审查委员会

使用人类组织进行体外的实验室研究同样会涉及受试者保护的问题。两种情况下会需要这种额外的监管。

首先，如果该组织是为了特定的研究取自于活体个体，此种情况需要提交伦理审查委员会审查。

---

[16] 流产或者堕胎之后捐献胚胎或者胎儿的组织并不直接使供体成为人类捐献者，只有在捐献者的个人信息被保留或者没有被完全隐藏的情况下才是。

whether the tissue is left over from a clinical procedure or is being obtained through a new intervention specifically for research, and whether the resulting tissue specimen has information attached to it that makes the donor's identity readily ascertainable. When tissue is collected through a physical intervention (such as a blood draw) specifically for research, the donor is a human subject, and an IRB oversees the recruitment of donors, the procedures used for collection, and the information provided to obtain consent.[16] However, as discussed below, merely giving consent to use of already excised tissue does not render the donor a human subject, unless the tissue has information that makes the donor's identity readily ascertainable.

Once the tissue has been obtained, it is available for laboratory research, subject to the usual rules for oversight of recombinant DNA research by IBCs. This pattern of regulation is the same regardless of whether genome editing will be carried out on the tissues.

Recombinant DNA Research and Institutional Biosafety Committees

Research with human tissues and cells that takes place entirely within a laboratory and does not involve either preclinical testing on nonhuman animals or clinical testing on humans is subject to regulations and requirements, many focused primarily on ensuring the safety of the laboratory environment for workers. For experiments subject to the *NIH Guidelines for Research involving Recombinant or Synthetic Nucleic Acid Molecules* (*NIH Guidelines*), there is a requirement that the research be reviewed and approved by an Institutional Biosafety Committee (IBC). The *NIH Guidelines* are applicable to all research that is conducted at or sponsored by an institution that receive NIH funding for such research, however many institutions follow the requirements of the *NIH Guidelines* even when they are not required. IBCs review nearly all forms of research utilizing recombinant (or synthetic) nucleic acid molecules at the local institutional level (e.g., university or research center). The IBCs ensure research is conducted in conformity with the biosafety provisions of the *NIH Guidelines*, and assess the research for potential risks to human health and the environment. This biosafety review is accomplished by assessing the appropriate physical and biological containment for the research and ensuring the researchers are adequately trained to conduct the work they are proposing safely.

An IBC is required to have at least five members with expertise in recombinant or synthetic nucleic

---

[16] Donating embryos or fetal tissue remaining after miscarriage or abortion does not render the donor a human subject unless identifying information about the donor is retained and insufficiently obscured.

虽然伦理审查委员会主要是保护临床研究中受试者的权利和利益的机构，收集组织用于研究也意味着是把受试者作为研究对象，即使之后使用该组织材料的工作也不会产生任何可以追踪到或以任何方式影响到捐献者的信息。

第二种情况发生在组织采集与采集对象不相关时，如本来会被废弃的手术组织。如果这种材料采集时采取了充分的匿名措施，就不会触发伦理审查委员会的审查。但是一旦受试者的身份信息可以被轻易确定，这时捐献者则被视为研究对象，审查也会随之而来，除非该工作享有部分或全部内容的知情同意豁免。

2017年生效的修订后的《普遍规则》对使用冻存人类样品的情况做出了修改，规定了由大多数联邦机构和部门资助或以其他方式受其管辖的人体课题的框架和要求[17]。该条例自2018年1月起实行，修订版条例覆盖了确定身份的组织的使用。

（1）允许从受试者处获得广泛的知情同意，（试图用于未知目的研究的知情同意）使用可识别的个人信息和生物标本用于储存、维护和二次使用。广泛的知情同意是一个研究者可选的替代方案，否则的话，就需要让机构审查委员会（IRB）放弃知情同意的要求，或获得特定研究的同意，以对未识别的信息和未识别的生物样本进行研究。

（2）可以根据研究的风险评估建立新的豁免类别。在特定的类别中，豁免研究可能被要求进行一定程度上的审查评估，以此来保证可识别的个人信息和生物样品信息的安全性[18]。

---

[17] FDA有自己的针对涉及人类样本的研究的管理体系。这一管理体系会有一些不同，如针对小风险研究中的知情同意豁免（21 CFR Part 50）。常规的规则也适用于以下美国机构资助的研究：国土安全部（Department of Homeland Security）；农业部（Department of Agriculture）；能源部（Department of Energy）；国家航空航天管理局（National Aeronautics and Space Administration）；商业部（Department of Commerce）；社会安全管理局（Social Security Administration）；国际发展署（Agency for International Development）；住房和城市发展部（Department of Housing and Urban Development）；劳工部（Department of Labor）；国防部（Department of Defense）；教育部（Department of Education）；退伍军人事务部（Department of Veterans Affairs）；环境保护署（Environmental Protection Agency）；卫生与公共服务部（Department of Health and Human Service）；国家科学基金（National Science Foundation）；运输部（Department of Transportation）。从历史上看，遗传学研究的资金主要来自美国卫生与公众服务部、国家科学基金会和能源部（Rine and Fagen, 2015）。尤其是，常规的规则也适用将其申请扩展到上述来源以外的其他资助来源的研究机构所进行的研究。

[18] 联邦政府保护受试人的政策：联邦登记册最后的规则（Federal Policy for the Protection of Human Subjects; Final Rule Federal Register），Vol. 82，no. 12（January 19, 2017），pp. 7149-7274.

---

acid molecule technology, at least two of whom are independent of the institution at which the research is being conducted. The *NIH Guidelines* encourage institutions to open IBC meetings to the public when possible and when consistent with the protection of privacy and proprietary interests (NIH, 2013c). The *NIH Guidelines* also require institutions to make IBC meeting minutes available to the public upon request.

Human Tissue Use and Institutional Review Boards

Laboratory-based research using human tissue may also trigger certain human subjects protections, even though all the work is done *in vitro*. Two situations trigger this additional level of regulation.

First, as noted above, if the tissue is being collected from a living individual, specifically for research, this interaction generally will be subject to oversight by an IRB. Although IRBs are designed primarily to protect the rights and welfare of research subjects in clinical investigations, the act of collecting tissue for research is considered to render the donor a research subject, even if subsequent work with the tissue will yield no information that could be traced to or in any way affect the donor.

The second situation may occur when tissue is collected without interaction with the person from whom it is derived, such as when surgical tissue that would otherwise be discarded is collected for use in research. If the tissue is sufficiently anonymized, the use of the tissue in research will not trigger IRB review. But if the donor's identity can be readily ascertained, the donor is considered a research subject, and IRB review is triggered unless the work is eligible for exemption or waiver of some or all elements of informed consent.

The rules will change with respect to research using stored human specimens upon the effective date of the January 2017 revisions to the "Common Rule" that sets out the framework and requirements for human subjects research that is funded by most federal agencies and departments, or is otherwise subject to its jurisdiction.[17] Effective as of January 2018, the revised rule covering use

---

[17] The FDA has its own set of regulations governing human subjects research, which can differ in some details (e.g., regarding waivers of consent in minimal risk research). Those regulations may be found at 21 CFR Part 50. The Common Rule applies to research funded by the following departments and agencies: Agency for International Development; Environmental Protection Agency; National Aeronautics and Space Administration; National Science Foundation (NSF); Social Security Administration; U.S. Department of Agriculture; U.S. Department of Commerce; U.S. Department of Defense; U.S. Department of Education ; U.S. Department of Energy (DOE); U.S. Department of Health and Human Services (HHS); U.S. Department of Homeland Security; U.S. Department of Housing and Urban Development; U.S. Department of Labor; U.S. Department of Transportation; and U.S. Department of Veterans Affairs. Historically, genetics research funding has come from DOE, HHS, and NSF, in particular (Rine and Fagen, 2015). The common rule also applies to research conducted at institutions that have voluntarily extended its application to research funded by sources other than those listed above.

使用匿名、去标识或编码的生物材料[19]的条例能允许更宽松的使用方式。如果是，会豁免大部分或是全部的审查监管。使之用于可识别的私人信息和可识别的生物样本的二次研究[20]，则在以下情况下，研究将免于全部或大部分审查委员会监督：

（1）确定的个人信息或确定的生物样品是公开的；

（2）研究者以不易确定受试者的身份的方式记录信息，研究者不与受试者接触或试图重新识别受试者；

（3）二次研究活动受到健康保险流通与责任法案管制（HIPAA）；或二次研究是由联邦实体或其代表实行的涉及使用联邦生成的非检索信息，前提是原始收集受特定联邦隐私保护并持续保护。

随之伦理审查委员会提出了一系列保护措施，旨在确保研究对象和社会风险（生理、心理和社会经济），以及可能存在的利益处于合理的平衡。此外，除非有豁免资格，否则需要从研究对象（在目前情况下，指提供组织样本的个人）或法律授权的代表获得知情同意书和自愿同意书才能继续研究。

人类配子和胚胎实验研究的附加规则

基因组编辑的基础科学研究可能需要对人类配子和胚胎进行实验，但不会将之移植到子宫进行进一步的发育（见第3章）。事实上，这种对胚胎的体外研究已经在中国进行（使用的是非生育性胚胎），并被瑞典和英国的相关监管机构批准（具有生育能力的胚胎）。这项工作能够帮助了解人类生殖细胞发展进程，同时作为众多实验方法之一，基因组编辑能够探索特定基因在胚胎中的作用（Irie et al., 2015）。

此类研究可能采取多种形式，但每种形式都涉及不同的伦理和法律问题。首先，它可能涉及体细胞组织编辑的方式，配子或许会受到影响。其

of identifiable tissue:

Allows the use of broad consent (i.e., seeking prospective consent to unspecified future research) from a subject for storage, maintenance, and secondary research use of identifiable private information and identifiable biospecimens. Broad consent will be an optional alternative that an investigator may choose instead of, for example, conducting the research on nonidentified information and nonidentified biospecimens, having an institutional review board (IRB) waive the requirement for informed consent, or obtaining consent for a specific study.

Establishes new exempt categories of research based on their risk profile. Under some of the new categories, exempt research would be required to undergo limited IRB review to ensure that there are adequate privacy safeguards for identifiable private information and identifiable biospecimens.[18]

The rules governing use of anonymous, de-identified or coded materials[19] will allow for even broader use. Research will be exempted from all or most IRB oversight if it is for secondary research use[20] of identifiable private information and identifiable biospecimens for which consent is not required, and when:

The identifiable private information or identifiable biospecimens are publicly available;

The information is recorded by the investigator in such a way that the identity of subjects cannot readily be ascertained, and the investigator does not contact subjects or try to re-identify subjects;

The secondary research activity is regulated under the Health Insurance Portability and Accountability Act (HIPAA); or

The secondary research activity is conducted by or on behalf of a federal entity and involves the use of federally generated nonresearch information provided that the original collection was subject to specific federal privacy protections and continues to be protected

With IRB review comes a set of protections focused on ensuring that the risks (physical, psychological and socioeconomic) and possible benefits to the

---

[19] 匿名组织在收集和存储的全过程没有保留任何个人的标识；消除识别的组织也将删除早些时候的标识；编码的组织利用编码遮蔽个人的标识。当研究人员不能简单破译编码时，个人的标识将不再是"随时都可确定的"。

[20] "二次研究的审查豁免是指在一次或者首次研究活动之外收集可以被标识的信息和生物样本。在此项豁免下的信息一般可以被研究者在同样的记录中发现，样本则一般在某一个组织样本库中（如医院存储临床病例样本的部门）" Federal Policy for the Protection of Human Subjects; Final Rule Federal Register, Vol. 82, no. 12 (January 19, 2017), pp. 7149-7274 at p. 7191.

[18] Federal Policy for the Protection of Human Subjects; Final Rule Federal Register, Vol. 82, no. 12 (January 19, 2017), pp. 7149-7274.

[19] Anonymous tissue is collected and stored without any personal identifiers at any time; de-identified tissue has earlier identifiers removed; coded tissue has identifiers that are obscured by virtue of coding. Where the investigators do not have easy access to a key to break the code, the tissue will no longer have a personal identity that is "readily ascertainable."

[20] "By "secondary research," this exemption is referring to re-using identifiable information and identifiable biospecimens that are collected for some other "primary" or "initial" activity. The information or biospecimens that are covered by this exemption would generally be found by the investigator in some type of records (in the case of information) or some type of tissue repository (such as a hospital's department for storing clinical pathology specimens)." Federal Policy for the Protection of Human Subjects; Final Rule Federal Register, Vol. 82, no. 12 (January 19, 2017), pp. 7149-7274 at p. 7191.

次，它可能涉及在体外或体内编辑的现有的配子或配子祖细胞，如精原干细胞。再次，它可能涉及编辑在受精过程中的卵细胞（如在卵母细胞胞浆内精子注射期间）或编辑已经受精的卵细胞（合子）或胚胎。

只要对配子和胚胎的编辑工作仍处于临床前阶段，即没有能够产生妊娠的移植，美国的管理监督和限制主要还是来自州胚胎研究法，或由联邦和其他资助者施加的限制。但如果涉及需要编辑的生殖材料产生妊娠的临床试验，那么此项研究将受到FDA管辖，并且每次在开展类似的试验之前需要通过实验性新药（IND）的申请（关于此类潜在的未来应用的讨论参见第5章）。

1994年，在美国由NIH人类胚胎研究小组发起了围绕人胚胎实验室研究的公共政策问题的广泛讨论，其目的是向NIH咨询委员会提供相应的建议。此次讨论产生了如下共识：首先，应视胚胎不同于普通人类组织，但如果需要将其应用于重要的、不能用争议较少的方法获得的科研领域，则仍是可以的。同时，专家组报告时呼吁尽量使用早期阶段的人类胚胎，应在符合研究需求的前提下使用最小数量的人类胚胎。此外，专家组也呼吁除了只在非常有限的情况下使用那些最初因繁殖而产生但随后不再需要而被丢弃的胚胎。另外，用于研究的捐赠胚胎也必须是以生殖为目的而得到的胚胎，并且需要获得知情同意书（NIH，1994）。虽然从技术上来说，此次报告涉及使用联邦基金研究人类胚胎的条件（随后被美国国会禁止[21]），但其建议获得了科学界对此类研究在伦理和道德标准上的认可。

目前，对人类研究的监管保护并不适用于体外的胚胎[22]。尽管如此，许多（虽然不是大多数）提供胚胎干细胞研究的机构已经实施了自愿监督措施（Devereaux and Kalichman，2013），同时国际干细胞研究学会最近通过了体外胚胎移植的指导方针，要求将监督委员会扩展到涉及人类胚胎的几乎所有研究，无论干细胞来源及资金来源如何（ISSCR，2016）。

一些对生殖细胞基因组进行编辑的临床前研

---

[21] 迪奇-威克修正案（The Dickey-Wicker Amendment）禁止绝大多数的美国联邦经费资助涉及制造或破坏人类胚胎的研究，或将胚胎置于损伤和有破坏风险的研究，除非是必要的增加其健康发育机会的研究。这一修正案在1996年后就被美国卫生与公众服务部、劳工部及教育部列入年度拨款法案中。
[22] 其他法规尤其是管理胚胎研究州级法律可能适用（见第5章）。

research subject and society are in reasonable balance. Furthermore, except when eligible for waiver, informed and voluntary consent is required from the research subject (in the present context, the person whose tissue is being used) or from a legally authorized representative.

Additional Rules Governing Laboratory Research on Human Gametes and Embryos

Basic science research on genome editing may entail experimentation on human gametes and embryos, with no intention of performing intrauterine transfer to establish a pregnancy in a woman (see Chapter 3). Indeed, such *in vitro* research on embryos has already proceeded in China (using nonviable embryos) and has been approved (with viable embryos) by the relevant regulatory bodies in Sweden and the United Kingdom. Work also is proceeding on understanding human germ cell development, research in which genome editing is one of many tools that can be used to explore the roles of specific genes (Irie et al., 2015).

This laboratory research might take a number of forms, each raising slightly different ethical and legal issues. First, it might involve editing somatic tissue in such a way that gametes would or might also be affected. Second, it might involve editing an existing gamete or gamete progenitor, such as a spermatogonial stem cell, *in vitro* or *in vivo*. Third, it might involve editing an egg in the process of fertilization (e.g., during intracytoplasmic sperm injection), or editing an already fertilized egg (zygote) or embryo.

As long as the work on gametes and embryos remains preclinical—i.e., there is no transfer for gestation—the regulatory oversight and limits in the United States derive from state embryo research laws or limitations imposed by federal or other funders. Should there be a clinical trial involving efforts to gestate the edited reproductive materials, the research would come under FDA jurisdiction, and approval of an Investigational New Drug (IND) application would be required prior to beginning each such trial (see Chapter 5 for discussion of such potential future applications).

In the United States, the public policy issues surrounding laboratory research with human embryos were debated extensively by the 1994 NIH Human Embryo Research Panel, which was convened to provide recommendations to the Advisory Committee to the NIH director. Its conclusions, reflect the view that embryos should be regarded as different from ordinary human tissue but nonetheless be used for some areas of research if in the service of important scientific knowledge that cannot be obtained with less controversial methods. In addition, the panel's report called for the use of human embryos at the earliest stages and in the smallest numbers

究，很可能会利用体外受精（IVF）所余留下的胚胎。虽然没有官方数据，但保守估计全球目前已经存储超过100万个胚胎，其中的大多数都产生于IVF并储存在美国（Lomax and Trounson，2013）。如前所述，美国联邦基金通常是禁止进行胚胎研究的，然而这项工作却得到了来自其他国家和个人来源的资金支持，这些研究往往采用类似1994年胚胎研究小组提出的政策。例如，加利福尼亚州近十年来一直使用州基金资助胚胎和胚胎干细胞研究，但仅限于在联邦资助有限的年份支持早期已经成系的胚胎干细胞系的研究。马里兰州、纽约州、新泽西州和康涅狄格州也因为不能获得联邦资助的研究项目建立了新的基金（NIH，2016c）。

用多能干细胞产生出的人类配子进行基因组编辑研究，不受管理胚胎研究的法律或基金政策的限制，除非此项研究中产生了为检测配子多能性的受精卵。单细胞受精卵在大多数相关的州和联邦的法律中都被认为是胚胎，许多基金申报和工作也受此限制。此外，这一步骤只是为研究目的制造胚胎（既不产生妊娠胚胎，也不发育到胎儿期），在美国这仍然是最具争议的胚胎研究形式。一些反对在研究中制造胚胎的人认为，受精卵的产生意味着一个新的、伦理上的人类已经存在了，即使这些胚胎仅作为研究目的使用，也是对人类生命的不尊重，并且可能具有潜在的重大弊端（NIH，1994，p.42）。在一些情况下，这种推理被扩展到包括用体细胞核移植（"克隆"）制备的全能细胞。即使那些对胚胎伦理无所谓的人，也可能对创造研究用的胚胎产生抵触（The Washington Post，1994；Green，1994）。

此外，胚胎研究小组还认为当"在自然条件下研究本身无法有效地进行时"或当"具有潜在的、出色的科学和治疗价值时"，制造胚胎是合理的（NIH，1994，P.45）。这似乎包括在体外产生配子的研究和避免线粒体疾病的技术，但在人类胚胎小组报告时，这两种技术还尚未进入应用领域。同时，为了检测已经编辑过的配子所需的基因组修饰而进行的研究，以及在IVF过程中（如胞质内精子注射）将基因组修饰过的组分连同精子一起引入，似乎都属于此类例外。而在那些允许对人类胚胎进行研究的国家中，即使实验目的和动机不符合NIG（1994）提出的规定，相关人员依然也会使用胚胎专门用于研究（联合国教科文组织，2004b）。

consistent with needs of the research. Except in very limited circumstances, the panel called for use of only those embryos that, although originally created in the course of a reproductive effort, now would otherwise be discarded. Donation to research would require the informed consent of those who had created the embryos for reproductive purposes (NIH, 1994). While technically the panel's report addressed conditions for federal funding of research that uses human embryos (which was subsequently prohibited by congressional action[21]), its recommendations came to be recognized within the scientific community as a more general evaluation of the ethics and acceptability of such research.

Regulatory protections for human research subjects do not apply to the *ex vivo* embryo.[22] Nonetheless, many (if not most) institutions housing embryonic stem cell research have put voluntary oversight measures in place (Devereaux and Kalichman, 2013), and the International Society for Stem Cell Research recently adopted guidelines calling for expanding these oversight committees to almost all research involving human embryos, regardless of whether stem cells will be derived and regardless of funding source (ISSCR, 2016).

Some preclinical research on germline genome editing would likely take advantage of embryos left over from reproductive attempts using *in vitro* fertilization (IVF). Although no official numbers are available, a conservative estimate indicates that more than one million embryos, most of them produced but ultimately not used for IVF, remain in storage across the United States (Lomax and Trounson, 2013), with many more being stored around the world. As noted earlier, U.S. federal funding for research on embryos generally is prohibited. The work can, however, be supported with funds from individual states and private sources, often with policies similar to those proposed by the 1994 embryo research panel. California, for example, has been funding embryo research and embryonic stem cell research for a decade using funds from a state bond issued during the years when federal funding was limited to a small number of older embryonic stem cell lines. Connecticut, Maryland, New Jersey, and New York also created funds for research that could not be federally funded (NIH, 2016c).

Genome-editing research that generates human gametes from pluripotent stem cells would not be governed by the laws or funding policies governing

---

[21] The Dickey-Wicker Amendment prohibits the use of most federal funds for research that involves creating or destroying embryos, and for research that puts embryos at risk of injury or destruction except when necessary to increase their chance for healthy development. The amendment has been attached to the annual appropriations bills for the Departments of Health and Human Services, Labor, and Education since 1996.

[22] Other rules, particularly state laws governing research using embryos, may apply (see Chapter 5).

其他国家对人类胚胎研究的监管

如前所述，美国的一些州已经有相关法律监管或完全禁止使用人类胚胎进行研究（NCSL，2016）。而在联邦政府的层面上，虽然有使用联邦基金进行研究的限制，但并没有完全禁止此类研究。

相比之下，在美国允许的许多研究，在英国受到更严格的管制，其中对人类配子和胚胎的研究需要接受人类受精和胚胎学管理局（Human Fertilisation and Embryology Authority）的审查，并且每个特定实验组都需要获得相应的许可（关于生殖系编辑的临床应用的讨论，参见第5章）。而在其他国家，如德国（DRZE，2016）、智利[23]、意大利（Boggio，2005）、立陶宛[24]和斯洛伐克[25]，无论在任何监管制度下，此类研究都不合法。

对此类研究的态度的不同，反映出许多国家对配子特别是对胚胎的研究一直存在着争议。对人类胚胎的法律和伦理状况存在以下不同观点：将其视为与任何其他人类组织一样使用；将其视为需要特别尊重的组织使用；将其视为与新生儿具有同样法律权利的组织。这些观点在各国之间或国家内部都存在着差异，并且受到当地宗教和民众的影响，从而造成目前这种不同区域内存在允许使用、监管使用或是完全禁止使用胚胎的状况。

虽然基因组编辑是在细胞中进行遗传修饰的新技术，但其在人胚胎研究中的使用依然存在着与过去类似的问题：胚胎的伦理状态，制备研究用胚胎或使用即将被丢弃的胚胎的可接受性，以及适用于在研究中使用胚胎的法律或限制（CIRM，2015；ISSCR，2016；NIH，2016）。本报告不涉及这些伦理论证，并且接受适用于每个国家的现行法律和监管政策。如果这些一般政策中的任何一个在未来发生改变，那么基因组编辑研究也将受到影响。

使用非人类动物的研究

美国于1966年颁布的《动物福利法》（7 USC § 2131）是关于在研究中使用动物的联邦法律，虽然其中包含一些并不像大鼠和小鼠那样常用的实验物

embryo research unless a fertilized egg would be made in order to test the gametes. A single-cell fertilized egg is treated as if it were an embryo for most relevant state and federal laws, and restrictions on the work or on the funding would apply. In addition, such a step would constitute making an embryo solely for research purposes (that is, without any intent to gestate the embryo and bring a fetus to term), and this has remained the most controversial form of embryo research in the United States. Some of those opposed to making embryos in research argue that fertilization brings a new, morally significant human being into existence, and that making embryos for research purposes is inherently disrespectful of human life and potentially open to significant abuses (NIH, 1994, p. 42). In some cases, this reasoning is extended to encompass totipotent cells made with somatic cell nuclear transfer ("cloning"). Even those who do not accord full moral status to an embryo might be wary of creating embryos for research (Green, 1994; *The Washington Post*, 1994).

On the other hand, the panel concluded that making embryos is justified when "the research by its very nature cannot otherwise be validly conducted" or when it is necessary for a study that is "potentially of outstanding scientific and therapeutic value" (NIH, 1994, p. 45). This would appear to include research on *in vitro*-derived gametes and on techniques for avoiding mitochondrial disease, neither of which were on the immediate horizon for human application at the time of the Human Embryo Research Panel report. The genome-editing research necessary to test edited gametes would seem to fall within this exception, as would the introduction of genome-editing components along with sperm during IVF procedures such as intracytoplasmic sperm injection. Among those countries that permit research on human embryos, rules differ on whether this exception also would permit making embryos specifically for research (UNESCO, 2004b).

Oversight in Other Nations for Research Using Human Embryos

As noted earlier, in the United States, a handful of states have laws governing or forbidding research using human embryos (NCSL, 2016). At the federal level, there is no prohibition on such research, although there are limits on the use of federal funds to perform the research.

By contrast, much of what is permitted in the United States would be more tightly regulated in the United

---

[23] Chile, Congreso Nacional, Sobre la investigacion cientifica en el ser humano, su genoma, y prohibe la clonacion humana, September 22, 2006, no. 20.120, art. 1, Witherspoon Council staff translation, http://www.leychile.cl/Navegar?idNorma=253478 [Spanish].
[24] Lithuania, Seimas, Law on Ethics of Biomedical Research, no. VIII-1679, May 11, 2000, amended July 13, 2004, no. IX-2362, art. 3, §2, http://www3.lrs.lt/pls/inter3/dokpaieska.showdoc_l?p_id=268769 [English, official translation].
[25] Slovakia, Health Care Act No. 277/1994, art. 42, 3(c), as quoted in UNESCO (2004b); Slovakia, Slovak Penal Code, art. 246a added in 2003, as quoted in UNESCO (2004b, p. 14).

种。但它规定了在研究中对多种物种的检测和维护方法，由美国农业部（USDA）动植物健康检查局执行，要求研究机构在当地建立一个动物护理和使用委员会，用来"监督和评估该机构对实验动物的护理和使用"，如确保对实验动物的物理保护和符合疼痛最小化的标准等。

如果基因组编辑研究中需要生产嵌合体，也需要满足 NIH 的基金守则中明确限制的某些组合（NIH，2015a）。NIH 最近已要求公众对人类干细胞研究指南中与嵌合体相关的规定提出修改意见（NIH，2016）。

人类基因组编辑的临床试验——伦理审查委员会的作用

目前，凡是涉及人类受试者的研究包括临床基因组编辑试验等，如果未经 FDA 许可不得开始（具体细节将在下面讨论）。同时还有其他三个机构，即伦理审查委员会、生物安全伦理委员会和重组 DNA 咨询委员会都有对基因组编辑临床试验的监督责任。

伦理审查委员会的审查和批准侧重于对临床研究的风险、收益及研究人员的招募方式。委员会负责审查由美国卫生和人类服务部支持或受 FDA 监管的所有涉及人类受试者的研究。同时委员会也要求研究项目满足以下条件：对研究的进行和支持需要由其他联邦机构制定共同规则；对研究的产物需要由 FDA 管理；对科研进程需要由研究人员自愿维护，否则将不受委员会约束。共同规则能够规范对人类个体的研究，同时一些联邦资助机构已经采用了关于胚胎研究的附加规则。如前所述，胚胎研究分别由一些州和联邦资金通过制定限制来单独调节。

伦理审查委员会有权批准或否决研究课题、人类受试者招聘计划和知情同意书，有时作为批准条件则需要修改实验方法。委员会还监督正在进行中的研究项目，如发现存在问题可以随时暂停研究，例如，不良事件作用的发生率或严重性。在这项工作中，委员会可能需要数据和安全监控部门的协助，在研究进行期间跟踪临时数据。同时，委员会也提供额外的专家和独立审查，以确保研究能够达到风险和潜在利益的合理平衡，并且根据在研究期间获得的数据，能够在受试者应聘前帮助其全面地了解研究的风险和收益。

虽然不同于个别制度政策或国家法律，联邦

Kingdom, where research on human gametes and embryos is subject to review by the Human Fertilisation and Embryology Authority and a license is required for each specific set of experiments. (See Chapter 5 for discussion of clinical use of germline editing.). In other countries, such as Chile,[23] Germany (DRZE, 2016), Italy (Boggio, 2005), Lithuania,[24] and Slovakia,[25] the research would not be legal under any regulatory regime.

This variation in governance approaches reflects the fact that research with gametes, and in particular with embryos, has been controversial in many countries. Views on the legal and moral status of the human embryo range from treating it the same as any other human tissue, to considering it a tissue deserving of some extra degree of respect, to viewing it as tissue that should be accorded the same respect or even the same legal rights as a live-born child. These views vary both among and within countries and reflect both religious and secular influences. The result has been public policies ranging from permissive, to regulated, to prohibitionist.

While genome editing is a powerful new technology for making genetic modifications in cells, its use in the context of research on human embryos raises issues essentially the same as those discussed in the past: the moral status of the embryo, the acceptability of making embryos for research or using embryos that would otherwise be discarded, and the legal or voluntary limits that apply to the use of embryos in research (CIRM, 2015; ISSCR, 2016; NIH, 2016). This report does not address those ethical arguments, and accepts as given the current legal and regulatory policies that apply in each country. If any of those general policies were to change in the future, genome-editing research would be affected as well.

Research Using Nonhuman Animals

The 1966 Animal Welfare Act (7 U.S.C. § 2131), the federal law covering the use of animals in research, regulates testing and maintenance of a number of species, although notably not some of those which are most commonly used, such as rats and mice. It is enforced by the U.S. Department of Agriculture's (USDA's) Animal and Plant Health Inspection Service, and at the local level requires that research institutions establish an institutional animal care and use committee "to oversee and evaluate all aspects of the institution's animal care and use program," such as ensuring that the standards for physical containment and pain minimization are met.

If genome editing research at any point were to require the creation of chimeric organisms, funding from NIH would come with rules limiting certain combinations (NIH, 2015a). NIH has recently requested public comment on proposed changes to provisions relevant to chimeras in its guidelines for human stem cell research, including

法规没有规定伦理审查委员会是否必须举行公开会议或向公众提供会议记录和其他文件，但是委员会的成员组成除了包含经过技术培训的专家之外，还必须包含至少一个非科学领域成员，以及至少一个不隶属于该机构的成员。此外，委员会还可以根据情况邀请具有特殊领域能力的人员协助审查复杂的问题。

联邦法规要求伦理审查委员会鉴定对研究对象的风险是否最小化，并且判断受试者的潜在利益和可能由研究产生的知识的重要性之间是否合理。委员会还需要确保对受试者的选择是公平的，受试者是在充分获得信息后自愿参与研究。在儿科方案中，风险耐受低。如果对儿童有益，研究可以在一方家长同意的情况下进行。但是如果研究并没有提供具有医疗受益的前景，那么在没有卫生和福利部秘书的特别干预的情况下，儿童可能就不会参与到任何比"最小风险危险一点"的项目中去。

当对胎儿进行研究时，某些联邦资助者坚持需要与风险程度有关的特殊规定，以及如何（以及向谁）通过规定（Subpart B 45 CFR 46）；虽然不是必需的，但使用其他资金的科研人员也可以采用这些相同的规定。这些规定指出，当研究胎儿对孕妇或是对胎儿自身能够产生利益时，平衡利益与风险后，应该使胎儿的风险在可能的程度上被最小化。如果没有这样的利益前景，胎儿的风险可能不会大于最小，而且研究的目的必须是开发无法通过任何其他试验手段获得的重要生物医学知识。当研究具有受益的前景时，只需要获得孕妇的同意便可进行。但如果研究具有仅对胎儿有益而不是对孕妇本身有益的话，就算研究可行，也需要获得父亲的同意。

人类受试者的自愿和知情权是对其的主要保护。卫生和福利部的规则中包括以下几项：

（1）对研究目的的解释、将要进行的步骤，以及说明是否有实验性步骤；

（2）对受试者或他人提供任何合理的、可预见的风险或利益的描述；

（3）说明适当的替代步骤；

（4）保密程度的说明（如果有的话）；

（5）对于涉及超过最小风险的研究，应说明在受伤情况下是否有任何补偿和医疗治疗；

（6）受试者必须是自愿参与的，拒绝参与研究

work that involves chimeras (NIH, 2016).

Clinical Trials of Human Genome Editing—The Role of IRBs

Clinical genome-editing trials—that is, studies involving human subjects—cannot commence without permission from the FDA, the details of which are discussed below. Along with FDA review, three other bodies—IRBs, IBCs, and the RAC—have clinical trial oversight responsibilities for genome editing.

IRB review and approval focuses on the risks and benefits of a clinical study and on the manner in which people are recruited for the study. It is required for any research involving human subjects that is supported by the U.S. Department of Health and Human Services (HHS) or regulated by the FDA. It is also required for research conducted or supported by any of the other federal agencies subscribing to the Common Rule, for research on products regulated by the FDA, and for research conducted by investigators at any institution that has voluntarily extended these protections to research that is otherwise not subject to these rules. The Common Rule addresses research with living individuals, and some federal funding agencies have adopted additional rules specifically with respect to research with fetuses. Research on embryos, as noted earlier, is regulated separately by some states and through federal funding restrictions.

IRBs have the authority to approve or deny approval for research protocols, human subject recruitment plans, and informed consent documents. They also may require modifications to a protocol as a condition of approval. IRBs also oversee amendments to ongoing studies and can suspend studies proving to be problematic—for example, due to the rate or severity of adverse events. In this task, IRBs may be assisted by data and safety monitoring boards, designed to track interim data while a study is ongoing. They provide additional expert and independent review to help ensure that a study continues to meet the standard for a reasonable balance of risks and potential benefits and that the information provided during initial recruitment of subjects remains a fair reflection of their risks and benefits as additional information is obtained during the study.

Federal regulations do not specify whether an IRB must hold open meetings or make its minutes and other documents available to the public; these are matters for individual institutional policies or state law. But an IRB, in addition to including experts with appropriate technical training, must include at least one member whose primary concern is in a nonscientific area and one lay member who is not otherwise affiliated with the institution. In addition, an IRB has the discretion to invite individuals with competence in special areas to assist in the review of

将不受到任何处罚,并且可以随时终止参与研究[26]。

首先在人类实验中很难遵守这些规定。因为根据相关定义标准,当研究从临床前模型转至人类干预时,很难评估其不确定性程度。然而,此类的试验必须进行。伦理审查委员会致力于确保研究对象不仅了解临床前的工作,并且了解存在的知识缺口,这些缺口将影响实验的预期结果。

联邦法案包括一项规定:伦理审查委员会不用考虑应用研究获得的知识将带来的大范围可能性后果(如研究对于国家政策所带来的影响)。因此,这项规定表明伦理审查委员会不具有单独拒绝申请的权利,因为研究所产生的知识成果及其所影响的政策可能具有社会争议性,如这类研究在未来应用时是否会引起争议?然而,这项规定允许伦理审查委员会拒绝将导致研究对象产生生理、心理或情感伤害的研究。

人类基因组编辑临床试验:重组 DNA 咨询委员会(RAC)的作用

20 世纪 60 年代末到 70 年代初,随着概念及技术的快速进步,导致了第一例人为的重组 DNA 分子的诞生(Berg and Mertz,2010)。由于潜在的应用,以及滥用重组 DNA 技术及其他已知和未知的风险是科学界和公众及政府部门所关心的,为此当时的 NIH 负责人 Donald Frederickson 在 1974 年创建了重组 DNA 咨询委员会(Recombinant DNA Advisory Committee,RAC)。重组 DNA 咨询委员会的成员组成被要求确保拥有广阔的公众视角,包括来自各行各业的成员,如科学家、临床医生、伦理学家、生物安全专家、神学家及公共代表等。随着时间的推移,重组 DNA 咨询委员会的成员及其责任已经随着科学发展和公众关注的变化而演变。

早期重组 DNA 咨询委员会的活动包括要求每个研究机构建立一个生物危害审查委员会(后期称为 IBC),用于审查风险及确保存在合适的安全措施。重组 DNA 咨询委员会的首要任务是制定针对重组 DNA 研究的准则,虽然缺乏法规的法律效力,但对防止转基因生物和材料的非预期释放或人类接触产生了巨大影响(Rainsbury,2000)。重组 DNA complex issues.

Federal regulations require an IRB to determine that risks to research subjects are minimized and are reasonable in relation to the potential benefits to the subjects and the importance of the knowledge that may be expected to result from the research. They are also required to ensure that selection of subjects is equitable and that subjects are freely volunteering for the research with sufficient information. In pediatric protocols, risk tolerance is lower. If benefit to the child is possible, the research may proceed with the consent of one parent and risk tolerance will be geared to the potential benefits. But if the research offers no prospect of medical benefit, the child may not be exposed to more than a "minor increment over minimal risk" absent special intervention by the secretary of HHS.

When research is done on fetuses, certain federal funders insist on special provisions related to the degree of risk that is permitted and to how (and from whom) consent must be sought (Subpart B 45 CFR 46); while not required, these same provisions may be adopted by investigators who use other funds. These provisions state that risk to the fetus is tolerated when it has been minimized to the extent possible and when it is balanced by the prospect of direct benefit for the pregnant woman or the fetus. If there is no such prospect of benefit, the risk to the fetus may not be greater than minimal, and the purpose of the research must be the development of important biomedical knowledge that cannot be obtained by any other means. Consent by the pregnant woman is sufficient when the research holds the prospect of benefit to her as well as the fetus. If the research holds the prospect of benefit only to the fetus and not to the pregnant woman herself, then paternal consent is also required, if feasible.

Requiring voluntary and informed consent is one of the key protections for human subjects. The elements, as listed in HHS regulations include among other items:

1. An explanation of the purposes of the research, the procedures that will be used and whether any procedures are experimental;

2. A description of any reasonably foreseeable risks or benefits to the subject or to others;

3. A disclosure of appropriate alternative procedures;

4. A statement describing the extent, if any, to which confidentiality will be maintained;

5. For research involving more than minimal risk, an explanation as to whether any compensation and medical treatments are available in case of injury; and

6. A statement that participation is voluntary, refusal to participate will involve no penalty and that the subject

---

[26] Federal Policy for the Protection of Human Subjects; Final Rule. Federal Register, Vol. 82, no. 12 (January 19, 2017), pp. 7149-7274 at p. 7266.

研究满足这个准则是获得 NIH 基金的条件之一，并且该准则适用于所有私立或公立研究所实施或赞助的研究（NIH，2013a）。许多其他的美国政府机构及私立研究所都要求获其基金资助的研究必须符合该准则（Corrigan-Curay，2013）。

起初，重组 DNA 咨询委员会审批所有在获得 NIH 资助的研究所中进行的有关转基因研究的提案，并就正式批准向 NIH 官方提出建议，技术上来说，官方批准来自于 NIH 的负责人，但必须建立在 RAC 决定的基础上（Freidmann et al.，2001）。之后，重组 DNA 咨询委员会发展了与 FDA 在审查上的相互合作。重组 DNA 咨询委员会对安全的关注开始拓展，包括为讨论社会和伦理问题提供场所。在 20 世纪 90 年代中期，FDA 拥有批准转基因研究方案的唯一权力，其中一些方案由重组 DNA 咨询委员会成员的初步审查，然后进行深度审查及公众探讨。还为同情性豁免程序的使用作出了规定（Rainsbury，2000；Wolf，2009）。

NIH 准则的附录 M 为"注意事项"文件，这里详细记录了人类转基因研究方案的提交、报告及 RAC 审查的相关要求（NIH，2013c）。NIH 准则提出"NIH 目前将不会接受有关生殖改变的研究提案"。关于胚胎的转基因，NIH 准则表示 NIH 可能愿意考虑这样的研究，但必须建立在 RAC 会议制定出针对于此的额外的临床前及临床研究的安全准则基础上。在 2015 年 4 月，NIH 的负责人发表声明，称"NIH 绝不资助任何在人类胚胎上使用基因组编辑技术的研究"。

在监管转基因研究的整个体系中，重组 DNA 咨询委员会提供专题讨论会以进一步针对方案的审查及公众讨论。但如上所述，IRB 将私下召开此类会议，会议将包括非科学领域成员。RAC 的公众属性是其作为 1972 年联邦咨询委员会法案（FACA）下的公共咨询委员会。为与 FACA 规章一致，重组 DNA 咨询委员会必须公开召开会议，并且提前通知时间和地点，提供会议记录，并允许公众参与（Steinbrook，2004）。

重组 DNA 咨询委员会同时也资助有关重组 DNA 研究的重要的科学和政策问题的公开座谈会（Friedmann et al.，2001），为科学、临床、伦理和安全专家及公众提供公开论坛，讨论转基因研究领域的新问题。重组 DNA 咨询委员会除了审查提案

may discontinue participation at any time.[26]

First-in-human trials make compliance with these provisions difficult, given that by definition, it is very difficult to assess the degree of uncertainty that pertains when research is moving from preclinical models to human interventions. Nonetheless, such trials must take place, and IRBs work to ensure that subjects understand not only what is known from preclinical work but also appreciate the existence of knowledge gaps that will affect the extent to which the outcome of the trials can be predicted.

The federal rules include a provision stating that an "IRB should not consider possible long-range effects of applying knowledge gained in the research (for example, the possible effects of the research on public policy) as among those research risks that fall within the purview of its responsibility." This provision therefore excludes from IRBs the power to withhold approval of a study solely because the knowledge it produces or the policies it affects may be socially controversial or because of fears that the study will represent the beginning of a slippery slope to future applications that are controversial. The provision does, however, allow IRBs to withhold approval of a study because it may cause physical, psychological, or emotional harm to the subjects.

Clinical Trials of Human Genome Editing: The Role of the Recombinant DNA Advisory Committee

The late 1960s and early 1970s saw the rapid progression of the concepts and technology that led to the first intentional creation of recombinant DNA molecules (Berg and Mertz, 2010). The RAC was established by then-NIH Director Donald Frederickson in 1974 in response to scientific, public, and political concerns about the potential use and misuse of recombinant DNA technologies, as well as the associated known and unknown risks. The proposed RAC membership included requirements designed to ensure a broader public perspective, such as a diverse membership that included scientists, clinicians, ethicists, biosafety experts, theologians and public representatives, among others, etc. Over time, the RAC's membership and responsibilities have evolved in response to scientific developments and shifting public concerns.

Early actions by the RAC included requiring that every research institution create a biohazard review committee (later renamed an IBC) to review risks and certify the presence of adequate safety measures. The major initial task of the RAC was the drafting of guidelines for recombinant DNA research that, while lacking the legal force of regulations, have had an enormous influence on practices for preventing the unintended release of or human exposure to genetically modified organisms and

及向研究监管机构提供信息外,这种透明体制意在优化个体研究方案的实施及在总体上促进转基因研究得到进步(O'Reilly et al.,2012)。通过这种方法,RAC作为一种重要渠道,为转基因领域的科学辩论、机构水平的监管、增加透明度、提升公众信任及信心服务。

2016年4月,《NIH准则关于涉及核苷酸分子的重组及合成研究的修正案》正式实施(NIH,2016a)。修改过的NIH准则,更多地反映了IOM研究下的提议(IOM,2014)。在修正案中,人类转基因试验必须满足NIH与监管机构(如IRB、IBC)的一致要求,这表明方案将显著受益于RAC审查,并且满足以下标准的一项或以上:

"实验方案若使用新的载体、基因结构及传递方法,将视为首次人类试验经历,因此代表着一种未知风险;

实验方案依赖于临床前安全数据,而此数据的获得使用了新的未知并未经证实其价值的临床前模型体系;

实验使用的载体、基因材料或传递方法可能带有并未广为人知的毒性,这样可能会增加监管机构严格评估方案的难度"(IOM,2014)。

如果NIH负责人认为人类转基因研究存在明显的科学、社会或伦理问题,则研究方案将也会面临重组DNA咨询委员会的审查。重组DNA咨询委员会已经审查了一些涉及三个主要基因组编辑技术的方案,在开发的早期阶段,某些人类基因组编辑方案有望满足这些标准。

在重组DNA咨询委员会支持下的公众参与

转基因研究的公众审查目的包括:①传播信息以助于其他科学家接受新的科学发现,并对他们的研究带来伦理上的思考;②提高公众意识,建立公众对于这些研究的信任度,允许研究的审查中出现公众声音(Scharschmidt and Lo,2006)。根据NIH的科学政策办公室(OSP)的要求,重组DNA咨询委员会审查的方案行使了许多功能(Corrigan-Curay,2013),包括:

(1)优化临床实验设计及增加实验对象的安全性,在某些案例中,增强研究者、卫生保健工作人员及密切联系的研究对象的生物安全保护是非常重要的。

material (Rainsbury, 2000). The guidelines are a term and condition of NIH funding, and are applicable to all recombinant DNA research that is conducted or sponsored by a public or private institution that receives NIH funding for any such research (NIH, 2013a). Many other U.S. government agencies and private institutions require that their funded research be conducted in accordance with the NIH Guidelines (Corrigan-Curay, 2013).

Initially, the RAC reviewed and approved all proposals for gene-transfer research protocols to be performed at institutions receiving NIH funds for recombinant DNA research, and advised the NIH director on the issuing of official approvals, as technically, official approvals came from the director of NIH, based on the RAC's decision (Freidmann et al., 2001). Over time, the interplay between RAC review and FDA review has evolved. The RAC's initial focus on safety broadened over time to include providing a venue for discussion of social and ethical issues. In the mid-1990s, the FDA assumed sole authority to approve gene transfer research protocols, with some protocols selected for in-depth review and public discussion by an initial review by RAC members. Provision was also made for a compassionate use exemption process (Rainsbury, 2000; Wolf, 2009).

Appendix M of the NIH Guidelines is a "Points to Consider" document that details the requirements for human gene-transfer protocol submission and reporting and review by the RAC (NIH, 2013c). The NIH Guidelines state that the "NIH will not at present entertain proposals for germline alteration." With regard to *in utero* gene transfer, the NIH Guidelines state that the NIH may be willing to consider such research but only after significant additional preclinical and clinical studies satisfy criteria developed at a RAC conference. In April 2015, the NIH Director issued a statement that "the NIH will not fund any use of gene editing technologies in human embryos."

Within the entire system of oversight for gene-transfer research, the RAC provides a forum for the in-depth review and public discussion of a protocol. The IRBs convene in private, although, as noted earlier, they do include nonscientists as members. The public nature of the RAC is due to its status as a public advisory committee under the Federal Advisory Committee Act (FACA) of 1972. To comply with FACA regulations, the RAC must hold open meetings, giving advance notice of the time and place; provide minutes; and allow for public participation (Steinbrook, 2004).

The RAC also sponsors public symposia on important scientific and policy issues related to recombinant DNA research (Friedmann et al., 2001), providing a public forum for scientific, clinical, ethics, and safety experts along with the public to discuss emerging issues in the field of gene transfer. Along with the RAC's protocol

（2）为了提升基因治疗研究的效率，将允许科学家建立一个公共基础平台，包含及时产生的新知识、透明的分析过程。

（3）为 FDA、NIH 人类研究保护办公室（OHRP）、IRB、IBC，以及其他对于批准基因治疗研究的实施极其重要的监管机构的审议提供信息。

目前的程序旨在实现高透明性。OSP 网站提供了有关协议和 RAC 大会讨论的信息，协议本身也会根据请求提供给公众（OBA，2013）。所有 RAC 与调查者之间的信函同时也作为研究方案的公共记录，可以提供给调查者、赞助者、伦理审查委员会、生物安全委员会、FDA 及 OHRP（NIH，2013a）。对于挑选出来用于深入审查及公众讨论的方案，当审查后，其调查者收到以 RAC 会议讨论的提议为基础的信件时，方案的注册程序方视为完成。该信件同时被送往相关的伦理审查委员会和生物安全委员会（NIH，2013b）。重组 DNA 咨询委员会会议的记录及网络广播可在重组 DNA 咨询委员会公共网页上获得。调查者及伦理审查委员会和生物安全委员会并未被要求必须遵循重组 DNA 咨询委员会的建议。但是，一个方案的批准需建立在其他监管机构信息的收集的基础上。在研究参与者可以开展临床试验之前，研究方案必须被相关的伦理审查委员会和生物安全委员会批准。这些机构通常依赖于重组 DNA 咨询委员会的建议来做出自己的决定，但是重组 DNA 咨询委员会的批准本质上并不要求进行研究（Wolf et al.，2009）。对于负责监管部门批准的 FDA 而言，当审查 IND 申请时，同样会考虑到重组 DNA 咨询委员会的观点（Takefman，2013）。

食品药品监管局对于研究性新药（IND）申请的审查

无论基金来源如何，FDA 是最终对基因组编辑产品进行监管和审批的机构。这些产品大多数被视为生物制剂而非医疗器械。因此，在被用于人类试验之前，它们的 IND 申请必须被 FDA 审查及批准。IND 针对基因治疗受到 CBER 下的组织中心和高级疗法（前身为生物制品评估和研究中心）管理。由于 FDA 与资助者的互动贯穿产品的整个生命周期，FDA 的审查将在 IND 前期至上市后的监管中一直存在。

review and mechanisms for informing institutional oversight bodies, this transparent system is intended to optimize the conduct of individual research protocols and to advance gene-transfer research generally (O'Reilly et al., 2012). In this way, the RAC serves as an important channel for scientific debate, informing institution-level oversight, increasing transparency, and promoting public trust and confidence in the field of gene transfer.

In April 2016, amendments to the NIH Guidelines for Research Involving Recombinant or Synthetic Nucleic Acid Molecules (NIH, 2016a) went into effect. Under the revised NIH Guidelines, which reflected many of the recommendations of an IOM study (IOM, 2014), individual human gene-transfer trials are limited to cases in which the NIH concurs with a request from an oversight body (such as an IRB or an IBC) that has determined that a protocol would significantly benefit from RAC review and has met one or more of the following criteria:

"The protocol uses a new vector, genetic material, or delivery methodology that represents a first-in-human experience, thus presenting an unknown risk.

The protocol relies on preclinical safety data that were obtained using a new preclinical model system of unknown and unconfirmed value.

The proposed vector, gene construct, or method of delivery is associated with possible toxicities that are not widely known and that may render it difficult for oversight bodies involved to evaluate the protocol rigorously" (IOM, 2014, p. 4).

Human gene-transfer protocols may also be reviewed by the RAC if the NIH director determines that the research presents significant scientific, societal, or ethical concerns. The RAC has reviewed several protocols involving the three major gene editing technologies, and certain human gene-editing protocols, at this early stage of development, would be expected to meet these criteria.

Public Engagement Under the Auspices of the Recombinant DNA Advisory Committee

Public review of protocols for gene-transfer research is intended (1) to disseminate information so that other scientists can incorporate new scientific findings and ethical considerations into their research, and (2) to enhance public awareness of and build public trust in such research, allowing for a public voice in the review of the research (Scharschmidt and Lo, 2006). According to NIH's Office of Science Policy (OSP), protocol review by the RAC serves many functions (Corrigan-Curay, 2013), including

- optimizing clinical trial design and increasing safety for research subjects, and in some instances strengthening biosafety protections necessary for researchers, health care workers, and close contacts

CBER 管理着大量的生物制剂，包括人类基因疗法产物，以及某些与转基因有关的医疗器械。FDA 将基因疗法产物定义为一种"通过转录或翻译转移的基因材料整合到宿主基因组来调节其作用的产品"，它们可以是核酸病毒或遗传工程微生物。迄今为止，依据《联邦食品、药品、化妆品法案》及修订后的《公共健康服务法》，FDA 审查的基因疗法产品的类型包括：非病毒性载体（质粒）、复制缺陷型载体（如腺病毒或腺病毒相关载体）、复制性溶瘤载体（如麻疹病毒、呼肠孤病毒）、复制缺陷型逆转录病毒及慢病毒载体，疱疹病毒载体，转基因微生物（如李斯特菌、沙门氏菌、大肠杆菌）及转基因细胞。

FDA 还设立了一个联邦咨询委员会，即细胞、组织、基因疗法咨询委员会，负责审查和评估拟用于移植的人类细胞、人类组织，转基因疗法和异种移植产品的安全性、有效性和适当性相关的数据，以及移植、输注和转移以治疗和预防人类疾病，并在各种条件下重建、修复或替换人体的组织[27]。

不论基金的来源如何，FDA 的程序适用于所有基因疗法的临床研究。在 FDA 审查 IND 及对随后研究进程主要阶段的审查期间（如从临床Ⅰ期到临床Ⅱ期阶段），RAC 初步的关于人类转基因研究的科学及伦理审查、新型应用的公共论述均会被列入考虑范畴内（Takefman, 2013）。与 RAC 不同，FDA 针对批准 IND 开始基因疗法临床试验的审查过程是不对公众开放的。为进入市场，产品需要接受生物制剂许可申请（BLA）（21-CFR 600-680）的批准，BLA 将关注于制造信息、标签、临床前和临床研究。申请 BLA 的过程将包括一些公众的参与。许多首创一类新药将进入咨询委员会阶段，其中将包含一些药学及科学专家，以及伦理学家、行业代表和病患代表。这些会议将代表着 FDA 对于一个新药的首次公众讨论。给那些关注会议的病患、医师及其他利益相关者和在机构网站上寻找会议记录的人提供途径。会议将提前公布于众，并且包括公众评论阶段。

FDA 将对研究团体通过"重点关注"文件的形式提供帮助，这代表了目前 FDA/CBER 人员对于重要的基因转移及基因治疗问题的思考（FDA,

- of research subjects;
- improving the efficiency of gene therapy research by allowing scientists to build on a common foundation of new knowledge emanating from a timely, transparent analytic process; and
- informing the deliberations of the FDA, the NIH Office of Human Research Protections (OHRP), IRBs, IBCs, and other oversight bodies whose approval is necessary for gene therapy research projects to be undertaken.

The current process aims to be highly transparent. The OSP website provides information about protocols and the public discussions at the RAC meetings, and the protocols themselves are made available to members of the public upon request (OBA, 2013). All correspondence between the RAC and investigators also is part of the public record for the protocol and is available to the investigators, sponsor(s), IRB(s), IBC(s), the FDA, and OHRP (NIH, 2013a). For protocols selected for in-depth review and public discussion, the protocol registration process is defined as complete when, following the review, the investigator receives a letter based on the recommendations discussed at the RAC meeting. The letter is also sent to the relevant IRB(s) and IBC(s) (NIH, 2013b). Minutes and webcasts of the RAC meetings are made available on the RAC's public website. Neither investigators nor IRBs or IBCs are required to follow any of the RAC's recommendations. Rather, a protocol's approval comes from a collection of other regulatory bodies. A protocol must be approved by the relevant IBC(s) and IRB(s) before research participants can be enrolled in a clinical trial. These bodies often rely on the RAC's recommendations in making their decisions, but the RAC's approval per se is not required for the research to move forward (Wolf et al., 2009). The FDA, the agency responsible for regulatory approval, also takes into account the views of the RAC when reviewing IND applications (Takefman, 2013).

U.S. Food and Drug Administration Review of Investigational New Drug (IND) Applications

Regardless of the funding source, the FDA is the agency ultimately responsible for the regulation and approval of genome-editing products. Most of these products will be viewed as biologic drugs rather than devices. Thus, before being used in human trials, they will need to have FDA review and approval of their IND application. INDs for gene therapy are regulated by the Office of Tissues and Advanced Therapies (previously the Office of Cellular, Tissue and Gene Therapies) within

---

[27] *Federal Register* 23309 (1986).

1991）。这些文件旨在指导调查者在准备他们的NID申请时，能够理解FDA对开发和测试的期望与要求。2015年，FDA发表了"细胞及基因治疗产物的早期临床试验设计的考虑因素"（FDA，2015b）。

为了确保所有监管部门的要求均能满足，FDA鼓励在产品开发的早期阶段，研究人员和机构人员间能够举行"前NID"会议，会议上可讨论关于临床试验设计的一些特殊问题。会议同样提供了机会来探讨药品科学及管理水平上的安全及/或潜在的临床保留问题[28]，例如，在儿童人群中研究转基因产品的计划。在会议上，研究者必须提供一份信息包，说明转基因产品的结构，拟用的临床症状、剂量和给药；提供临床前及临床的研究说明和数据摘要包含的化学、制造、控制（CMC）信息；规定了会议的预期目标（FDA，2000）。

对于某些类型方案——包括涉及转基因产品的方案——考虑到重组DNA蛋白来自于细胞系，有必要讨论一些特殊问题，如细胞特性是否胜任、潜在的细胞系污染、外源性物质的移除或灭活，或者产品的潜在抗原性。在向NID提交申请前，调查者需要考虑并解决源自前NID会议的FDA准则。

一般来说，当审查NID申请书时，FDA将平衡临床试验参与者所获得的利益与承担的风险（Au et al.，2012；Takefman and Bryan，2012）。一旦研究人员已提交NID，FDA将在30天内决定是否允许其继续进行，或者将其用于临床等待以便从提交者那里获取更多的数据。申请应该包括产品制造、安全性、质量检测、纯度和效力，以及临床前、药理学和毒理学测试的详细资料。对于基因治疗产品的安全性检测有着特殊的要求，包括：①对于体外转导细胞、载体、转基因存在的潜在异常免疫反应；②载体或转基因的毒性，包括载体对生殖细胞、睾丸和卵巢组织的影响；③传递过程中的潜在风险（FDA，2012b）。

申请书中的临床方案部分信息应包括：临床Ⅰ期、Ⅱ期及/或Ⅲ期研究，起始剂量，剂量递增，

CBER. The review follows a regulatory framework in which the FDA and the sponsor interact throughout the product's life cycle, from pre-IND to postmarketing surveillance.

CBER regulates a range of biologics, including human gene therapy products, and certain devices related to gene transfer. The FDA defines gene therapy products as products that "mediat[e] their effects by transcription and/or translation of transferred genetic material and/or by integrating into the host genome …and [that] are administered as nucleic acids, viruses, or genetically engineered microorganisms." The general types of gene therapy products reviewed by the FDA to date, pursuant to its authority under the Federal Food, Drug and Cosmetic Act (Public Law 75-717) and the Public Health Service Act (Public Law 78-410) as amended, are nonviral vectors (plasmids), replication-deficient viral vectors (e.g., adenovirus, adeno-associated virus), replication-competent oncolytic vectors (e.g., measles, reovirus), replication-deficient retroviral and lentiviral vectors, cytolytic herpes viral vectors, genetically modified microorganisms (e.g., *Listeria*, *Salmonella*, *E. coli*), and *ex vivo* genetically modified cells.

The FDA also maintains a federal advisory committee, the "Cellular, Tissue and Gene Therapies Advisory Committee," that reviews and evaluates available data related to the safety, effectiveness, and appropriate use of human cells, human tissues, gene-transfer therapies, and xenotransplantation products that are intended for transplantation, implantation, infusion, and transfer in the treatment and prevention of a broad spectrum of human diseases and in the reconstruction, repair, or replacement of tissues for various conditions.[27]

The FDA process applies to all gene therapy clinical research, regardless of source of the funding. During the FDA's review of INDs and its subsequent review of major steps in the research process (e.g., movement from phase I to phase II studies), any RAC preliminary scientific and ethical review of human gene transfer, as well as its public discussion of novel applications, is taken into account (Takefman, 2013). Unlike RAC review, the FDA's review process for granting an IND to begin a gene therapy clinical trial is closed to the public. To go on the market, products need to receive approval of their Biologic Licensing Application (BLA) (21 CFR 600-680), which focuses on manufacturing information, labeling, and preclinical and clinical studies. The process for approving the BLA may include some public participation. Many first in class products are taken to an advisory committee, which typically includes members with medical and scientific expertise, as well as ethicists, industry representatives, and patient representatives. These meetings often represent the FDA's first public

---

[28] 临床保留是一项推迟推荐的临床研究或暂停正在进行的临床研究。发布临床保留的条件包括研究对象存在的不合理风险，或发现削弱研究方案及研究者信心的信息（Clinical Holds and Requests for Modification，21 CFR，Sec. 312.42［April 1, 2016］）。

给药途径，给药方案，患者人群的定义（详细记录以及排除标准），安全监测计划。它同样包含了关于研究设计、临床程序描述、实验室检测或其他方法检测产品效益的信息。由于基因治疗的产品中的载体及转基因可能在研究对象的整个生命周期中都存在，FDA已经发布了持续观察研究对象的迟发性反应的指南（FDA，2006）。

联邦法规要求，许多临床试验信息必须刊登于 ClinicalTrials.gov 网页上，这个政府部门的数据库含有大量的临床试验数据及相似的网址。其适用于许多被 FDA 管理的临床试验药物（包括生物制剂产品），以及设备产品。该法规自 2017 年 1 月起生效，这里有一个扩展的表格记录了额外结果数据，可以帮助患者找到试验[29]。目标是为了完善实验设计，阻止重复不成功试验，提升依据基础及药物和设备开发的证据基础和效率，并建立公众信任度。

依据法定指令，在 NID 阶段，FDA 的审查几乎没有透明度，包括机构是否考虑授予特殊产品 NID。一旦 FDA 颁发了产品的许可证，就会在网站上公布针对该产品的临床、药理及其他方面的技术审核（例如，Zhu 和 Rees［2012］报道的 Ducord，一种脐带来源的干细胞产品，用于某些移植过程）。尽管在这些公布的技术审核方法里对专利信息会有所保留，但临床审核还是提供了大量关于这些实验的信息[30]。这些信息里有关于实验设计的早期讨论，以及赞助商是否符合某些伦理道德和良好的试验操作标准评估方面的总结。必要时，FDA 可以聘请细胞、组织和基因治疗咨询委员会，让公众也参与到对产品的广泛应用性的迫切问题中。

2008 年发起的 FDAS Sentined 计划，建立了一个风险识别和分析系统，利用电子健康护理数据对药品、生物制品和仪器设备投入市场后的安全性进行监测，这是对不良事件报告系统的补充。通过 Sentined，FDA 可以通过一个能同时维护患者隐私的流程，从电子健康记录、保险索赔数据、注册中心等来源获取信息。CBER 在哨点发起了几个项目，

discussion of a new medical product, providing access to information for patients, physicians and other stakeholders who observe the meeting, and to those who use the meeting transcripts made available on the agency website. Meetings are publicly announced in advance, and include public comment periods.

FDA offers assistance to the research community in the form of "Points to Consider" documents that present the current thinking of FDA/CBER staff about important issues in gene transfer and gene therapy (FDA, 1991). These documents are intended to guide investigators in understanding FDA perspectives and requirements for development and testing as they prepare their IND applications. In 2015 the FDA released "Considerations for the Design of Early-Phase Clinical Trials of Cellular and Gene Therapy Products" (FDA, 2015b).

To ensure that all regulatory requirements are met, the FDA encourages a "pre-IND" meeting between investigators and agency officials early in the protocol development process at which specific questions related to the planned clinical trial design are discussed. The meeting also provides an opportunity for the discussion of various scientific and regulatory aspects of the medical product as they relate to safety and/or potential clinical hold issues,[28] such as plans for studying the gene-transfer product in pediatric populations (FDA, 2001). For the meeting, the investigator must submit an information package that describes the structure of the gene-transfer product, its proposed clinical indication, dosage and administration; provides preclinical and clinical study descriptions and a data summary; includes chemistry, manufacturing, and controls (CMC) information; and specifies objectives expected from the meeting (FDA, 2000).

For certain types of protocols—including those involving gene-transfer products—it is sometimes necessary to discuss special issues regarding recombinant DNA proteins from cell-line sources, such as the adequacy of characterization of cells, potential contamination of cell lines, removal or inactivation of adventitious agents, or potential antigenicity of the product (FDA, 2015b). An investigator is expected to consider and address FDA guidance resulting from the pre-IND meeting before submitting an application for an IND.

As a general rule, when reviewing IND submissions, the FDA balances potential benefits and risks to participants in the clinical trials (Au et al., 2012; Takefman and Bryan, 2012). Once the investigator has submitted the IND, the FDA has 30 days either to allow it proceed or to put it

---

[29] *Federal Register* 64981-65157.
[30] FDAAA（2007）第 916 部分要求 FDA 网站公布关于产品许可的一些信息。http://www.fda.gov/BiologicsBloodVaccines/GuidanceComplianceRegulatoryInformation/ProceduresSOPPs/ucm211616.htm（accessed February 2, 2017）.

[28] A clinical hold is an order to delay a proposed clinical investigation or suspend an ongoing investigation. Conditions for issuing a clinical hold include unreasonable risk to research subjects or discovery of information that undermines confidence in the investigators or the study protocol (*Clinical Holds and Requests for Modification*, 21 CFR, Sec. 312.42 [April 1, 2016]).

旨在改进对疫苗和其他生物制品颁布许可后的安全性监督。除了监督之外，其他投入市场后的质量控制措施还包括登记册、特殊患者信息册和正规Ⅳ期实验要求。欧盟也有自己的投入市场后的监测和控制方法，细节上不太一样，但目的都相似（Borg et al.，2011）。

一旦药物通过 FDA 批准，其处方使用方法也可能会与它所批准和标注的有所不同。如前所述的这种标注外使用是合法的，也是卫生服务提供者在他们认为该方式适合他们的患者时采取的一种普遍做法。这可能意味着，在不同的医疗条件下，产品的使用方法会与其批准的不同（例如，批准用于一种癌症却用于另一种癌症），在不同的形式下或用于不同类型的患者，其剂量也会不同。标注外使用有助于药物批准后，医生的自由裁断和信息的有效使用，也维持了投入市场后的安全监测。在美国，一些医学领域，如儿科学（AAP，2014）和癌症护理（American Cancer Society，2015），已知有很高的标注外使用率。

有许多方法可以加速产品的监管过程，包括快速跟踪、突破性治疗、加速批准和优先审核（FAD，2015a）。用于加速审核的方法，包括：与 FAD 工作人员进行更早、更频繁和更深入的协商；放宽关于材料提交的规定；改变研究所需的终点；或在已经提交申请的其他产品之前进行审核。

在 2016 年 12 月签署的《21 世纪治愈法》中[31]，加速审查的这一条款已扩大到了再生医学和其他细胞治疗产品。该法批准了一个"再生医学疗法"，这是基于合理预期的替代终点和包括临床对照试验范围以外的更广泛来源的证据而提出的。批准后措施仍可以包括未来试验要求，以及监督、患者信息手册、登记册和其他风险缓解措施。这一过程在某种程度上类似于日本针对再生医疗产品采用的"有条件批准"方法，尽管它缺乏任何当上市后风险减轻和临床实验承诺未得到履行时自动撤销批准的触发机制。

人类基因组编辑所开发的治疗方法也包含在这一扩大的范畴里，一些符合"再生医学疗法"的定

on clinical hold while more data are obtained from the sponsor. The application includes details on product manufacturing, safety and quality testing, and purity and potency, as well as preclinical, pharmacological, and toxicological testing. Safety testing required specifically of gene therapy products includes (1) potential adverse immune responses to the *ex vivo* transduced cells, the vector, or the transgene; (2) vector and transgene toxicities, including distribution of the vector to germ cells in testicular and ovarian tissues; and (3) potential risks of the delivery procedure (FDA, 2012b).

The clinical protocol section of the application includes information about phase I, II, and/or III studies, including starting dose, dose escalation, route of administration, dosing schedules, definition of patient population (detailed entry and exclusion criteria), and safety monitoring plans. It also includes information regarding study design, including description of clinical procedures, laboratory tests, or other measures to be used to monitor the effects of the product. Because vectors and transgenes of gene therapy products may persist for the lifetime of the research subject, the FDA has issued guidance on observation of subjects for delayed adverse events (FDA, 2006).

Federal regulations require that information about many clinical trials be posted at ClinicalTrials.gov, the government's database for information about a large proportion of clinical trials, or a similar site. This applies to many clinical trials of drug products (including biological products) and device products that are regulated by the FDA. Effective as of January 2017, there is an expanded registry with additional results data to help patients find trials.[29] The goals are to enhance trial design, to prevent duplication of unsuccessful trials, to improve the evidence base and efficiency of drug and device development, and to build public trust.

In accordance with statutory mandates, however, there is little or no transparency in FDA reviews during the IND stage, including whether the agency is considering an IND for a specific product. But once the FDA has approved a license for a product, it may post the clinical, pharmacological, and other technical reviews of the product on its website (see, for example, information for Ducord, an umbilical-cord–derived, stem cell product for use in certain transplantation procedures, as reported by Zhu and Rees [2012]). Although proprietary information is redacted from these posted reviews, the clinical reviews provide considerable information about the trials.[30] They

---

[31] 21st Century Cures Act，Public Law No. 114-255，HR 34，114th Cong. (2015-2016)（https://www.congress.gov/bill/114th-congress/house-bill/ 34/text?format=txt）. 或 http://www.fda.gov/BiologicsBloodVaccines/Cellular GeneTherapyProducts/ucm537670.htm（accessed January 30，2017）.

[29] *Federal Register* 64981-65157.
[30] Section 916 of the Food and Drug Administration Amendments Act (2007) requires posting of certain information about a BLA approval on the FDA website. See SOPP8401.7 Action Package for Posting http://www.fda.gov/ BiologicsBloodVaccines/GuidanceComplianceRegulatoryInformation/ ProceduresSOPPs/ucm21 1616.htm (accessed February 2, 2017)

义（包括细胞治疗、治疗组织工程产品、人体细胞和组织产品）并满足对"严重或危及生命的疾病或条件"的需求的标准的治疗方法，有资格获得各种加速机制。自新法通过以来，FAD 一直在努力执行这些规定，同时也在考虑立法中规定的符合再生医学疗法定义的产品的范围等问题[32]。如该法中所写，满足对严重或危及生命的疾病的需求的标准，可能不包括其改善的预期用途[33]。

FDA 与 NIH 及 RAC 的相互影响

1999 年，参与转基因试验的 Jesse Gelsinger 的死亡，引起了人们对转基因试验的高度关注（Shalala，2000；Steinbrook，2002）。针对此事件，NIH 采取了措施协调对负面事件的报道，并扩大了公众获取有关人类转基因试验信息的途径，如建立基因改造临床研究信息系统（GeMCRIS）（NIH，2004），该数据库于 2004 年投入使用，包括在 NIH 注册的人类转基因实验的注册信息。GeMCRIS 主要包括正在进行的研究的医疗条件、进行实验的机构、进行实验的研究人员、正在使用的基因产品和基因产品的传递方法及研究方案的摘要等信息。

RAC 与 FDA 监督转基因研究的方法之间存在差异。作为美国唯一对生物医学产品进行监管的联邦监管机构，FDA 在评估基因转移产品时，从第一次在人体中使用到商业分销，在其整个使用寿命期间，都将重点放在安全性和有效性上。由于存在专利信息，根据法定条款规定，FDA 的监管中有许多步骤都是保密的（Wolf et al.，2009）。相反，RAC 能够解决由转基因和基因治疗研究引起的更广泛的科学、社会和伦理问题，而且与 IRB 不同，RAC 可以在个别方法的审核过程中解决这些更广泛的问题（NIH，2016b，Sec. IV-C-2-e）。此外，RAC 是由不经政府聘用的专家公开进行审查的（Wolf et al.，2009）。

为了促进各机构之间的交流，《RAC 章程》

---

[32] The FDA has also released notice of a new category, "regenerative advanced therapy," along with stipulations for the designation. (http://www.raps.org/Regulatory-Focus/News/2017/01/20/26651/FDA-Begins-Accepting-Regenerative-Therapy-Applications-for-RAT-Designation/).

[33] 21st Century Cures Act, Public Law No. 114-255, HR 34, 114th Cong. (2015-2016) (https://www.congress.gov/bill/114th-congress/house-bill/34/text?format=txt; accessed January 30, 2017).

may summarize early-stage discussions about trial design and assessments of whether sponsors conformed to certain ethical and good trial practice standards. When necessary, the FDA can engage its Cellular, Tissue and Gene Therapies Advisory Committee to receive public input on a pressing issue of broad applicability.

The FDA's Sentinel Initiative—launched in 2008 to establish a national risk identification and analysis system using electronic health care data to monitor the safety of drugs, biologics, and devices after they have reached the market—complements the Adverse Event Reporting System. Through Sentinel, the FDA can access information from electronic health records, insurance claims data and registries, and other sources using a process that also maintains patient privacy. CBER has launched several projects within Sentinel aimed at improving postlicensure safety surveillance of vaccines and other biologics. In addition to monitoring, other postmarket quality control measures include registries, special patient information pamphlets, and requirements for formal phase IV studies. The European Union has its own tools for postmarket monitoring and control, different in detail but similar in purpose (Borg et al., 2011).

Once FDA has approved a drug, it may be prescribed for uses that differ from those for which it was approved and labeled. As noted earlier, such off-label prescribing is legal and a common practice of health care providers when they deem it medically appropriate for their patient. This may mean use of the product for a different medical condition from that for which it was approved (for example, approved for one kind of cancer and used for another), or its administration at different doses, in different forms, or to different categories of patients. Off-label prescribing allows for physician discretion and the efficient use of information following a drug's initial approval, while still maintaining postmarket surveillance for safety. In the United States, some areas of medicine, such as pediatrics (AAP, 2014) and cancer care (American Cancer Society, 2015), are known to have a high rate of off-label use.

There are a number of mechanisms by which products may follow an accelerated regulatory pathway, including Fast Track, Breakthrough Therapy, Accelerated Approval, and Priority Review (FDA, 2015a). The means used to accelerate the review range from earlier, more frequent, and more intensive consultations with FDA staff; to easing rules for the submission of materials; to changing the endpoints required in the study; to conducting the review before that of other products for which the applications were submitted earlier.

This provision for accelerated review was expanded to include regenerative medicine and other cell therapy

要求一名 CBRE 的成员作为 RAC 的无投票权的联邦代表（NIH，2011）。NIH 和 FAD 还协调了对负面事件的报道。

其他国家的监管细节

正如前 FAD 理事 Robert Califf 所指出的，"科技进步不应局限于国家界限，我们必须了解国际同行新的观点"。为了促进技术信息的交流并确定监管协调的领域，FAD 积极参与到国际药品监管机构论坛及其基因治疗项目组（Califf and Nalubola, 2017）。

其他各国对基因治疗的监管方法与美国相似（见附录 B），特别是都集中于上市前风险和利益评估。例如，韩国有一种针对基因治疗的监管方法与美国非常相似，但它包括一个有条件批准制度，允许使用不健全的证据。英国和美国一样，有严格的上市前风险和利益审核，涉及配子或胚胎的疗法需要更严格的监管方式（见延伸内容 2-2）。欧盟和美国一样，也允许使用标注外方法，但其对包括一些基因治疗产品的"先进治疗药品"有更多的质量控制方法（George，2011）。日本也使用一种类似于美国的《仪器设备法规》的针对基因治疗产品的系统，对新产品按照预期风险水平进行前瞻性分类和管理。新加坡还采用了一种基于风险的方法，即采用以下标准：操作是实质性的还是最低限度的；预期用途是同源还是异源[34]；是否会与药物、设备或其他生物制品结合等。这些标准与许多美国在确定组织是否应该遵守移植药物规定或细胞治疗产品市场规定时使用的标准类似（Charo，2016b）。延伸内容 2-2 以英国为例，阐明了美国和其他国家监管方法之间的差异。

总结和建议

基因组编辑在预防、改善或消除许多人类疾病方面具有巨大的前景，然而，伴随这一承诺，需要进行道德上负责任的研究和临床应用。

本章对美国现有监管结构的讨论，为实验室

---

[34] FDA 将同源用途定义为，通过具有与供体相同的基本功能并能在受体中行使功能的 HCT/P，对受体的细胞或组织进行修复、重建、替换或补充（21 CFR 1271.3（c）），这些细胞或组织也能用于自体移植。

products in the 21st Century Cures Act,[31] signed into law in December 2016. The act allows for approval of a "regenerative-medicine therapy" based on surrogate endpoints reasonably expected to predict clinical outcomes and on evidence provided by a wider range of sources, including those outside the realm of controlled clinical trials. Postapproval measures can still include requirements for further trials, as well as surveillance, patient information brochures, registries, and other risk mitigation measures. This process resembles to some extent the "conditional approval" mechanism adopted in Japan for regenerative-medicine products, although it lacks any trigger that automatically withdraws approval if postmarket risk mitigation and clinical trial commitments are not fulfilled.

The therapies being developed with human gene editing were not excluded from this new expanded category, and some might be eligible for a variety of accelerating mechanisms if they meet the definition of "regenerative-medicine therapy" (which "includes cell therapy, therapeutic tissue engineering products, human cell and tissue products"), as well as the criterion of having the potential to fulfill an unmet need for a "serious or life-threatening disease or condition." Since passage of the new law, the FDA has been working on implementing these provisions, and is considering a number of issues, including the scope of the products that meet the definition of regenerative medicine therapy, as specified in the legislation.[32] As written, though, the criterion of fulfilling an unmet need for a serious or life-threatening disease would seem to exclude intended uses for enhancement.[33]

The Interplay Between the FDA and the NIH RAC

Concern about the conduct of gene-transfer trials reached a new level of intensity after the 1999 death of Jesse Gelsinger, a participant in one such trial (Shalala, 2000; Steinbrook, 2002). In response, NIH took steps to coordinate reporting of adverse events and expand public access to information regarding human gene transfer trials, for example, through the creation of the Genetic Modification Clinical Research Information System

---

[31] 21st Century Cures Act, Public Law No. 114-255, HR 34, 114th Cong. (2015-2016) (https://www.congress.gov/bill/114-congress/house-bill/34/text?format=txt; accessed January 21, 2017). See also http://www.fda.gov/BiologicsBloodVaccines/CellularGeneTherapyProducts/ucm537670.htm (accessed January 30, 2017).

[32] The FDA has also released notice of a new category, "regenerative advanced therapy," along with stipulations for the designation. (http://www.raps.org/Regulatory-Focus/News/2017/01/20/26651/FDA-Begins-Accepting-Regenerative-Therapy-Applications-for-RAT-Designation/)

[33] 21st Century Cures Act, Public Law No. 114-255, HR 34, 114th Cong. (2015-2016) (https://www.congress.gov/bill/114-congress/house-bill/34/text?format=txt; accessed January 30, 2017).

延伸内容2-2

其他的监管制度示例
Example of an Alternative Governance Regime

英国是对基因治疗有着其他监管制度的一个例子。对于体细胞基因治疗，其监管方法与美国没有什么不同，但涉及配子和胚胎时，其具有更集中和强化的管理方法。

对于体细胞基因治疗，英国生物技术顾问系统也参与到基因治疗咨询委员会和转基因生物健康安全执行科学咨询委员会里。英国临床试验法规要求在进行基因治疗的临床试验前，必须获得药品和保健产品监管机构的批准。同时也需要获得基因治疗咨询委员会的批准，附加法令条例也有对生产细胞治疗产品设备的质量控制的管理（Bamford et al.，2005）。公众的参与也来自于这些监管机构和专业团体。例如，英国基因和细胞疗法协会每年开放一次公共参与日，会让各个阶层的学生、患者、医生和科学家在一起讨论。和美国一样，基因治疗一旦在临床上可用，就允许标签外使用。

对于生殖系基因治疗问题，英国有更加集中和垂直整合的监管制度，对于产品是否可以生产、何时可以生产、由哪些专家或者机构生产有更严格的管理。美国的系统，会先有一个正式的"研究"阶段，然后再批准商业运营（即在市场上的临床应用），并且在使用方法上给予了专业人员自由判断的权力；英国的系统没有这些单独的分类，限制了医生在使用配子或胚胎治疗时的自由裁定权。这个严格的监督系统可以跟踪每个用于研究或治疗的胚胎的命运。

为了实施这一系统，英国创建了人类受精和胚胎管理局，这是一个对使用卵细胞和精子的治疗方法与涉及人类胚胎的治疗或研究独立的监管机构。它制定标准，然后颁发许可证到特定的诊所进行明确的监管。要获得许可证，该诊所必须满足安全和质量保证标准，为患者提供咨询，利用新技术监测出生情况及孩子的健康情况，并提供可以进行持续特定监测的人员和系统。除了诊所和研究中心（涉及指定个人），HFEA也颁

The United Kingdom serves as an example of an alternative governance regime for gene therapy. For somatic gene therapy, its approach is not unlike that in the United States, but it has more centralized and intensive regulatory control over therapies that involve gametes and embryos.

For somatic gene therapy, the U.K.'s biotechnology advisory system involves interplay between the Gene Therapy Advisory Committee and the Health and Safety Executive Scientific Advisory Committee on Genetically Modified Organisms (Contained Use). The U.K. Clinical Trials Regulations require that before clinical trials of gene therapy are conducted, approval must be obtained from the Medicines and Healthcare Products Regulatory Agency. Approval from the Gene Therapy Advisory Committee is also required, and additional laws and regulations govern quality control at facilities producing the cell-based therapy products (Bamford et al., 2005). Public input is sought by these regulatory agencies and by professional societies. For example, each year the British Society for Gene and Cell Therapy runs a Public Engagement Day to bring students at all levels, patients, caregivers, and scientists together for discussion and debate. Once available clinically, gene therapies may be used off-label, as in the United States.

With respect to the possibility of germline editing, the U.K. has a more focused and vertically integrated regulatory regime, with tighter controls over whether and when a procedure can be done, and by which professionals and clinics. Unlike the U.S. system, which has a formal "investigational" phase of defined uses followed by an approval for commercial marketing (that is, clinical use in the market), and which then allows professionals wide latitude in how to use it, the U.K. system foregoes these separate categories and limits physician discretion in the area of therapies using embryos or gametes. It is a rigorous system of oversight that can track the fate of every embryo used for research or treatment.

To implement this system, the U.K. created the Human Fertilisation and Embryology Authority, an independent regulator of treatments using eggs and sperm, and of treatment or research involving human embryos. It develops standards, and then issues licenses to specific clinics to proceed with specific interventions. To be licensed, clinics must meet safety and quality assurance standards, offer counseling to patients, monitor birth outcomes and the well-being of children conceived through the new technologies, and generally provide personnel and systems that allow for ongoing compliance

发针对特定项目或治疗的许可证，可以是更广泛的常规完善的程序，或个别特殊情况。例如，关于植入前遗传诊断的许可证是针对特定疾病的。对于有经验者，这项政策可以放宽，研究中心现在也可以获得针对一系列遗传疾病的许可证，但这些仍然必须在HFEA的批准名单上，"标注外使用"仍然不允许用于涉及配子和胚胎的应用。

研究的管理、临床试验前测试、临床试验、涉及人类基因组编辑的潜在医学用途，以及对美国与其他国家监管制度之间差异的了解，提供了一个起始框架。

不同国家的产品监管制度非常相似，都着重于对上市前风险和利益的平衡。但是在有条件批准的有效性、其他对细胞治疗产品加速审核方法，以及在胚胎或配子管理方面仍存在一些差异。在英国临床护理中，标注外使用方法通常是允许的，但对涉及胚胎或配子的治疗进行更全面的监控。

综上所述，美国监管系统虽然还有改进空间，但已经足以用于监督人类基因组编辑研究和产品批准。第3~7章描述了其他方面还可以做出的努力。

建议2-1：关于人类基因组编辑的监督制度、深入研究和临床应用均应遵循以下原则：
（1）福利提升原则
（2）透明性原则
（3）妥善关注原则
（4）科学诚信原则
（5）尊重个人原则
（6）公平原则
（7）跨国界合作原则

(GeMCRIS) (NIH, 2004). This database, which became operational in 2004, includes summary information on human gene-transfer trials registered with NIH (2004). Included in the GeMCRIS summaries is information about the medical conditions under study, institutions where trials are being conducted, investigators carrying out these trials, gene products being used, and routes of gene product delivery, as well as summaries of study protocols.

Differences remain between the RAC's and the FDA's

monitoring. In addition to the clinic/research center (with named individuals involved), HFEA issues licenses for the specific project or treatment. The latter can be on a broad basis for common, well-established procedures, or specifically for each individual case. For Preimplantation Genetic Diagnosis, for example, licenses were originally for specific diseases. With experience this was relaxed and centers can now obtain licenses for a range of genetic diseases, but these still must be on the HFEA's approved list, and so "off-label use" is not allowed for applications involving gametes and embryos.

approach to oversight of gene-transfer research. The FDA, as the sole federal regulatory agency for biomedical products in the United States, focuses on safety and efficacy when evaluating gene-transfer products, from the first time they are used in humans through their commercial distribution (Kessler et al., 1993) and over the lifetime of their use. FDA regulation includes many steps that, by statutory provision, are confidential because of the presence of proprietary information (Wolf et al., 2009). In contrast, the RAC is able to address broader scientific, social, and ethical issues raised by gene-transfer and gene therapy research, and—unlike IRBs—the RAC is permitted to address these broader issues in its review of individual protocols as well (NIH, 2016b, Sec. IV -C-2-e). In addition, RAC review is conducted publicly by experts who are not employed by the government (Wolf et al., 2009).

To encourage communication between the agencies, the RAC charter calls for a member of CBER to be one of the nonvoting federal representatives to the RAC (NIH, 2011). NIH and the FDA also have harmonized reporting of adverse events.

GOVERNANCE IN OTHER NATIONS

As noted by former FDA Commissioner Robert Califf, "[s]cientific advances do not adhere to national boundaries and therefore it is critical that we understand the evolving views of our international counterparts." To that end, the FDA actively participates in the International Pharmaceutical Regulators' Forum and its Gene Therapy working group, for the purpose of exchanging technical information, and identifying areas for regulatory coordination (Califf and Nalubola, 2017).

The regulatory pathways for gene therapy in other jurisdictions are similar to those in the United States in important ways (see Appendix B), particularly with respect to the centrality of premarket risk and benefit assessment. For example, gene therapy in South Korea has a pathway very similar to that in the United States except that it includes a system of conditional approval that allows for use with less robust evidentiary bases. The United Kingdom has rigorous premarket risk and benefit review, as in the United States, but singles out therapies involving gametes or embryos for more intensive regulation (see Box 2-2). The European Union has additional layers of quality control for "advanced therapy medicinal products", which would include some gene therapy products, although as in the United States, off-label use would be permissible (George, 2011). And Japan uses a system for gene therapy products that resembles the U.S. device regulations, in which new products are sorted prospectively by level of anticipated risk and regulated accordingly. Singapore also has adopted a risk-based approach, with such criteria as whether the manipulation is substantial or minimal; whether the intended use is homologous or nonhomologous[34]; and whether it will be combined with a drug, a device, or another biologic. These criteria resemble many of those used by American authorities in determining whether tissues should be subject to rules governing transplant medicine or rules governing the marketing of cell-therapy products (Charo, 2016b). Box 2-2 illustrates the differences between the United States and other regulatory regimes by describing the example of the United Kingdom.

CONCLUSIONS AND RECOMMENDATION

Genome editing holds great promise for preventing, ameliorating, or eliminating many human diseases and conditions. Along with this promise, however, comes the need for ethically responsible research and clinical use.

The existing U.S. regulatory structures discussed in this chapter provide a starting framework for governance of laboratory research, preclinical testing, clinical trials, and potential medical uses involving human genome editing in the United States, as well as for an understanding of differences between the U.S. system and the regulatory infrastructures of other nations.

There is considerable similarity in the structures for product regulation among different jurisdictions, with an emphasis on premarket balancing of risk and benefit. Some differences exist in the availability of conditional approval or other accelerated approval mechanisms for cell-therapy products, as well as in the management of embryos and gametes. In clinical care, off-label use is commonly permitted, again with the notable exception of the more comprehensive controls on therapies involving embryos and gametes in the United Kingdom.

Overall, while capable of improvement, the structure of the U.S. regulatory system is adequate for overseeing human genome editing research and product approval. Specific areas in which additional effort might be made are identified in Chapters 3-7.

RECOMMENDATION 2-1. The following principles should undergird the oversight systems, the research on, and the clinical uses of human genome editing:
 1. **Promoting well-being**
 2. **Transparency**
 3. **Due care**
 4. **Responsible science**
 5. **Respect for persons**
 6. **Fairness**
 7. **Transnational cooperation**

---

[34] FDA defines homologous use as the repair, reconstruction, replacement, or supplementation of a recipient's cells or tissues with an HCT/P that performs the same basic function or functions in the recipient as in the donor (21 CFR 1271.3(c)), including when such cells or tissues are for autologous use.

3
基因组编辑技术的基础研究应用
Basic Research Using Genome Editing

近年来在基因及基因组编辑技术上的显著进步引起了学术界的广泛兴趣,并在基础及应用研究的诸多领域产生了重大影响。大约在 60 多年前,人们已经发现地球上所有的生命都是由 DNA 编码来遗传给下一代的,而生物学领域的加速发展极大地提高了人们对 DNA 的认识及操控能力。本章回顾了涉及人类基因组编辑的基础研究的各种类型及目的,先描述了基因组编辑的基本工具和基因组编辑技术的快速进展,接着详述了在推进了解人类细胞和组织、人类干细胞、疾病和再生医学,以及哺乳动物繁殖和发育的基础实验研究中,基因组编辑技术是如何应用的。最后我们总结了本研究相关的伦理及监管问题。整个章节对涉及基因组编辑的基础研究中的关键术语和概念进行了定义;延伸内容 3-1 定义了这些术语中最基本的含义。

基因组编辑基本方法

所有的生命体,从细菌到植物再到人类,尽管它们基因组的大小和基因的数量千差万别,但都是以相似的机制来编码和表达基因的。因此,了解任何一种生命形式都是了解所有其他生命形式的巨大信息来源,并且能够提供跨物种的见解和应用,这对推进基因及基因组编辑技术的研发尤其具有价值。最早期的分子生物学研究是利用细菌及病毒来进行的。它们相对简单和易于分析的特性是建立遗

The recent remarkable advances in methods for editing the DNA of genes and genomes have engendered much excitement and activity and had a major impact on many areas of both basic and applied research. It has been known for 60 years that all life on earth is encoded in the sequence of DNA, which is inherited in each succeeding generation, but accelerating advances have greatly enhanced understanding of and the ability to manipulate DNA.

This chapter reviews the various types of and purposes for basic laboratory research involving human genome editing. It begins by describing the basic tools of genome editing and the rapid advances in genome-editing technology. The chapter then details how genome editing can be used in basic laboratory research aimed at advancing understanding of human cells and tissues; of human stem cells, diseases, and regenerative medicine; and of mammalian reproduction and development. Ethical and regulatory issues entailed in this research are then summarized. Throughout the chapter, key terms and concepts germane to basic research involving genome editing are defined; Box 3-1 defines the most foundational of these terms.

THE BASIC TOOLS OF GENOME EDITING

All living organisms, from bacteria to plants to humans, use similar mechanisms to encode and express genes, although the sizes of their genomes and their numbers of genes differ greatly. Hence, understanding of any form of life is immensely informative with respect to understanding all other forms, and provides insights and applications that obtain across species—a fact that

延伸内容3-1

基 本 术 语
Foundational Terms

以下术语是理解任何 DNA 相关研究的基础。

DNA 是由四种（A，T，C，G）相似而重复的单元组成的长多聚物，其中英文字母代表不同的核苷酸碱基单元。碱基通过彼此特异性配对（A 配 T，C 配 G）形成两条互补链组成的众所周知的 DNA 双螺旋。编码基因的 DNA 片段能作为模板被复制（转录）成被称为 RNA 的第二种核苷酸多聚物。一些 RNA 通过与其他 RNA 配对来影响其功能。另一些 RNA 则参与组成细胞活动所必需的结构，这些细胞活动包括以一些 RNA 为模板来编码蛋白质。蛋白质是另一种由 20 种不同氨基酸单元组成的多聚物；由于 RNA 经过编码变成了不同的"语言"——"被写成"氨基酸而不是核苷酸，因此以 RNA 为模板编码蛋白质的过程被称为翻译。这些蛋白质聚合物折叠成复杂的三维结构来形成构建模块进而形成人体细胞，并且执行生命体的各种功能。DNA 到 RNA 的转录和 RNA 到蛋白质的翻译合称为基因表达，这一过程被严密调控以确保基因能因时因地适量表达。因此，单个细胞的功能是由其基因的表达来决定的。生物体内完整的一组基因被称为基因组。人的大部分细胞包含两个拷贝的完整的人基因组，每个拷贝包含 30 亿对碱基对，能够编码大约 2 万个编码蛋白质的基因，以及调节基因表达的控制元件。人们可以将基因组看成是"代码"或"软件"，将 RNA 和蛋白质及其形成的结构看成是细胞和生物体的"硬件"。

The following terms are foundational to understanding any research involving DNA.

DNA is a long polymer of similar repeating units of four types (A,T,C,G), where the letters denote distinct units called nucleotide bases. The bases pair specifically with each other (A with T and C with G) to form the well-known double helix of DNA with its two complementary strands. Segments of DNA sequence encode genes that can be copied (transcribed) from the DNA into a second type of nucleotide polymer called RNA. Some of these RNAs act by pairing with other RNAs to affect their functions, while others contribute to structures necessary for cellular activities, including the copying of some of the RNA molecules to encode proteins. Proteins are polymers of a different type of unit, called amino acids, of 20 different types; hence the copying of RNA to form proteins is known as translation since the copying is into a different "language"—"written" in amino acids rather than nucleotides. These protein polymers fold into complex three-dimensional shapes that form the building blocks of the cells that make up the human body and perform the myriad functions of living organisms. The combination of transcription from DNA to RNA and translation of RNA to protein is known as gene expression, and is tightly regulated so that genes are expressed at the appropriate times and places and in the correct amounts. Thus, the functions of individual cells are dependent on the genes they express. The complete set of genes in an organism is called its genome. Most human cells contain two complete copies of the human genome, each comprising 3 billion base pairs and encoding approximately 20,000 genes encoding proteins, plus the regulatory elements that control their expression. One can think of the genome as the "code" or "software" and RNA and proteins and the structures they form as the "hardware" of cells and living organisms.

传密码和基因表达基础的关键。更复杂生物的平行研究也是建立在对细菌研究的进展之上。到 19 世纪 60 年代中期，细菌、植物和动物共享着许多基本分子机制的事实已昭然若揭。细菌中的关键发现揭示了它们对抗病毒的保护机制。这其中包含了所谓的限制性内切核酸酶——用来剪切入侵病毒的

has been particularly invaluable in the development of methods for editing genes and genomes.

The earliest studies in molecular biology were on bacteria and their viruses. Their relative simplicity and ease of analysis were key in establishing the basis of the genetic code and the expression of genes. Parallel research on more complex organisms built on the advances in these studies of bacteria, and by the mid-1960s, it was

DNA 并"限制"病毒生长的蛋白质。这项发现使科学家能以可预测、可重复的方式切割 DNA，并将切割片段重新组装到重组 DNA 中。

到 20 世纪 70 年代中期，重组 DNA 技术为有效地组合 DNA 提供了有力手段，并且在生物技术领域得到广泛应用。然而，这种潜力同时也带来对这些新方法应用的风险性质疑。鉴于这些担忧，一些科学家及相关人士于 1975 年在艾斯罗马（Asilomar）召开了会议，讨论了可能需要哪些预防措施来监管这项新技术，并制定了一套指南来规范该研究的范围与实行。这些指南的衍生条例至今仍然监管着重组 DNA 的研究，有些已经被纳入官方监管系统。事实上，人们最极端的顾虑并没有发生。如今，重组 DNA 技术在全世界广泛应用，并且在科学认知和医学发展（包括许多有价值的药物和治疗）方面使人类受益匪浅，与此同时，生物技术产业也已是世界经济蓬勃发展的一部分。

利用重组 DNA 技术发展出来的诸多技术之一是能够将 DNA 引入细胞中并在其内表达，即所谓的转基因。此方法在实验室基础研究中已经得到了广泛应用（详见附录 A）。虽然整个过程并不高效，但外源 DNA 引入到细胞内后很大程度上会随机整合到细胞的基因组上并表达外源的 RNA 和蛋白质（取决于插入方式和位点）。重组 DNA 技术中一个关键性的进步是能特异性切割 DNA 的分子工具的发展，从而使靶向修改基因和基因组的 DNA 序列成为可能。研究发现，双链断裂（DSB，即 double-strand break 的英文首字母全称）可以有目的地用能特异性切割 DNA 位点的核酸酶（归巢核酸酶，有时也叫巨大核酸酶，最初是在酵母中发现）来制造出来（Roux et al., 1994a, b; Choulika et al., 1995）。基于这些突破性的发现，在接下来的 20 年里，多种不同类型的、能靶向特异位点的核酸酶已被研究出来并适用于靶向 DNA 切割（Carroll, 2014）。

双链断裂在 DNA 复制、辐射或化学损伤中可以自然产生，细胞也已经进化出了末端再连接机制（一种被称为非同源性末端连接或 NHEJ 的过程；NHEJ 即 Nonhomologous End Joining 的英文首字母全称）来修复它们。但是，这种再连接通常不太完美，修复过程中会引入小片段的插入和缺失。这样的插

clear that bacteria, plants, and animals shared many fundamental molecular mechanisms. Key discoveries in bacteria uncovered some of their mechanisms for protection against viruses, including so-called restriction endonucleases, proteins bacteria use to cleave the DNA of infecting viruses and "restrict" their growth. This discovery allowed scientists to cut DNA in predictable and reproducible ways and to reassemble the cut pieces into recombinant DNA.

By the mid-1970s, it was evident that recombinant DNA offered a powerful means of combining DNA in productive ways, with promising applications in biotechnology. However, this potential also raised questions about whether the application of these novel methods might entail some risk. In light of those concerns, a group of scientists and others convened a meeting at Asilomar in 1975 to consider what precautions might be needed to oversee this new technology and established a set of guidelines to regulate the containment and conduct of the research. The descendants of those guidelines still regulate recombinant DNA research to this day, some of them incorporated into official regulatory systems. In practice, the most extreme concerns did not eventuate. Today, the use of recombinant DNA methods is widespread worldwide and has yielded enormous benefits to humankind in terms of scientific understanding and medical advances, including many valuable drugs and treatments, and the biotechnology industry is now a thriving part of the world economy.

Among methods developed through the use of recombinant DNA technology is the ability to introduce DNA into cells where it can be expressed—a so-called transgene. This method is widely used in fundamental laboratory research (see Appendix A for more detail). When such exogenous DNA is introduced into a cell, it can insert into the DNA of the cell's genome largely at random and, depending on how and where it is inserted, can be expressed as RNA and protein, although this overall process is not very efficient. A key advance was the development of techniques for generating molecular tools that could be used to cut the DNA of genes and genomes in specific places to allow targeted alterations in the DNA sequence. It was found that double-strand breaks (DSBs) could be deliberately generated by nucleases that cut DNA at defined sites (homing nucleases, sometimes also called meganucleases, originally discovered in yeast) (Choulika et al., 1995; Roux et al., 1994a,b). In the succeeding 20 years, based on these groundbreaking discoveries, several additional types of nucleases that can be targeted to specific sites were developed and adapted for use in targeted DNA cleavage (Carroll, 2014).

Such double-strand breaks also occur naturally during DNA replication or through radiation or chemical damage,

入和缺失（indels，即 insertions and deletions 的缩写）会破坏被切割基因的 DNA 序列并常常使其失活。这种靶向切割和非同源性末端连接的非精准修复提供了一种使基因或基因调控元件失活的方法。虽然插入缺失突变通常只有一至数个核苷酸的长度，但在有些情况下能包含几千个碱基对。通过在两个不同位点同时制造两个双链断裂，然后再利用非同源性末端连接来连接这两个位点，基因组编辑技术还能被用于制造特定染色体缺失和移位。这些位点可以是在同一染色体（产生缺失）或不同染色体上（产生移位）。

双链断裂的修复过程中，如果提供额外的、与被切割 DNA 具有相同序列（即同源）的 DNA 片段，就能达成更精准的编辑。同源修复也见于正常的细胞修复机制中，这些机制能被发掘来做精准的 DNA 修改。若将与被切割序列有细微差异的同源 DNA 引入细胞，差异序列能被插入基因或基因组，该过程被定义为同源定向修复（HDR，即 homology-directed repair 的英文首字母全称）。同源定向修复也能用于在基因组精确位置插入不同长度的全新序列（如一个或多个基因）。与非同源性末端连接不同，同源定向修复介导的基因组编辑使科学家能够同时预期编辑发生的位置及修改后 DNA 的大小和序列。因此，同源定向修复的原理和文档编辑非常类似，因为它能精确地改变文档中的字母。

锌指核酸酶（ZFNs，即 zinc finger nucleases 的英文首字母全称）和转录激活因子样效应物核酸酶（TALENs，即 transcription activator-like effector nucleases 的英文首字母全称）是广泛发展的应用于基因及基因组编辑的两种靶向核酸酶。两种酶在正常时的基本功能是结合相对较短的 DNA 序列的蛋白质。锌指是多细胞生物用来调控基因表达的蛋白质的 DNA 结合结构域（它们往往结合锌作为结构的一部分，并因此得名）。分子生物学家能改造锌指使其能识别不同的 DNA 短序列，还能将之与核酸酶偶联来切割 DNA。因此，锌指结构被用于靶向基因和基因组上的特定序列，而与之偶联的核酸酶则用来切割 DNA 的两条链来产生双链断裂。ZFNs 已被开发用于基因编辑且已进入临床试验——例如，在艾滋病患者体内尝试产生对 HIV 病毒的抗性（Tebas et al., 2014）。TALENs 的工作原理与 ZFNs 类似，它们利用的是一种最初在感染植

and cells have evolved mechanisms for repairing them by rejoining the ends (a process known as nonhomologous end joining or NHEJ). However, this rejoining often is not perfect, and small insertions and deletions can be introduced during the repair. Such insertions and deletions (indels) can disrupt the sequence of the DNA and often inactivate the gene that was cut. This targeted cleavage and inaccurate repair through NHEJ provide a means of inactivating genes or gene-regulatory elements. Although the resulting indels are usually one or a few nucleotides long, in some cases they can consist of thousands of base pairs. Genome editing through NHEJ can also be harnessed to create defined chromosomal deletions or chromosomal translocations by simultaneously creating two DSBs at different sites, followed by rejoining at those two sites. These sites can be either on the same chromosome (producing a deletion) or on different chromosomes (producing a translocation).

More precise editing can be achieved if, during the breakage-repair process in the cell, an extra piece of DNA is provided that shares sequence (that is, is homologous) with the cleaved DNA. Such homologous repair also is used by normal cellular repair mechanisms. These mechanisms can be exploited to make precise changes. If homologous DNA slightly different in sequence from the cleaved sequence is introduced into the cell, that difference can be inserted into the sequence of the gene or genome, a process termed homology-directed repair (HDR). HDR can also be used to insert a novel sequence (e.g., one or more genes) of variable length at a precise genomic location. In contrast to NHEJ, HDR-mediated genome editing allows scientists to predict both where the edit will occur and the size and sequence of the resulting change. Thus, HDR-mediated editing is very much like editing a document because precise changes in the characters can be made.

Two types of targeted nucleases that have been widely developed for use in editing genes and genomes are zinc finger nucleases (ZFNs) and transcription activator-like effector nucleases (TALENs). Both rely on proteins whose normal function is to bind to specific relatively short DNA sequences. Zinc fingers are segments of proteins used by multicellular organisms to control the expression of their genes by binding to DNA (they also typically bind zinc as part of their structure; hence their name). They can be engineered by molecular biologists to recognize different short DNA sequences and can be joined to nucleases that cleave DNA. Thus, the zinc fingers target specific sequences in genes and genomes, and the attached nucleases cleave the DNA to generate a double-strand break by cleaving both strands of the DNA. ZFNs have been developed for gene editing and are in clinical trials—for example, in attempts to confer resistance to the

物的细菌中发现的DNA识别蛋白（转录激活样效应子或TALEs，即transcription activator-like effectors的英文首字母全称）。TALE蛋白的DNA识别序列由一系列重复单元组成，每个单元识别单个DNA碱基对。改造TALEs比改造锌指更为简便，也能同样偶联DNA核酸酶来产生TALENs。用TALENs改造用于治疗急性淋巴性白血病的淋巴细胞的临床前应用已经发表（Poirot et al., 2015）。

综上，这些工具已是成熟的、可用于基因治疗的基因编辑方法，许多相关的安全和监管问题已经得到了解决（见第4章）。然而，设计靶向特异位点的TALENs、ZFNs更甚，所要用到的蛋白质改造工程依然在技术上极具挑战性，并且非常耗时和花费巨大。

最近五年里，人们见证了一个全新体系的兴起，它被称为CRISPR/Cas9（CRISPR代表聚簇固定间隔短回文重复，即clustered regularly interspaced short palindromic repeats的英文首字母全称）（Doudna and Charpentier, 2014; Hsu et al., 2014）。短的效仿CRISPR体系制造的RNA序列与Cas9（CRISPR结合蛋白质9，一种RNA定向核酸酶）或其他类似的核酸酶配对，就能快捷地被编程用于编辑特定序列的DNA。与以往的方法相比，CRISPR/Cas9体系在更简便经济的同时还非常高效。CRISPR/Cas9与TALEs类似，最初也是在细菌中发现的，是保护细菌对抗入侵病毒的免疫系统的一部分（Barrangou and Dudley, 2016; Doudna and Charpentier, 2014）。CRISPR/Cas9最显著的特点是用可与目的DNA碱基互补配对的RNA序列代替蛋白质结构域来识别特定DNA序列。

如在2012年第一次被改造成功时那样（Jinek et al., 2012），细菌核酸酶Cas9可以结合被称为**向导RNA**的量身打造成可以识别任何所选DNA序列的单个RNA序列。这种双因子体系可以通过向导RNA结合到所选DNA位点，并用Cas9核酸酶切割DNA。由于合成任何所需序列的RNA十分简单，制造CRISPR/Cas9靶向核酸酶非常直观——这套体系能被快捷地编程用于靶向任何基因组的任何序列。已有用于选择合适的向导RNA的程序，虽然不是所有的向导RNA都一样有效，但测试若干向导RNA并从中找出有效的几个既不困难也不昂贵。这样易于设计且具有非凡特异性和高效性的

HIV virus in AIDS patients (Tebas et al., 2014). TALENs work similarly to ZFNs, also using DNA recognition proteins (transcription activator-like effectors or TALEs) originally identified in bacteria that infect plants. The DNA recognition sequences of TALE proteins are made of repeating units, each of which recognizes a single base pair in the DNA. TALEs are simpler and easier to engineer than are zinc fingers and can similarly be joined to DNA-cleaving nucleases to yield TALENs. The preclinical application of TALENs to engineer lymphocytes for the treatment of acute lymphoblastic leukemia was recently reported (Poirot et al., 2015).

Thus, these tools are already well-established approaches to the use of genome editing for applications in gene therapy, and many of the associated safety and regulatory issues have already been addressed (see Chapter 4). However, the protein engineering required to design site-specific versions of TALENs and, even more so, of ZFNs, remains technically challenging, time-consuming, and expensive.

The last 5 years have seen the development of a completely novel system, known as **CRISPR/Cas9** (CRISPR stands for clustered regularly interspaced short palindromic repeats,) (Doudna and Charpentier, 2014; Hsu et al., 2014). Short RNA sequences modelled on the CRISPR system, when paired with **Cas9** (CRISPR associated protein 9, an RNA-targeted nuclease), or alternatively with other similar nucleases, can readily be programmed to edit specific segments of DNA. The CRISPR/Cas9 system is simpler, faster, and cheaper relative to earlier methods and can be highly efficient. CRISPR/Cas9, like TALEs, was originally discovered in bacteria, where it functions as part of an immunity system to protect bacteria from invading viruses (Barrangou and Dudley, 2016; Doudna and Charpentier, 2014). The key distinguishing feature of CRISPR/Cas9 is that it uses RNA sequences instead of protein segments to recognize specific sequences in the DNA by complementary base pairing.

As first reengineered in 2012 (Jinek et al., 2012), the bacterial nuclease Cas9 binds a single RNA sequence known as a guide RNA tailored to recognize any sequence of choice. This two-component system can bind to the chosen site in DNA via the guide RNA and cleave the DNA using the Cas9 nuclease. Since it is simple to synthesize RNA of any desired sequence, generation of CRISPR/Cas9 targeting nucleases is straightforward—the system is readily programmed to target any sequence in any genome. Programs exist for choosing suitable guide RNAs, and while not all guides work equally well, testing a number of guides to find effective ones is not difficult or expensive. This ease of design, together with the remarkable specificity and efficiency of CRISPR/Cas9 has

CRISPR/Cas9 彻底改变了基因组编辑领域，对基础研究的发展有重大意义，并能广泛应用到生物技术、农业、昆虫控制和基因治疗等领域。

图 3-1 总结了基因组编辑的三种方法：ZFN、

revolutionized the field of genome editing and has major implications for advances in fundamental research, as well as in such applications as biotechnology, agriculture, insect control, and gene therapy.

Figure 3-1 provides a summary of the ZFN, TALEN,

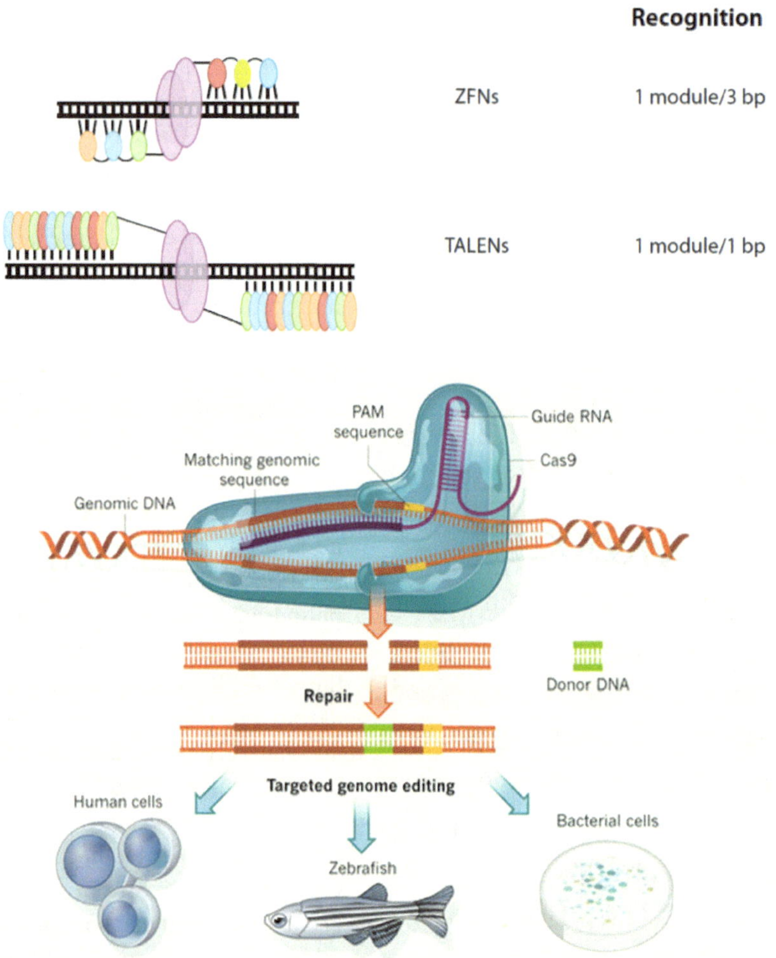

图 3-1 基因组编辑技术
FIGURE 3-1 Methods of genome editing.
上图：锌指核酸酶（ZFNs）：彩色模块代表锌指，每个模块被设计成识别 DNA 中三个相邻的碱基对；这些模块与能在 DNA 中形成双链断裂的 FokI 核酸酶二聚体偶联。
Top: Zinc finger nucleases (ZFNs): The colored modules represent the Zn fingers, each engineered to recognize three adjacent base pairs in the DNA; these modules are coupled to a dimer of the FokI nuclease that makes a double-stranded cut in the DNA.
中图：转录激活因子样效应核酸酶（TALENs）：每个彩色模块识别 DNA 中的单个碱基对；这些模块与能在 DNA 中形成双链断裂的 FokI 核酸酶二聚体偶联。
Middle: Transcription activator-like effector nucleases (TALENs): The colored modules each recognize a single base pair in the DNA; these modules are coupled to a dimer of the FokI nuclease that makes a double-stranded cut in the DNA.
下图：CRISPR/Cas9。需要两种来源于聚簇固定间隔短回文重复（CRISPR）区域的因子。核酸酶如 Cas9（蓝色）在能结合到基因组上邻近短的前间区序列邻近基序（PAM）（黄色）的 20 个碱基序列上的向导 RNA（紫色）的协助下定位到 DNA 上特定的序列，并在 DNA 上制造双链断裂。
Bottom: CRISPR/Cas9: Two components derived from the clustered regularly interspaced short palindromic repeat (CRISPR) region are needed. A nuclease such as Cas 9 (blue) is targeted to a specific site on the DNA by the guide RNA (purple), which binds a 20-base sequence in the genome adjacent to a short protospacer adjacent motif (PAM) sequence (yellow) and targets a double-strand cut in the DNA.
所有三种技术所造成的 DNA 断裂都能通过非同源性末端连接来修复或被一段同源 DNA（绿色）序列指导修复，后者在打靶生物体（可以是任意物种）的基因组上产生替换。
In all three cases, the DNA cut can be repaired by nonhomologous end joining of the ends or by repair directed by a stretch of homologous DNA (green), producing alterations in the genome of the target organism, which can be from any species. (For more detail, see Appendix A.)
来源：上图和中图（Beumer and Carroll, 2014）；下图（Charpentier and Doudna, 2013）。
SOURCES: top and middle (Beumer and Carroll, 2014); bottom (Charpentier and Doudna, 2013).

TALEN 和 CRISPR。如上所述，这些基因组编辑方法已被广泛应用于生物科学的各种领域，从细胞和实验动物的实验室基础研究，到改良农作物和牲畜的农业应用，再到人类健康的研究及越来越多的临床应用。农业应用已经在美国国家科学院的其他研究中得到解决（见第 1 章），潜在的临床应用是本报告随后章节的内容。本章重点关注在实验室基础研究中使用基因编辑。

这项研究探讨了基因组编辑技术在培养的细胞和多细胞实验生物（如小鼠、果蝇、植物）中的使用及优化的基础问题。这样的探索性基础研究对于改善未来基因组编辑的应用是必不可少的。在实验室研究中，基因组编辑技术也为以下领域的研究提供了强有力的工具：细胞的基本功能、新陈代谢过程、免疫学和对病理性感染的抵抗力，以及诸如癌症和心脏病之类的疾病。这些研究由标准实验室安全机制来监督。除了这些应用，本章也评论了用类似方法对人生殖细胞进行不带生育目的、只是纯粹实验室基础研究的可能性。通过对人胚胎和生殖细胞的研究，或通过对干细胞的起源和体外培养维持的改良，这些工作可以为人早期发育和成功繁殖提供有价值的见解，也可能推动临床上的进步。

基因组编辑技术的飞速发展

CRISPR/Cas9 的发展是基因和基因组编辑科学中的革命性标志，随着基础科学飞速发展，其他基于 CRISPR 的系统正被开发并配置于多个不同的目的。不同类型细菌用的 CRISPR 系统稍有区别，虽然 CRISPR/Cas9 系统因其简易性在目前得到广泛应用，能使基因编辑更为灵活的替代系统也在开发中（Wright et al.，2016；Zetsche et al.，2015）。

在此项技术中最亟须解决的问题是由 CRISPR 指导的核酸酶介导的 DNA 断裂的特异性和高效性。虽然向导 RNA 能够识别大约 20 个碱基对并因此提供了极大的特异性（在大约 1×10^{12} 个碱基对中才可能会遇到一个精确匹配——一万亿分之一——相当于几百个哺乳动物基因组），但是还有很小的可能会发生所谓的脱靶事件，即核酸酶在非目的位置造成切口，尤其是当向导 RNA 与目标

and CRISPR methods of genome editing. As mentioned, these genome-editing methods are being widely applied across a broad range of biological sciences, from fundamental laboratory research on cells and laboratory animals; to applications in agriculture involving improvements in crop plants and farm animals, to applications in human health, both at the research level and, increasingly, in clinical applications. Agricultural applications have been addressed in other studies by the U.S. National Academies of Sciences, Engineering, and Medicine (see Chapter 1) and potential clinical applications are the subject of subsequent chapters of this report. The focus in this chapter is on basic laboratory research using genome editing.

This research addresses fundamental questions concerning the use and optimization of genome-editing methods both in cultured cells and in experimental multicellular organisms (e.g., mice, flies, plants). Such basic discovery research is essential for improving any future applications of genome editing. Applications of genome editing in laboratory research also have added powerful new tools that are contributing greatly to understanding of basic cellular functions, metabolic processes, immunity and resistance to pathological infections, and diseases such as cancer and cardiovascular disease. These laboratory studies are overseen by standard laboratory safety mechanisms. In addition to these applications this chapter reviews the potential for using similar approaches in basic research on human germline cells, not for the purposes of procreation but solely for laboratory research. This work will provide valuable insights into the processes of early human development and reproductive success, and could lead to clinical benefits, directly as a result of work with human embryos and germline cells or through improvements in the derivation and maintenance of stem cells *in vitro*.

RAPID ADVANCES IN GENOME-EDITING TECHNOLOGY

The development of CRISPR/Cas9 has revolutionized the science of gene and genome editing, and the basic science is advancing extremely rapidly, with additional CRISPR-based systems being developed and deployed for multiple different purposes. Different species of bacteria use somewhat different CRISPR systems, and although the CRISPR/Cas9 system is currently the most widely used because of its simplicity, alternative systems being developed and will provide increased flexibility in methodology (Wright et al., 2016; Zetsche et al., 2015).

Among the issues that need to be addressed going forward are the specificity and efficiency of the DNA cleavage mediated by CRISPR-guided nucleases. While

靶点稍有不同的 DNA 序列相结合时。某些早期实验提示脱靶事件的可能性也许会很高，但随着方法的改良，以及该技术越来越多地被用于正常细胞而不是培养的细胞系，脱靶率实际上很低。人们在提高 Cas9 切割特异性方面已经取得了很多进展，也已开发出了监控脱靶切割率的方法（详见附录 A）。

该技术的另一个重要进步是修改 CRISPR/Cas9 系统来避免 DNA 的切割。例如，去掉 Cas9 的核酸酶活性，从而使这种"死的"Cas9（dCas9）与向导 RNA 的复合体仍能被向导 RNA 靶向结合到特定 DNA 位点，但不会切割 DNA（Qi et al., 2013）。偶联其他具有不同活性的蛋白质到 dCas9 上，可以对 DNA 及相关蛋白做不同的修饰。因此，人们可以通过设计各种 CRISPR/Cas9、ZFN 或 TALE 的变体来打开或关闭相邻的基因、改变单个碱基、改变染色体上与 DNA 相互作用的染色体蛋白，从而改变基因的表观遗传调控（Ding et al., 2016; Gaj et al., 2016; Konerman et al., 2015; Sander and Joung, 2014）。这些非切割型变体都不能切割 DNA，从而降低了恶性脱靶事件的概率。很多其他的修饰也被引入来增强特异性和降低脱靶事件（详见附录 A）。新近报道的可被编程的 RNA 介导的切割 RNA 的核酸酶 CRISPR/C2c2（Abudayyeh et al., 2016; East-Seletsky et al., 2016）能够敲降特定基因的 RNA 拷贝而不影响基因本身。这项发展使未来非遗传性的或可逆的基因编辑成为可能。

从这个简短的概述中可以看出，飞速发展的多样化 RNA 引导的基因编辑体系开创了操纵基因表达和功能的无数手段。近期报道的在很多不同类型细胞，包括人多能干细胞及小鼠体内的可诱导基因敲降、敲除的方法（Bertero et al., 2016）进一步拓展了这些方法的潜力。这些和其他技术上的进步，使该技术在世界范围内快速成为分子生物学的基本工具，为过去 40 年基因编辑工具的积累又增添了新的一员。这些技术的应用使得研究基因在细胞和实验动物（如酵母、小鼠及许多其他物种）中的功能及增进对生命的进一步理解变得前所未有的容易。它们还能被用于研究干细胞的起源和分化，为再生医学提供基础性的见解；开发人类疾病模型，增进对疾病进程的理解，实现对人类细胞的离体（*ex vivo*）药物测试。

the roughly 20-base sequence recognized by the guide RNA provides a great deal of specificity (an exact match should occur by chance in approximately $1 \times 10e12$ base pairs—1 in a trillion—the equivalent of several hundred mammalian genomes), so there is some small potential for so-called **off-target events**, in which the nucleases make cuts in unintended places, especially if the guide RNA binds to DNA sequences that are slightly different from the intended target. Some early experiments suggested that off-target events might occur at a significant rate, but as the methods have been improved and as their application has increasingly been in normal cells rather than cultured cell lines, the frequency of off-target cleavages appears to be very low. Advances have been achieved in the specificity of Cas9 cleavage (Kleinstiver et al., 2016; Slaymaker et al., 2016) and methods have been developed for monitoring the frequency of off-target cleavage. (See Appendix A for more detail.)

Another significant advance has occurred in the development of methods for modifying the CRISPR/Cas9 system so that DNA cleavage is avoided. For example, the nuclease function of Cas9 can be inactivated so that a complex of guide RNA and such a "dead" Cas9 (dCas9) will target a specific site via the guide RNA but will not cleave the DNA (Qi et al., 2013). By coupling other proteins with different activities to the dCas9, however, different sorts of modifications can be made to the DNA or its associated proteins. Thus, it is possible to design variants of CRISPR/Cas9, ZFN, or TALE that will turn on or turn off adjacent genes, make single-base changes, or modify the chromatin proteins that associate with DNA in chromosomes and thus modify the epigenetic regulation of genes (Ding et al., 2016; Gaj et al., 2016; ; Konerman et al., 2015; Sander and Joung, 2014). All of these noncleaving variants fail to cleave DNA, thus reducing the potential for deleterious off-target events, and many other modifications are being introduced to enhance specificity and reduce off-target events (see Appendix A for further detail). Most recently, CRISPR/C2c2, a programmable RNA-guided, RNA-cleaving nuclease, has been described (Abudayyeh et al., 2016; East-Seletsky et al., 2016) that could be used to knock down specific RNA copies of genes without affecting the gene itself. This development raises the future possibility of nonheritable or reversible editing.

As can be seen from this brief survey, the rapidly developing versatility of these RNA-guided genome-editing systems is opening up numerous means of manipulating the expression and function of genes. A recent report of methods for inducibly knocking down or knocking out genes in a multiplex fashion in many cell types, including human pluripotent stem cells, as well as in mice (Bertero et al., 2016) further expands the potential

人类细胞和组织基础实验研究进展

基础生物医学研究能够发现更多基因组编辑的途径和机制,这为人类药物发展提供了重大机遇。实验室中对人类细胞、组织、胚胎和配子的基因组编辑研究,为增进对人类基因功能、基因组重排、DNA 修复机制、人类早期发育、基因与疾病的联系、癌症及其他有严重遗传基础的疾病的认识开辟了重要途径。用基因组编辑技术操纵基因及基因表达使人们能了解基因在人类细胞行为中的功能,包括基因功能障碍是如何导致疾病的。例如,对培养的人类细胞进行基因组编辑可以模拟癌症或遗传病中发生的病理变化,为了解病理缺陷的分子基础提供了理想的疾病培养模型。这类实验室研究还能开发解决这些缺陷的方法,比如在细胞培养中测试潜在的药物。现在所有这类方法在使用上都已经比几年前更为简便。

在受精卵发育为能植入女性子宫的胚胎之前的早期胚胎中,有一部分胚胎细胞被称为胚胎干(ES,即 embryonic stem 的英文首字母全称)细胞。这些 ES 细胞在培养中能增殖,具有分化发育成几乎所有类型体细胞的潜力,但缺乏自行发育成胎儿的潜力,因而具有许多科技优势。如今,利用成体细胞产生干细胞也成为了可能。成体细胞可以被重编程转变为能分化发育成多种类型细胞的状态,减少了从早期胚胎中获取干细胞的需求。这种方式获得的多能细胞被称为诱导多能性干(iPS,即 induced pluripotent stem 的英文首字母全称)细胞。诱导多能性干细胞能在体外培养,并被诱导分化成许多不同类型的细胞,如神经元、肌肉或皮肤细胞及其他种类细胞。过去几十年对干细胞的了解及其如何使用方面的进步是再生医学领域的基础。再生医学旨在修复或替换人体组织中的受损细胞,或在受伤或疾病后产生新的组织。虽然这些更多的是属于临床实践的领域,且本章不包含被遗传改变的细胞在人体中的应用(见第 4 章),但这仍然是科学家在实验室从事人和动物干细胞基础研究的重大理由。

基因组编辑技术对于在人类 ES 和 iPS 细胞内产生多种遗传修饰非常有用。在高效的基因组编辑工具发明之前,这些细胞已被证实对利用同源重组等常规手段的遗传修饰有很大抗性。虽然同源重组

of these methods. These and other advances have rapidly rendered these methods basic tools of molecular biology worldwide, adding to the existing toolkit assembled over the past 40 years. These methods are now being applied to study with unprecedented ease the functions of genes in cells and in experimental animals, such as yeast, fish, mice, and many others, to enhance understanding of life. They also are being used to investigate the derivation and differentiation of stem cells, providing fundamental insights relevant to regenerative medicine, and to develop culture models of human disease both to advance understanding of disease processes and to enable testing of drugs on human cells *ex vivo*.

BASIC LABORATORY RESEARCH TO ADVANCE UNDERSTANDING OF HUMAN CELLS AND TISSUES

Basic biomedical research aimed at discovering more about the mechanisms and capabilities of genome editing offers significant opportunities to advance human medicine. Genome-editing research conducted on human cells, tissues, embryos, and gametes in the laboratory offers important avenues for learning more about human gene functions, genomic rearrangements, DNA-repair mechanisms, early human development, the links between genes and disease, and the progression of cancer and other diseases that have a strong genetic basis. Manipulation of genes and gene expression by genome editing allows one to understand the functions of genes in the behavior of human cells, including why they malfunction in disease. For example, editing of cultured human cells to model the changes that arise in cancer or in genetically inherited diseases provides culture models of those diseases with which to understand the molecular basis of the resulting defects. Such laboratory studies also allow the development of means of combating those defects, such as the testing of potential drugs in cell culture. All of those approaches are much easier now than they were just a few years ago.

Certain cells derived from an early embryo, after fertilization but prior to the developmental stage at which it would implant in a woman's uterus, are referred to as embryonic stem (ES) cells. These ES cells have scientific advantages because they can reproduce in cell culture and have the potential to form all the different body cell types while lacking the potential themselves to develop into a fetus. It is now also possible to create stem cells by manipulating adult somatic cells to convert them to a state in which they, too, have the ability to form multiple cell types, reducing the need to take stem cells from an early embryo. These are referred to as induced pluripotent stem (iPS) cells. Such pluripotent stem cells can be cultured *in vitro* and induced to develop into many

对编辑小鼠 ES 细胞是有效的，但在人类细胞中运用同源重组获得靶向重组子的效率极其低下。使用 CRISPR/Cas9 可以大大提高重组效率，从而能够快速产生有标记的报告细胞系，使跟踪分化的途径、寻找相互作用的蛋白质、筛选适宜的细胞系、研究细胞中单个基因和通路的功能，以及许多其他应用成为可能（Hockemeyer and Jaenisch，2016）。例如，在特定基因中进行精确靶向突变或修正的能力，使人们能够在相同遗传背景下产生带有不同特定疾病等位基因的人类 ES 细胞系（Halevy et al.，2016），用于研究这些疾病基因造成的后果。相反，基因组编辑也能在患者特异的 iPS 细胞系中靶向修正致病突变，产生遗传匹配的对照细胞系。这种修饰后的干细胞系主要用于实验和临床前研究，探索特定疾病进程，并检测治疗这些疾病的潜在药物。将来，这些编辑过的干细胞系可能可以用于各种形式的、基于体细胞的细胞治疗（见第 4 章）。

哺乳动物繁殖与发育基础实验研究进展

生殖细胞是指具有参与形成新个体的能力，且能将自身遗传信息传递给子代个体的一类细胞，包括卵母细胞和精母细胞，及其子细胞卵子和精子。卵子和精子经过受精形成的胚胎，早期称为合子（受精卵）和囊胚，具有分化为包括体细胞和新的生殖细胞在内的个体所有细胞种类的潜能。随着胚胎继续发育，细胞功能逐步特化，进而分化为特定种类细胞，如神经系统细胞、皮肤细胞或肠细胞等。

在生殖和发育的过程中，发生在配子（卵子和精子）、卵母或精母细胞、非常早期的胚胎阶段的细胞中的遗传信息的改变，将遍及生命体所有后继产生的细胞，并能遗传给后代。与上文强调的一样，本章只关注基因编辑技术在实验室中的应用，而非人体临床或以植入妊娠为目的的胚胎上的应用。尽管如此，了解人发育过程涉及的细胞种类及其功能，不仅对研究者对科学问题的研究有帮助，而且对围绕人细胞和胚胎应用的实验室基础研究的伦理通告、规章制度及社会舆论等同样具有重要意义。

生殖干细胞和生殖前体细胞的基因编辑

同体细胞一样，在小鼠受精卵（合子）、早期胚胎细胞、胚胎干细胞或精原干细胞中的基因修饰

different cell types, such as neurons, muscle or skin cells, and many others. Advances over the last several decades in understanding stem cells and how they can be used form the foundation for the field of regenerative medicine, which seeks to repair or replace damaged cells within human tissues or to generate new tissues after disease or injury. Although these are increasingly areas of clinical practice, and the application of genetically altered cells in humans is not covered in this chapter (see Chapter 4), there are nevertheless a number of important reasons why scientists aim to undertake basic investigations in human and animal stem cells in the laboratory.

Genome-editing methods have been extremely useful in generating a variety of genetic modifications in human ES and iPS cells. Before the advent of efficient genome-editing tools, these cells had proven resistant to genetic modification with the standard tools of homologous recombination that had been used effectively in mouse ES cells. Using those tools in human cells resulted in very low frequencies of targeted recombination. Improvements in efficiency resulting from the use of CRISPR/Cas9 have enabled rapid generation of tagged reporter cell lines, making it possible to follow differentiation pathways, look for interacting proteins, sort appropriate cell types, and investigate the functions of individual genes and pathways in cells, among many other applications (Hockemeyer and Jaenisch, 2016). For example, the ability to make precisely targeted mutations or corrections in specific genes has made possible the generation of human ES lines with different specific disease alleles on the same genetic background (Halevy et al., 2016) for use in research on the consequences of such disease genes. Conversely, genome editing also allows the targeted correction of disease mutations in patient-specific iPS cell lines to generate genetically matched control lines. Such modified stem cell lines are used primarily to conduct experimental and preclinical studies, to investigate specific disease processes, and to test drugs that could be used to treat such diseases. In the future, such edited stem cell lines could be used for various forms of somatic cell–based therapies (see Chapter 4).

BASIC LABORATORY RESARCH TO ADVANCE UNDERSTANDING OF MAMMALIAN REPRODUCTION AND DEVELOPMENT

Germline cells are cells with the capacity to be involved in forming a new individual and to have their genetic material passed on to a new generation. They include precursor cells that form eggs and sperm, as well as the eggs and sperm cells themselves. When fertilization occurs to create an embryo, the earliest stages of this embryo, referred to as the zygote (fertilized egg) and

已证明可行。在这些情况下，基因修饰的效果可以直接在胚胎或培养的细胞中研究。基因编辑方法有多种，可用于编辑的细胞来源也有多个，以下细胞系都被认为是生殖系的一部分或有能力成为生殖系：

- 正常囊胚分离获得的胚胎干细胞；
- 从体细胞核转移（somatic cell nuclear transfer，SCNT）[35] 形成的胚胎中分离获得的细胞；
- 体细胞重编程诱导为类似 ES 细胞的诱导多能性干细胞。

在小鼠中，以上来源的干细胞系均可通过基因编辑改造并移植到囊胚或桑葚胚中，在体内形成生殖系传递。移植后产生的胚胎称为嵌合体，由移植的外源干细胞和胚胎自身细胞共同组成。小鼠或大鼠的精母细胞经过培养和基因编辑后移植到受体小鼠或大鼠睾丸，它们能分化产生具有至少在体外能使卵细胞正常受精的精子的细胞（见附录 A 和 Chapman et al., 2015）。所有这些情况下所获得的胚胎都可以被移植回子宫来完成妊娠，从而建立起能稳定携带这些遗传修改的小鼠系。这种方法为探索基因组中所有基因的功能，以及建立人类疾病啮齿类动物模型提供了前所未有的机遇。在小鼠受精卵（Long et al., 2014；Wu et al., 2013）、胚胎干细胞或精母干细胞（Wu et al., 2015）中对疾病基因进行突变修复，再通过生殖系传递产生遗传修正了的小鼠的原理验证性实验也有报道。

以非生殖目的为前提，将基因组编辑技术应用于相对应的人类细胞系在基础研究中具有巨大潜力。对人类胚胎早期发育的深入了解意义重大，不仅能回答人类胚胎早期发育过程中的诸多问题，还能够对相当一部分临床疾病的病理研究及预防或治疗提供莫大的帮助。以下是基因编辑技术在人类细胞中的一些应用。

辅助生殖技术的进步

人类在生殖技术及遗传疾病的胚胎植入前基因诊断（preimplantation genetic diagnosis，PGD）方面获得

blastocyst, have the potential to divide and form all the cells that will make up the future individual, including somatic (body) cells and new germ cells. As the embryo continues to develop, its cells differentiate into specific cell types that become increasingly restricted in their functions (for example, to form specialized cells such as those in the nervous system, skin, or gut).

During reproduction and development, genetic changes made directly in gametes (egg and sperm), in egg or sperm precursor cells, or in very early embryos would be propagated throughout the future cells of an organism and would therefore be heritable by subsequent generations. As emphasized above, this chapter focuses exclusively on the use of genome-editing technologies in the laboratory, and not on clinical applications in humans or in embryos for the purposes of implantation to initiate pregnancy. Nevertheless, it is important to understand which cell types are involved in human development and their functions, because this information informs researchers' decisions about how to study particular scientific questions and informs ethical, regulatory, and social discussions around when and why it may be useful to use human cells, including embryos, in basic laboratory research.

Genome Editing of Germline Stem Cells and Progenitor Cells

It is already possible in mice to genetically modify the genome in a fertilized egg (the zygote), in individual cells of the early embryo, in pluripotent ES cells, or in spermatogonial stem cells, just as in somatic cells. In all these cases, the effects of the genetic modifications can be studied directly in the embryo or in cells in culture. There are a number of ways to undertake these genetic manipulations and a number of cell types in which they can be conducted. The cell types below are all considered part of the germline or have the capacity to contribute to the germline:

- embryonic stem cells derived from normal early embryos (blastocyst stages),
- cells from early embryos produced after somatic cell nuclear transfer (SCNT),[35] and iPS cells obtained by reprogramming somatic cells into an ES cell-like state.

In mice, these cell types can all be manipulated experimentally through genome editing. Stem cells of the types listed above can contribute to the germline *in vivo* after they are introduced into mouse embryos at the morula or blastocyst stage. This process generally

---

[35] SCNT，somatic cell nuclear transfer，体细胞核转移技术，指将从另一细胞（如已经经过基因修饰的体细胞）取出的"供体"细胞核移植到去核卵细胞中的技术。世界上第一只克隆哺乳动物"多利"羊即通过核移技术产生。

[35] SCNT is a technique in which the original nucleus of an egg cell is removed and replaced with a "donor" nucleus taken from another cell (for example, from a somatic cell that has undergone genome editing). This is the technique that was used to create Dolly, the first cloned mammal obtained from an adult cell.

的成功深深依赖于体外受精（in vitro fertilization, IVF）和从人类受精卵到囊胚期胚胎体外培养技术。然而，用于确保体外培养的胚胎正常并能在植入后完成妊娠的方法依然缺乏。绝大多数胚胎研究都是在小鼠胚胎上进行的，而小鼠胚胎发育与人类胚胎有类似的地方，也有差异很大的地方（见延伸内容3-2）。甚至人类胚胎的体外培养条件很大程度上都是建立在培养小鼠胚胎的基础上，相较于别的物种，体外培养的人类胚胎出现非整倍体的概率偏高[36]。这些非整倍体经常以嵌合的形式出现，即在胚胎中细胞与细胞之间各有不同（Taylor et al., 2014）。至于非整倍体是如何产生的，以及是否与体外培养环境有关尚不清楚。体外胚胎培养也有可能导致表观遗传[37]层面的改变，这种表观异常可能会影响到正常发育过程甚至在成体后致病（Lazaraviciute et al., 2014）。体外早期胚胎研究有助于研究者们更好地了解调控人类胚胎发育早期的细胞和分子路径，以及人类胚胎能成功发育的体外培养条件。后者能为提高体外受精成功率打下基础。

综上，人和小鼠胚胎发育存在明显的差异，这使得人们无法通过研究小鼠来精确地推断人胚胎发育的进程。事实上，这一不足之处也影响了人们优化体外受精技术，获取最佳多能性干细胞及其他干细胞来模拟人类疾病和用于未来的再生治疗。因此，在司法允许对人类胚胎进行研究的地区，植入前的人胚胎体外试验研究具有相当大的吸引力，这些研究主要涉及受精过程的发生、胚胎发育过程中相关基因的激活、细胞谱系发生、表观遗传调控如X染色体的失活等，并和已有的小鼠研究结果进行异同比较。

另外，人胚胎发育的体外研究同样有助于找出导致人类自然妊娠早期高流产率（10%~45%不等，视孕妇年龄而定）及不孕的原因。同样，更好地了解精子发育过程对解决男性不育也至关重要。多能性干细胞来自于早期胚胎并能在培养中产生胚胎干细胞。对人胚胎发育的深入研究有助于对多能性的起源和调控的理解，并以此为理论指导获得更好的干细胞来用于再生医学。体外人胚胎研究带来的潜

---

[36] 染色体数不是通常单倍体数的精确倍数。
[37] 表观遗传，指通过对基因组DNA或染色体中结合在DNA上的蛋白质和RNA进行化学修饰来调控基因表达。

creates an embryo that is a chimera, in which some cells are derived from the stem cells introduced into the embryo, and some are formed from the initial embryonic cells. Mouse or rat spermatogonial stem cells can be cultured and their genomes edited, and the cells can then be introduced into recipient mouse or rat testes, where they can give rise to sperm able to fertilize oocytes, at least *in vitro* (see Appendix A and Chapman et al., 2015). In all of these cases, when the resultant embryos are transferred back into the uterus to complete pregnancy, it is possible to establish lines of mice carrying the genetic alterations. These approaches provide unprecedented opportunities to explore the functions of all the genes in the genome and to develop rodent models of human diseases. Proof-of-principle experiments also have been reported in which disease-related genetic mutations have been corrected in mouse zygotes (Long et al., 2014; Wu et al., 2013), embryonic stem cells, or spermatogonial stem cells (Wu et al., 2015) and then transmitted though the germline to produce genetically corrected mice.

The application of genome-editing technologies to the equivalent human cell types holds considerable potential value for fundamental research without any intent to use such manipulated cells for human reproductive purposes. Improved knowledge of how an early human embryo develops also is valuable in its own right, and because such knowledge can help answer questions about humans' own early development, as well as facilitate understanding and potential prevention or treatment of a wide range of clinical problems. A number of these applications are described below.

Improvements in Assisted Reproductive Technology

The success of human reproductive technologies and preimplantation genetic diagnosis (PGD) of inherited diseases has been, and continues to be, dependent on *in vitro* fertilization (IVF) and on culturing of human embryos from the zygote to the blastocyst stage. However, tools for ensuring that an individual embryo in culture is normal and capable of completing pregnancy remain limited. Most embryo research has been conducted on mouse embryos, which are similar to human embryos in certain respects but significantly different in others (see Box 3-2). Even the conditions in which human embryos are kept in culture are based largely on those established for mouse embryos. High rates of aneuploidy[36] are found in cultured human embryos relative to other species. This aneuploidy is often mosaic—that is, it varies among cells in the embryo (Taylor et al., 2014)—

---

[36] Having a chromosome number that is not an exact multiple of the usual haploid number.

延伸内容3-2

小鼠和人胚胎发育差异
Differences Between Mouse and Human Development

近年来，人们在理解促使受精卵发育到囊胚——胚胎发育过程中最早开始形成不同细胞类型的阶段——的关键事件方面取得了显著进展。多数此类研究都是以小鼠胚胎为研究模型。然而，小鼠和其他啮齿类动物，与人或多数其他哺乳动物胚胎发育之间存在一系列重大差异（见下图）。因此研究这些差异是什么、如何产生、为什么如此显著仍是今后的研究重点。在实验室中，对人类细胞及早期胚胎进行基因组编辑，为探讨上述问题提供了有力工具。

从受精卵发育到囊胚，小鼠需要3~4天，而人需要5~6天。囊胚（直径约0.1mm）约含100个细胞，由三类细胞组成；外面一层为滋养外胚层，中间包着由原始内胚层和上胚层细胞组成的内细胞团。小

In the past few years, considerable progress has been made in understanding the events that allow the zygote to develop into the blastocyst, the earliest stage of embryonic development in which different cell types are formed. Much of this research has been conducted on mouse embryos. However, it is clear that a number of important differences exist between the development of mouse and rodent embryos and those of humans and most other mammals (see the figure below). Significant research remains necessary to understand what these differences are, how they arise, and why they are significant. Genome editing in human cells and early embryos in the laboratory provides an important tool to help address these questions.

A mouse blastocyst takes 3-4 days to develop, whereas a human blastocyst takes 5-6 days. The blastocyst has about 100 cells, is about 1/10th of a millimeter in diameter, and contains only three cell types; there is an outer layer called *trophectoderm* that encloses an inner cell mass consisting

小鼠与人囊胚植入后初期发育的比较
Comparison of blastocyst and early postimplantation development between mouse and human.

尽管两物种的囊胚期（左图）看起来很相似，发育到较晚的植入期阶段（右图）两者却存在显著差异，尤其是胚外组织。
Whereas the blastocysts of the two species (left-hand figures) look very similar, later stages at the time of implantation (right-hand figures) show significant differences, particularly in the extraembryonic tissues.
(A) 小鼠囊胚的滋养外胚层在植入后会经历一个由上胚层（EPI，蓝色）的FGF4信号刺激的增殖时期来形成胚外外胚层（绿色）和绒膜锥（浅绿色）。直至胎盘发育的后期，只有限的小鼠滋养层细胞会侵入母体子宫。PE = 原始内胚层；TE = 滋养外胚层。
(A) The trophectoderm of a mouse blastocyst after implantation undergoes a proliferative phase stimulated by FGF4 signals from the epiblast (EPI, blue) to form the extraembryonic ectoderm (green) and ectoplacental cone (light green). There is only limited invasion of the maternal uterus by mouse trophoblast cells until much later in placental development. PE = primitive endoderm; TE = trophectoderm.
(B) 植入后，人囊胚的滋养外胚层与上胚层并不紧密接触，但会侵入子宫内膜，随后会形成绒毛膜绒毛。
(B) The trophectoderm of the human blastocyst does not stay in close contact with the epiblast after implantation but invades into the endometrium, where it will later form the chorionic villi.
来源：Rossant, 2015.
SOURCE: Rossant, 2015.

鼠、人及哺乳动物胚胎发育的前几天主要分化出几种辅助胚胎在子宫中正常存活所必需的细胞类型——胚盘和卵黄囊——前者来源于滋养外胚层，后者来源于原始内胚层（Cockburn and Rossant，2010）。上胚层细胞属于多能性细胞，在发育过程中生成包括生殖细胞在内的完整胚胎（Gardner and Rossant，1979）。

除上图中总结的差异外，这两类物种早期胚胎发育的调控机制也存在明显不同，导致人们对小鼠胚胎囊胚期三种细胞形成的信号通路和下游基因调控通路了解得比较清楚（Frum and Ralston，2015）。在囊胚期，每个细胞的命运便被限定在这三种之一。小鼠早期胚胎发育的分子层面研究有助于了解细胞多能性相关调控机制，再加上体外胚胎干细胞（ES）研究所获得的信息，共同促成了里程碑式的诱导性多能性干细胞（iPS）的产生，即利用早期胚胎中表达的因子将成体细胞重编程到多能性状态的细胞（Takahashi and Yamanaka，2006；Takahashi et al.，2007）。iPS 细胞技术的出现展示了基础实验研究可以推动一个领域——在这里是再生医学——的重大进步。iPS 细胞能够形成用于研究或治疗疾病的各类细胞，大幅度减少了胚胎细胞的使用。自体来源的 iPS 细胞还可以最小化使用异体细胞带来的免疫排斥反应。

与小鼠发育研究相比，人们对人胚胎发育到囊胚期过程涉及的分子调控机制与细胞动向了解很少。已有研究试图确定细胞谱系限定发生的时间，但是由于实验中可用的胚胎数量受限，该研究结果并不如小鼠中那么精确。缺少合适的研究方法也是人胚胎发育研究面临的一大挑战，而基因组编辑工具的发展能够弥补这些缺陷。

小鼠和人胚胎植入后初期阶段的差异体现在形态学上和其他目前可能还不清楚的地方。研究表明，人胚胎细胞谱系限定的发生较小鼠的晚，可能要到囊胚发育完全之后。新的技术可以做到在少量组织甚至单细胞层面上进行基因表达分析，为科学家探究调控谱系发育的分子机制提供有利帮助。一种检测单个细胞中全基因组基因表达的方法——单细胞 RNA-seq，已被用于早期人类胚胎分析（Blakeley et al.，2015；Petropoulos et al.，2016）。这些数据清晰地显示，人和小鼠在发育过程中的基因表达谱

of *primitive endoderm* and *epiblast* cells. Mice, humans and all mammalian embryos spend their first few days of development making mostly the cell types needed to survive in the uterus, the placenta and the yolk sac, which are derived from trophectoderm and primitive endoderm respectively (Cockburn and Rossant, 2010). The epiblast cells are the pluripotent cells that give rise to the entire embryo including its germ cells (Gardner and Rossant, 1979).

In addition to the differences summarized in the figure above, there are clearly other differences in the control of development of the early embryos of the two species. In a mouse embryo, the signaling pathways and downstream gene regulatory pathways that drive the formation of the three cell types of the blastocyst are fairly well understood (Frum and Ralston, 2015). Each cell is restricted to one of these three fates by the blastocyst stage. Information from the molecular analysis of events in the early mouse embryo has helped in understanding the underlying principles of the establishment of cellular pluripotency. Together with knowledge gained from studying embryonic stem (ES) cells, this information aided in the momentous development of induced pluripotent stem (iPS) cells in which adult cells are reprogrammed to pluripotency using factors known to be expressed in the early embryo (Takahashi and Yamanaka, 2006; Takahashi et al., 2007). The ability to create iPS cells represents an example in which basic laboratory research contributed to critical advances in a field—in this case regenerative medicine. Scientists can use these cells to generate cells to study or treat disease, greatly reducing the need to use cells derived from embryos. Because iPS cells can be created from a person's own cells, they also minimize the immune rejection that can occur if cells arising from one individual are used in another.

In contrast to understanding of mouse development, much less is known about the cellular and molecular events of blastocyst formation in human embryos. Some experiments on the timing of cell-lineage restriction have been performed, but not with as much precision as in mice because of the restricted number of embryos that are available for research. Compounding the challenges of studying development in human embryos has been the absence of appropriate methods. The development of genome-editing tools can help address this gap.

The early postimplantation stages of mouse and human development differ morphologically and perhaps in other ways not currently understood. Current data suggest that cell-lineage restriction occurs later in a human embryo than in a mouse embryo, probably not until after the blastocyst is fully developed. New technologies for analyzing gene expression in small amounts of tissue, and even from single cells, are helping scientists to gain insights into the molecular pathways controlling lineage

既有共同点也存在差异性,其中包括一些已知在小鼠植入前胚胎发育中起关键作用的因子。从这些数据中能够推测出人胚胎植入前发育过程中所需的因子,但不能确定其重要性。基因组编辑技术如CRISPR/Cas9可以帮助我们确定这些基因在人类胚胎植入前的胚胎发育中的作用。

小鼠与人囊胚期形成上的差异也被认为是导致从两个物种获得的ES细胞在性状上显著不同的原因(Rossant,2015)。尽管小鼠和人的ES细胞都是从囊胚分离的多能性细胞并且表达一些相同的关键多能性因子,但两种细胞之间在性状上仍存在许多不同,如它们自我更新依赖不同的生长因子。人ES细胞的许多特性与小鼠发育到植入后初期阶段的上胚层细胞更相似,这就产生了一个疑问:人类胚胎是否经历了在小鼠囊胚和ES细胞中观察到的所谓的"原始态(naïve)"阶段(Huang,2014;Pera,2014)。

除了从囊胚多能细胞中得到的ES细胞,小鼠干细胞还可以从滋养外胚层和胚外内胚层中分离各自的祖细胞获得,称为TS和XEN细胞(Rossant,2015)。这些干细胞对于研究细胞分化的分子机制,以及这些细胞系的特性十分重要,而后者与正常和疾病状态下的胎盘及卵黄囊生物学特性相关。然而,从人囊胚分离得到同等的胚外干细胞类型的尝试却至今没有成功(Hayakawa et al.,2014)。这些细胞系具有重要意义,尤其是考虑到不同哺乳动物胎盘类型在进化上的巨大差异。这些差异包括组成人类胎盘的细胞类型与小鼠和其他哺乳动物存在的实质性差异。例如,侵入并与子宫内膜(子宫内膜细胞)直接相互作用的合胞体滋养层细胞在类人猿的胎盘上已经存在,但在人类中却有其独特性。因此,依靠研究小鼠细胞获得的知识去理解人胎盘的正常发育是行不通的。同样,试图通过类似方法研究人体胎盘或胎盘与母体结合失败导致流产,或何时胎盘侵入子宫过深导致绒毛膜癌等的病理原因的方案同样不可行。

development. Single-cell RNA-sequencing, a method to look at genome-wide gene expression in single cells, has been applied to early human embryos (Blakeley et al., 2015; Petropoulos et al., 2016). From these data it is already apparent that there are both similarities and significant differences between humans and mice in the developmental profiles of gene expression, including in some of the key genetic drivers known for mouse preimplantation development. It is possible to speculate about which of these genes are required to drive human preimplantation development, but it is not yet known how critical they are. Genome-editing methods such as CRISPR/Cas9 will allow determination of the roles of those genes expressed specifically during human pre-implantation development.

Differences between mouse and human blastocyst formation also are thought to underlie the significant differences in the properties of ES cells derived from these two species (Rossant, 2015). Although both mouse and human ES cells are derived from the pluripotent cells of the blastocyst and share expression of some key pluripotency genes, they have many different properties, including their dependence on different growth factors for their self-renewal. Many of the properties of human ES cells are more similar to those of epiblast cells found in early postimplantation stages of mouse development, leading to the question of whether human embryos actually go through the same so-called "naïve" state of pluripotency observed in mouse blastocysts and ES cells (Huang et al., 2014; Pera, 2014).

In addition to ES cells obtained from the pluripotent cell lineage in the blastocyst, it is possible in mice to derive stem cells corresponding to progenitors of both trophectoderm and extraembryonic endoderm, termed TS and XEN cells, respectively (Rossant, 2015). These stem cell types are valuable for understanding molecular mechanisms of differentiation, as well as the properties of these cell lineages, which are relevant to placental and yolk sac biology in health and disease. However, attempts to derive equivalent extraembryonic stem cell types from human blastocysts have not yet been successful (Hayakawa et al., 2014). Such cell lines would be of great value, especially given the considerable evolutionary divergence seen in placental types among mammals. This divergence includes substantial differences in the types of cells comprising the human placenta compared to mice and other mammals. For example, syncytiotrophoblast cells that invade and directly interact with the endometrium (the cells lining the uterus/womb), are present in the placenta of great apes and may even have unique properties in humans. As a result, it is not possible to rely on knowledge gained from studying mouse cells to understand normal development of the human placenta. Similarly, it is not possible to understand pathologies in which the placenta or placental interaction with the mother fails, which can cause miscarriage, or when the placenta invades too vigorously into the uterus, which can lead to choriocarcinoma.

在价值不仅体现在胚胎干细胞，植入前的早期胚胎已决定了后期发育为卵黄囊和胎盘的细胞类型。在妊娠期间，卵黄囊和胎盘紧密地联系着母体和胎儿，为胎儿提供营养和其他因素，确保胎儿能够存活。它们的缺陷会导致流产、早产或出生缺陷等。对卵黄囊和胎盘发生过程的研究将有助于治疗不孕症、预防早期流产和先天畸形的发生。虽然目前人们对这些进程一无所知，但对这些胚外细胞类型的研究也能为早期植入后胚胎动向提供线索。本章中涉及的有关人胚胎实验室研究可行性总结见表 3-1。

but how it arises and how it relates to *in vitro* culture conditions are not well understood. There is also concern that epigenetic[37] abnormalities might occur in human embryos *in vitro* (Lazaraviciute et al., 2014), which might compromise development or health, even later in life. Research on early-stage human embryos in culture should enable scientists to better understand the cellular and molecular pathways that control early human embryo development and the conditions under which human embryos in culture can develop successfully. This knowledge could in turn help improve IVF outcomes.

表 3-1 人胚胎实验室研究的必要性

体外研究	临床效果
体外受精研究	促进体外受精 (IVF) 和植入前基因诊断 (PGD) 技术的提高 有效避孕
人早期胚胎培养条件优化	促进体外受精和植入前基因诊断技术的提高 有助于找出导致流产和先天畸形的原因
胚外组织发生研究（卵黄囊和胎盘）	有助于找出导致植入失败和流产的原因
多能性干细胞的分离和体外分化	建立用于试验检测药物和其他治疗方法的人类疾病体外模型 获取用于体细胞基因/细胞疗法和再生医学的改良细胞
精子和卵子发生研究	发现治疗不孕不育的新方法

TABLE 3-1 Reasons for Laboratory Studies of Human Embryos

In Vitro Studies	Clinical Outcomes
Studies of fertilization *in vitro*	Improvements in *in vitro* fertilization (IVF) and pre-implantation genetic diagnosis (PGD) Possible improvements in contraception
Improved culture of early human embryos	Improvements in IVF and PGD Insights into reasons for miscarriages and congenital malformations
Development of extraembryonic tissues (yolk sac and placenta)	Insights into reasons for failures in implantation and for miscarriages
Isolation and *in vitro* differentiation of pluripotent stem cells	*In vitro* models for human diseases for experimental testing of drugs and other therapies Improved cells for somatic gene/cell therapies and for regenerative medicine
Investigations of sperm and oocyte development	Possible novel approaches to infertility

探索人胚胎发育

利用 CRISPER/Cas9 和类似技术进行基因编辑是开展这类研究所需的关键性工具之一。对发育过程中基因调控的整体研究可通过 CRISPER/Cas9 介导的特定目的激活或失活来实现。事实上，随着 CRISPER/Cas9 技术效率的不断提升，在受精卵中敲除[38]基因并直接研究基因缺失对胚胎发育的影响将变得可行。这些实验都不涉及人类妊娠，因此

All of the differences between humans and mice discussed above mean that it is not possible to accurately infer developmental events in human embryos from studying mice. This limitation has practical consequences for the development of improved IVF technologies, as well as for the ability to derive the best pluripotent or other stem cells for modeling of human disease and for future regenerative therapies. Thus, there is considerable interest in experimental investigation of preimplantation human development in culture, in jurisdictions where such research on human embryos is permitted. The goals

---

[38] 敲除：即基因敲除，指一种遗传工程技术，针对某个已知序列但功能未知的序列，改变生物的遗传基因，令特定的基因功能丧失作用，从而使部分功能被屏障，并可进一步对生物体造成影响，进而推测出该基因的生物学功能。

[37] The term "epigenome" refers to a set of chemical modifications to the DNA of the genome and to proteins and RNAs that bind to DNA in the chromosomes to affect whether and how genes are expressed.

都不会产生可遗传的生殖系修改。所有的实验都只能是体外实验，实验结果主要在发育最初的1~6天的囊胚期阶段进行分析。某些情况下，人们会对在人胚胎发育较晚期的一些阶段，特别是在胚胎植入子宫后的一些早期阶段中改变某些特定的基因所产生的效果感兴趣。目前，许多国家都允许将人胚胎体外培养到胚层形成前的那个阶段（受精后第14天或胚胎产生原条结构时）。改良的、能让植入期人胚胎在体外继续发育的体外培养体系正在开发中。最近有研究表明，这些改良体系可用于研究胚外结构的发生和上胚层形成"胚盘"的过程——这些过程只在人类发育中发生而不见于小鼠发育（Deglincerti et al.，2016；Shahbazi et al.，2016）。改良后的细胞培养体系，结合更先进的、可用于分析基因功能的基因编辑技术，在深入理解早期人胚胎发育基本进程中被寄予厚望。为了解决这些基础生物学问题，目前世界上已经有至少两个研究团队（分别在英国和瑞典）拿到了使用CRISPER/Cas9在人胚胎中进行基因编辑的许可。

此类研究成果为提高体外受精和胚胎植入成功率及降低流产率具有重要指导意义，反之，也可以为寻找成功避孕的新方法提供依据。此类研究也能用于改良分离和维持干细胞的技术，并有助于分化分离用于基础研究、疾病和损伤后治疗的特定类型细胞。运用基因编辑在人早期胚胎中的研究对确定这些技术是否适用于最终的临床应用具有借鉴意义，即人们希望从基础研究中获取足够的理论指导来确定制造可遗传的，最好是非嵌合的基因组改变的可行性（见第5章）。因为可用于实验的人胚胎宝贵而相对稀缺，所以在实验研究中采用最有效的方法显得尤为重要。因此，本研究对涉及影响人胚胎中基因编辑效率的因素进行了探讨，其中包括：

- 导入细胞的基因组编辑元件的类型和形式；
- 是否使用Cas9或其他核酸酶；
- 基因组编辑元件的导入方式，如DNA、mRNA、蛋白质或RNA-蛋白质复合物；
- 是否在基因编辑中使用单个、成对的或多个向导RNA；
- 是否需要DNA模板及模板的大小；
- 基因编辑最佳时机的选择，如胚胎二细胞时期进行编辑是否能获得所需信息、有无必要

of this work are to understand the fundamental events of fertilization, activation of the embryonic genome, cell lineage development, epigenetic events such as X-inactivation, and others, and how these events compare and contrast with what is understood from studying mice.

Similar research also could provide insights into the reasons for the high rates of early pregnancy loss in natural human pregnancies (10 to 45 percent, depending on the age of the mother), as well as the causes of infertility. Better understanding of sperm development would be crucial in addressing issues of male infertility. Pluripotent stem cells arise from the early embryo, and these cells can generate ES cells in culture. Better understanding of human embryonic development would provide insights into the origins and regulation of pluripotency and how to translate that knowledge into improved stem cells for regenerative medicine. The potential benefits of such research are not limited to embryonic stem cells. Cell types that give rise to the yolk sac and the placenta also are determined in the early embryo prior to implantation. The yolk sac and placenta establish the crucial links with the mother during pregnancy and provide nutrients and other factors that enable the embryo to survive. Defects in these tissues can compromise a pregnancy, leading to miscarriage, premature birth, or postnatal abnormalities. Better understanding of how the yolk sac and placenta originate would help in improving techniques for overcoming infertility and preventing early miscarriage, as well as understanding and preventing congenital malformations. These extraembryonic cell types also provide cues that pattern the early postimplantation embryo, although almost nothing is known about these processes in humans. These possibilities and others discussed in this chapter are summarized in Table 3-1.

Understanding of Human Development

Genome editing by CRISPR/Cas9 and similar techniques has a key place in the tool set needed to undertake such experiments. CRISPR/Cas9-guided activation or inactivation of specific target pathways could be used to understand overall gene regulation in development. Indeed, as the efficiency of CRISPR/Cas9 continues to increase, it should be possible to use genome editing to knock out[38] genes in zygotes and study the effects directly in genetically altered embryos. None of these experiments would involve human pregnancies, so none could result in heritable germline modifications. They would all be *in vitro* experiments, with results being analyzed primarily at the blastocyst stage in the first 1-6 days of development.

---

[38] A gene is said to be "knocked out" when it is inactivated because the original DNA sequence has either been replaced or disrupted.

选择一细胞期,还是最好在体外受精时随精子细胞一起导入;
- 是否允许嵌合式胚胎的出现,需要注意的是有些情况如对细胞命运进行追踪的研究中嵌合式胚胎是有利的,而在另外一些情况如研究某些分泌蛋白基因功能则需要避免嵌合式胚胎的出现;
- 实验中如何检测和优化改良版核酸酶如Cas9,或特定修复机制的抑制剂(例如,同源指导修复实验中,可能会需要一个有效的非同源性末端连接通路的抑制剂[Howden et al.,2016])。

探索生殖细胞发生和不孕症

来自成年小鼠睾丸的精原干细胞系(SSC,即spermatogonial stem cell 的英文首字母缩写)为我们研究体外和移植回睾丸后体内精子发生过程提供了丰富的细胞资源。人们可以改变这些细胞的遗传信息,研究其对精子发生过程及移植回小鼠体内后对其后代的影响。在体外利用CRISPR/Cas9 技术修正这些干细胞中基因突变的方法也已有报道(Wu et al.,2015)。该研究组成员使用CRISPR/Cas9 对小鼠精原干细胞中一个可导致白内障的基因突变进行了修正,将改造后的细胞移植回小鼠睾丸内,然后收集圆形精子细胞并注射到卵母细胞胞质中(ICSI,即intracytoplasmic sperm injection 的英文缩写)完成体外受精并获得胚胎。最终得到的胚胎在体内正常发育后生成的后代中,基因修复效率达到百分之百。类似的实验在包括恒河猴(Hermann et al.,2012)在内的其他物种中也有报道。稳定的人精原干细胞系还未见报道,但可以预见其在理解男性不育症和探索与年龄相关的高基因突变率的研究中会是一个重要工具。精母干细胞系体外研究对恢复男性癌症患者在放射性治疗和化疗后的生育能力方面应该有很大帮助,如果将培养改造后的精原干细胞移植回睾丸或用于体外受精,则可能会导致这种修改经生殖系传递传给后代。

小鼠胚胎干细胞经过体外人为诱导能够得到卵子和精子前体细胞,这些胚胎干细胞来源的卵子可以用正常精子进行受精,而精子也可以通过卵胞浆内单精子注射使正常卵子受精(Hayashi et al.,2012;Saitou and Miyauchi,2016;Hikabe et al.,

In some cases, there could be interest in exploring the effects of altering specific genes at the next stages of human development, notably the early stages after the embryo would implant in a uterus. At present, culture of human embryos up to the stage just prior to germ-layer formation (at 14 days after fertilization or the formation of the "primitive streak") is permitted in many countries. Improved culture systems that allow human embryos to develop in culture during the implantation period are being developed. Recent results suggest that these systems could be used to study the elaboration of extraembryonic structures and of the epiblast into an "embryonic disc"—processes that occur in humans in ways not found in mice (Deglincerti et al., 2016; Shahbazi et al., 2016). These improved cell culture systems, combined with better ways of analyzing gene function using genome editing, can be expected to lead to better understanding of the fundamental processes of early human development. Already at least two research groups (in the United Kingdom and Sweden) have received regulatory permission to carry out CRISPR/Cas9 experiments on human embryos, aimed at addressing these kinds of fundamental biological questions.

Knowledge gained from such studies is expected to inform and improve IVF procedures and embryo implantation rates and reduce rates of miscarriage. Conversely, the same studies may lead to novel methods of contraception. Such research also should lead to better ways of establishing and maintaining stem cells from these early embryonic stages, which could facilitate efforts to derive cell types for studies and treatments of disease and traumatic injury. Knowledge gained from these laboratory studies using genome-editing methods in early human embryos should also provide information about the suitability of these methods for any eventual potential clinical use. That is, basic research can be expected to inform an understanding of the feasibility of making heritable, and preferably nonmosaic, changes in the genome (see Chapter 5). Because human embryos that can be used in research are a valuable and relatively scarce resource, it will be important to ensure that the most efficient methods are used for these laboratory studies of their basic biology. Thus, it is likely that in the course of this research, various technical issues associated with improving the use of genome-editing methods in human embryos will be addressed. Relevant questions include the type and form of genome-editing components to be introduced; whether to use Cas9 or an alternative nuclease;

- what method to use to introduce the genome-editing components—e.g., as DNA, mRNA, protein, or ribonucleoprotein complex;
- whether to use single guide RNAs, pairs, or multiple guide RNAs as part of the editing machinery;
- the size of the DNA template and whether such a

2016；Zhou et al.，2016）。尽管最近有两篇文献报道了将人胚胎干细胞诱导为早期生殖细胞前体，但人的生殖细胞至今仍然无法从多能性干细胞诱导获得（Irie et al.，2015；Sasaki et al.，2015）。运用基因编辑技术对生殖细胞发生过程相关基因功能进行研究也凸显了人与鼠生殖细胞发生过程的显著差异。由于人与小鼠精子发生过程不尽相同，因此从研究小鼠精子发生所获得的知识不能完全适用于人的同类进程上。由此看来，对人类细胞的研究对解决人体生命过程中的疑问至关重要。如果能像在小鼠中一样从人的多能干细胞分化得到单倍体生殖细胞，将为研究人生殖细胞发生和找出不孕症的原因打开新的大门，还将为用可遗传的基因组修饰来治疗基因遗传疾病提供新的可能。

基础研究中的伦理和监管问题

如第 2 章中详细介绍的那样，实验室进行的体细胞基础科学研究将受制于对实验人员及环境的安全管理，包括公共机构生物安全委员会对重组 DNA 相关的工作进行的特别审查。如果细胞和组织来自可辨识的、还存活的个体，则需考虑捐赠者的许可和隐私，而且在大多数情况下试验方案将至少受到机构审查委员会的一些审查，但总的来说没有新的伦理问题出现。

相对于以上研究，胚胎研究更加备受争议。如前所述，使用能存活的胚胎来进行研究在美国少数州是非法的（NCSL，2016），而且虽然多数州允许胚胎研究，但是对胚胎有风险的研究大多不能得到联邦卫生与公众服务（HHS）部门的基金资助，这是因为从 20 世纪 90 年代以来迪奇-威克修正案（公共法律第 114-113 号 H 部 V 题第 508 节）一直是 HHS 拨款程序的一部分，包括用于 2017 年的资助法案（见第 2 章）[39]。它指出：

（a）本法案提供的基金不得用于：

（1）为了研究目的制造人胚胎；

（2）根据公共卫生服务法案（42 USC 289g(b)）的 45 CFR 46.204(b) 和第 498(b) 条，会破坏、丢弃人胚胎，或明知人胚胎受伤和死亡的风险大于子宫内胎儿研究所允许的风险的研究。

---

[39] 同时在 S. 3040 和 H.R.5926 中。

template is required;

- the optimal timing for genome editing, i.e., whether information can be obtained by using two-cell embryos, whether it necessary to use one-cell embryos, or whether it is best to introduce the reagents along with the sperm during *in vitro* fertilization;
- whether mosaicism can be tolerated, keeping in mind that it may be an advantage for certain experiments, as when cell fate is to be followed, but may need to be avoided in other cases, such as when investigating a gene whose product is a secreted protein; and
- how to test and improve modified versions of nucleases such as Cas9 or inhibitors of certain repair mechanisms (e.g., an effective inhibitor of nonhomologous end joining may be needed if the experiment demands homology-directed repair [Howden et al., 2016]).

Understanding of Gametogenesis and Infertility

In mice, the generation of spermatogonial stem cell (SSC) lines from the adult testes has provided a rich source of cells with which to study the process of spermatogenesis *in vitro* and *in vivo*, after regrafting to the testes. It is possible to alter these cells genetically and study the impact of the changes on the process of spermatogenesis itself or, in mice, the impact on the offspring. It is also possible to correct genetic mutations in the stem cells *in vitro* using CRISPR/Cas9. Proof of principle for such an approach has been published (Wu et al., 2015). This work used CRISPR/Cas9 editing in mouse SSCs to correct a gene mutation that causes cataracts in mice. The edited SSCs were transferred back to mouse testes, and round spermatids were collected for intracytoplasmic sperm injection (ICSI), a form of *in vitro* fertilization, to create embryos. Resulting offspring were correctly edited at 100 percent efficiency. Similar experiments have been conducted using SSCs from other species, including macaques (Hermann et al., 2012). Stable human SSC lines have not yet been reported, but would clearly be an important tool for understanding male infertility and for exploring such issues as the higher rate of mutations associated with age. This is an active area of research because it may enable restoration of fertility in male cancer patients after radiation or chemotherapy. The ability to grow and manipulate human SSCs would, however, raise the possibility of generating human germline alterations if the cells were grafted back to the testes or used in IVF.

Related issues arise from experiments in which both oocytes and sperm progenitors have been generated from mouse ES cells. ES-derived oocytes can be fertilized by

(b) 就本章节而言，术语"人胚胎"包括从该法案颁布之日，根据45CFR46未作为人类受试者来保护的任何生物体，包括来源于受精、孤雌生殖、克隆或其他以任何方式从人的配子或二倍体细胞获得的生物体。

这种州和联邦法律结合的结果使得胚胎研究在美国大部分地区合法，但一般得不到HHS基金资助。

另外，根据国家科学院关于胚胎干细胞研究的建议，来自干细胞研究督查委员会的对使用人胚胎的实验研究的法外监督被广泛采用（IOM，2005；NRC and IOM，2010）。最近，国际干细胞研究学会（成员来自包括美国在内的世界各地的研究者）发表指南呼吁将自愿的干细胞研究监督委员会转变成人胚胎研究监督（"EMRO"）审查委员会，来监督所有涉及植入前人体发育的各阶段、人胚胎或胚胎衍生细胞，或者需要在体外制造可用于受精或产生胚胎的人类配子（ISSCR，2016）的研究。这些由诸多科学家、伦理学家及公众成员组成的独立的多学科委员会，将负责审查研究者的提案和资质上的各类细节。所倡议的委员会将在道德框架内评估研究目标来确保研究以透明和可靠的方式进行。项目提案应包括对替代方案的讨论，提供使用所需人体材料的合理性解释，包括对要使用的植入前胚胎的数量、所提出的方法，以及为何在人类而不是动物模型系统中进行实验的正当理由。

总结和建议

人类基因组编辑相关（不与患者接触）的实验研究遵循与其他人类组织体外基础实验相同的监管途径，所引起的伦理问题也有已有伦理准则和监管制度予以监督。

这不仅包括体细胞研究工作，也包括在司法允许的区域捐赠和使用人类配子及胚胎用于研究目的。尽管有人对其中一些规则体现的政策存在异议，但规则依然有效。为了解决与人类生殖繁殖相关的重大科学和临床问题，对人类配子和祖细胞、胚胎及多能性干细胞的研究仍将持续。这些研究对于不以可遗传的基因组编辑为目的的科学和医疗用途是必需的，然而，如果将来有此类意图，这些研究也可提供重要的信息和技术。

normal sperm, and ES-derived spermatids can fertilize eggs by ICSI (Hayashi et al., 2012; Hikabe et al., 2016; Saitou and Miyauchi, 2016; Zhou et al., 2016). Human gametes have not yet been generated successfully from pluripotent stem cells, although two recent papers report the generation of early germ cell progenitors from human ES cells (Irie et al., 2015; Sasaki et al., 2015). Through the use of genome-editing methods, this work also highlighted significant differences between mice and humans in the genes involved in specification of primordial germ cells. There is evidence as well that knowledge gained from studying later stages of spermatogenesis in mice may not always be applicable to the same process in humans. These findings reflect the role of research on human cells in answering questions about human biology. If human haploid gametes could be generated from human pluripotent cells, as they can be in mice, it would open up new avenues for understanding gametogenesis and the causes of infertility. It would also open up possibilities for using heritable genome modifications to address health problems that originate from genetic causes.

ETHICAL AND REGULATORY ISSUES IN BASIC RESEARCH

As described in more detail in Chapter 2, basic science research performed in the laboratory on somatic cells will be subject to regulation focused on safety for laboratory workers and the environment, including special review by institutional biosafety committees for work involving recombinant DNA. Few new ethical issues are raised, although if the cells and tissues come from identifiable living individuals, donor consent and privacy will be a concern, and in most cases the protocols will be subject to at least some review by institutional review boards.

Research with embryos is more controversial. As noted earlier, research using viable embryos is illegal in a small number of U.S. states (NCSL, 2016), and while permitted in most states, research that exposes embryos to risk generally may not be funded by the U.S. Department of Health and Human Services (HHS) funds; this is due to the Dickey- Wicker Amendment (Public Law No: 114-113 Division H Title V Section 508), which has been adopted repeatedly since the 1990s as part of the HHS appropriations process, including in the bills introduced for 2017 funding (see Chapter 2).[39] It states:

(a) None of the funds made available in this Act may be used for—

(1) the creation of a human embryo or embryos for research purposes; or

(2) research in which a human embryo or embryos are

---

[39] Sec. 508(a) in both S.3040 and H.R.5926.

建议 3-1 现有的监管设施，以及审查与评估对人类细胞和组织进行基因组编辑的基础实验室研究的流程，应该用来评估以后人类基因组编辑的基础实验室研究。

destroyed, discarded, or knowingly subjected to risk of injury or death greater than that allowed for research on fetuses *in utero* under 45 CFR 46.204(b) and section 498(b) of the Public Health Service Act (42 U.S.C. 289g(b)).

(b) For purposes of this section, the term "human embryo or embryos" includes any organism, not protected as a human subject under 45 CFR 46 as of the date of the enactment of this Act, that is derived by fertilization, parthenogenesis, cloning, or any other means from one or more human gametes or human diploid cells.

The effect of this combination of state and federal law is to make embryo research legal in most of the United States but generally not eligible for HHS funding.

Additional, extralegal oversight of laboratory research using human embryos comes from the stem cell research oversight committees that were widely adopted pursuant to recommendations of the National Academies regarding embryonic stem cell research (IOM, 2005; NRC and IOM, 2010). Recently, the International Society for Stem Cell Research, whose membership includes investigators from around the world as well as the United States, adopted guidelines calling for the transformation of these voluntary stem cell research oversight committees into human embryo research oversight (EMRO) review committees that would oversee "all research that (a) involves preimplantation stages of human development, human embryos, or embryo-derived cells or (b) entails the production of human gametes *in vitro* when such gametes are tested by fertilization or used for the creation of embryos" (ISSCR, 2016). The review would include details of the proposal and the credentials of the researchers under the auspices of these independent, multidisciplinary committees of scientists, ethicists, and members of the public. The proposed committees would assess research goals "within an ethical framework to ensure that research proceeds in a transparent and responsible manner. The project proposal should include a discussion of alternative methods and provide a rationale for employing the requested human materials, including justification for the numbers of preimplantation embryos to be used, the proposed methodology, and for performing the experiments in a human rather than animal model system."

CONCLUSIONS AND RECOMMENDATION

Laboratory research involving human genome editing—that is, research that does not involve contact with patients—follows regulatory pathways that are the same as those for other basic laboratory *in vitro* research with human tissues, and raises issues already managed under existing ethical norms and regulatory regimes.

This includes not only work with somatic cells, but also the donation and use of human gametes and embryos for research purposes, where this research is permitted. While there are those who disagree with the policies embodied in some of those rules, the rules continue to be in effect. Important scientific and clinical issues relevant to human fertility and reproduction require continued laboratory research on human gametes and their progenitors, human embryos and pluripotent stem cells. This research is necessary for medical and scientific purposes that are not directed at heritable genome editing, though it will also provide valuable information and techniques that could be applied if heritable genome editing were to be attempted in the future.

RECOMMENDATION 3-1. Existing regulatory infrastructure and processes for reviewing and evaluating basic laboratory genome-editing research with human cells and tissues should be used to evaluate future basic laboratory research on human genome editing.

4

体细胞基因组编辑
Somatic Genome Editing

应用基因编辑技术对人类体细胞基因组进行编辑来治疗遗传性疾病已经进入临床试验。体细胞可以参与到机体内除生殖细胞系以外的各个组织中，这意味着，与可遗传的生殖系编辑不同（详见第 5 章），这种对体细胞的改造只影响被治疗的个体，而不会遗传给下一代。基因治疗，即对体细胞进行遗传改造，并不是一个全新的治疗概念[40]，在过去几十年里，基因治疗在临床疾病治疗方面的应用已经取得了长足的进展（Cox et al., 2015; Naldini, 2015）。尽管截至 2016 年年底，只有两个基因疗法获得批准（Reeves, 2016），但是数百个早期和少数晚期的临床试验正在进行当中（Mullin, 2016）。现行的基因治疗方案是以广泛的、来自不同细胞和非人动物模型的实验室研究成果为基础，建立起来的一套在活细胞或生物体内对基因进行添加、删除或者改造的方法。最近，基因编辑技术的发展，特别是基于核酸酶的编辑工具的开发，极大地增强了基因疗法的应用前景（详见第 3 章）。

本章从内容上可分为四部分。第一部分先叙述了人类体细胞基因组编辑背景知识，定义了相应关键术语，接着对基因组编辑相比于传统基因疗法和早期方案的优势进行了小结，并且简短回顾了基于核酸酶的基因编辑用到的修复方法——同源性的和

The use of human genome editing to make edits in somatic cells for purposes of treating genetically inherited diseases is already in clinical trials. Somatic cells contribute to the various tissues of the body but not to the germline, meaning that, in contrast with heritable germline editing (discussed in Chapter 5), the effects of changes made to somatic cells are limited to the treated individual and would not be inherited by future generations. The idea of making genetic changes to somatic cells, referred to as gene therapy, is not new,[40] and considerable progress has been made over the past several decades toward clinical applications of gene therapy to treat disease (Cox et al., 2015; Naldini, 2015). Hundreds of early-stage and a small number of late-stage trials are underway (Mullin, 2016), although only two gene therapies have been approved as of late 2016 (Reeves, 2016). Existing technical approaches to gene therapy are based on the results of extensive laboratory research on individual cells and on nonhuman organisms, establishing the means to add, delete, or modify genes in living cells or organisms. Prospects for future applications of gene therapy have recently been greatly enhanced by improvements in genome-editing methods, particularly the development of nuclease-based editing tools (see Chapter 3).

This chapter begins by providing background information on human somatic cell genome editing, including definitions of key terms. It then summarizes the advantages of genome editing over traditional gene therapy and earlier approaches, and briefly reviews the repair methods— homologous and nonhomologous—used

---

[40] 基因治疗致力于置换有缺陷的基因或添加新的基因来治愈疾病或提高抵抗疾病的能力。

[40] Gene therapy denotes the replacement of faulty genes or the addition of new genes to cure or improve the ability to fight disease.

非同源性的。第二部分是对人类体细胞基因组编辑潜在应用的讨论。第三部分依次对基因组编辑面临的科学技术问题、伦理问题和监管问题进行了分析。本章以结论和推荐意见结束。更多有关基因编辑技术的科学和技术细节见附录A。

研 究 背 景

基因、基因组和遗传变异体

所有人类都拥有两套遗传自其父母的基因，每一套基因均被称为一个基因组，并被包装于23条染色体之中。人类单倍体（单一）基因组的长度约为30亿（$3×10^9$）个碱基对，每个体细胞（二倍体）中的两套遗传继承的基因组编码了人一生中细胞和机体装配及运行所需的信息。尽管被统称为人类基因组，但是每一套基因组之间在许多位点并不相同（大约每1000个碱基对中有一个，或是总共约有300万个位点不同），而这些基因组间的不同使得每个人类个体都是独特的（The 1000 Genomes Project Consortium，2015）。这些碱基差异大多对个体不具有或者只有微小的影响，但是其中一些变异会影响基因的表达和／或功能。人类基因组约含有20 000个编码蛋白质的基因，这些蛋白质实际参与构建了人类细胞及机体，另外还有其他众多的DNA元件，它们负责调节每一个基因表达的时间、地点及水平（Ezkurdia et al.，2014）。一些基因变异能够改变它们所编码的蛋白质的性质，另一些基因组变异则能影响基因的表达。这些基因变异影响头发或眼睛的颜色、血型、身高、体重和其他诸多个人特征，不过人类特征多数是由多个基因相互作用决定的。此外，其他影响因素如饮食、运动、教育和环境，可以通过与个人的遗传组分相互作用而产生重大影响。

基因组序列中的许多变异来源于细胞分裂期进行DNA复制（拷贝）时产生的碱基对序列的改变（可以比喻为印刷改变）。这些突变以一定的速率持续发生着，尽管细胞存在校对和修正（编辑）这些突变的机制，但是还有一些突变逃脱了细胞的校对程序而存留下来。此外，DNA突变的频率会因为辐射（如阳光中的紫外线、宇宙射线、X射线）或者环境中化学物的影响（如香烟烟雾和其他致癌物）而增加。正如前文所述，这些突变很多

for nuclease-based genome editing. Next is a discussion of potential human applications of somatic cell genome editing. Scientific and technical considerations and ethical and regulatory issues are then examined in turn. The chapter ends with conclusions and recommendations. Additional scientific and technical detail on methods for genome editing are provided in Appendix A.

BACKGROUND

Genes, Genomes, and Genetic Variants

All humans contain two sets of genes inherited from their parents; each of these sets of genes is called a genome and is packaged into 23 chromosomes. The haploid (single) human genome is around 3 billion (3×10^9) base pairs long, and the two inherited genomes in each somatic cell (diploid) encode the information required for the assembly and functioning of a person's cells and body throughout life. Although people speak of the human genome, each genome differs from any other at many positions (around 1 in 1,000 base pairs, or about 3 million positions), and these genetic differences contribute to what makes individual humans unique (The 1000 Genomes Project Consortium, 2015). Many of these variations probably have little or no effect, but some affect the expression and/or functions of genes. Within the human genome lie approximately 20,000 genes that encode proteins, the molecules that actually build human cells and bodies, plus many other DNA elements that control when, where, and how much each gene is expressed (Ezkurdia et al., 2014). Some variants in genes can change the properties of the proteins they encode, while other genomic variants can affect the expression of genes. Such variants influence the color of hair or eyes, blood type, height, weight, and many other individual features, although most human traits are affected by interactions among multiple genes. Furthermore, other influences, such as diet, exercise, education, and environment, have major impacts by interacting with a person's genetic makeup.

Many of the variations in genomic sequences arise from alterations in the sequences of base pairs that arise during replication (copying) of the DNA during cell division (one can think of them as typographical changes). These alterations occur continually at a certain rate, and although cells have mechanisms for proofreading and correcting (editing) such changes, some escape the proofreading process and persist. Furthermore, the frequency of DNA alterations can be increased by radiation (e.g., by ultraviolet rays in sunlight or by cosmic or X-rays) or by environmental chemicals (e.g., cigarette smoke and other carcinogens). As mentioned, many of these variants have little or no effect, but others have positive or deleterious effects. This process of variation in

对人体不产生影响或是只有微小的影响，但是另一部分突变就有可能产生有利或有害的影响。基因组里的这些突变过程在人类进化为独立的物种之前就已经存在并持续至今。物种的进化正是依赖于这些突变的持续发生——那些有利的突变被选择性保留，有害的则被去除。然而，某特定突变对生物体是有利还是有害可根据具体情况而改变，这或许就是决定是否将编辑突变应用到临床中的一个考虑因素。

遗传性疾病

人们致力于将基因编辑的新进展用于可能的临床应用的一个主要动力是基于它为治疗和预防人类疾病提供新途径的可能性。基因编辑的一种可能用途是治疗数千种已知的遗传性疾病[41]。某些有害突变遗传自父母的一方或双方，而另一些有害突变并不遗传自父母，而是在胚胎期原发产生的。遗传模式随突变的性质不同而不同。如果导致基因功能丧失的突变遗传自一个亲本，来自另一亲本的正常基因通常足以提供所需的功能而使这种突变往往没有明显表型，这种遗传模式遗传学家称之为隐性遗传。在杂合状态下，隐性突变基因与正常基因同时存在于受精卵（合子）阶段及后继的幼儿和成人阶段，该突变通常（但非总是）对人体不产生或者只产生微弱的影响。也就是说隐性有害基因突变通常不会导致相应的疾病，除非分别来自父母双方的两个等位基因均带有这种突变。如果父母都处于杂合状态，每人都携带一个拷贝的有害突变体，他们的每个子女将有 25% 的概率同时遗传到两个拷贝的该突变体——也就是所谓的纯和状态。这种情况下两个等位基因都没有正常功能，因而会导致遗传性疾病的发生。这一现象有许多例子，如以气泡男孩症为代表的某些重症联合免疫缺陷病、镰状红细胞贫血症和泰 - 萨克斯症。

有的变异体，哪怕以单拷贝的形式存在且仍保留有一条正常的等位基因，也能导致疾病，这种突变被称为显性突变，它们在杂合状态下也会产生有害影响。一个明显的例子是亨廷顿舞蹈症，这种疾病中一个拷贝的显性突变就可以导致迟发型疾病发生。

human genomes has been going on since before humans evolved as a separate species and continues to this day. Evolution relies on this continual generation of variants—those that are advantageous are selected for, whereas those that are deleterious are selected against. Whether a particular variant is advantageous or deleterious, however, can vary with the context and may be a consideration in deciding whether to edit variants for clinical benefit.

Genetically Inherited Diseases

One primary impetus for interest in possible clinical applications of the recent advances in genome editing is the possibility that they provide new avenues for treating and preventing human disease. One such possible use is in the treatment of genetically inherited diseases, thousands of which are known.[41] Certain deleterious variants can be inherited from one or both parents, while others can arise de novo in the embryo rather than being inherited from either parent. The pattern of inheritance varies with the nature of the variant. If a variation that causes loss of function in a gene is inherited from one parent, it often has no evident effect, because the unaltered variant inherited from the other parent is sufficient to provide the function needed. Geneticists refer to this mode of inheritance as recessive. Recessive gene variants usually (but not always) have little or no effect in the so-called heterozygous state, when two different variants are present in the fertilized egg (zygote) and in the subsequent child and adult. That is, a person generally will not have the disease caused by a recessive deleterious gene variant unless that variant is inherited from both parents. If both parents are heterozygous, each having one copy of a deleterious variant, each of their children will have a 25 percent chance of inheriting two copies of that variant—the so-called homozygous state. In that case, there is no functional variant is available, and the consequence may be a genetically inherited disease. Many examples of this phenomenon exist (e.g., certain forms of severe-combined immunodeficiencies, such as bubble boy disease, as well as sickle-cell anemia and Tay-Sachs disease).

Other variants may actually produce medical problems even when present in a single copy despite the presence of a functional gene variant. Such variants, called dominant, produce deleterious effects even in the heterozygous state. A clear example is Huntington's disease, in which a single copy of a dominant disease-causing variant produces late-onset disease.

Some inherited diseases, such as certain forms of hemophilia, which affect blood clotting, involve genes that are present on the X chromosome (so-called X-linked).

---

[41] OMIM, https://www.omim.org（访问于 2017 年 1 月 10 日）；遗传联盟，http://www.diseaseinfosearch.org/（访问于 2017 年 1 月 25 日）.

[41] OMIM, https://www.omim.org (accessed January 10, 2017); Genetic Alliance, http://www.diseaseinfosearch.org (accessed January 25, 2017).

还有一些遗传疾病，如影响血液凝结的血友病中的某些类型，疾病相关基因定位于 X 染色体（所谓的 X-连锁）。因为男性只有一条 X 染色体，而女性有两条，故而男性中的单个异常的 X 染色体连锁的血友病基因将导致疾病显现，而仅具有一个有害变体的女性将是变异基因的携带者且通常没有出血症状（所谓的沉默携带者）。

如前文提到的，一些基因突变体是有利还是有害取决于其所处的环境，这增加了人们了解遗传疾病的复杂性。最著名的例子就是镰状红细胞贫血症，它是由一种基因突变引起的，而这个基因编码的是红细胞中运载氧的血红蛋白。如果镰状血红蛋白变异体从父母双方遗传得到（纯合子），将会导致血红蛋白在特定条件下聚集，使得红细胞变成镰刀状从而阻碍血液循环，引发多种障碍、剧烈的疼痛及正常组织功能损害。仅遗传到一个镰刀形基因变体的杂合个体（杂合子）几乎没有任何疾病的迹象，因为他们携带镰状细胞变体并能将其传递给下一代，故被称为携带者。事实证明，这种变体的杂合状态使突变基因携带者对感染其红细胞的疟疾寄生虫具有一定的抗性。也就是说，在疟疾盛行的区域，镰状细胞变体为携带者的生存提供了显著优势，因而在进化中被选择保留，这也是为什么相对于其他区域，镰状细胞变体携带者更普遍地分布于疟疾流行的地域，如非洲、印度和地中海地区。基于杂合优势的这种平衡选择的其他例子并不少见，杂合优势可以平衡因遗传了两种疾病相关的变体所产生的不利影响。

最后值得一提的是，大多数人类疾病被认为受多种基因的遗传变异的影响，每种变异对疾病进展仅具有微小的影响。因此，尽管应用人类基因组编辑技术来治疗遗传性疾病的前景在某些情况下极被看好，如治疗那些确定是由单一基因突变而引起的遗传疾病，但是对于大多数人类疾病来说，其前景并不乐观。

基因编辑较传统基因治疗和早期方案的优势

基因治疗是指将外源基因导入到细胞中以达到改善疾病状况的目的，利用病毒能够进入细胞这一特性，病毒载体能最有效地完成外源基因的导

Because men have only one X chromosome, whereas women have two, a single abnormal X-linked hemophilia gene in a man will lead to the disease being manifest, whereas women with just one deleterious variant will be carriers of the altered gene, usually without having bleeding symptoms (so called silent carriers).

Adding to the complexity of understanding genetic disorders is the observation, noted above, that some variants may be either deleterious or advantageous depending on the context. Probably the best known example is sickle-cell disease, which is caused by a variation in one of the genes encoding hemoglobin, the protein that carries oxygen in red blood cells. If the sickle hemoglobin variant is inherited from both parents (homozygous), it causes the hemoglobin protein to aggregate under certain conditions, leading to deformation of the red blood cells into a sickled shape that interferes with blood circulation, causing multiple difficulties and much pain and impairment of normal tissue functions. Heterozygous individuals (heterozygotes) who inherit just one sickle gene variant have few if any signs of disease and are known as carriers since they carry the sickle-cell variant and can pass it on to their children. It turns out that heterozygosity for this variant makes carriers somewhat resistant to malaria parasites that infect their red blood cells. That is, the sickle-cell variant provides a significant survival advantage in areas where malaria is present, and for that reason has been selected for and is relatively prevalent in such areas such as Africa, India, and the Mediterranean, where carriers are more common than in other areas. There are other examples of such balanced selection based on heterozygous advantage, balanced against the disadvantage of inheriting two disease-associated variants.

Finally, it is important to note that most human diseases are thought to be affected by genetic variants in multiple genes, with each variant having only a minor effect on disease progression. Thus, while the prospect of human genome editing to treat genetically inherited diseases has great appeal in some cases—for example, those in which a single gene can be clearly identified as causal—that is not true of the majority of human diseases.

ADVANTAGES OF GENOME EDITING OVER TRADITIONAL GENE THERAPY AND EARLIER APPROACHES

Gene therapy is the introduction of exogenous genes into cells with the goal of ameliorating a disease condition. This is most efficiently done using viral vectors that take advantage of a virus's natural ability to enter cells. The viral vectors are used to introduce a functional transgene and compensate the malfunction of an inherited mutant gene (gene replacement) or to instruct

入。病毒载体能被用于导入功能性基因并补偿遗传性突变基因导致的功能障碍（基因替代）或赋予被修饰细胞一项新的功能（基因添加）。这些载体上还包含用于驱动转基因表达的外源转录调节序列（启动子）。由于病毒载体的承载容量有限，转基因及启动子必须经过修改，而不能是它们自然存在于基因组中的形式，这就可能导致转入的基因不能完全重现其生理表达模式。根据所选择的载体和靶细胞的类型，基因修饰可分为短暂性、长期性和永久性三大类。永久性基因修饰是通过将慢病毒或γ-逆转录病毒载体插入到宿主细胞的基因组（整合）而实现的。然而，由于这种插入是半随机性的，它可能会影响插入位点或附近基因的功能和表达，因而具有潜在风险（插入突变）。目前，由于病毒载体特别是慢病毒和重组腺相关病毒（rAAV）的改良，人们在基因治疗方面取得了巨大进展，并在临床中对这些策略进行了深入的研究。虽然据报道大多数使用这种方法治疗的患者疗效显著（Naldini, 2015），但人们仍需要更灵活精准的遗传修饰，以进一步提高基因治疗的安全性并将其应用于治疗更多疾病，而靶向基因编辑使得这一切变为可能。

直到十年前，试图利用基因修改来治疗遗传疾病的尝试（又叫基因打靶），都是在培养的细胞中转入一段携带目标序列的DNA模板，然后让其随机插入到基因组中，或依赖罕见的同源重组整合到基因组中预定位置。DNA模板通常使用重组质粒（小的环状DNA），或利用天然具有进入细胞的能力的病毒载体转入细胞内。那些极少数获得目标序列的细胞需要再进行遗传筛选和克隆扩增。尽管有其局限性，但基因打靶作为一种实验工具，其重要性体现于同源重组的方法已经被广泛应用于改造酵母、脊椎动物细胞系乃至小鼠，使人们可以从遗传角度进一步剖析多种生物过程（Mak, 2007; Orr-Weaver et al., 1981）。

按照以前的这些策略，基因打靶的成功率从使用DNA质粒的10^{-6}（100万个细胞中有1个）到使用病毒载体（如rAAV）的10^{-2}~10^{-3}（100~1000个细胞中有1个）不等。然而，当科学家采用在改造DNA时用核酸酶在基因组的目标位点造成双链DNA断裂（DSB）的策略，基因打靶的成功率急剧上升（Carroll, 2014; Jasin, 1996）。这个基于核酸

a novel function in the modified cells (gene addition). The vectors also include exogenous transcriptional regulatory sequences (promoter) to drive transgene expression. Because viral vectors have a limited cargo capacity, both the transgene and the promoter have to be modified from the natural version present in the genome and may thus fail to properly recapitulate physiological expression patterns. According to the choice of vector and type of target cells, the genetic modification may be transient, long-lasting or permanent. Permanent modification is achieved using lentiviral or gamma-retroviral vectors that physically insert into the genome of the infected cells (integration). However, because insertion is semi-random, it may affect the function and expression of genes at or nearby the insertion site, thus representing a potential risk (insertional mutagenesis). Currently, tremendous progress is being made in gene therapy because of improved viral vectors, particularly lentiviral and recombinant adeno-associated viruses (rAAV), and these strategies are being intensively investigated in the clinic. However, despite the fact that remarkable benefits are being reported in most treated patients (Naldini, 2015), more flexible and precise genetic modifications, such as those made possible by targeted genome editing, are needed to further improve the safety of gene therapy and broaden its application to the treatment of more diseases and conditions.

Until the past decade, attempts to use genome modification in the treatment of genetically inherited disease, also called gene targeting, were performed by introducing a DNA template carrying the desired sequence into a cell population in culture, and then either allowing insertion at a random location or relying on rare homologous recombination events to incorporate that template sequence at an intended location in the genome. The DNA template generally was introduced into the cell using such systems as recombinant plasmids (small circular pieces of DNA) or viral vectors, which take advantage of a virus's natural ability to enter cells. The rare cells that acquired the desired sequence then had to be genetically selected and clonally expanded. Despite the limitations of this approach, the importance of gene targeting as an experimental tool is reflected in the broad use of homologous recombination to modify yeast, vertebrate cell lines, or even mice to genetically dissect a wide range of biological processes (Mak, 2007; Orr-Weaver et al., 1981).

The frequency of successful gene targeting using these older strategies ranged from 10^{-6} (1 in 1 million cells) for plasmid DNA to 10^{-2}-10^{-3} (1 in 100 to 1 in 1,000 cells) using viral vectors (such as rAAV). When scientists modify DNA with a nuclease that makes a double-strand break (DSB) at a desired location in the genome, however, the frequency of successful genome editing increases dramatically (Carroll, 2014; Jasin, 1996). Nuclease-

酶的靶向基因修饰系统是本文所探讨的基因编辑技术的根本。利用这个基于核酸酶的编辑系统，人们现在可以先切断 DNA，随后要么通过非同源性末端连接的 DNA 双链断裂（DSB）修复策略引入小片段的插入或缺失，要么通过效率相对偏低的同源重组在靶位点处插入新序列，从而实现高达 100% 的改造基因组上目标序列。由于基因重组效率的显著提高，科学家和临床医生得以考虑扩大基因编辑技术的应用，其中包括应用于疾病的治疗。

基因编辑技术的灵活性

基于核酸酶的基因编辑技术包含多种可以修改细胞内 DNA 序列的方法。根据 DNA 修改位置和目的的不同，基因编辑可以达成多种结果。利用基因编辑技术可以达到以下目的：

(1) 对基因的编码序列进行靶向基因破坏（失活）（基因破坏）；
(2) 精准替换一个或多个核苷酸（例如，将遗传变体原位替换为野生型或另一等位基因变体）；
(3) 将编码蛋白的转基因靶向插入到预定位点；
(4) 靶向改造一些调节基因表达水平的非蛋白编码的遗传元件（如启动子、增强子和其他 DNA 调控元件）[42]；
(5) 在基因组特定位置引入大片段缺失。

基因编辑技术的安全性和有效性

早期使用的基因置换质粒由于可以在整个基因组中随机插入，故存在插入突变的风险，尽管现在使用的最新一代插入质粒可以降低这种风险，但基于核酸酶的基因编辑却可能消除这种风险。此外，利用基因编辑技术对遗传突变进行原位基因修复，不光能恢复突变基因的功能，还能重建对其表达的生理调控。相较于采用人工构建的启动子来驱动治疗性转基因的表达，基因编辑技术提供了一种更为安全和有效的修复策略。这种随机插入的转基因或许不能准确重现其生理表达模式，并有可能受到插入位点的强烈影响，从而导致转基因细胞群体中基

---

[42] 小片段插入或者缺失的产生可以使一个元件失活；缺失更大的目标片段可以去掉整个元件；可以在元件内替换特定核苷酸；或者遗传元件可以被插入到基因组精确的位点上。

---

based systems that make targeted genetic alterations are at the root of the genome-editing technologies discussed in this report. With nuclease-based editing systems, it is now possible to cut and, consequently, modify up to 100% of the desired target sequence in the genome, either by small insertions or deletions introduced by the non-homologous end-joining DSB repair, or by relying on homologous recombination to introduce a new sequence at the target site, albeit with a somewhat lower efficiency. These dramatic improvements in efficiency have enabled scientists and clinicians to consider using genome editing for a greatly expanded range of applications, including application to the treatment of diseases.

Flexibility

Nuclease-based genome editing encompasses various methods for altering the DNA sequence of a cell. This editing can achieve several types of results, depending on where in the DNA the edits are made and for what purpose. Changes that can be made with genome editing include

- targeted disruption (inactivation) of the coding sequence of a gene (gene disruption); precise substitution of one or more nucleotides (e.g., in situ conversion of a genetic
- variant to wild type or to another allelic variant);
- targeted insertion of a transgene into a predetermined site for protein-coding genes; targeted alterations made to non–protein-coding genetic elements that regulate gene
- expression levels (e.g., promoters, enhancers, and other types of regulatory elements)[42]; and
- creation of large deletions at chosen genome locations.

Safety and Effectiveness

Nuclease-based genome editing may abrogate the risk of insertional mutagenesis inherently associated with prior gene-replacement vectors that integrate quasi-randomly throughout the genome, although late-generation integrating vectors used today may mitigate this risk. In addition, in situ gene correction of inherited mutations using genome editing reconstitutes both the function and the physiological control of expression of the mutant gene. This provides a safer and more effective correction strategy than gene replacement, in which expression of the therapeutic transgene is driven by a reconstituted artificial promoter. Such randomly inserted transgenes may fail to

---

[42] Small insertions or deletions can be created to inactivate an element; larger defined deletions can be created to remove entire elements; specific nucleotide substitutions can be made in the element; or new genetic elements can be inserted into precise locations in the genome.

因表达出现异质性。事实上，离体基因编辑技术的首批潜在应用之一可能是干细胞介导的原发性免疫缺陷的修复——应用基因编辑技术可以降低早期转基因方法中治疗基因的异位表达或组成型表达带来的导致癌变或机能障碍的风险。如果临床相关细胞中的基因打靶效率能高到满足治疗要求，那么基因编辑技术可以在安全性方面完全胜过基因替换（传统基因治疗），前提是脱靶效应不会造成癌症相关基因的修改，从而带来相似的风险。

基因编辑另一个潜在的广泛应用是将基因表达盒精确地靶向整合到基因组中的安全位点，这些精心挑选的位点不仅有利于转基因的稳定表达，还能保证插入到该位点的片段足够安全，不会对邻近基因产生有害影响。这种方法在确保治疗基因可预测的稳定表达的同时，避免了由意外插入引起的原癌基因激活这样的癌变风险。靶向整合到安全位点和原位突变修复都具有广泛应用于干细胞相关治疗的潜能，只要靶细胞经得起临床应用前的大量体外选择培养和扩增。随着对生长和分化不同类型细胞的培养条件的改良，特别是结合多能性细胞的分化技术，人们可以预期，基因编辑技术的应用将变得更加广泛（Hockemeyer and Jaenisch, 2016）。

基因编辑技术可实现靶基因中断

相对于传统的基因治疗策略而言，基因编辑的一个独特应用是靶向基因破坏。事实上，利用锌指核酸酶（ZFNs, zinc finger nucleases 的英文首字母缩写）的基因破坏技术已经在进行临床测试，测试结果表明其对 T 细胞有一定效果（Tebas et al., 2014），并且这种方法最近被拓展到造血干细胞（HSCs, hematopoietic stem cells 的英文首字母缩写）。这些临床实验旨在破坏细胞因子受体CCR5（5 型趋化因子受体）的表达。CCR5 是 HIV 感染的辅助受体，但并不是 T 细胞功能的必要成分，中断它的表达可以使感染了 HIV 的患者的 T 细胞对 HIV 病毒感染产生抗性[43]。原则上，基因破坏技术也可用于消除引起显性遗传疾病的基因突变。

reproduce the physiological expression pattern faithfully, and they can be strongly influenced by the insertion site, giving rise to substantial variegation of expression among a population of transduced cells. Indeed, one of the first potential applications of *ex vivo* genome editing may well be stem-cell-mediated correction of primary immunodeficiencies—an improvement over prior transgenic approaches in which ectopic or constitutive expression of the therapeutic gene posed a risk of cancerous transformation or malfunction. If on-target editing frequencies of clinically relevant cell types are high enough to be therapeutically useful, genome editing may eventually outperform gene replacement (traditional gene therapy) in terms of safety, provided that off-target changes do not pose similar risks by modifying genes associated with cancer.

Another potential broad application of genome editing is precisely targeted integration of a gene expression cassette into a so-called safe genomic harbor, chosen because it is conducive to robust transgene expression and allows a safe insertion that does not have a detrimental effect on adjacent genes. This approach may ensure predictable and robust expression of a therapeutic gene without the risk of oncogenesis caused by inadvertent insertional activation of an oncogene. Targeted integration into a safe harbor and in situ correction of mutations are both potentially widely applicable to stem-cell-based therapies as long as the targeted cells are amenable to extensive *in vitro* culture selection and expansion prior to clinical use. One can envisage increasing application of these types of genome editing as the ability to grow and differentiate different types of cells in culture improves, particularly in conjunction with differentiation from pluripotent cells (Hockemeyer and Jaenisch, 2016).

Gene Disruption

A unique application of genome editing relative to standard gene therapy strategies is targeted gene disruption. Indeed, clinical testing of gene disruption using zinc finger nucleases (ZFNs) is already under way, with some indication of benefit for T-cells (Tebas et al., 2014), and this approach has recently been extended to hematopoietic stem cells (HSCs). These trials aim to disrupt expression of a cytokine receptor, C-C chemokine receptor type 5 (CCR5), which also functions as a coreceptor for HIV infection and is not essential for T-cell function, thus making the T-cells of an HIV-infected individual resistant to viral infection.[43] Gene disruption

---

[43] 利用锌指核酸酶技术中断 CCR5 基因表达进行治疗的临床实验有 6 个。其中 3 个已经完成，一个在进行中，剩下两个正在招募参与者。更详细的资料，请参看 https://www.clinicaltrials.gov/ct2/show/NCT02500849?term=zinc+finger+nuclease+CCR5&rank=1（访问于 2017 年 1 月 10 号）。

[43] There are six clinical trials involving the use of ZFNs to disrupt expression of CCR5. Three of these trials have been completed, one is ongoing, and two are currently recruiting participants. For more information, see https://www.clinicaltrials.gov/ct2/show/NCT02500849?term=zinc+finger+nuclease+CCR5&rank=1 (accessed January 10, 2017).

基因编辑技术的亲民性

在过去 5~10 年间，已开发和改进了多种核酸酶平台，其他一些类似的平台也可能会很快被开发出来。自 2012 年发展起来的 CRISPR/Cas9 核酸酶平台，让科学研究、临床应用和患者群体对基因编辑的应用前景持有明显乐观态度，它的应用将使基因编辑平民化，使更多的实验室可以使用该技术。因此，CRISPR/Cas9 的应用已经使人们意识到基因编辑可以作为一种治疗工具，同时也引起了与其使用相关的伦理和监管问题的考虑（Baltimore et al.，2015；Corrigan-Curay et al.，2015；Kohn et al.，2016）。其实这些问题并不是新产生的，也不是只针对 CRISPR-Cas9 系统，其中许多问题在早期基因治疗和基因编辑应用的背景下就已经被提出和讨论过。

基于核酸酶的同源与非同源重组编辑修复技术

基于核酸酶的基因编辑依赖人工改造的核酸酶结合并切割基因组上的特定靶序列，从而产生 DNA 双链断裂或只切开 DNA 单链的"切口"。细胞自身对 DNA 损伤的修复主要有两种方式：①非同源性末端连接法（nonhomologous end-joining，NHEJ），该方法在修复过程中常常使基因或遗传元件失活；②基于同源性的修复机制，一般称为同源定向修复（homology directed repair，HDR）（参见第 3 章）。

非同源性末端连接法介导的基因组编辑会在断裂位点引起碱基插入或缺失（"indel"），从而改变所编辑的基因的序列。值得注意的是，尽管非同源性末端连接法介导的基因组编辑能够通过产生 DNA 断裂或切口精确定位，但却无法预测单个细胞中修改后的 DNA 序列或长度，也无法预测出一群细胞之间碱基插入缺失突变（indel）的异同。

同源指导修复介导基因组编辑以已有 DNA 为模板，在基因组上精准地替换一至数个碱基，或插入一段序列（如一个或多个基因）。不同于非同源性末端连接，同源定向修复介导的基因编辑使研究者能预测编辑发生的位点和改变后的DNA 序列及长度。因此，同源定向修复介导的编辑类似于文档编辑，因为它能精确地改变 DNA 序列。

could, in principle, also be used to eliminate a dominant disease-causing gene variant.

Accessibility

Multiple nuclease platforms have been developed or improved in the last 5-10 years, making it likely that additional such platforms will be developed in the near future. The CRISPR/Cas9 nuclease platform, developed just since 2012, has generated significant optimism among research, clinical, and patient communities and has democratized genome editing, making it usable by many more laboratories. As a result, CRISPR/Cas9 has raised awareness of genome editing as a therapeutic tool and motivated consideration of the ethical and regulatory issues associated with its use (Baltimore et al., 2015; Corrigan-Curay et al., 2015; Kohn et al., 2016). These issues are not new, however, nor are they specific to the CRISPR-Cas9 system; many of them have already been confronted and addressed in the context of earlier gene therapy and genome editing applications.

HOMOLOGOUS AND NONHOMOLOGOUS REPAIR METHODS USED FOR NUCLEASE-BASED GENOME EDITING

Nuclease-based genome editing relies on the design of an artificial enzyme a nuclease to bind a specific target sequence in the genome where it creates either a DNA double-strand break or a DNA single-strand cut known as a "nick." The cell usually repairs the break through one of two major mechanisms: (1) nonhomologous end-joining (NHEJ), which frequently inactivates the gene or genetic element during the repair process; or (2) homology-based mechanisms, generically described as homology directed repair (HDR). (See also Chapter 3.)

Genome editing by NHEJ creates an insertion or deletion ("indel") at the break site that alters the sequence of the edited gene. Importantly, while genome editing by NHEJ is precisely located by where the DNA break or nick is produced, it is not possible to predict the size or sequence of the resulting change in a single cell or the variability of the changes (indels) among a group of cells.

In genome editing by HDR, a DNA template is used either to create one or more nucleotide changes, perhaps to match a known human reference sequence, or to insert a novel sequence (e.g., one or more genes) at a precise genomic location. In contrast to NHEJ, HDR-mediated genome editing allows scientists to predict both where the edit will occur and the size and sequence of the resulting change. Thus, HDR-mediated editing is analogous to editing a document because it enables precise changes in DNA sequence.

人类体细胞基因编辑技术的潜在应用

基因编辑技术应用可从几个方面进行分类：

- 基因编辑针对的**细胞或组织类型**：①体细胞或组织，这种修饰不会遗传给后代；②生殖细胞或生殖前体细胞，这种修饰可以遗传给后代；③受精卵，这种修饰会导致新个体的体细胞和生殖细胞都发生改变（本章主要讨论体细胞编辑，生殖系基因编辑在第 5 章讨论）。
- 基因编辑的**实施位置**：①离体实验，即在体外试管中对细胞或者组织进行编辑，然后将其移植回个体体内；②体内实验，即直接在人体内编辑。
- 基因编辑的**具体目的**：如治疗或者预防疾病，基因编辑技术可将能致病的 DNA 突变修正为正常序列；或是引入额外的或新的性状，应用基因编辑技术将一个基因序列改变为人基因组中原本不存在的序列。
- 基因编辑的**属性**：①对致病性突变或者高风险等位基因变体进行简单修饰；②对一个内源基因进行基因破坏、异位表达或过表达等更复杂的修饰；③通过基因修饰获得新的功能，如强化某个生物学效应，或者建立对疾病或病原体的抗性。

这些修饰的目的可以是治疗或者预防某种疾病，也可以是修改（理论上说，甚至可以完全新造出）被编辑细胞或者组织的某种性状。值得注意的是，通过基因编辑强化某种细胞特性（如超量分泌某种蛋白质，或者抵抗病毒感染）也可用于疾病治疗。这种以治疗疾病为目的的细胞特性强化，应当与以在人身上获得某种想要的或全新的性状为目的的强化区分开来（这是第 6 章中将要详细讨论的内容）。

表 4-1 列举了一些有可能通过体细胞基因编辑技术治疗的疾病类型。尽管不够全面，但足以证明该技术的广泛应用前景。

有关基因编辑如何应用于疾病治疗最典型的案例，是通过同源重组将导致镰状红细胞贫血症

POTENTIAL HUMAN APPLICATIONS OF SOMATIC CELL GENOME EDITING

Genome-editing applications can be categorized based on several general features:

- **Which cells or tissue(s) are modified**—in particular, whether the modification is made in somatic cells or tissues, which do not contribute to future generations; in a germ cell or germ cell progenitor, which can result in heritable changes passed to future children; or in a zygote, in which case both somatic and germ cells would be modified. (The focus here is on somatic editing; germline editing is discussed in Chapter 5.);
- **Where the editing takes place**—in the test tube, followed by return of the cells or tissues to the individual (*ex vivo*), or directly in the person's body (*in vivo*);
- **The specific goal(s) of the modification**—for example, to treat or prevent disease or to introduce additional or new traits. These goals may be achieved by modifying a pathogenic DNA variant to a known nonpathogenic variant present in human reference sequences, or by modifying a gene to a sequence other than one that is a known existing human sequence; and
- **The precise nature of the modification**—simple modification of a disease-causing mutation or risk-associated allelic variant, or more a complex change, such as disruption or ectopic/overexpression of an endogenous gene or addition of a novel function that augments a biological response or establishes resistance to a disease or pathogen.

The intent of each of these modifications could be to treat or prevent a disease but could also be to modify (or, in principle, even create novel) phenotypic traits in the treated cells or tissues. It is important to note, for example, that one can use genome editing to achieve enhancement of a cellular property (e.g., secreting supranormal amounts of protein or resisting a viral infection) with the intent of curing a disease. Such cellular enhancement with intent to modify disease course needs to be distinguished from the concept of enhancement aimed at creating a desired or novel organismal feature in humans (a topic discussed in detail in Chapter 6).

Table 4-1 provides examples of the types of human diseases that might be treated using somatic cell genome editing. Even though this list is not comprehensive, it highlights the broad range of potential applications.

Clear examples of how genome editing might be applied to cure disease are to use homologous recombination to

表 4-1 体细胞编辑潜在的治疗应用举例

疾病类型	遗传性/传播模式	离体或在体	NHEJ 或 HDR 介导	发展阶段	常用策略
镰状红细胞贫血症	常染色体隐性遗传	离体 (HSPC)	HDR	临床开发	修复为非致病性等位基因
镰状红细胞贫血症/β 地中海贫血症	常染色体隐性遗传	离体 (HSPC)	NHEJ	临床前研究	引入胎儿血红蛋白
X 染色体连锁的重症免疫缺陷 (SCID-X1)	X 染色体连锁隐性遗传	离体 (HSPC)	HDR	临床开发	基因敲入全长或者部分互补 DNA（cDNA）以修复下游致病性突变
X 染色体连锁的超 IgM 抗体综合征	X 染色体连锁隐性遗传	离体 (T 细胞)	HDR	临床前向临床转化	基因敲入全长 cDNA 以修复下游致病性突变
B 型血友病	X 染色体连锁隐性遗传	在体 (肝)	HDR	临床试验*	通过强启动子表达凝血因子
囊性纤维化	常染色体隐性遗传	在体 (肺)	HDR	开发中	修复为非致病性等位基因
HIV	病毒感染	离体 (T 细胞和 HSPC)	NHEJ	临床试验	使获得对 HIV 感染的抗性
HIV	病毒感染	离体	HDR	开发中	使持续分泌抗 HIV 因子
癌症免疫治疗	NR	离体 (T 细胞)	NHEJ 或 HDR	临床概念性验证阶段	通过基因编辑获得更强效的癌细胞特异性 T 细胞
杜氏肌营养不良症 (DMD)	X 染色体连锁隐性遗传	在体	NHEJ	临床前研究	删除 DMD 的致病突变基因使之转变为症状较轻的贝克型肌营养不良
亨廷顿舞蹈症	常染色体显性遗传	在体	NHEJ	开发中	删除致病性的三联体重复序列
神经退行性疾病	包含各种类型	离体或者在体	HDR	概念性阶段	使细胞分泌神经保护因子

* 临床试验的最新信息可以参考 clinicaltrials.gov 网站信息。

注释：HDR = homology-directed repair (同源指导修复)；HSPC = hematopoietic (blood) stem and progenitor cells (造血干/祖细胞)；NHEJ = nonhomologous end joining (非同源性末端连接)；NR = not relevant (不相关)，基因编辑只针对 T 淋巴细胞，目的是杀死癌细胞。

TABLE 4-1 Examples of Potential Therapeutic Applications of Somatic Cell Editing*

Disease	Inheritance/Transmission Pattern	*Ex vivo* or *In vivo*	NHEJ or HDR Mediated Editing	Stage of Development	General Strategy
Sickle-Cell Disease	Autosomal recessive	*Ex vivo* (HSPC)	HDR	Clinical development	Edit to non-disease-causing variant
Sickle-Cell Disease/β-Thalassemia	Autosomal recessive	*Ex vivo* (HSPC)	NHEJ	Preclinical	Induction of fetal hemoglobin
Severe Combined Immunodeficiency X-linked (SCID-X1)	X-linked recessive	*Ex vivo* (HSPC)	HDR	Clinical development	Knock-in of full or partial complementary DNA (cDNA) to correct downstream disease-causing variants
X-Linked Hyper IgM Syndrome	X-linked recessive	*Ex vivo* (T-cells)	HDR	Preclinical–clinical development	Knock-in of full cDNA to correct downstream disease-causing variants
Hemophilia B	X-linked recessive	*In vivo* (liver)	HDR	Clinical trial*	Express clotting factor from a strong promoter
Cystic Fibrosis	Autosomal recessive	*In vivo* (lung)	HDR	Discovery	Edit to non-disease-causing variant
HIV	Viral infection	*Ex vivo* (T-cells and HSPC)	NHEJ	Clinical trial	Engineer resistance to HIV
HIV	Viral infection	*Ex vivo*	HDR	Discovery	Engineer constitutive secretion of anti-HIV factors
Cancer Immunotherapy	NR	*Ex vivo* (T-cells)	NHEJ or HDR	Conceptual through clinical trial	Engineer more potent cancer-specific T-cells by genome editing
Duchenne's Muscular Dystrophy (DMD)	X-linked recessive	*In vivo*	NHEJ	Preclinical	Deletion of pathologic variant to convert DMD to milder Becker's muscular dystrophy
Huntington's Disease	Autosomal dominant	*In vivo*	NHEJ	Discovery	Delete disease-causing expanded triplet repeat
Neurodegenerative Diseases	Various	*Ex vivo* or *In vivo*	HDR	Conceptual	Engineer cells to secrete neuroprotective factors

*Current information on clinical trials is available at clinicaltrials.gov.

NOTE: HDR = homology-directed repair; HSPC = hematopoietic (blood) stem and progenitor cells; NHEJ = nonhomologous end joining; NR = not relevant, the edits are to lymphocytes designed to kill the cancer.

（Dever et al., 2016; DeWitt et al., 2016）的疾病等位基因修改回野生型β血红蛋白表达序列，以及修复导致重症联合免疫综合征的遗传缺陷（Booth et al., 2016）。另一项更微妙的采用基因编辑技术修复致病性突变的研究，是将野生型 mRNA 的互补性 DNA 拷贝或者 cDNA 插入一个内源基因位点，以修复下游基因突变（Genovese et al., 2014; Hubbard et al., 2016; Porteus, 2016）。例如，以肝脏作为靶器官，通过转基因在一部分肝细胞的清蛋白基因启动子下游插入凝血因子，可改善血友病小鼠模型的出血性状。

一些基因疗法的潜在应用需要制造基因破坏，前提是引入核酸酶后不会因细胞毒性或免疫排斥导致被处理细胞的死亡。这些应用包括中断某些神经退行性疾病相关的显性突变，如去除导致亨廷顿舞蹈症的三联体重复序列（Malkki, 2016）；在杜氏肌营养不良症（DMD）中，通过删除或强行跳过带有致病突变的外显子，重建有功能的抗肌萎缩蛋白（Long et al., 2016; Nelson et al., 2016; Tabebordbar et al., 2016）。其他例子还包括通过破坏一个内源基因的抑制子，恢复表达一个胎儿基因，来补偿有缺陷的成体基因，如目前研究者们尝试通过在红细胞谱系里中断 BCL11A 基因的表达，恢复红细胞中胎儿血红蛋白的表达，以补偿重症地中海贫血症中成体β血红蛋白表达的缺失；或者以此抵消镰状红细胞贫血症中镰状β血红蛋白突变产生的影响（Hoban et al., 2016）。在 T 细胞免疫治疗中，基因组编辑一个很有前景的应用是用来破坏 T 细胞中一至数个基因，使它们不会对抗、抵消或抑制引入的外源细胞表面受体的活性，而这些表面受体可以引导 T 细胞去攻击肿瘤相关的抗原（Qasim et al, 2017）。这些策略的实施可以强化当前的细胞免疫治疗，并有希望攻克目前细胞免疫疗法在大多数实体瘤中的应用限制。

基因编辑的设计和应用涉及的科技考量

各种类型的基因编辑都需要考虑特定的参数，而这些参数共同决定了基因编辑工具的效率和潜在的细胞毒性。这些科技方面的考量，不但能够说明如何以及为何要采用某种特殊手段以达到研

change the variant that causes sickle-cell disease back to the sequence that encodes wild-type β hemoglobin (Dever et al., 2016; DeWitt et al., 2016) or correct the deficits in severe combined immune deficiencies (Booth et al., 2016). A more subtle use of genome editing to correct a disease-causing variant is to insert the wild-type DNA copy of the mRNA (complementary or cDNA) into an endogenous locus to correct downstream mutations (Genovese et al., 2014; Hubbard et al., 2016; Porteus, 2016). Concerning the liver as a target organ, it has been shown that targeted insertion of a clotting factor transgene downstream of the promoter of the albumin gene in a fraction of hepatocytes may rescue the hemophilia bleeding phenotype in mouse models (Anguela et al., 2013; Sharma et al., 2015).

Several potential applications of genome therapy entail causing gene disruption, provided that the delivery of the nuclease does not lead to loss of the treated cells because of toxicity or immune rejection. Among these applications are the disruption of dominant mutations and expanded triplet repeats in some neurodegenerative diseases, such as Huntington's disease (Malkki, 2016), and the reconstitution of a functional dystrophin in Duchenne's muscular dystrophy by deletion or forced skipping of the exon carrying the disease-causing mutation (Long et al., 2016; Nelson et al., 2016; Tabebordbar et al., 2016). Other examples include disruption of an endogenous gene repressor to rescue expression of a fetal gene compensating for a defective adult form, as is currently being attempted by disrupting expression of BCL11A in the erythroid lineage; to rescue fetal globin expression to compensate for the lack of expression of adult beta-globin in thalassemia major; or to counteract the sickling beta globin mutant in sickle-cell anemia (Hoban et al., 2016). In T-cell immunotherapy, a promising application of genome editing is single or multiplex disruption of genes that may antagonize, counteract, or inhibit the activity of exogenous cell-surface receptors introduced into T-cells to direct them against tumor-associated antigens (Qasim et al., 2017). These strategies can strongly potentiate current cell-based immunotherapy strategies, possibly overcoming current barriers that limit efficacy in most solid tumors.

SCIENTIFIC AND TECHNICAL CONSIDERATIONS ASSOCIATED WITH THE DESIGN AND APPLICATION OF GENOME-EDITING STRATEGIES

All types of genome editing involve consideration of certain parameters that together determine the efficacy and potential toxicity of a genome-editing tool. These scientific and technical considerations inform how and why a particular approach is chosen to meet a research or therapeutic goal; they also impact the nature of the data that will be available for the regulatory evaluations that

究和治疗的目的，而且它们还能影响监管评价所需的数据性质，而这些评价又是潜在的临床前试验、临床试验、评估和对这些方法的持续监管所需要的。

核酸酶平台的选择

核酸酶的选择包括：选择基于蛋白质-DNA识别（如巨大核酸酶、ZFNs或TALENs）还是基于碱基配对识别（如CRISPR/Cas9）的平台类型，以及选择如何设计和构建靶向特定基因序列的组件。当使用锌指和TAL效应因子制备DNA结合蛋白结构域时，针对每种特异序列的结合结构域均可进行大量工程操作和改进，这就使得对整个平台的性能和特异性难以做出一个广泛适用的综合性预测。也就是说，对于ZFNs和TALENs而言，对一种特异核酸酶的性能（活性和特异性）优化，不一定适用于另一种核酸酶。

与之不同，当基于RNA的核酸酶如Cas9的技术发展起来后，对平台本身的通用改进适用于每一个特定的靶序列。因为不同CRISPR-Cas9系统的唯一主要区别是向导RNA，所以对一个Cas9核酸酶的优化方案通常适用于其他核酸酶。这种特性暗示针对一种临床应用设计的基因编辑系统可以快速简便地应用于其他的病例。

传递策略：离体和在体基因编辑

基因编辑操作途径有离体和在体两种。在离体编辑中，由于先是在实验室对细胞进行操作，所以可在将已编辑细胞施用给患者之前对其进行一系列的检测。然而，由于离体基因编辑是在体外实施，这种方法只适用于某些特定的细胞类型。相比之下，体内编辑适用于其他更多的细胞和组织类型，但为了在原位修饰靶细胞，这种方法需要将基因编辑工具直接注射到患者的体液（如血液）、体腔或器官中，给应用带来了额外的安全和技术方面的挑战。

离体基因编辑将一部分靶细胞分离到体外进行操作，然后将改造后的细胞移植回体内。细胞来源可以是自体的或同种异体的：自体来源的细胞来自于同一个体，而异体来源的细胞来自于免疫匹配的供体。无论细胞来源于患者自身还是免疫匹配的供体，注入患者体内的细胞通常具有干细胞

will be required for potential preclinical testing, clinical trials, review, and ongoing oversight of these methods.

Choice of Engineered Nuclease Platform

The choice of nuclease includes the platform type, which can be based on protein-DNA recognition (e.g. meganucleases, ZFNs, or TALENs) or on nucleic acid base-pairing recognition (e.g., CRISPR/Cas9), and the design and generation of the components that target the intended genomic sequence. When developing protein-based DNA-binding domains that are made using zinc fingers and TAL effectors, extensive engineering and improvement are possible for each specific sequence-binding domain, such that it is difficult to make a general prediction on the performance and specificity of the overall platform. That is, for ZFNs and TALENs, optimization of performance (activity and specificity) often requires work for each nuclease that may or may not translate to another nuclease.

In contrast, when RNA-based nucleases such as Cas9 are developed, general improvements are made to the platform itself and should translate to each specific target sequence. Because the only major difference among CRISPR-Cas9 systems is the targeting guide RNA, optimization of one Cas9 nuclease often will generalize to improved performance of other nucleases. This fact has implications for the ease or speed with which genome-editing systems designed for one clinical application could be adapted to target others.

Delivery Strategy: *Ex Vivo* and *In Vivo* Genome Editing

Genome editing can be carried out ***ex vivo*** or ***in vivo.*** In *ex vivo* editing, it is possible to conduct a number of checks on the edited cells before they are administered to a patient because the cells are first manipulated in the laboratory. *Ex vivo* editing, which occurs outside the body, is suitable only for certain cell types, however. By contrast, *in vivo* editing allows other types of cells and tissues to be edited, but poses additional safety and technical challenges because it involves administering the genome-editing tool directly into a patient's body fluids (e.g., blood), body cavities, or organs in order to modify targeted cells in situ.

Ex vivo genome editing can be performed by isolating and manipulating a population of the intended target cells outside the body and then transplanting those cells into an individual. The source of cells can be autologous or allogeneic: autologous cells are derived from the same individual, while allogeneic cells are derived from an immunologically matched donor. Whether the cells are sourced from the same patient or a matched donor, the administered cells often have stem-cell-like properties, which may allow their self-renewal and long-term maintenance *in vivo*, as well as repopulation of the treated tissue with their genetically modified progeny. In some

特性，这可以允许其在体内进行自我更新和长期维持，并分化产生可以重建所治疗组织的细胞。在一些情况下，可以在移植回患者体内之前将这些细胞在体外诱导分化为特定类型的细胞或细胞谱系。此外，被编辑细胞可以是已分化的体细胞，如对存活时间或长或短的免疫效应细胞进行离体扩增和遗传修饰，来增强它们对抗肿瘤或病原体的活性。目前人们已对多种体细胞进行了分离、遗传修饰和移植，包括造血干细胞和前体细胞、成纤维细胞、角质细胞（皮肤干细胞）、神经干细胞及间充质基质/干细胞。随着科学知识和技术的进步，治疗所涉及的细胞种类很可能会继续增加，从而扩大离体基因编辑可能的潜在应用。在体基因编辑中，需要将编辑组件包括切割DNA的核酸酶，以及将编辑定位于基因组特定位置的向导RNA（以CRISPR/Cas9为例）导入到细胞中。如果要使用同源定向修复，则还需要一个同源重组模板。在体基因编辑的靶细胞可以是长时间存活的组织特异性细胞，如肌纤维细胞、肝细胞、中枢神经系统神经元或视网膜感光细胞，也可以是稀少的组织特异性干细胞、其他难以获取和移植的细胞类型。然而，相较于离体基因编辑，在体基因编辑在很多技术层面仍存在很大挑战，包括如何有效地将基因编辑工具导入到体内正确的细胞中、如何确保基因组的靶向位点被成功编辑，以及如何将由于脱靶带来的错误降到最低等。

其他注意事项

一系列与离体和在体基因编辑相关的其他科技考量促进了人基因编辑体系的发展。

相关类型细胞的分离

为了进行离体基因编辑，首先需要从特定组织中分离，或通过多能性干细胞诱导产生相关类型细胞，然后在离体实验条件下对细胞进行扩增和修饰，最后再将细胞移植到患者体内，使它们能整合和/或发挥其特定的生物学功能。离体基因编辑有几个优点：只有靶向细胞暴露在基因编辑试剂中；有多种可供选择的、适用于不同类型细胞的试剂递送平台；细胞用于治疗前可对其进行鉴定、纯化和扩增。目前，这种操作仅适用于少数细胞类型，包括最终会分化为皮肤、骨骼、肌肉、血液和神经元的细胞。

approaches, the cells can be treated in culture to induce commitment or differentiation toward a desired cell type or lineage before being administered to the patient. Otherwise, the edited cells can be differentiated somatic cells, such as short-lived or long-lived immune effector cells that are expanded and genetically modified *ex vivo* to enhance their activity against a tumor or infectious agent. Several somatic cell types have been isolated, genetically modified, and transplanted, including blood-forming hematopoietic (blood) stem and progenitor cells, fibroblasts, keratinocytes (skin stem cells), neural stem cells, and mesenchymal stromal/stem cells. This list likely will grow as scientific knowledge and techniques improve. An expanded repertoire of cell types has the potential to increase the range of possible *ex vivo* genome-editing applications.

In *in vivo* genome editing, the editing machinery that needs to be delivered to the cells includes the nuclease that cuts the DNA and, in the case of CRISPR/Cas9, the guide RNA that targets the editing to a specific genomic location. If HDR is intended, a homologous template is also required. Target of *in vivo* genome editing may include long-lived tissue-specific cells, such as muscle fibers, liver hepatocytes, neurons of the central nervous system, or photoreceptors in the retina, but may also include rare, tissue-specific stem cells and other types of cells that cannot easily be harvested and transplanted. Relative to *ex vivo* approaches, however, *in vivo* approaches pose greater challenges with respect to efficient delivery of the genome-editing machinery to the right cells in the body, ensuring that the correct location in the genome has been successfully edited, and minimizing errors resulting from off-target editing.

Additional Considerations

A number of additional scientific and technical considerations related to both *ex vivo* and *in vivo* genome editing inform the development of human genome-editing systems.

Ability to Isolate the Relevant Cell Types

To carry out *ex vivo* genome editing, it is necessary first to isolate the relevant cell types from an appropriate tissue source or to generate them from pluripotent stem cells, and then to grow and modify them *ex vivo* and finally administer them to the patient so that they can engraft and/or deliver the intended biological activity. There are several advantages to the *ex vivo* strategy: only the intended cells are exposed to the editing reagents, there is a wide choice of delivery platforms that can best be fitted to each cell type and application, and it is possible to characterize and even purify and expand the edited cells before administration. Currently this process has been established for only a few cell types, including cells that

随着能分离的原代细胞和多能性干细胞诱导分化细胞种类的增加，离体增殖和成功将这些细胞移植回患者体内的技术增强，离体基因编辑潜在的应用范围将进一步扩大。

离体基因编辑策略有许多离体细胞培养中常见的短板，包括需要长期体外培养、从少量甚至单个起始细胞扩增，两者均有风险，可能会积累突变，导致细胞增殖能力的下降。这些问题与基因编辑尤其相关，因为引发基因编辑所必需的双链DNA断裂可能触发细胞凋亡（细胞死亡）、分化（细胞类型改变）、细胞衰老（老化）和增殖停滞（细胞停止分裂）等细胞反应，而所有的类似反应对细胞扩增和干性维持都是不利的。

由于大多数治疗应用都需要注入大量的细胞，这些短板成为离体基因编辑的主要应用障碍。要克服这些障碍，人们需要进一步完善细胞培养体系，增进对基因随机突变累积导致的安全风险的理解，以及找到可靠评估这些问题的方法。另外，在利用多能性细胞体外诱导分化为治疗用细胞的过程中，对细胞分化命运的控制及不同分化阶段细胞的纯化也是应用前需解决的问题，如果将未成熟细胞移植到体内，会存在诱发肿瘤或不能成功整合到受体组织的风险。尽管存在上述短板，离体基因编辑仍具有明显的优点，可在将细胞移植到患者体内之前对所需修改进行选择并验证其准确性。

控制基因编辑工具在体内的分布

在体基因编辑需要考虑的其他因素与负责递送编辑元件平台的选择有关，因为这种选择影响了基因编辑工具的作用范围、作用时间及体内的生物分布。这一考量对潜在的效率、急性和慢性毒性、免疫原性，甚至是否存在对生殖细胞无义编辑的风险具有重要影响。对靶向基因位点的有效编辑通常需要细胞内核酸酶的高表达，不过这种高表达可以是瞬时的，以防止过高的毒性和脱靶活性。尽管在体外培养细胞中做到短时高表达基因编辑核酸酶相对容易，但要想在体内做到这点则更具挑战性。最后，在体内环境下进行基因编辑可能会导致生殖细胞或原生殖细胞发生意外的（非故意的）基因修饰。因此，临床前在体基因编辑的发展应该考虑到会导致可遗传给子孙后代的生殖细胞修饰的风

will eventually give rise to skin, bone, muscle, blood, and neurons. The range of possible *ex vivo* genome-editing applications will expand with the development of scientific knowledge about how to isolate additional primary cell types and derive other cell types from pluripotent cells, grow the cells *ex vivo*, and ultimately successfully and safely transplant them back into patients.

Ex vivo genome-editing strategies, have a number of expected limitations, which are common to all attempts at culturing cells *ex vivo*. These limitations include the need for prolonged culture and expansion from a few cells or even a single founder cell, both of which entail the risk of accumulating mutations, as well as incurring replicative exhaustion. This issue is particularly relevant for genome editing because inducing double-stranded breaks to DNA, as is required for initiating the process, may itself trigger such cellular responses as apoptosis (cell death), differentiation (changing cell type), cell senescence (aging), and replicative arrest (cells stop dividing). All of these cellular responses are detrimental to cell expansion and maintenance of pluripotency.

These limitations represent significant hurdles to *ex vivo* genome editing because most therapeutic applications require substantial numbers of cells for infusion. Overcoming these hurdles will require better ways to culture cells, better understanding of the safety risks associated with genomic accrual of random mutations in these settings, and reliable assays for assessing such events. Additional hurdles are the ability to fully control the commitment and differentiation of cells in culture and their purification from the source pluripotent cells. This is an important consideration because administration of immature cells may be associated with a risk of tumorigenesis or failure to integrate functionally within the tissue. Despite these limitations, *ex vivo* genome editing has the advantage that cells with the desired alteration can be selected and the accuracy of the alterations validated before transplantation to the patient.

Ability to Control Biodistribution of the Genome-Editing Tool

Additional considerations for *in vivo* genome editing are linked to the choice of the delivery platform for the editing machinery because this choice impacts the extent, time course, and *in vivo* biodistribution of the genome-editing tool. This consideration has major implications for potential efficacy, acute and long-term toxicity, and immunogenicity and even the risk of unintentional editing of germ cells. Efficient editing of the intended genomic site usually requires a high level of intracellular nuclease expression, even though this often can be for only a short time to prevent excess toxicity and off-target activity. Whereas short-term, high expression of the genome-editing nuclease can be obtained relatively easily for

险，并在临床试验的患者中将这种潜在风险降到最低。

一般来说，如果可以证明基因编辑工具在被编辑细胞中无残留，在注入时不以活性形式被引入，则被离体基因编辑过的细胞导致生殖系传递的风险很低。在这些情况下，可能没必要对生殖系传递进行非临床研究。另一方面，对于在体注入基因编辑工具，则需要对其在生殖腺中的潜在生物学分布和对生殖细胞基因组的基因编辑活性进行评估。这些参数将会极大地受到递送基因编辑工具的平台的选择、递呈时机及路线的影响。当使用病毒载体递呈核酸酶时，临床前研究可能需要顾及通过动物和人类研究积累起来的关于这些病毒载体是否能接触到生殖系细胞的知识。在类人猿等动物模型上进行的临床前研究，可用来监测在体内的生物学分布和核酸酶在包括生殖腺在内的脱靶组织细胞中的活性。在非临床模型中研究潜在的生殖系传递建议依据决策树的方法：一个阳性发现触发下一阶段的研究。例如，先研究基因编辑工具和/或证明它活性的基因标记（插入缺失突变，indels）是否存在于生殖腺中，然后确定它们是否存在于判断为阳性的生殖腺中分离出的生殖细胞中，以及这种发现是偶然的还是持续出现，最终分析上述遗传修饰是否能够传递到可生育的实验动物后代中。可以通过设计分子实验来追踪预期或替代的核酸酶酶切位点上插入缺失突变的发生，前提是被研究物种基因组中存在上述位点，并与核酸酶有足够的亲和力使结果在实验的灵敏度范围内。就像已经在几种基因治疗产品中发现的那样，在替代动物中进行该研究时存在诸多局限，包括已有方法灵敏度过低、载体生物分布，以及能否接触到生殖腺细胞上存在物种特异性差异，通常测试雌性生殖系传递比测试雄性困难。由于存在上述局限性，因此无论非临床生物分布数据的结果如何，通常都推荐正在接受在体基因治疗临床试验的患者采取避孕措施（如果有意义或适用），避孕持续时间至少为预期清除体液中所给予的载体所需要的时间，并通常延长至包括至少一个精子发生周期，在男性体内为64~74天。在此期间，可在不同的时间点进行精液测试，如果结果为阳性，则应继续避孕并通知相应的监管机构。另外，目前尚没有针对女性生殖系传递的非侵入性监测手段。

cells cultured *in vitro*, it is more challenging *in vivo*. Finally, in an *in vivo* setting there could be unintentional (inadvertent) modification of the germ cells or primordial germ cells; therefore, pre-clinical development of *in vivo* editing should address the risk of modification of germ cells resulting in heritable changes that could be passed along to future generations and minimize this potential risk in humans enrolled in clinical trials.

In general, the risk of germline transmission associated with the administration of *ex vivo* gene edited cells is likely to be low if one can show that the editing reagents do not remain associated with the treated cells and are not shed in active form at the time of administration. In these conditions, non-clinical studies of germline transmission may not be necessary. On the other hand, *in vivo* administration of editing reagents would require assessment of their potential biodistribution to the gonads and activity on germ cell genomes. These parameters will be strongly influenced by the delivery platforms used and the timing and route of delivery. When viral vectors are used to deliver the nuclease, the preclinical studies might take into consideration accumulating knowledge from animal and human studies concerning the potential of these vectors to reach germline cells. Preclinical studies in animal models such as non-human primates could be designed to monitor both the biodistribution of the vector/vehicle as well the activity of the nuclease in cells from off-target tissues, including the gonads. A suggested approach to studying the potential of germline transmission in such nonclinical models can follow a decision tree, in which a positive finding triggers the next level of investigation. One could start from investigating the presence of the reagent and/or genomic signs (indels) of its activity in the gonads, then identifying their actual occurrence in germ cells isolated from the positive gonads, the transient or sustained occurrence of this finding and, eventually, the transmission of the genetic modification to the viable progeny of the treated animals. Molecular assays could be designed to track the occurrence of indels at the intended or surrogate nuclease target sites, provided that such sites exist in the genome of the species used for the study with sufficient affinity for the nuclease to support the sensitivity of the assay. Many limitations exist when conducting such studies in surrogate animal species, as already discovered for several gene therapy products, including the low sensitivity of the available assays, species-specific differences in vehicle biodistribution and access to the gonadal cells, and the general difficulties of testing transmission to the female vs. male germline. Because of these limitations, regardless of the outcome of non-clinical biodistribution data, contraceptive measures are usually recommended (if meaningful or applicable) for patients undergoing *in vivo* gene therapy clinical trials at

最小化转染载体或基因组编辑蛋白引起的免疫反应

目前，基因编辑相关蛋白和核苷酸的体内导入主要通过以下两种方式来完成。第一种方式以化学共轭物为基础（脂类和／或糖类复合物）。尽管这种方法已经在肝细胞等特定细胞类型的基因编辑中取得了一定效果（Yin et al.，2014），但在许多组织细胞中仍存在基因表达时间短且相对低效的缺点，而且这种方式可能会使患者体内与治疗无关的细胞受到来自核酸酶的潜在细胞毒性损害。另一种方式通过病毒载体使蛋白质稳定地、组织特异性地表达，但这种方法导入的因子表达时期长，引起免疫反应的可能性较大。例如，自我互补型的 rAAV8 载体（scAAV）就被改造为一种介导工程核酸酶持续表达的工具。持续性的核酸酶表达会增加产生 DNA 损伤和基因毒性的风险，以及后继的患者细胞大面积（尽管可能很慢）死亡和恶性转化的风险。此外，目前所有的基因编辑工具中都含有衍生于常见微生物病原体的蛋白元件，这有可能引起被治疗个体的初级或次级免疫反应。如同在病毒载体介导的治疗研究中充分报道的一样，在治疗中病毒载体蛋白可能引起免疫应答，从而导致被编辑细胞快速而完全的清除，使治疗无效化。机体原始免疫和抗原表达的时长及程度的提升均可加大编辑后细胞被清除的风险。

非分裂细胞中的基因组编辑

离体和在体基因编辑的另一巨大障碍在于向神经元之类的有丝分裂后细胞中靶向插入一段 DNA 序列的做法并不可行，因为有丝分裂后细胞中同源重组活性极低或完全丧失。相反，非同源性末端连接（NHEJ）在非分裂细胞中同样具有活性，所以主要通过在非分裂细胞中制造插入缺失突变来使基因失活。然而，通过对方法的调整，非同源性末端连接（NHEJ）也可用于在特定位点的基因插入（例如，Maresca et al.，2013；Suzuki et al.，2016a）。据最新报道，有一种不依赖于同源性的基因靶向插入（HITI，即 Homology-Independent Targeted gene Integration 的缩写）的技术已经在体外和体内试验中成功应用于分裂细胞（如干细胞），以及最重要的是用于非分裂细胞（如神经元）中 DNA 序列的靶向敲入（Suzuki et al.，2016a）。

least for the expected time of clearance of the administered vector/vehicle from the body fluids, and usually extended to encompass at least one cycle of spermatogenesis, which is approximately 64-74 days in men. Testing of semen can be done at various time points during this time interval and, if samples are positive, should continue and the respective regulatory authorities be notified. On the other hand, there are currently no non-invasive means to monitor women for germline transmission.

Ability to Limit Immune Response to Delivery Vectors or Genome-Editing Proteins

In vivo delivery of proteins and nucleic acids is currently done with either of two types of platforms. The first is based on chemical conjugates (lipo- and/or glyco- complexes) that provide short-lived but relatively inefficient expression across multiple different tissue types, although advances have been achieved in targeting specific cell types, such as liver (Yin et al., 2014). This approach can expose therapeutically irrelevant cell types in the patient to the potential toxicity of the nuclease. The second type of platform relies on viral vectors that can provide robust and tissue-specific expression, but they also are frequently long-lived and more likely to provoke an immune response. Self-complementary rAAV8 vectors (scAAV), for example, have been shown to mediate continued expression of the engineered nuclease. Sustained nuclease expression increases the risk of DNA damage and genotoxicity, with subsequent potential risk of widespread (albeit possibly slow) cell death or malignant transformation of the patient's cells. Moreover, all current formulations of editing machinery contain elements that are derived from proteins of common microbial pathogens, which could trigger primary or secondary immune responses in treated individuals. As has been well documented in viral gene therapy studies, immune recognition of viral vector proteins may lead to rapid and complete clearance of cells that have received the editing machinery, which eliminates the benefit of the treatment. The risk of clearance of the edited cells is exacerbated by preexisting immunity and by the extent and duration of expression of the antigen.

Ability to Make Genome Edits in Nondividing Cells

Another major hurdle for both *ex vivo* and *in vivo* editing is that targeted insertion of a DNA sequence into postmitotic cells, such as neurons, is not feasible because of their low or absent homologous recombination activity. In contrast, NHEJ, which is active in nondividing cells, has been harnessed mainly for the generation of indels to inactivate a gene. However, NHEJ can, with modifications to the methods, be used to generate site-specific gene insertions (e.g., Maresca et al., 2013; Suzuki et al., 2016a). Most recently, it was reported that one of these methods, Homology-Independent Targeted gene Integration, or

在一些小鼠模型中显示出其可行性及潜在的治疗性作用后，在体基因编辑的应用被寄予厚望，然而，至少就目前的应用方式来看，将其转化为临床应用仍存在巨大挑战。再者，考虑到动物模型中免疫应答众所周知的难预测性，稳定的核酸酶表达可能仍然不是临床应用的首选途径，尽管一些动物模型对其表现出很好的耐受性。

关于基因组编辑的活性和特异性评估

每一种靶向核酸酶都能通过对 DNA 的切割效率和特异性来表征。切割效率能通过对打靶位点测序，相对容易地检测出来。特异性则反映了核酸酶上靶活性与脱靶活性的比值。检测特异性的手段同样有多种，各具优缺点（详见附录 A）。虽然全基因组测序是分析单细胞和克隆的黄金标准，但这种测序的深度还不足以用来评估细胞群体的脱靶谱。

人类基因组编辑的脱靶率与自然突变率的比较

需要特别注意的是，要对基因组编辑方法的特异性进行精确评估，就必须将在基因组编辑过程中产生的突变与整个生命期间自发发生的突变区分开来。正常基因组复制的自然错误频率在基因组不同位点之间是不一样的，但平均每个碱基在每轮 DNA 复制中的错误频率约是 10^{-10}。人类每个细胞约含有 60 亿个 DNA 碱基对，所以即便在自然错误率非常低的情况下，平均每轮的细胞复制将能产生大约一至数个原发（de novo）突变。因此，随着细胞的增殖，细胞中的碱基突变数量以这样的速率自然地累积着。除了这种背景突变频率以外，很大一部分的 DNA 损伤是来源于正常环境的直接接触，如环境辐射、氧化应激及其他 DNA 损伤剂。虽然对于基因组编辑用的核酸酶所产生的突变率与自发突变率之间还没进行过直接比较，但是这类分析的结果很可能会取决于所研究核酸酶的特定种类，以及检测用的细胞类型。核酸酶技术的错误率将被继续改进，如果还没有，那么在将来的某个时候，其错误率将低于自发突变频率。

检测不同递送平台下的效率与特异性

在基因编辑应用中，包含核酸酶和靶向序列的系统必须被传送到细胞内。由于递送平台的选择决

HITI, allows targeted knock-in of DNA sequences in dividing cells (e.g., stem cells), and most importantly, in non-dividing cells (e.g., neurons) both *in vitro* and *in vivo* (Suzuki et al., 2016a).

In vivo gene editing is a highly sought-after application that has been shown to be feasible and potentially therapeutic in some mouse models. Substantial challenges to its translation to the clinic remain, however, at least in the current modalities of administration. Also considering the well-known difficulties in predicting immune response in animal models, stable expression of nucleases, despite being apparently well tolerated in some animal models, may not be the preferred route to clinical development.

Assessment of the Activity and Specificity of Genome Editing

Each targeting nuclease can be characterized by the *efficiency* and *specificity* of DNA cleavage. Efficiency can be relatively easily measured (by sequencing the targeted site. Specificity reflects on-target versus off-target site activity, which also can be measured by various assays, each with advantages and disadvantages (see Appendix A for details). While whole-genome sequencing could be the gold standard for analyzing single cells or clones, the depth of this sequencing is not sufficient to assess the off-target spectrum in populations of cells.

Comparing Off-Target Editing Rate with the Natural Mutation Rate of the Human Genome

It is important to note that accurate assessment of the specificity of a genome-editing approach requires that mutations created by the genome-editing process be distinguished from those that occur spontaneously throughout a life span. The natural error frequency of normal genome replication varies among sites in the genome but is approximately 10^{-10} per base per round of DNA replication. Because each human cell contains approximately 6 billion DNA base pairs, even the naturally low error rate means that DNA replication can be expected to generate, on average, approximately one or a few de novo mutations in each round of cellular replication. Thus as cells proliferate, they naturally accumulate mutations at this rate. In addition to this background mutation frequency, a significant amount of DNA damage results from normal environmental exposures such as radiation, oxidative stress, and DNA-damaging agents in the environment. A direct comparison between the mutation frequency generated by a genome-editing nuclease and the spontaneous mutation frequency has not yet been conducted, but results from this type of analysis are likely to depend on the specific nuclease in question and on which cell type is examined. The error rate of nuclease technologies continues to improve and

定了基因编辑系统的表达范围、水平和时长，因此它不仅会影响在给定实验条件下基因编辑的效率和特异性，还进一步决定了其毒性和免疫原性谱。此外，所选递送平台（DNA、RNA或者蛋白质形式；递送机制）的一些固有特征也影响了其潜在毒性（参见附录A中的表A-1）。这些毒性效应通常是由靶向细胞针对外源分子的先天免疫反应所造成，而且DNA，尤其是DNA质粒，常常会比RNA和蛋白质引起更剧烈的免疫反应。针对病毒的先天免疫反应可能随病毒和细胞的类型不同而有所不同：通常先天免疫体系对人类体细胞中的腺相关病毒和慢病毒的反应非常弱（Kajaste-Rudnitski and Naldini, 2015），但某些类型的免疫细胞除外，如树突状细胞和巨噬细胞，这类细胞具有大量齐全的内置病毒感应器，从而能够引发干扰素分泌和炎症反应（Rossetti et al., 2012）。另外，试剂的纯度和组成（质粒还是线性化DNA、RNA中的碱基突变，以及是否用高效液相色谱组分[HPLC]纯化组分）也可以起到很重要的作用。

最后，预期的靶序列和相关序列在基因组中出现的频率，以及靶向位点处的局部染色质环境，也能影响基因编辑方法的效率和特异性。上述所有的因素都可能随着所处理细胞的类型和操作方式（离体还是在体）的差异而产生不同的影响。此外，上靶和脱靶的比率也会受到靶向细胞类型的固有生物学表征的影响，包括在细胞周期不同阶段、DNA损伤反应及修复能力的差异。

评估效率和特异性的临床前研究

在人类基因编辑应用的发展阶段，临床前研究被用来确定每种核酸酶编辑系统的活性和特异性。这类临床前研究的设计受靶细胞和实验条件选取的影响，故而实验结果应该被视为相对，而不是绝对有用的。另外，必须附加说明的是，大多数这类临床前研究是在大量细胞群体的基础上测量活性和特异性的，这些细胞的核酸酶表达情况并不均一。因为上靶与脱靶活性之间的比率会随着核酸酶表达水平的差异而有所不同，所以对于那些核酸酶表达水平更高的细胞来说，上靶与脱靶活性比率应该是不太有利的，这是因为在此情况下，上靶活性会被饱和，而在脱靶位点的活性会变得更加明显。另一方面，对于那些核酸酶表达水平较低的细胞，因为活

may at some point, if it is not already, be less than the spontaneous mutation frequency.

Measuring Efficiency and Specificity for Each Delivery Platform

For genome-editing applications, the system (nuclease and targeting sequences) must be delivered inside cells. Because the choice of delivery platform determines the extent, level, and time course of expression of the genome-editing machinery, it affects the efficiency and specificity displayed in a given set of experimental conditions and furthermore determines the toxicity and immunogenicity profile. In addition, several intrinsic features of the chosen delivery platform (DNA, RNA, or protein; delivery mechanism) also influence its potential toxicity (see Table A-1 in Appendix A). These effects usually are due to normal innate target-cell responses to exogenous molecules, and they often are stronger for DNA especially DNA plasmids than for RNA or proteins. The innate responses to viruses may vary with virus and cell type: usually they are very low for AAV or lentivirus in human somatic cells (Kajaste-Rudnitski and Naldini, 2015), with the exception of some immune cell types, such as dendritic cells and macrophages, which have a large complement of built-in viral sensors and may trigger interferon and inflammatory responses (Rossetti et al., 2012). The purity and composition of reagents (plasmid versus linearized DNA, mutant bases in RNA, high performance liquid chromatography [HPLC] purification of components) also can play a significant role.

Finally, the frequency with which the intended target sequence and related sequences occur in the genome and the local chromatin environment at the target site also can influence the efficiency and specificity of a genome-editing approach. All the factors mentioned above are likely to vary according to the treated cell type and modality (*ex vivo* versus *in vivo*). Moreover, the ratio of on- to off-target activity also is affected by the intrinsic biology of the targeted cell type, including differences in cell-cycle status, DNA-damage responses, and repair capability.

Preclinical Studies to Assess Efficiency and Specificity

In the development of human genome editing applications, preclinical studies are undertaken to establish the activity and specificity of each editing nuclease system. The design of these preclinical studies is influenced by the choice of target cells and experimental conditions, and the results should be viewed as providing relative rather than absolute values. An additional caveat is that most of these preclinical studies measure nuclease activity and specificity over a large population of cells, among which nuclease expression will vary. Because the ratio of on- to off-target activity also varies with nuclease

性主要体现在上靶位点，所以会表现出更有利的活性比率。这一方面的考量表明，上靶和脱靶比率的剂量依赖性应当作为验证基因编辑方法过程的一部分予以考虑。

随着科技的进步，对核酸酶特异性的评估手段将被不断改进。从攥写本文（2016年底）时的业务角度来看，以下是进行此项评估的几项合理方法：

- 使用生物信息学和无偏向性的筛选来识别潜在的脱靶位点（参见附录A）。
- 使用对细胞系和原代靶细胞类型的深度测序来确定在上靶和脱靶位点的插入缺失突变频率（验证）。
- 评估已验证的脱靶位点的潜在生物学效应，去掉那些能在预测得具有生物学效应的位点产生脱靶活性的核酸酶。值得注意的是，迄今为止，所鉴定到的大多数脱靶位点都位于基因组的非蛋白编码区，这使得它们的功能重要性难以被评估。
- 使用能够检测全染色体完整性的方法，如核型分析、单核苷酸多态性（single-nucleotide polymorphism，SNP）阵列，以及染色体易位测定。但是这些分析方法因为灵敏度相对较低，应用上有局限性。
- 使用感兴趣的目的细胞的多种功能验证来检验克隆优势的风险，并评估基因编辑制造过程的真实可行性、效率及毒性。

值得注意的是，在开发用于临床使用的基因组编辑方法时，进行全面的、足够灵敏的和特异性的研究，以求捕获所有可能的脱靶位点，可能是没必要或者说不可行的。例如，正在进行的标准基因治疗已经表明，无法控制的慢病毒插入相对于双链断裂的非同源性修复来说能造成更具破坏性的改变，但它在一些细胞类型和组织中却相对安全并被良好地耐受，即便在引入大量插入（每个患者插入高达 10^8 或者 10^9 个拷贝数）时也是如此。更进一步的考量是脱靶活性具有序列依赖性。大多数早期的旨在确认靶向效率和特异性的临床前实验都是在非人类生物尤其是小鼠中进行，然而，人类和小鼠的基因组有着极大差异，因此，在小鼠或者其他啮齿类动物基因组中对工程核酸酶特异性的评估结果，对

expression level, the cells with higher nuclease expression may have a less favorable ratio since the on-target activity will saturate, while activity at off-target sites becomes more evident. On the other hand, cells with lower expression may exhibit a more favorable ratio because activity is evident mainly at the intended target site. This consideration suggests that the dose dependence of on- and off-target rates be considered as part of the process of validating a genome-editing approach.

Assessment of nuclease specificity will continue to evolve as scientific knowledge and techniques improve. From an operational standpoint as of this writing (late 2016), however, the following represents a reasonable approach to conducting this assessment:

- Use both bioinformatics and unbiased screens to identify potential off-target sites (see Appendix A).
- Use deep sequencing of both cell lines and the primary target cell type to determine the frequency of indels at both on- and off-target sites (validation).
- Evaluate validated off-target sites for potential biological effects, and eliminate nucleases that generate off-target activity at sites that could be predicted to have biologic effects. It should be noted that most off-target sites identified to date lie in non-protein coding regions of the genome, making their functional importance difficult to assess.
- Use assays that measure gross chromosomal integrity, such as karyotyping, single-nucleotide polymorphism (SNP) arrays, and translocation assays. These assays are limited in being relatively insensitive.
- Use diverse functional assays of the target cells of interest to measure risk of clonal dominance and to assess the actual feasibility, efficiency, and toxicity of the genome-editing manufacturing process.

It is important to note that to develop a genome-editing approach for clinical use, it may not be necessary or feasible to conduct comprehensive efficiency and specificity studies performed at high-enough sensitivity to capture all possible off-target edits. Ongoing work in standard gene therapy, for example, has indicated that uncontrolled lentiviral insertions, which cause even more disruptive changes than non-homologous repair of a double-strand break, may be relatively safe and well tolerated in several types of cells and tissues. This is true even when large numbers of insertions (up to 10^8 or 10^9 per patient) are introduced. A further consideration is that the off-target activity is dependent on the sequence. Much of the early preclinical testing aimed at establishing targeting efficacy and specificity has been carried out

于预测该方法在人类中运用的效果作用有限。

总结

总的来说，基因编辑已被整合到体细胞基因治疗方法中，并且这种应用有着上升趋势。基因编辑策略与其他治疗方法存在着竞争关系，这些治疗方法包括：小分子疗法；生物制剂治疗；竞争最激烈的基因编辑以外的基因治疗方法，如用于基因置换的慢病毒载体和重组腺相关病毒载体。因此，各种治疗策略最终将需要在效力、风险、成本和可行性方面进行相互评估。

体细胞基因编辑产生的伦理和监管问题

在大多数方面，体细胞基因编辑的开发得益于基因治疗强大的技术知识基础，并由现行的监管体系和伦理规范监督。这些监管体系和伦理规范在全世界范围内促进了目前体细胞和基因治疗在研究和临床上的应用，其中包括澳大利亚、中国、欧洲各国、日本和美国（见第2章）。这些管理体系包括各式各样的临床前模型和研究设计，用以支持应用编辑过的细胞进行治疗的临床开发，以及从人类临床试验到最后市场营销的路线图。

美国的监管系统

如第2章所述，在美国开展体细胞基因编辑的临床试验需要向食品药品监督管理局（Food and Drug Administration，FDA）递交研究性新药（IND）应用申请并获得批准，其临床方案将需要得到机构审查委员会（IRB）的批准并接受其持续的审查（FDA，1993）。另外，美国国家卫生研究院（National Institutes of Health，NIH）的重组DNA顾问委员会（RAC）针对该临床试验会撰写评述，报道FDA和IRB的相关评审意见，为公众讨论提供平台。正如第2章所述，尽管在研究到何种阶段细胞疗法方可市场化、何种状况下可以撤回产品的规定上存在差异，世界上其他国家也有着类似的监管程序。

临床使用是否被批准很大程度上取决于确定当按照标示和预期来使用时，何时疗效将高于风险（Califf，2017）。人们越来越多地通过一个有组

in nonhuman organisms, especially mice. However, the genomes of humans and mice are sufficiently divergent that assessment of the specificity of engineered nucleases in the genomes of mice or other rodents may have somewhat limited predictive power for the same genome-editing approach in humans.

Summary

In summary, genome editing is already being incorporated into somatic gene therapy approaches, and such applications are likely to increase. Genome-editing strategies are in competition with other therapeutic approaches, including small molecule therapies; biologics; and most notably other gene therapy approaches, such as lentiviral vectors and rAAV vectors used for gene replacement. In the end, therefore, each strategy will need to be evaluated against the others in terms of efficacy, risk, cost, and feasibility.

ETHICAL AND REGULATORY ISSUES POSED BY SOMATIC CELL GENOME EDITING

In most respects, somatic cell genome editing will be developed with the benefit of gene therapy's robust base of technical knowledge, and within the existing system of regulatory oversight and ethical norms that have facilitated the current research and clinical development of somatic cell and gene therapy around the world, including Australia, China, Europe, Japan, and the United States (see Chapter 2). These regulatory systems include a wide range of preclinical models and study designs to support the clinical development of therapies based on edited cells, as well as a roadmap for first-in-human clinical testing and eventual marketing.

Regulatory Oversight in the United States

As described in Chapter 2, clinical testing of somatic cell genome editing could not begin in the United States without the Food and Drug Administration's first having approved an Investigational New Drug (IND) application, and the clinical protocol would require IRB approval and ongoing review (FDA, 1993). In addition, the reviews by the National Institutes of Health's Recombinant DNA Advisory Committee (RAC) inform the deliberations of the FDA and IRBs, and provides a venue for public discussion. Other countries have similar pathways, as described in Chapter 2, albeit with some variations in the stage of research at which a cell-based therapeutic can be marketed and the terms under which it can be withdrawn.

The question of approval for clinical use hinges largely on identifying when benefits may be expected to outweigh risks when used as labeled and as intended (Califf, 2017). Clinical trial data is increasingly reviewed within

织的机构来评估临床实验数据以确定临床需求、替代治疗方案、不确定的领域和风险管理渠道[44]。FDA 前任专员 Robert Califf 说道："FDA 产品审查小组必须权衡科学和临床证据，考虑利益相关者和公众在看待疗效和对治疗风险的容忍性上有观点上的冲突。他们必须综合考虑替代疗法的存在性和有效性、疾病的严重程度、受影响患者的风险耐受程度，以及上市后的数据是否具有提供更多认知的潜能。FDA 的决策需要在高质量的证据和尽早的应用、利益和风险、保护美国民众利益和鼓励可能改善健康的创新之间寻求合适的平衡点"（Califf，2017）。

　　一种基因疗法能否被批准也可能取决于在它进入临床使用后，它的治疗风险和治疗效果能否被细致地监测。在这点上，FDA 发布了一个有影响力（但没有约束力）的基因疗法临床试验指南，该指南与基因编辑临床试验有关（FDA，2006）。基因疗法并不总是需要长期随访的，例如，载体序列，整合和潜伏性相关的临床前数据证明了它们的长期风险非常低。但是当存在长期风险时，"基因治疗临床试验则必须进行长期随访观察，以减轻那些风险"。如果没有这样的长期随访观察计划，风险将会过高，（很可能）临床试验将不会获得批准。对于有必要进行长期随访的试验，指南建议试验后要有 15 年的接触、观察和身体检查（但如果出于诸如载体的持久性，或受试者预期只能短期生存等考量，这个随访期限可以适当缩短）。在招募受试对象时，受试者必须是自愿的，且知情同意长期随访，虽然他们具有随时退出的权利，但还是希望他们能遵守协议。

　　基因疗法一旦获得 FDA 批准用于特定人群和适应证，它将受到上市后的监测和对不良事件的报道。如果基因疗法显示出不安全性或无治疗效果，将会被给予特殊警告或被要求完全撤回产品。此外，如果基因疗法有着可能会导致审批被废除的重大安全隐患，则可能需要市场风险评估和缓解措施（REMS），比如要求医生技能娴熟或要求患者进行登记注册。

a structured framework that identifies need, alternatives, areas of uncertainty, and avenues for risk management.[44] According to former FDA Commissioner Robert Califf, "FDA product review teams must weigh scientific and clinical evidence and consider conflicting stakeholder and societal perspectives about the value of benefits and the tolerability of risks. They must consider the existence and effectiveness of alternative treatments, disease severity, risk tolerance of affected patients, and potential for additional insight from postmarket data. Such decisions require seeking the appropriate balance between high-quality evidence and early access, between benefit and risk, between protecting the US public and encouraging innovation that may improve health outcomes" (Califf, 2017).

Approval of a gene therapy may depend upon how carefully risks and benefits can be monitored once it enters clinical use. On this topic, the FDA has issued an influential (though non-binding) guidance for gene therapy trials that would have relevance to genome editing trials (FDA 2006). Long-term follow-up is not always required, for example when preclinical data on such things as vector sequence, integration and potential for latency demonstrates that long-term risks are very low. But when long-term risks are present, "a gene therapy clinical trial must provide for long-term follow-up observations in order to mitigate those risks." Without such a plan for long-term follow-up observations, the risks would be unreasonable and (presumably) the trial not approvable. Where merited, the guidance suggests a 15-year period of post-trial contact, observation and physical exams (though this can be shortened based on factors such as vector persistence, or when subjects are predicted to have only short-term survival). Prior to enrolling, subjects must give voluntary, informed consent to long-term follow-up, and while they may withdraw at any time, it is hoped they will comply.

Once approved by the FDA for particular populations and indications, gene-based therapies would be subject to postmarket monitoring and adverse event reporting, and special warnings added or the products withdrawn completely if shown to be unsafe or ineffective. In addition, postmarket risk evaluation and mitigation strategies (REMSs), such as requiring physicians to have special proficiency or requiring patients to be entered into a registry, could be required if significant safety concerns would preclude approval absent these extra controls.

Off-label use of cells subjected to genome editing would be legal in the United States, in Europe, and in other countries, and is probably to be expected with

---

[44] 药物管制决策中利益风险评估的结构化方法，PDUFA V 计划（2013—2017 财政年度）。2013 年 2 月草稿。http://www.fda.gov/downloads/ForIndustry/UserFees/PrescriptionDrugUserFee/UCM329758.pdf（访问于 2017 年 1 月 30 日）。

[44] Structured Approach to Benefit-Risk Assessment in Drug Regulatory Decision-Making, PDUFA V Plan (FY 2013-2017). Draft of February 2013. http://www.fda.gov/downloads/ForIndustry/UserFees/Prescription-DrugUserFee/UCM329758.pdf (accessed January 30, 2017).

在美国、欧洲和其他国家，标示外使用基因编辑过的细胞是合法的，而这种标示外使用对于不同患者群体或不同严重程度的疾病情况[45]来说是预料之中的（例如，如果某一疗法准许应用于治疗成人，则它的应用很可能会拓展到标示外的儿科群体）。对于标示外基因疗法应用的预期，使人们推测这项技术已不受控制地扩展到不安全、不明智、不必要或不公平的应用中。诚然，标示外的使用，虽然是药物创新的一个重要方面，但是有时会导致不合理的应用。与许多药物相比[46]，编辑后细胞的特异性导致它们比起一般药物来说只能应用于有限的领域。人们或许会认为针对肌营养不良患者的基因编辑细胞疗法会使那些想让肌肉变得更强壮的健康人对其产生兴趣，其他细胞疗法却不易使人们产生类似的联想，至少在近期内如此。这一点与人们对基因疗法超出恢复或维持普通健康的用途的忧虑（在第6章中讨论）尤其相关，因为编辑后细胞的特异性将使得它在目前不太可能会有这样的应用。

体细胞基因编辑转向临床试验需要面对一些传统基因疗法已经面对了的技术挑战。离体基因编辑策略立足于修改人类细胞类型，因而只需在体外培养模型中或者异种移植入免疫缺陷小鼠当中进行检测。这些研究分析改造后细胞在体内的活力、生物学分布和生物学功能，包括自我更新、多能性和克隆形成等干细胞的所有关键特征。在体试验策略则可能需要在非人灵长类动物中进行临床前的毒性和生物学分布试验，包括提供生殖细胞有没有发生任何意外改变的证据。事实上，基因治疗领域已经明确规定会导致生殖细胞被意外修改的在体试验方案是不被允许的。然而，值得注意的是，大多数检测生殖系传递的方法灵敏度低下，因此在考虑临床开发和管理时，可能需要设法管理这种一定程度上的不确定性。

美国和欧洲的监管机构以及国际协调会议（ICH）公布了几个指导文件，以阐明在非临床研究中调查和解决基因治疗产品意外整合到生殖细胞的风险的一般原则，并提供将参加临床试验的人的潜

---

[45]FDA最近举行了一次公开听证会，就未经批准或非标示使用的医疗产品（包括细胞疗法）问题，讨论了其有关与制造商沟通的法规和政策（FDA，2016a）。
[46]通讯，FDA，2016年12月15日。

respect to patient populations (for example, if approved for adults, use might well be extended off-label to pediatric populations) or for varying degrees of severity of the disease indication.[45] The prospect of off-label use has led to speculation about uncontrolled expansion of the technology into uses that are unsafe, unwise, unnecessary, or unfair. And it is true that off-label use, while an important aspect of innovative medicine, can at times lead to uses that lack a rigorous evidentiary basis. But the specificity of these edited cells may limit the range of off-label uses for unrelated indications more than is the case with many drugs.[46] While one might imagine a cell therapy based on genome editing for muscular dystrophy being of possible interest to those with healthy muscle tissue who wish to become even stronger, other examples are more difficult to envision, at least for the near future. This point is of particular relevance to concerns about uses that go beyond restoration or maintenance of ordinary health (discussed in Chapter 6) because the specificity of edited cells will make such applications less likely at this time.

Several technical challenges faced in moving somatic genome editing toward clinical testing have already been met by conventional somatic gene therapy. Concerning *ex vivo* strategies, they are based on modifying human cell types and thus can be tested only in *in vitro* culture models or upon xenotransplant of the modified cells into immunocompromised mice. These studies interrogate cell viability, biodistribution, and biological function *in vivo*, including self-renewal, multipotency, and clonogenicity, all crucial features of stem cells. *In vivo* strategies may require preclinical testing of toxicity and biodistribution in nonhuman primates, including evidence that unintentional modification of the germline does not occur. Indeed, the field of gene therapy has determined that *in vivo* approaches that would lead to unintentional modification of the germline should not be permitted. Note, however, that most assays of germline transmission have low sensitivity, and thus a certain degree of uncertainty may have to be managed in considering clinical development and regulation.

Several guidance documents have been published by regulatory authorities in the U.S. and Europe and by the International Conference on Harmonisation (ICH) to illustrate the general principles for investigating and addressing the risks for inadvertent germline integration of gene therapy products in non-clinical studies, and provide considerations to minimize this potential risk in humans enrolled in clinical trials (EMEA, 2006; FDA, 2012a; ICH, 2006). Such guidelines may be suitably

---

[45] The FDA recently held a public hearing to discuss its regulations and policies on manufacturer communications about unapproved or off-label uses of medical products, including cell-based therapies (FDA, 2016a).
[46] Communication, FDA, December 15, 2016.

在风险降到最低的考虑因素（FDA，2012a；EMEA，2006；ICH，2006）。这个指南可以适当地运用到体细胞基因编辑策略的临床前研究设计中去。

为了加快再生医学的发展，一种新的公私合作伙伴关系已经正式推出。目前，国际标准协调机构已被创立用于"推进加工、测量和分析技术以支持细胞、基因、组织工程和再生医学产品，以及基于细胞的药物开发产品的全球性应用。制定标准来创造一个更加统一规范的环境，为将来在国际上协调监管提交的各机构提供帮助"[47]。这些国际标准包括对细胞的遗传修饰的标准，具体提到了对基因编辑脱靶效应的测定（Werner and Plant，2016）。

根据方法和适应证来对体细胞基因编辑进行管理

对未来体细胞基因编辑应用的伦理和监管评估可能取决于基因编辑的方法和预期治疗的适应证。与传统的基因治疗一样，体细胞基因编辑可应用于将基本的基因突变恢复为不与疾病相关的正常序列，这将使得一部分靶细胞恢复正常功能。体细胞基因编辑也可以用于改造细胞，使得它的表型与正常细胞不同，从而能够更好地抵抗或预防疾病，例如，它可以改造细胞使细胞产生高于正常量的蛋白质，或者令细胞产生对病毒感染的抗性。基因组编辑的离体和在体方法均可用于治疗或预防疾病。此外，基因组编辑还可用于改变与疾病无关的性状（参见第6章）。

无论用于评估人类体细胞基因编辑应用的最终制度是什么，至关重要的是，监管监督机制具有足够的法律权威和执行能力来识别与阻止未经授权的应用。到目前为止，现有的机构能够阻止未经授权的基因治疗应用，而当前的监管制度对此提供了关键性指导。人类基因组编辑疗法可能比传统基因治疗更加难以控制，因为技术的进步使得编辑步骤更容易操作，即便如此，操作细胞和将编辑后细胞输送到患者体内这两个过程的实施仍然无法离开高质量的实验室和医疗设施，而在这些设备里通常能确保对人类基因组编辑疗法进行到位的管理监督。

adapted to design preclinical studies of somatic genome editing strategies.

In an effort to speed the development of regenerative medicine, a new public–private partnership has been launched. The International Standards Coordinating Body was established "to advance process, measurement, and analytical techniques to support the global availability of cell, gene, tissue-engineered, and regenerative medicine products, and cell-based drug discovery products. Creating standards creates a more uniform compliance environment and addresses and assists in future efforts for harmonization internationally of the regulatory framework for submissions across the globe."[47] The sectors of activity include genetic modification of cells, with specific mention of standards for measuring off-target events in genome editing (Werner and Plant, 2016).

Regulating Somatic Genome Editing by Approach and Indication

An ethical and regulatory assessment of future somatic genome-editing applications may depend on both the technical approach to the editing and the intended indication. Like traditional gene therapy, somatic genome editing could be used to revert an underlying genetic mutation to a variant not associated with disease, which would result in a fraction of the targeted cells regaining normal function. Somatic genome editing also could be used to engineer a cell so that its phenotype differed from that of a normal cell and was better able to resist or prevent disease. For example, a cell could be changed so that it made above-normal amounts of a protein, or so that it was resistant to a viral infection. Both *ex vivo* and *in vivo* approaches to genome editing could be applied to treat or prevent a disease. In addition, genome editing could be used to alter a trait not associated with disease (see Chapter 6).

Regardless of the final framework used to assess human somatic cell genome-editing applications, it is vital that the regulatory oversight mechanisms have sufficient legal authority and enforcement capability to identify and block unauthorized applications. To date, the existing structures have been successful in preventing unauthorized applications of gene therapy and the current framework provides guidance on key elements. Although human genome editing may be somewhat more difficult to control than traditional gene therapy because technical advances have made the editing steps easier to perform, the cellular manipulations and delivery of edited cells to the patient continue to demand high-quality laboratory and medical facilities, which generally will ensure that regulatory oversight is in place.

---

[47] http://www.regenmedscb.org（2017年1月10日访问）。

[47] See http://www.regenmedscb.org (accessed January 10, 2017).

防止未成熟或未经验证的基因编辑应用

干细胞/再生医疗领域，不受监管的治疗尤其成问题，世界范围都有的不法团体毫无科学依据地宣称他们可以进行干细胞治疗，从而从孤注一掷的患者身上获利（Enserink，2016；FDA，2016b；Turner and Knoepfler，2016）。出现这种情况的一部分原因是过去的一些对再生医疗近期前景过于乐观的陈述误导了公众，一部分原因是存在一些不受监管的地区，另外还有一部分原因（至少在美国）来自对政府管理机构的抵制。在美国，联邦法院已经批准了FDA（食品药品监督管理局）对基因操作细胞使用的管理规定，但这个问题依然困扰着人们[48]。编辑后的细胞（尤其是那些取自患者然后回输到患者的细胞）可能也会造成这样的困惑：到底它是一个受监管的医疗产品，还是仅仅是一次医疗的实施？所有的这些，都需要监督管理机构从一开始就进行明确。总的来说，监管机构需要法律权威、领导层承诺和政策支持来行使他们的法律权力，叫停那些使用未经监管部门审查和批准的人类基因编辑产品的治疗方法的营销（Charo，2016b）。至于干细胞治疗，尽管意大利关闭了一个诊所的事件证明了监管部门要这样做所需要的法律和政治权力水平（Margottini，2014），但是FDA的缺少执行力度的现状还是令人关注（Turner and Knoepfler，2016）。

胎儿中的基因编辑特殊注意事项

在特定情况下，最有效或者唯一的方法就是试图在分娩之前编辑胎儿的体细胞。符合这些特定情况的疾病包括多系统疾病、特别早发的疾病（如果在出生后才介入干预则太迟了，不能使儿童受益），以及从技术角度来讲需要面临极大挑战的疾病。此外，由于胎儿发育具有极大的可塑性，在胎儿时期进行编辑在某些情况下可能比产后编辑更有效，比如试图恢复影响大脑中每个神经元的致病突变。

一般来说，治疗性的编辑过程可以通过离体操作实现：先从胎儿获取细胞，在体外进行编辑，后再移植回胎儿体内。目前，已建立的分离和移植自体胎儿细胞的方法只能用于有限的细胞种类，在将来，适用的细胞种类可能会有所增加。

Preventing Premature or Unproven Uses of Genome Editing

The issue of unregulated therapy has been particularly problematic in the field of stem cell/regenerative medicine, with rogue entities around the world making scientifically unfounded claims about stem cell therapies and profiting from desperate patients (Enserink, 2016; FDA, 2016b; Turner and Knoepfler, 2016). In part this is due to some of the past unduly optimistic statements about the near-term prospects of regenerative medicine, in part to the presence of unregulated jurisdictions, and in part to some resistance—at least in the United States—to the regulatory authority of the government. In the United States, federal courts have confirmed the FDA's jurisdiction over the use of manipulated cells, but this is still the subject of some confusion.[48] Edited cells—particularly those taken from a patient and then returned to that patient—may engender the same confusion about whether this is a regulated product or merely the practice of medicine, and the regulatory authority needs to be made clear from the outset. Overall, then, regulatory bodies need the legal authority, leadership commitment, and political support to apply their legal powers to halt the marketing of therapies that use human genome-editing products that have not undergone regulatory review and approval (Charo, 2016b). With regard to stem-cell therapies, there has been considerable concern about the absence of vigorous use of enforcement powers by the FDA (Turner and Knoepfler, 2016), although Italy's experience with closing down one clinic has illustrated the level of legal and political power needed to do this (Margottini, 2014).

Special Considerations Associated with Genome Editing in Fetuses

In certain situations, either the most effective or the only approach would be to attempt to edit the somatic cells of a fetus prior to delivery. Diseases for which these special circumstances might apply include those that are multisystemic or have an extremely early onset that would make postnatal intervention too late to benefit the child or are extremely challenging from a technical standpoint. In addition, because of the tremendous developmental plasticity of the fetus, fetal editing might be more effective than postnatal editing in certain circumstances. An example would be attempting to revert a disease-causing variant that affects every neuron in the brain.

In a more general sense, the therapeutic editing process could be carried out *ex vivo* in a scenario in which cells could be harvested from the fetus, edited outside the body, and then transplanted back into the fetus. Currently, established

---

[48] U.S. v. Regenerative Sciences, 741 F. 3d 1314 (D.C. Cir 2014).

胎儿的治疗性编辑也可以通过在体的方式实现，在这种情况下，编辑工具将被运送到胎儿体内并在原位改造细胞。如上所述，在发育早期对致病基因变体的原位校正比产后的在体编辑更有效，因为产后许多器官系统发育已经完全。目前已有子宫内干细胞治疗的尝试（成功例子有限）(Couzin-Frankel，2016；Waddington et al.，2005)，所以随着一些医学新兴领域的出现，有关子宫内疗法的概念已经经历了一些伦理分析。国际胚胎移植和免疫学会已经成立，该组织每年举办年度会议来评估胎儿基因治疗的前景和发展[49]。

虽然胎儿基因编辑有潜在的优势，但至少有两个特殊伦理问题需要解决：关于受试者同意操作的特殊规则（见第2章）；造成生殖细胞或生殖细胞祖/干细胞的修改导致生殖细胞产生可遗传变异的风险增加。

关于受试者知情同意的管理，现有的监督机制已经解决了关键的问题。胎儿手术已经用于临床护理，而子宫内胎儿基因疗法正吸引着越来越多人的兴趣（McClain，2016；Waddington，2005）。相对于产后或成人干预，胎儿干预的风险/疗效评估发生了偏移，当无法预期对未来的儿童有医疗益处的时候，胎儿可能遭受的风险程度受到严格限制。而当存在上述医疗益处的可能性时，更常见的风险/效益平衡标准才可以适用。在决定对胎儿进行手术前需要清楚，孕妇具有知情同意的伦理和法律权力。在美国和其他国家，产妇的同意是必要条件（Alghrani and Brazier，2011；O'Connor，2012），当研究同时针对母方健康时，只需母方同意就足够了[50]。然而在美国，NIH（国家卫生研究所）资助的研究受限于45 CFR第46部分、B部分的特殊规定，如果研究仅仅是对胎儿有利，则父方的同意（如果父方在的话）也是必需的。在美国，哪怕不是由国家卫生研究院资助的许多研究项目也采用了这些相同的规则。

第二个问题是对胎儿进行在体体细胞编辑时，如何评估生殖细胞是否也在无意中被编辑了。生殖系细胞发育的一个关键特性是，能产生生殖细胞的原始生殖细胞在关键的发育阶段会与体细胞隔

---

[49] http://www.fetaltherapies.org（2017年1月30日访问）。
[50] 涉及孕妇或胎儿的研究，45 CFR，Sec. 46.204。

methods for isolating and transplanting autologous fetal cells are available for a limited number of cell types, but the range of cell types is likely to increase in the future.

Therapeutic editing in fetuses also could be performed *in vivo*, in which case the editing machinery would be delivered to the fetus to modify cells in situ. As noted above, the in situ correction of a disease-causing variant early in development has the potential to be more effective than postnatal *in vivo* editing, when many organ systems are more fully developed. *In utero* stem cell therapy has been tried (with limited success) (Couzin-Frankel, 2016; Waddington et al., 2005), so the general concept of *in utero* therapy with emerging areas of medicine has already undergone some ethical analysis. And an International Fetal Transplantation and Immunology Society has been formed, which holds annual meetings to review prospects and progress for fetal gene therapy.[49]

Although fetal genome editing has potential advantages, at least two special ethical issues would need to be addressed: special rules for consent (see Chapter 2) and the increased risk of causing heritable changes to the germline by causing modification of germ cells or germ cell progenitor/stem cells.

With regard to consent, key issues have been addressed by existing oversight mechanisms, fetal surgery has already been used in clinical care, and *in utero* fetal gene therapy is attracting increasing interest (McClain, 2016; Waddington, 2005). The risk/benefit calculation is shifted relative to a postnatal or adult intervention, with the degree of risk to which a fetus can be subjected being strictly limited when there is no prospect of medical benefit to the future child. When such benefit is possible, however, the more usual standards for risk/benefit balance apply. Decisions about fetal surgery have been made with the understanding that the pregnant woman has the ethical and legal authority to give informed consent. In the United States, as in other countries, maternal consent is required (Alghrani and Brazier, 2011; O'Connor, 2012), and when research is aimed at maternal health as well, maternal consent alone is sufficient.[50] In the United States, however, NIH-funded research is subject to special regulations set forth at 45 CFR Part 46, Subpart B, and paternal consent (if available) also is required if the research holds out the prospect of benefit solely to the fetus. Even when not funded by NIH, many studies in the United States employ these same rules.

A second issue is the challenge of assessing whether unintended germline editing has occurred if *in vivo* somatic editing is attempted in a fetus. A key feature of germline cell development is that the primordial cells that will give rise to germ cells are sequestered from

---

[49] See http://www.fetaltherapies.org (accessed January 30, 2017).
[50] *Research Involving Pregnant Women or Fetuses*, 45 CFR, Sec. 46.204.

离。在发育早期阶段，在生殖系与体细胞的隔离发生或结束前，生殖细胞可能与想改造的体细胞靶标一样能被有效地编辑。结果就是，相比胚胎发育后期阶段的干预，在胚胎发育早期阶段进行基因编辑存在着更高的生殖系细胞意外被编辑的风险。对生殖细胞或生殖细胞祖细胞是否发生编辑的评估只可能在产后进行，但这时就太晚了，结果已经不可更改。

总结和建议

一般来说，有大量民众支持使用基因疗法（可扩展至利用基因修饰的基因疗法）治疗和预防疾病与残疾。

人类体细胞基因编辑在多种疾病的治疗和预防方面，以及改善目前正在使用或正处于临床试验的基因治疗技术的安全性、疗效和效率方面有着很大的应用前景。虽然基因编辑技术的优化仍在继续，但这项技术最应该被应用于人类疾病与残疾的治疗和防治，而不是其他次要的目的。

因基因治疗而发展起来的相关伦理规范和调控制度可适用于对体细胞基因编辑治疗的监管。体细胞基因编辑临床试验的管理评估方法与其他药物疗法相类似，包括风险最小化、对参与者承担的治疗风险相对于可能取得的治疗效果是否合理的分析、确定所招募的参与者是否在自愿及知情同意的情况下参加。同时，监管制度还需要具有法律权威和执法能力，可以阻止基因编辑在未经授权或未发展成熟的情况下被使用。另外，监管当局需要对所用技术的具体技术方面的知识进行不断地更新，至少监管当局在进行评估的时候，除了考虑基因编辑系统的技术体系，还需要了解申请中所涉及的临床应用，用以权衡预期风险和疗效。因为脱靶问题会随着平台技术、细胞类型、靶向基因组序列和其他相关因素的变化而变化，所以对于体细胞基因编辑特异性（例如，可被接受的脱靶率）而言，目前还没有一个特定的标准。

建议4-1：现有用于审查和评估治疗或预防疾病及残疾的体细胞基因疗法的监管基础设施和流程可以用于评估基于基因编辑技术的体细胞基因治疗。

somatic cells at key developmental points. Before this sequestration of germline and somatic cells occurs or has been finalized in early development, germline cells might be edited as efficiently as would be the desired somatic cell targets. As a result, there could be a higher risk of unintentional edits to germline cells early in fetal development compared with performing the same intervention later in fetal development. It might be possible only to assess postnatally whether editing of germ cells or germ cell progenitors had occurred, at which time it would be too late to change the outcome.

CONCLUSIONS AND RECOMMENDATIONS

In general, there is substantial public support for the use of gene therapy (and by extension, gene therapy that uses genome editing) for the treatment and prevention of disease and disability. Human genome editing in somatic cells holds great promise for treating or preventing many diseases and for improving the safety, effectiveness, and efficiency of existing gene therapy techniques now in use or in clinical trials. While genome-editing techniques continue to be optimized, however, they are best suited only to treatment or prevention of disease and disability and not to other less pressing purposes.

The ethical norms and regulatory regimes already developed for gene therapy can be applied for these applications. Regulatory assessments associated with clinical trials of somatic cell genome editing will be similar to those associated with other medical therapies, encompassing minimization of risk, analysis of whether risks to participants are reasonable in light of potential benefits, and determining whether participants are recruited and enrolled with appropriate voluntary and informed consent. Regulatory oversight also will need to include legal authority and enforcement capacity to prevent unauthorized or premature applications of genome editing, and regulatory authorities will need to continually update their knowledge of specific technical aspects of the technologies being applied. At a minimum, their assessments will need to consider not only the technical context of the genome-editing system but also the proposed clinical application so that anticipated risks and benefits can be weighed. Because off-target events will vary with the platform technology, cell type, target genome sequence, and other factors, no single standard for somatic genome-editing specificity (e.g., acceptable off-target event rate) can be set at this time.

RECOMMENDATION 4-1. Existing regulatory infrastructure and processes for reviewing and evaluating somatic gene therapy to treat or prevent disease and disability should be used to evaluate somatic gene therapy that uses genome editing.

建议 4-2：当前，监管机构只能授权或批准与治疗或预防疾病和残疾相关的临床试验。

建议 4-3：监管当局应该在按预期使用的风险和收益的背景下，评估拟议的人类体细胞基因编辑应用的安全性和有效性，认识到脱靶事件可能会随着平台技术、细胞类型、目标基因的位置，以及其他因素的变化而变化。

建议 4-4：在监管当局考虑是否授权对超出治疗或预防疾病和残疾以外的情况进行体细胞基因编辑临床试验之前，监管当局应该先进行透明而全面的公共政策辩论。

RECOMMENDATION 4-2. At this time, regulatory authorities should authorize clinical trials or approve cell therapies only for indications related to the treatment or prevention of disease or disability.

RECOMMENDATION 4-3. Oversight authorities should evaluate the safety and efficacy of proposed human somatic cell genome-editing applications in the context of the risks and benefits of intended use, recognizing that off-target events may vary with the platform technology, cell type, target genomic location, and other factors.

RECOMMENDATION 4-4. Transparent and inclusive public policy debates should precede any consideration of whether to authorize clinical trials of somatic cell genome editing for indications that go beyond treatment or prevention of disease or disability.

5

可遗传性基因组编辑
Heritable Genome Editing

当准父母知道可能会将严重遗传性疾病传递给他们的孩子时,可提供一个潜在的方法[51]让自己的孩子不受该疾病的影响,这是许多此类父母共同的愿望(例如,Chan et al., 2016; Quinn et al., 2010)。数以千计的遗传性基因疾病是由单个基因的突变引起的。[52]尽管许多遗传性基因疾病是罕见的,但是统计下来也影响了相当大一部分(5%~7%)的人口。由于这些严重疾病的遗传,单个家庭在情感上、经济上或其他方面都承受着相当大的负担,一些家庭可能会愿意使用生殖系编辑来减轻负担。基因组编辑技术的最新进展已经使最终在人类生殖系中应用这些技术的可能性成为现实。如本文其他部分所述,基因组编辑技术的改进正在提高基因组编辑的效率和准确性,同时还降低了脱靶事件的风险。因为生殖系基因组编辑是可遗传的,即其影响可以是多代的,因此,它的潜在的受益和危害都会倍增。此外,自主生殖系遗传改变的概念引起了有关这种形式的人工干预是否是明智和适当的激烈争论,以及对这一技术可能产生的人文效应的担忧。如下所述,这些担忧包括生殖系编辑会削弱人类尊严和多

For prospective parents known to be at risk of passing on a serious genetic disease to their children, heritable genome editing[51] may offer a potential means of having genetically related children who are not affected by that disease—a desire shared by many such parents (e.g., Chan et al., 2016; Quinn et al., 2010). Thousands of genetically inherited diseases are caused by mutations in single genes.[52] While individually, many of these genetically inherited diseases are rare, collectively they affect a sizable fraction of the population (about 5-7 percent). The emotional, financial, and other burdens on individual families that result from transmission of such serious genetic disease can be considerable, and for some families could potentially be alleviated by germline editing. Recent advances in the development of genome-editing techniques have made it realistic to contemplate the eventual feasibility of applying these techniques to the human germline. As discussed elsewhere in this report, improvements in genome-editing techniques are driving increases in the efficiency and accuracy of genome editing while also decreasing the risk of off-target events. Because germline genome edits would be heritable, however, their effects could be multigenerational. As a result, both the potential benefits and the potential harms could be multiplied. In addition, the notion of intentional germline genetic alteration has occasioned significant debate about the wisdom and appropriateness of this form of human intervention, and speculation about possible cultural effects of the technology. As discussed below, these include concerns about diminishing

---

[51] "生殖系编辑"是指对生殖系细胞的所有操作,有原始生殖细胞(PGCs)、配子祖细胞、配子、受精卵和胚胎(参见第3章)。"可遗传生殖系编辑"是生殖系编辑的一种形式,包括妊娠期注入编辑过的物质,意图产生能将"编辑"传递给子孙后代潜力的个体,这些将在本章中讨论。它们的区别在于有意图而不是技术干预本身,这在两种情况下都非常相似。

[52] 查看 https://www.omim.org(accessed January 3, 2017)和 http://www.diseaseinfosearch.org(accessed January 3, 2017)。

[51] "Germline editing" refers to all manipulations of germline cells; primordial germ cells [PGCs], gamete progenitors, gametes, zygotes, and embryos) (see Chapter 3). "Heritable germline editing," a form of germline editing that includes transfer of edited material for gestation, with the intent to generate a new human being possessing the potential to transmit the "edit" to future generations, is discussed in this chapter. The distinction turns on intent rather than on the technological intervention, which is highly similar in both cases.

[52] See https://www.omim.org (accessed January 3, 2017) and http://www.diseaseinfosearch.org (accessed January 4, 2017).

样性，减弱对自然世界重要性的认知，以及在改变人类和自然时缺乏自控和自谦（Skerrett，2015）。类似的争论也发生在一系列相关的技术中，例如，线粒体置换技术，通过使用来自供体的健康线粒体来避免线粒体DNA携带的遗传病。因为卵子中的线粒体是通过母系世代传下来，虽然不改变细胞核中的DNA，但也导致了遗传性基因的变化。线粒体置换技术已在墨西哥（Hamzelou，2016）和乌克兰（Coghlan，2016）使用，并已在英国被授权但尚未被使用（HFEA，2016a）。而美国科学院最近研究的建议是，在严格遵守标准和受到监督的前提下，在美国允许进行线粒体置换临床试验（IOM，2016）（见延伸内容5-1）。2017年年初，有报道称在乌克兰有一个孩子在使用该技术后出生，在这个案例中是用于具有不孕不育的情况，并不符合HFEA或美国科学院报告制定的标准（Coghlan，2017）。

the dignity of humans and respect for their variety; failing to appreciate the importance of the natural world; and a lack of humility about our wisdom and powers of control when altering that world or the people within it (Skerrett, 2015). A similar debate is already under way regarding a related set of techniques—mitochondrial replacement—in which genetic disease carried by mitochondrial DNA is avoided by using healthy mitochondria from a donor. Because mitochondria in the egg are passed down maternally through the generations, the effect of these techniques is to make a heritable genetic change, albeit one that does not change the DNA in the nucleus. Mitochondrial replacement has been used in Mexico (Hamzelou, 2016) and Ukraine (Coghlan, 2016) and has been authorized although not yet used in the United Kingdom (HFEA, 2016a). A recent National Academies study led to the recommendation that mitochondrial replacement be permitted to proceed to clinical trials in the United States provided it is subject to strict criteria and oversight (IOM, 2016) (see Box 5-1). In early 2017, there were reports of a child born after use of the technique in the Ukraine, in this case for an infertility-related condition that would not have met the criteria laid out by either the HFEA or the National Academies report (Coghlan, 2017).

延伸内容 5-1

线粒体置换技术（MRT）
Mitochondrial Replacement Techniques

线粒体置换技术（MRT）已经引起了对可遗传基因修饰的新讨论。我们所有的人都是由细胞核中的DNA来编码我们大部分的特征，但还有非常少量的DNA存在于我们细胞内的线粒体中。很多疾病是由突变的线粒体通过母本卵子遗传给后代的。用其他卵子的正常线粒体代替这些缺陷型线粒体大大降低了后代患遗传疾病的风险，并且满足了准父母对拥有一个亲生后代的渴望。儿子不能将供体线粒体传给自己未来的孩子，但女儿可以通过她们经过修饰的卵子做到，这就使其成为一种潜在可遗传的生殖系改变形式。

2016年，IOM委员会建议采用审慎的、渐进的方法，对仅限于涉及已知的患严重疾病的风险的情况进行试验。同时建议只在雄性胚胎上使用这个技术，使捐赠的任何影响都将仅发生在第一代孩子身上。通过进一步的研究和安全性的证据，此技术也可以扩大到包含雌性胚胎的可遗传的MRT形式。英

A technique for mitochondrial replacement already has engendered fresh discussion of heritable genetic modification. All people have DNA in the nucleus that encodes most of our traits, but also a very small amount of DNA in the mitochondria within our cells. A variety of diseases are caused when mutated mitochondria are passed from parent to child through the mother's egg. Replacing these defective mitochondria with normal mitochondria from another woman's egg allows for conception of children with vastly reduced risk of inherited disease while satisfying prospective parents' desire for genetically related offspring. Sons could not pass along the donated mitochondria to their own future children, but daughters could, through their now-modified eggs, thus rendering this a potentially heritable form of germline alteration.

A 2016 IOM committee recommended pursuing a cautious, incremental approach, with trials limited to situations involving a known risk of passing along serious disease. It also recommended initially limiting the technique's use to male embryos, so that any effects of the donation would be experienced only by the first generation of children. With additional research and evidence of

国人类受精和胚胎学管理局（HFEA）也审查了 MRT 流程并得出结论：此技术已经成熟到既可以应用于雄性胚胎，也可用于雌性胚胎，虽然到目前为止此技术流程已获得批准，但并没有被执行。

本章首先回顾了可遗传基因组编辑的潜在应用和替代方法；接着依次描述科学和技术、伦理和社会问题，以及与这些应用相关的潜在风险；之后本章转向可遗传基因组编辑的规则；最后一节提出总结和建议。

潜在应用方向和替代方法

预防遗传性疾病的传播

对于是否应该使用生殖系基因组编辑来预防遗传性疾病的传播有不同的意见。生殖系基因组编辑不是达到这个目的的唯一方法。其他选择包括：考虑不生孩子；领养一个孩子；使用捐赠的胚胎、卵子或精子。然而，这些选择不允许父母双方与他们的孩子有遗传关联，而这对许多人是非常重要的。或者，体外受精（IVF）与胚胎着床前基因诊断（PGD）可以用于识别受影响的胚胎，使父母可以选择只植入那些诊断无突变的胚胎。这个选择并不是没有潜在的风险和成本，并且它也涉及丢弃受影响的胚胎，有些人会觉得不可接受。人们也可以通过进行胎儿的胚胎着床前基因诊断，选择性地将受影响的胎儿流产来避免将基因突变传给下一代。但与胚胎着床前基因诊断一样，一些人觉得终止妊娠是不可接受的，无论孩子的预测是否健康。

在这些情况下，对于那些意识到他们有遗传突变风险的人，使用可遗传的生殖系基因组编辑为他们提供了一个让自己孩子没有突变担忧的潜在方法。这种形式的编辑可以用在配子（卵子、精子）、配子前体或早期胚胎，但需要注意的是，体外受精（IVF）的过程将来可能被要求来产生用于后续的基因组修饰的胚胎。在大多数情况下，胚胎着床前基因诊断可用于筛选未受影响的胚胎来植入。

safety, it concluded, this could be expanded to a heritable form of MRT that includes female embryos. The United Kingdom's Human Fertilisation and Embryology Authority has also reviewed the MRT procedures, and concluded that they are robust enough to proceed with either male or female embryos, although to date the procedure has been approved but not yet performed.

This chapter begins by reviewing potential applications of and alternatives to heritable genome editing. It then describes in turn scientific and technical issues, ethical and social issues, and potential risks associated with these applications. The chapter then turns to the regulation of heritable genome editing. The final section presents conclusions and a recommendation.

POTENTIAL APPLICATIONS AND ALTERNATIVES

Preventing Transmission of Inherited Genetic Disease

Opinions differ as to whether germline genome editing should be used to prevent the transmission of inherited genetic diseases. Germline genome editing is not the only way to accomplish this goal. Other options include deciding not to have children; adopting a baby; or using donated embryos, eggs, or sperm. These options, however, do not allow both parents to have a genetic connection to their children, which is of great importance to many people. Alternatively, *in vitro* fertilization (IVF) with preimplantation genetic diagnosis (PGD) of the embryos can be used to identify affected embryos so that parents can choose to implant only those embryos that are free of the diagnosed mutation. This option is not without potential risks and costs, however, and it also involves discarding affected embryos, which some find unacceptable. One can also avoid transmitting genetic mutations to the next generation by using prenatal genetic diagnosis of the fetus followed by selective abortion of affected fetuses. But as with PGD, some people find it unacceptable to terminate an ongoing pregnancy regardless of the predicted health of the future child.

In these situations, for those who are aware they are at risk of passing on such a mutation, the use of heritable germline genome editing offers a potential avenue to having genetically related children who are free of the mutation of concern. This form of editing could be done either in gametes (eggs, sperm), in gamete precursors, or in early embryos, but it is important to note that IVF procedures would be required to generate embryos for subsequent genomic modification. In most cases, PGD could be used to identify unaffected embryos to implant.

However, there are some situations where all or a majority of embryos will be affected, rendering PGD difficult or impossible. For example, dominant late-onset genetic

然而，有一些遗传疾病会影响全部或大部分胚胎，这使得胚胎着床前基因诊断变得困难或者不可能。例如，显性晚发性遗传疾病，如亨廷顿舞蹈症，在一些隔离人群中，如果亲本之一是突变纯合型的，也会导致足够高的发生率。在这种情况下，所有的胚胎都会携带导致儿童疾病的显性致病等位基因。胚胎着床前基因诊断在这种情况下没有用。在其他人群中，准父母双方都携带相同的突变基因的机会很大，导致某种致病突变的频率也可能足够高。例如，肿瘤抑制基因 *BRCA1* 和 *BRCA2*，它们即使在单拷贝中遗传（因为缺失了正常的基因拷贝）也能增加乳腺癌和卵巢癌的患病风险；Tay-Sachs 病和其他由于继承了两个拷贝的隐性突变导致的早发性溶酶体贮积病。在这些情况下，将只有 1/4 概率的胚胎不携带致病突变。那些未受影响的胚胎可以通过胚胎着床前基因诊断鉴定，但是潜在可用于植入的胚胎数目将会显著减少。还有对于由某一基因中两个不同突变的配对引起的疾病（称为"共显性"），或者由两个或以上基因的特定等位基因组合造成的疾病，胚胎着床前基因诊断也更加困难。

随着医学的进步，严重隐性基因疾病如囊性纤维化、镰状细胞贫血、地中海贫血和溶酶体贮积病的存活率不断提高，这可能导致准父母双方均为纯合突变的概率增加，这一可能性不能被忽视。社会环境和这些人面临的医疗压力经常把他们聚集到一起，互相影响成为关系密切的社群。类似的社群在允许发育到繁殖年龄（如软骨发育不全、成骨不全）的常染色体显性遗传疾病的患者中也会形成，而这将再次增加传播疾病等位基因的可能性。随着通过传统的和体细胞基因组编辑疗法治疗儿童和成人严重遗传病能力的改善，解决准父母可能将这些疾病遗传给子女的问题将会变得越来越迫切。这种情况可能会增加患者和携带者对使用基因组编辑技术来避免有害基因传递给他们后代的兴趣。

当女性携带会引起卵母细胞在发育期间或产后损失的突变，如脆性 X 综合征、BRCA-1(de la Noval, 2016; Oktay et al., 2015) 等时，会导致生育能力受损，这可能带来另外的问题。在希望避免遗传病传播的女性中，除了遗传突变，外部因素如癌症治疗和环境中化学物质也会减少女性的卵巢储

diseases, like Huntington's disease, can occur at high enough frequency in some isolated populations that one parent will be homozygous for the mutation. In such situations, all embryos would carry the dominant disease-causing allele that would cause the disease in the children. PGD is not useful in this situation. In other populations, the frequency of particular disease-causing mutations may be high enough that there is a significant chance that both prospective parents will be carriers of mutations in the same gene. Examples include the tumor suppressor genes, BRCA1 and BRCA2, which increase the risk of breast and ovarian cancer even when inherited in a single copy (because of loss of the unaffected copy of the gene) and Tay-Sachs disease and other early-onset lysosomal storage diseases that are caused by the inheritance of two copies of recessive mutations. In these situations, only one in four of the embryos would be free of a disease-causing mutation. Those unaffected embryos could be identified by PGD, but the number of embryos potentially available for implantation would be significantly reduced. There are also examples of diseases that are caused by pairing of two different mutations in a given gene, known as "co-dominance," and combinations of specific alleles of two or more genes, in which PGD becomes more difficult.

As the survival of people with severe recessive diseases like cystic fibrosis, sickle-cell anemia, thalassemia, and lysosomal storage diseases improves with advances in medical treatments, the possibility cannot be dismissed that there will be an increase in the number of situations in which both prospective parents are homozygous for a mutation. The societal and medical pressures faced by these people often bring them together into social groups where they are more likely to interact and develop close relationships. Similar associations can develop among patients with autosomal dominant genetic diseases that allow development to reproductive age (e.g., achondroplasia, osteogenesis imperfecta), again increasing the likelihood of transmitting disease alleles. As our ability to treat children and adults with serious genetic diseases improves through both conventional and somatic genome editing therapies, there may be a growing need to address concerns potential parents might have about passing along these diseases to their children. Such situations may well increase interest among carriers and affected individuals in using genome editing techniques to avoid passing on deleterious genes to their children and subsequent generations.

There can be an additional problem in the case of mutations that compromise fertility, which is the case for women who carry mutations such as Fragile X, BRCA-1 (de la Noval, 2016; Oktay et al., 2015) and others that cause the loss of oocytes during development or postnatally. Beyond inherited mutations, external factors like cancer treatments and environmental chemicals can also reduce ovarian reserve in women who wish to avoid

备。在这些情况下，女性每个超排卵周期中可用于筛选的胚胎会变得更少，并且，通过未受影响的胚胎（通过 IVF 和 PGD）来怀孕的机会也低于那些没有突变的女性。结果就是，这些女性可能需要多个超排卵周期，伴随更多的风险、不适和成本，以获得未受影响的胚胎。

在所有这些情况下，若使用可遗传基因组编辑来矫正突变的技术变得安全而有效（如在配子祖细胞中），那么这种方法对于准父母来说可能会超越目前仅有的胚胎着床前基因诊断技术而变成首选。上述情况所涉及的人数可能很少，但面对这些艰难抉择的担忧却是真实的。

治疗影响多种组织的疾病

一些遗传性疾病影响特定的细胞类型或组织，如特定类型的血细胞。这些疾病可以通过体细胞基因组编辑来治疗，事实上，一些治疗方法已经开始使用（见第 4 章）。然而，体细胞基因组编辑不太适合治疗影响多种组织的其他遗传疾病，因为它可能无法针对疾病的所有方面，或可能在受感染的组织中难以达到足够数量的细胞以改善症状。目前已经研究了体细胞基因组编辑但可能不是完全有效的实例包括囊性纤维化，其影响多种上皮组织（肺、肠和其他器官组织），以及肌肉营养不良，其可以影响多种类型的肌肉组织如心肌，以及其他组织如大脑。想拥有自己孩子的患病夫妇可能是生殖系基因组编辑的未来候选人，因为编辑缺陷基因生殖系可以使所有组织受益。

杜氏肌营养不良症（Duchenne muscular dystrophy，DMD）是应用生殖系基因组编辑面临的挑战的一个指导性实例。DMD 是一个 X 染色体连锁疾病，在男性出生中患病率约为 1/3600。症状开始出现在出生后的最初几年内，而且逐渐恶化。DMD 患者的平均预期寿命约为 25 年。DMD 是由基因组中最大基因之一——肌萎缩蛋白突变引起的，包含多个重复类似片段。肌萎缩蛋白基因的大小和重复令它易于突变，从而使得这种遗传性疾病相对常见。体细胞基因组编辑方法已发展到能够用来去除肌肉前体中肌萎缩蛋白基因的有害突变，而这种体细胞基因组编辑方法将只能用于改善病症，不会纠正所有组织中的症状。

一旦这些体细胞编辑方法进行了临床试验，人

transmission of a genetic disease.

In these cases, women have fewer embryos available to screen from each cycle of superovulation, and the chance of establishing a pregnancy with an unaffected embryo (via IVF and PGD) is lower than it is for women without these mutations. As a result, affected women might require multiple superovulation cycles, with their attendant risks, discomforts, and costs, to identify an unaffected embryo.

In all of these situations, if it were safe and efficient to use heritable genome editing (for example in gamete progenitors) to correct the mutation, this alternative might be preferred by prospective parents who otherwise would be considering PGD. The number of people in situations like those outlined above might be small, but the concerns of people facing these difficult choices are real.

Treating Diseases That Affect Multiple Tissues

Some genetically inherited diseases affect specific cell types or tissues, such as particular types of blood cells. These diseases can be treated by somatic genome editing, and indeed, some of these treatments are already being used (see Chapter 4). However, somatic genome editing is less well suited to treating other genetic diseases that affect multiple tissues because it may be unable to target all aspects of the disease, or may have difficulty reaching a sufficient number of cells in the affected tissues to ameliorate symptoms. Examples of conditions for which somatic genome editing is already being investigated but may not be fully effective include cystic fibrosis, which affects multiple epithelial tissues (tissues of the lungs, gut, and other organs), and muscular dystrophies, which can affect multiple muscle types, including heart muscle, as well as other tissues such as brain. Couples with diseases who want to have genetically related children might be future candidates for germline genome editing because editing the defective gene in the germline could have therapeutic benefit in all tissues.

Duchenne muscular dystrophy (DMD) is an instructive example of the challenges faced by the application of germline genome editing. DMD is an X-linked disease that affects about 1 in 3,600 male births. Symptoms begin to appear within the first few years after birth and progressively worsen. The average life expectancy for a person with DMD is about 25 years. DMD is caused by mutations in one of the largest genes in the genome, dystrophin, which contains multiple repeating similar segments. Both the size of the dystrophin gene and its repeats predispose it to mutation, making this genetic disease relatively common. Somatic genome editing approaches are already being developed to remove the deleterious alterations in the dystrophin gene in muscle precursors. Such somatic genome editing approaches will ameliorate the condition but are not expected to correct the symptoms in all tissues.

Once those somatic editing approaches have been tested

们可以想象尝试利用生殖系编辑来校正所有组织中的缺陷。然而，知道自己携带 DMD 突变的女性可以利用 PGD 来避免生下一个遗传突变的孩子。此外，1/3 的 DMD 病例是由于起始突变，这只能在出生后才能被识别，因此不适合生殖系基因组编辑。目前体细胞编辑方法似乎比生殖系编辑对这类疾病更为有用，但是，基因组编辑方法和干细胞生物学的进步可能会改变这种情况。

相关的科学和技术问题

实际上，在受精卵和早期胚胎中使用基因组编辑仍存在相当大的技术困难。虽然目标的效率和准确性可以非常高，并且有相当的理由来支持脱靶效应会大大减少（见第 3 章和附录 A），仍必须要确保只有具有正确靶向等位基因的胚胎可以重返子宫来完成妊娠。

嵌合体

如果在受精卵或早期胚胎中进行基因组编辑，早期胚胎中的一些细胞将有很大机会无法编辑获得预期的（甚或任何的）特性。这种情况被称为"嵌合体"，它对生殖系基因组编辑在受精卵或胚胎上的应用提出了一个严峻的挑战。通过 PGD 检测编辑过的胚胎是不是嵌合型并不能确保植入胚胎含有正确编辑，因为单个细胞可能不反映胚胎的其他细胞的基因型，而获取多个细胞进行测试则会破坏胚胎。

嵌合体的影响在一定程度上取决于目的基因。如果目的基因编码蛋白与必需的细胞功能相关，嵌合体就会是一个严重的问题；但如果基因编码一个分泌因子（如生长激素或促红细胞生成素）或编码导致分泌所需的分子（如凝血因子），则只在部分细胞中校正基因可能就足够了。此外，由于出生孩子的生殖系也可能嵌合，只编辑部分细胞可能不能为后代解决问题。但 PGD 可以提供更好的机会找到无疾病的胚胎，或允许培养和选择编辑的精原干细胞（见下文有关可遗传编辑的潜在替代路线的部分），从而使那些孩子有一个不被影响的后代。总的来说，目前，嵌合体的问题将会严重阻碍人类生殖系基因组编辑在受精卵和早期胚胎中的临床应用，虽然最近的进展表明这个障碍将最终变得可以

clinically, one might imagine trying to use germline editing to correct the defect in all tissues. However, women who know they are carriers for a DMD mutation could use PGD to avoid having an affected child. Furthermore, one-third of DMD cases are due to de novo mutations, which would not be recognized until after birth and thus are not amenable to germline genome editing. Somatic editing approaches currently appear to be more useful than germline editing for this disease However, the pace of advances in genome editing methods and stem cell biology may alter that situation.

SCIENTIFIC AND TECHNICAL ISSUES

Practically speaking, considerable technical difficulties remain to be overcome in applying genome editing to zygotes and early embryos. Although the efficiency and accuracy of targeting can be extremely high, and there are sound reasons for believing that off-target effects can be greatly reduced (see Chapter 3 and Appendix A), there still would be a need to ensure that only embryos with correctly targeted alleles would be returned to the uterus to complete pregnancy.

Mosaicism

If genome editing were performed in a zygote (fertilized egg) or an early embryo, there would be a significant chance that some of the cells in the resulting early embryo would not have the desired (or even any) edits. This situation is called "mosaicism" and it presents a significant challenge to the application of germline genome editing on zygotes or embryos. Screening of an edited embryo by PGD to test for mosaicism would not ensure correct editing of the implanted embryo because a single cell may not reflect the genotype of the other cells of the embryo, and removal of multiple cells for testing would destroy the embryo.

The impact of mosaicism depends to some extent on the gene being targeted. Mosaicism is a serious problem if the gene of interest encodes a required cellular function, but if the gene encodes a secreted factor (e.g., growth hormone or erythropoietin), or leads to the secretion of a required molecule (such as a blood clotting factor), then correcting the gene in only a subset of cells may be sufficient. Furthermore, because the germline in the resulting child may also be mosaic, editing only a subset of cells may not solve the problem for succeeding generations. But it may offer a better chance of finding a disease-free embryo after PGD, or allow culture and selection of edited spermatogonial stem cells (see section on potential alternative routes to heritable edits, below), thereby enabling those children to have unaffected offspring. Overall, at present, the issue of mosaicism would present a serious impediment to the clinical application of human germline genome editing in zygotes or early embryos, although recent progress suggests that this impediment may

克服（Hashimoto et al., 2016）。

可遗传编辑的潜在替代方法

编辑胚胎基因组不是实现遗传基因组修饰的唯一潜在方法。直接修饰配子（卵子和精子）或它们受精之前的前体可以克服嵌合体，也潜在地允许在体外受精之前预选适当的靶向配子。

目前存在若干潜在的途径来进行配子基因组编辑，其中一些已经在小鼠中使用，其他的仍然有待进一步发展。例如，精原干细胞（其将产生精子）可以通过活检睾丸来分离，编辑培养，检测正确的基因编辑，然后重新植入睾丸；或者，通过诱导多能性干细胞（iPS）产生精子或卵母细胞祖细胞，这样可以通过在干细胞中进行基因组编辑来达到目的。检测正确靶向的克隆，在体外或体内产生精细胞或者是精子，用于供体卵子的受精。这种技术在小鼠和其他哺乳动物，包括非人灵长类动物中取得了重大进展（Hermann et al., 2012；Hikabe et al., 2016；Shetty et al., 2013；Zhou et al., 2016）。在未来，将这种方法发展到可以确保精确和有效地校正致病变异体并非不可能（参见第3章和附录A中的进一步讨论）。如果对人类精子和卵子祖细胞进行基因组编辑成为现实，未来人类的遗传生殖系基因组编辑将要发生极大的变化。

人类基因库的影响

另一个需要考虑的问题是，一些引起严重遗传疾病（如镰状细胞贫血）的基因，已经经过正向选择以维持人口中的致病等位基因，因为当其只存在一个拷贝（杂合子）时，会产生一些抵抗传染病的保护机制。对于一些其他致病变体也是如此，有一些证据表明囊性纤维化可能就是其中之一，虽然还没有确定（Poolman and Galvani, 2007）。这样的例子已经使一些人质疑通过可遗传生殖系编辑移除致病变异基因是否将会显著改变人类基因库。如前所述，人类生殖系编辑治疗如果被批准，其病例数将会非常小，并且在可预见的未来几乎没有机会对基因库有显著的影响。前文中也提到任何可遗传生殖系编辑应该限于变成自然产生的序列［即将有害的致病变体（突变）等位基因转变为常见的非致病性DNA序列］，最小化后代基因修饰的意外风险（参见第6章关于以基因强化目的进行基因编辑的讨论

eventually become surmountable (Hashimoto et al., 2016).

Potential Alternative Routes to Heritable Edits

Editing the embryo genome is not the only potential way to achieve heritable genome modification. Approaches that directly modify the genomes of the gametes (eggs and sperm) or their precursors before fertilization could overcome problems of mosaicism and would potentially allow preselection of appropriately targeted gametes before *in vitro* fertilization.

There are a number of potential routes to gamete genome editing, some of which are already in use in mice and others of which remain to be fully developed. For example, spermatogonial stem cells (which will give rise to sperm) could be isolated by biopsy from testes, edited in culture, tested for correct gene editing, and then reimplanted into the testes. Alternatively, generating sperm or oocyte progenitors via pluripotent induced pluripotent stem (iPS) cells would allow genome editing to occur in the stem cell population. Correctly targeted clones could be identified and used to generate spermatids or perhaps sperm, either *in vitro* or *in vivo*, and used to fertilize donor eggs. Significant progress on such technologies is being made in mice and other mammals, including nonhuman primates (Hermann et al., 2012; Hikabe et al., 2016; Shetty et al., 2013; Zhou et al., 2016). A future in which this kind of approach could be extended to ensure precise and effective correction of a disease-causing variant in gametes is not unrealistic (see further discussion in Chapter 3 and Appendix A). The future prospects for heritable germline genome editing in humans will change dramatically if genome editing in progenitors of human eggs and sperm becomes a reality.

Effect on the Human Gene Pool

Another consideration is that some genes that cause serious genetic diseases, like sickle-cell anemia, have been subject to positive selection to maintain the disease-causing allele in the population because it produces some protection against infectious disease when present in one copy (heterozygous). The same might be true for some other disease-causing variants and there is some evidence suggesting that might be the case for cystic fibrosis, although that is not yet established (Poolman and Galvani, 2007). Such examples have led some to question whether heritable germline editing to remove disease-causing variant genes might significantly alter the human gene pool. As discussed earlier, the numbers of cases of human germline editing to treat disease, if it were to be approved, would be very small and there is little chance of any significant effects on the gene pool in the foreseeable future. It has also been proposed that any heritable germline editing should be restricted to making changes that occur naturally in the human population (i.e., converting deleterious disease-causing variant (mutant) alleles to a common non-pathogenic DNA sequence), to

论）。将致病变体改变为已知的非致病性突变序列是目前所有预想的治疗应用的前提，因此任何这种为了治疗目的的生殖系基因组编辑，预期对人类基因库的影响都很小。

选择合适基因靶点的能力

最后，目前我们对人类基因、基因组、遗传变异和基因与环境的相互作用知识的掌握程度能否足以保证生殖系基因组编辑安全进行的疑问越来越多。虽然目前我们的知识对于一些基因编辑是足够的，但在许多情况下还不够。目前此技术仍存在不确定性，例如，为什么与阿尔茨海默病的风险增加明显相关的 APOE4 等位基因，在人类基因库中有这样高的出现频率。一种理论是，它可能在其他方面具有优势，类似于镰状细胞突变的杂合优点提供对抗疟疾的保护。基因如 APOE4 不会是生殖系基因组编辑的良好候选基因，因为它可以提供一些免受丙型肝炎感染肝损伤的保护（Kuhlman et al., 2010; Wozniak et al., 2002），同时其毒害作用不会完全渗透。随着大规模的项目将基因组序列与健康细节、环境和生活方式联系起来，积累基因组-环境相互作用的知识，如英国的十万基因组计划和美国的精准医疗计划。随着对基因组进化的认识，以及基因编辑和干细胞技术的改进，未来编辑可遗传生殖系用来改善人类健康和福祉的可能性是一个需要持续谨慎考虑的问题。每一个潜在的目标基因都需要在科学上和伦理上进行仔细评估，只有充分了解的基因，才能进行生殖系基因组编辑。

以纠正引发疾病性状为目的的种系编辑带来的伦理监管问题

近半个世纪前，Bernard Davis 发表了一篇评论，预见性地概述了今天仍在进行的关于基因研究的承诺、风险和路线图的讨论，包括进行生殖系可遗传性状的改造（Davis，1970）。他开篇便呼吁"客观地评估通过各种手段修改人类基因模式的前景"；继而又谨慎地提出，人们必须记住"最有趣的人类特征，与智力、气质和身体结构相关的特征，是高度多基因的"，因此依赖于大量基因与环境之间复杂的相互作用。直到今天该观点仍然是正确的，但

minimize risk of unexpected effects of the modification in generations to come (see also Chapter 6 for discussion of gene editing for enhancement purposes). Changing a disease-causing mutation to a known existing non-pathogenic sequence would be the case in any currently envisioned therapeutic applications and thus the effect of any such germline genome editing changes for therapeutic purposes is expected to have minimal effect on the human gene pool.

Ability to Select Appropriate Gene Targets

Finally, the issue arises of whether current knowledge of human genes, genomes, and genetic variation and the interactions between genes and the environment is sufficient to enable germline genome editing to be performed safely. While our knowledge is arguably sufficient for some genes, in many cases it currently is not. There is uncertainty, for example, about why the APOE4 allele, which clearly correlates with increased risk of Alzheimer's disease, is present in the human gene pool at such a high frequency. One theory is that it may confer an advantage in other respects, similar to the heterozygous advantage of sickle-cell mutations that confer protection against malaria. A gene such as APOE4 would not be a good candidate for germline genome editing because it may confer some protection against liver damage by hepatitis C infection (Kuhlman et al., 2010; Wozniak et al., 2002) and also the fact that its deleterious effects are not fully penetrant. Knowledge of genome–environment interactions will improve over time as large-scale projects linking genomic sequences with details of health, environment, and lifestyle are carried out such as the 100,000 Genomes Project in the United Kingdom and the Precision Medicine Initiative in the United States. As understanding of the genome progresses and genome editing/stem cell technologies improve, future possibilities for editing the heritable germline to improve human health and well-being will need to be the subject of ongoing, careful consideration. Each potential target gene would need to be evaluated carefully on both scientific and ethical grounds, and only well-understood genes would be suitable candidates for germline genome editing.

ETHICS AND REGULATION OF EDITING THE GERMLINE TO CORRECT DISEASE-CAUSING TRAITS

Nearly half a century ago, Bernard Davis published an essay that presciently outlined discussions still underway today, about the promise, the risks and the roadmaps for genetic research, including research on making heritable changes in the germline (Davis, 1970). He began his article with a call "to assess objectively the prospects for modifying the pattern of genes of a human being by various means," and continued with a caution: that one must keep in mind that "the most interesting human traits—relating to intelligence, temperament, and physical structure—are

更重要的是提高对控制复杂性状基因调控通路的认识，以及持续评估生殖系基因组编辑的潜在受益和风险。

平衡个人受益和社会风险

关于可遗传性生殖系基因编辑技术辩论的一个焦点是，需平衡给个体（如准父母和孩子）带来的潜在受益与其可能产生的个体的风险，以及对社会及人文的损害。这是一个复杂的伦理学问题，很大程度是因为个人的利益和风险更加直接、更加具体，而对人文效应的担心毫无疑问更加空泛。此外，虽然对过去的技术革新的考察可以帮助预测人文变化，但仍然具备不确定性，因为由新技术带来的任何人文变化都需要时间来显现，故不能直接摆出论据，因此，这种伦理学辩论会变得很困难。

在美国，对社会和文化问题的适当关注是在公民权利和法律判决的框架下进行的，通常是考虑人身自由度的束缚或区别限制是否在某些州合理且在必要的限制制度范围内。

这些挑战的结果通常较少取决于支持或反对这项技术的实质性论据或者个人选择，而更多地是由维持限制的正当性程度所决定的。当普通自由受到限制时，法院将只会支持政府限制的纯粹理性立场。但如果自由是根本的，如那些在《人权法案》中特别指出的或法院认定为基本权利的，政府限制则需要一个更有说服力和精心构造的理由。但后者的界定是不确定的。因为法院认定不存在于人权法案中的基本权利时，其方法和合法性存在长期争议。生育权利属于这一争议领域，使得对生育活动一个或多个方面的政府限制进行质疑时，不知道何种正当性程度适合（Murray and Luker，2015）。

父母受益

最可能从可遗传性生殖系基因组编辑获益的个体是，想要有一个正常的自己的孩子但又害怕遗传给孩子某种疾病的父母（和那个孩子）。这种愿望显而易见：当许多准父母在面对生一个自己的孩子或者生出的孩子可能患有遗传性疾病风险的抉择时，他们将选择冒险生出可能患病的孩子（Decruyenaere et al.，2007；Dudding et al.，2000；

highly polygenic" and therefore depend upon large numbers of genes interacting in complex ways with the environment. This is still true, but more is being learned every year about the genetic regulatory circuits that control complex traits, and there is an ongoing need to consider the potential benefits and risks of germline genome editing.

Balancing Individual-Level Benefits and Societal-Level Risks

One of the challenging characteristics of debates concerning heritable germline editing is that they require a balancing possible benefits that accrue primarily to individuals (such as prospective parents and children) against not only risks to the individuals, but also against possible harms that are discussed at a social and cultural level. This is a complicated ethical analysis, in no small part because the individual benefits and risks are more immediate and concrete, whereas concerns about cultural effects are necessarily more diffuse. In addition, although examination of past technological innovations can help make predictions about the cultural changes, these remain necessarily speculative, because any cultural changes resulting from a new technology take time to develop. Thus, the ethical debates become difficult because the arguments can fail to engage each other directly.

In the United States, appropriate consideration of social and cultural concerns is usually resolved within the context of civil rights jurisprudence and legal decisions, which compare the burden on individual liberties or the discriminatory impact of those burdens to whether there is a rational or compelling need for these particular state restrictions.

In these challenges, the outcomes are often determined less by the substantive arguments for and against the technology or an individual's choice, and more by the level of justification required to uphold the restrictions. When ordinary liberties are restricted, a mere rational basis for governmental restrictions will be upheld by the courts. If the liberties are fundamental, such as those specifically identified in the Bill of Rights or otherwise deemed fundamental by the courts, then a much more compelling and well-crafted justification for restricting fundamental liberties. The contours of the latter category can be uncertain, however, because of enduring debates surrounding the methodology and legitimacy of judicial determinations that some rights are fundamental despite their absence from the Bill of Rights. Procreative rights fall within this disputed area, making it harder to predict which level of justification will be needed in case of challenge to government restriction on one or more aspects of procreative activities (Murray and Luker, 2015).

Parental Benefits

The possible benefits of heritable germline genome editing accrue most immediately to individuals: the

Krukenberg et al., 2013）。如果是针对已经研究透彻的基因进行编辑，且所做的改变是编辑为已知的、常见的、非致病性序列，那么生殖系基因组编辑很多时候可能是比移植前基因诊断（PGD）更有效或更易接受的选项。它使父母受益，让出生的孩子享受更好的健康。对于可能导致早逝的疾病，生殖系基因组编辑可以让生出的孩子有一个更正常的寿命。

在美国，允许生殖系基因组编辑符合父母自主权最广泛的法律和人文解释。生育一个自己的孩子的愿望可能来自多方面，从想让自己或祖先外貌特征反映在孩子身上的愿望，到对满足家系传承、达到某种形式的不朽的生物学联系的信仰（Rulli, 2014）。在不同的国家和文化中，限制生殖系基因组编辑技术可能被视为限制父母的自主权。事实上，有些人认为他们有宗教或历史的义务来生育自己的孩子。还有一些人不同意这种父母自主权的看法，他们认为生殖系编辑迈出了将孩子产品化的一步，是对自身无法避免的缺陷的不能容忍和不能满足父母期望的结果（Sandel, 2004）。有些人会说，满足生育自己的孩子的愿望不是一种真正的受益，因为尤其是在收养、同性婚姻、配子捐献、代孕和继父母越来越普遍时，它代表了被某些人认为已经过时的亲属和家庭观念（Franklin, 2013）。

在美国，生育自由相关法律案例可以保证个人拥有可以按照自己的意愿选择何时要孩子，以及怎样养育孩子的权利[53]。相关案例聚焦在父母选择符合自己偏好的方式养育孩子并塑造其性格的权利；不被政府强制节育的权利；用避孕方法避免孕育的权利；在胎儿尚不能生活于子宫外的时候终止妊娠的身体自主控制权；以及在胎儿可以生活于子宫外的时候终止妊娠的保护自身健康的权利。在一个由《美国残疾人法案》法定解释的相关案例中，美国最高法院承认生育是一项主要的生命活动[54]。这些案例中的普遍内容包括实现怀孕的方法，以及使用相同技术降低那些孩子患疾病与残疾风险的权利。

然而，美国的宪法法案并不直接涉及摧毁子宫

prospective parents who want to have an unaffected genetically related child (and that child) but fear passing along a disease. The desire for genetic relation is evidenced by the fact that many prospective parents, faced with the choice between foregoing genetically related children or risking the birth of a child with a genetic illness, will choose to risk having an affected child (Decruyenaere et al., 2007; Dudding et al., 2000; Krukenberg et al., 2013). If an edit is made in a gene that is well-understood, and the change is a conversion to a known, common, non-pathological sequence, then germline genome editing might offer an option that is at times more effective or acceptable than PGD. It would offer benefits to the parents and allow for the birth of a child who will enjoy better health. In the case of some disorders that are lethal at a young age, it allows for the birth of a child with the prospect of a more ordinary lifespan.

Access to germline genome editing would be consistent with the broadest legal and cultural interpretations of parental autonomy rights in the United States. The desire to have genetically related children may arise from a variety of factors, ranging from a wish to see one's self or one's ancestors reflected in the appearance of the children to a belief in the need for a biological linkage in order to satisfy a sense of lineage, continuity or even some form of immortality (Rulli, 2014). Precluding access to this technology could be regarded as limiting parental autonomy, depending upon the country and the culture. Indeed, some people feel they have a religious or historical mandate to have genetically related children. There are others who do not share this view of parental autonomy, and see germline editing as a step toward seeing children as constructed products, and an increasing intolerance of their inevitable imperfections and failures to live up to parental expectations (Sandel, 2004). And some would argue that satisfying the desire for genetically related children is not an unalloyed benefit, as it can be seen as reifying what some view as outdated notions of kinship and family, precisely at a time when adoption, same-sex marriage, donor gametes, surrogacy and step-parenting are being normalized (Franklin, 2013).

In the United States, procreative liberty is grounded in legal cases that relate to the right to have children at the time one wishes, and with considerable latitude in rearing practices.[53] Relevant cases focus on a right to rear children and shape their characters largely to fit parental preferences; to not have the state involuntarily sterilize persons; on the right to use contraception to avoid conceiving; on the right to control one's body even if it entails terminating a pre-viable pregnancy; and on the right to preserve one's health even if it entails terminating a viable pregnancy. In a related case

---

[53] *Meyer v Nebraska*, 262 U.S. 390 (1923); *Pierce v Society of Sisters*, 268 U.S. 510 (1925); *Farrington v. Tokushige*, 273 U.S. 284 (1927); *Prince v. Massachusetts*, 321 U.S. 158 (1944); *Wisconsin v Yoder*, 406 U.S. 205 (1972).
[54] *Bragdon v. Abbott*, 524 U.S. 624 (1998).

外胚胎的权利，也不涉及移植前基因诊断（PGD）或试管受精（IVF）的合法性。上述观点能在多大程度上解释生育自由仍未知，父母权利和生殖的相关案例也未明确地支持这种解释（Nelson，2013）。就像所有自由一样，生育自由既可以被视为一种保护当事人免受政府禁令的被动权利，又可以作为一种要求政府促进选择或提供服务的主动权利。在美国，生育的被动权利保护父母免于政府对生育选择（如使用避孕方法）和父母酌情权（如教育教学语言的选择）等关键方面的禁令。但是，保护患者的健康和受益的合理的技术规范，甚至涉及相关的宪法权利，都理应得到全面批准。这些并未涉及要求政府资助甚至批准新的生殖技术的主动权利。

潜在风险

生殖系基因组编辑可能的收益需要与各种潜在的风险平衡。正如本章前面所讨论的，目前基因组编辑技术仍然面临技术挑战，只有克服这些技术挑战，才能对人类生殖系进行基因组编辑，这些需要谨慎和仔细审查任何尝试临床试验的提案。

非预期后果

担忧之一是人工干预可能产生意想不到的后果，这种担忧有两个不同的方面：其一，在体细胞基因组编辑过程中可能存在一定的脱靶效应。如本报告其他部分所述（详见附录 A，以及其他章节），这个技术问题受到很多科学关注。尽管通过对基因组编辑技术的改进，脱靶事件的发生率正在降低，评估脱靶率的方法正在被开发，其中一些方法已经批准用于体细胞治疗，但是还没有达到可以授权在生殖系基因编辑的临床试验上的要求。在任何临床使用生殖系基因组编辑试验之前，证明基因编辑不会导致任何显著增加的非预期变异是非常必要的。这种变异的可接受程度需要比体细胞基因修饰带来的变异要低，但是，考虑到生殖系基因组编辑被接受还有一段时间，降低和评估脱靶事件毫无疑问还有极大的改进空间，并且，体细胞基因组编辑的经验也将帮助评估变异率的接受范围。

无论采用什么样的标准来评判体细胞基因编辑应用中的脱靶效率，对生殖系基因组编辑的要求都会

concerning statutory interpretation of the Americans with Disabilities Act, the U.S. Supreme Court acknowledged that procreation is a major life activity.[54] The broad view of these cases would include methods of achieving pregnancy and a right to use the same technologies to reduce the risk of disease and disability in those children.

However, the constitutional law cases in the United States do not directly address a right to destroy an *ex utero* embryo, nor do they address PGD or the legality of IVF. The expansive view above remains untested with respect to how broadly it construes procreative liberty, and related cases on parental rights and reproduction do not clearly support this interpretation (Nelson, 2013). Procreative liberty—like all liberties—can be viewed either as a negative right that protects parties from governmental prohibitions or as a positive right that obligates government to facilitate choices or provide services. In the United States a negative right analysis of procreation protects parents from government prohibitions on key aspects of reproductive choice (such as use of contraceptives) and parental discretion (such as choice of language of educational instruction). But reasonable regulation of a technology for the protection of the health and wellbeing of those affected is entirely permissible, even where claims of constitutional rights can be made. And this has never been extended to a positive right to demand that government fund or even approve new reproductive technologies.

Potential Risks

Balanced against the possible benefits of germline genome editing are a variety of potential risks. As discussed earlier in the chapter, in its current state genome editing technology still faces technical challenges that would need to be overcome before it can be applied to human germline genome editing and these require caution and careful review of any proposals to proceed to clinical trials.

Unintended Consequences

One concern is that human intervention may have unintended consequences. In the case of germline genome editing, this concern has two distinct components. The first is the possibility of off-target effects of the editing process, as in the case of somatic genome editing. As discussed elsewhere in this report (in detail in Appendix A but also in other chapters), this technical question is receiving a lot of scientific attention. Although improvements in genome editing technology are reducing the incidence of off-target events and methods are being developed for assessing their rate, with some approaches already approved for somatic therapies, they have not yet reached the point where clinical trials in germline gene editing could be authorized. Before any clinical trials using germline genome editing could be approved it would be necessary to demonstrate that the

更加严格。定义那些受到基因编辑影响的人也要包括其未来的后代，他们并没有决定作为被研究或是治疗的对象，然而副作用可能通过逐代繁衍被放大。这都决定了要采取更加保守的方法来达到风险-受益之间的平衡。

担忧之二是，即使基因编辑的脱靶率为零，基因编辑本身仍然存在一些不确定的后果。就生殖系基因组编辑而言，其目的是将一种已知造成疾病的基因变异转变成另外一种不产生病理表型的变异，结果是编辑成在人群中已经广泛存在无严重后果的基因拷贝，因此不太会产生一些意想不到的后果。而另一方面，靶向编辑的非预期后果的问题确实存在，若编辑成基因组中并不普遍存在的 DNA 序列，则产生了所谓的"基因强化"，在生殖系编辑中，这些改变会影响后代，所以更令人担忧，这个问题在第 6 章将进一步讨论。

长期随访

与处理任何新的技术一样，生殖系基因组编辑也需要仔细监控临床试验方案，注意监控脱靶事件及基因编辑的效率和正确性。与常规临床试验不同，生殖系基因组编辑需要对后续世代进行长期的、有目的性的随访研究。这种随访研究需要研究受这种治疗影响的孩子，尽管这些孩子在当初参与研究时并没有决定权。这一类数据对于决定这项技术是否达到治疗目的非常重要（Friedmann et al.，2015）。即使是那些自愿参与到研究项目中的人，也不能被强迫参与长期随访，因此需要加以不断地鼓励。异种移植及一些药物和器械试验的经验表明，这种鼓励能够成功。尽管研究后代比研究志愿者更具挑战，但是其他生殖技术相关的经验表明，足够数量的后续研究还是可以做到，这将帮助得出许多长期效应的结论（Lu et al.，2013）。

社会影响

有些人认为，生殖系编辑有很多理由可以获得支持，而不仅仅是父母的选择。有一种观点认为，对生殖细胞的修饰能够为目前处于劣势的人群和他们的后裔创造一个公平的竞争环境（Buchanan et al.，2001）。另一些人认为生殖系基因组编辑对公共健康有益，因为它可能在某种程度上减少许多恶

editing procedures would not lead to any significant increase in unintended variants. The required level of such variants would necessarily be lower than for somatic genome editing but, given that germline genome editing will not occur for some time, the technology for minimizing and assessing off-target events will undoubtedly have improved significantly. In addition, experience with somatic genome editing will have refined understanding of what might be considered acceptable or unacceptable rates of unintended variants.

Whatever standards are developed for somatic applications, there will be less tolerance for off-target effects in germline applications. By definition those affected by the edits (future offspring) did not make the decision to be subjects of research or attempts at therapy, and adverse effects might be multiplied by reverberation across generations. Both factors lead to a more conservative approach to the risk-benefit balance.

A second concern is that the intended genome edits themselves might have unintended consequences, even in the absence of off-target effects. In the case of germline genome editing to convert a well understood disease-causing variant gene to a widely occurring non-pathological variant, the editing change would be to a version of the gene that is known not to have deleterious consequences. These are broadly present in the population already, so the chances that the edit will have some unexpected effects are small. On the other hand, the question of unintended consequences of a targeted edit does arise in the context of edits to make a change to a DNA sequence that is not already prevalent in the population, as would be the case of some so-called "enhancements." In germline editing, the concern is magnified because the alteration could affect descendants, which is discussed further in Chapter 6.

Long-Term Follow-Up

As with any new procedure, carefully monitored clinical trial protocols would be required for germline genome editing, with attention to monitoring off-target events as well as the efficiency and correctness of the specific edit. Unlike conventional clinical trials, germline genome editing trials would likely require long-term prospective follow-up studies across subsequent generations. This follow-up would entail study of the future children affected by the intervention, none of whom would have been party to the initial decision to participate in a research trial. Data of this type would be important for determining whether the techniques had achieved their goals (Friedmann et al., 2015). Even those who have volunteered to be research subjects cannot be compelled to participate in long-term follow-up. Nonetheless, encouragement is permitted. Experience from xenotransplantation, and some drug and device trials, shows that this encouragement can be successful. And despite the particular challenge of studying offspring rather than

性疾病，如神经节苷脂沉积病和亨廷顿病。即便如此，也应注意在历史上，滥用和强制优生（在下文"人类尊严和对人类优生的恐惧"部分中有讨论）与善意的公共健康和卫生运动错综复杂的关系，这就是为什么讨论公共卫生福利往往引起一些怀疑和不安。一些当代的超人类主义者指出，人体本身是有缺陷的，容易生病，需要大量的睡眠，具有各种认知极限，并最终死亡。他们认为通过改善人类物种，使人类变得更耐病、更道德和更智能是有意义的（Hughes，2004）。一些人，如哲学家约翰·哈里斯认为，在某些情况下，从基因方面强化我们自己是道义上的义务（Harris，2007）。但这些都是关于基因强化的论点，而不是恢复或维持普通的健康。第6章将更详细地讨论这个问题。

一个"天然"的人类基因组与适当的人为干预

在反对生殖系基因组编辑的社会和文化的论证中，人们更支持"天然"的基因组。尽管对农业和医学的人工干预已得到广泛接受，但有些人认为人类基因组是不同的，不应该对其进行操纵，这归结于它自然属性的某些方面，可以定义为"正常"、"真实"，或是由非人力干预的力量决定（Nuffield Council，2015）和"人性"（Machalek，2009；Pollard，2016）。然而，人类基因组并不完全是"人类"，它包括了尼安德特人（Neanderthal）和丹尼索瓦人（Denisovan）的DNA（Fu et al.，2015；Pollard，2016；Vernot et al.，2016）。它也不存在任何单一的、静止的状态。如第4章所述，每次细胞分裂时，DNA序列就会发生许多变化，环境污染（如辐射）和化学物质（天然的和合成的）也会导致序列变化。此外，减数分裂和受精过程在每个个体中产生了一种新型的基因变体，结果是个体之间甚至在单个个体的细胞内基因组DNA序列都存在显著变化（Kasowski et al.，2010；Zheng et al.，2010）。每个人（除了同卵双胞胎之外）都开始于独特的基因组——实际上两个基因组随着细胞分裂而多样化，每个基因组是"天然的"，没有一个人的基因组由全人类共享。

这种担忧转变成人们应该以谦逊的态度对待人类基因组，人类应该认识到智慧和科学的极限，甚至人为干预比自然过程更危险或更不可预测。这种担心通常用术语"扮演上帝"来表示，它认为人类

those who consented to the research, experience with other reproductive technologies suggests that follow-up can be carried out in numbers sufficient to permit conclusions about many possible long-term effects (Lu et al., 2013).

Societal Effects

Some have argued that germline editing can be justified on grounds that go beyond mere parental choice. There is a line of thought that germline modification could be used to create a level playing field for those whose traits now put their children and descendants at a disadvantage (Buchanan et al., 2001). Others see a public health benefit in access to germline genome editing because it might somewhat reduce the prevalence of many devastating diseases, such as Tay-Sachs and Huntington's disease. That said, it is important to note that the history of abusive and coercive eugenics (discussed in section on human dignity and the fear of eugenics, below) is intertwined with previous, undoubtedly well-intentioned public health and hygiene movements, which is one reason why discussions of public health benefits often engender some skepticism and unease.

Some contemporary transhumanists point out that the human body is flawed in that it easily becomes diseased, requires a great deal of sleep, has various cognitive limitations, and eventually dies. They suggest that it would make sense to improve the human species by making it more resistant to disease, more moral, and more intelligent (Hughes, 2004). Some, such as philosopher John Harris, say that in certain cases there is a moral obligation to enhance ourselves genetically (Harris, 2007). But these are arguments about enhancement, not the restoration or maintenance of ordinary health. That topic is addressed in more detail in Chapter 6.

A "Natural" Human Genome and the Appropriate Degree of Human Intervention

Among the social and cultural arguments against germline genome editing are positions that support a preference for a "natural" genome. Although there is wide acceptance of human intervention in agriculture and medicine, some hold the view that the human genome is different and should be free of intentional manipulation because of some aspect of its naturalness, whether defined as "normal," "real," or otherwise determined largely by forces other than human intervention (Nuffield Council, 2015) and some aspect of its "humanness" (Machalek, 2009; Pollard, 2016). However, the human genome is not entirely "human," as it includes Neanderthal and Denisovan DNA (Fu et al., 2015; Pollard, 2016; Vernot et al., 2016). Nor does it exist in any single, static state. As reviewed in Chapter 4, each time a cell divides, numerous changes in the DNA sequence occur, and environmental insults such as radiation and chemicals (both natural and synthetic) also produce sequence changes. Moreover,

不像上帝一样全能，能够在基因组中安全地进行任何改变（并且预测改变会达到预期目的）。甚至从非神学角度来探讨该问题时也使用了该术语来表示对环境或人类基因组的人为操控应该加以适当限制（President's Commission，1983）。

在某种程度上，这个论点隐含地接受这样的论断：自然和进化的力量对基因组的改变是比人类干预更好或至少出现更少问题的方法。然而，人类基因组中的自然变异是偶然产生的，在进化过程中通过始祖效应和根据选择压力（如气候、营养和传染病）进行选择或淘汰，其中一些可能与现代世界完全不相关。因此，虽然重要的是认识到人类理解的极限并谨慎行事，这并不一定意味着社会应该放弃任何人类干预。

总的来说，从科学的角度来看，关于生殖系基因组编辑可能带来的收益和风险的一些结论可以得到相当程度的确定，而另一些结论仍然不确定，需要进一步的调查和社会辩论来呼吁对这些结论保持慎重的态度。评估功效和风险的可能性是临床试验的核心，可以将其视为了解人类知识极限的表现。

除了对风险和不确定性的科学评估，人类对自然界的适当干预程度，长期以来在精神和宗教领域也进行了讨论。在当代西方，基督教传统对今天宗教多样性和世俗文化产生了极大的影响，这些思想一直在辩论关于改善或管理自然的任务中，哪些是人类的领域或义务，又有哪些是上帝的（Cole-Turner，1993；Vatican，2015）。这种思想反映了各种传统和圣经故事中的信仰，包括圣弗朗西斯的歌曲创造（Canticle of Creation）和一些美国原住民的信仰体系。

另一方面，在犹太教传统中，有明确的义务来建立和发展有利于人类的世界，这种建立和发展被看成是上帝与人类之间的积极合作，而不是相互干预（Steinberg，2006）。同样，许多穆斯林和佛教徒认为基因工程只是众多减少疾病的痛苦的干预措施之一（HDC，2016；Inhorn，2012；Pfleiderer et al.，2010）。问题在于，在自然界及人类自身中，什么程度的人为干预是合适的或者是被允许的。这是宗教和非宗教人士提出的一个精神和实践问题，虽然有时更经常是由前者提出（Akin et al.，2017）。即使在宗教信徒中，不同信仰的信徒也会有不同程度的兴趣或关注点（Evans，2010）。

meiosis combined with fertilization creates in each individual a novel assortment of gene variants. The result is significant variation in genomic DNA sequence among individuals (Kasowski et al., 2010; Zheng et al., 2010) and even within the cells of a single individual. Every human (other than monozygotic twins) begins with a unique genome—actually two genomes that subsequently diversify as cells divide, each of which is "natural. There is no single human genome shared by all of humanity.

The concern then devolves to the view that the human genome should be treated with a sense of humility and that humanity should recognize the limits of wisdom and science, and even that human intervention is more dangerous or more unpredictable than natural processes. This concern often is expressed by the term "playing God," which captures the notion that humans lack a god-like omniscience that would be required to make any changes in the genome safely (and to predict that such changes would actually serve the intended purpose). Even those approaching the issue from a nontheological point of view may use the term to represent a more general notion about appropriate limits of intentional human control of the environment or of the human genome (President's Commission, 1983).

To some extent, this argument implicitly accepts the thesis that the forces of nature and evolution are a better—or at least less problematic—source of genome alteration than human intervention. However, the natural variations in human genomes arise by chance and are selected for or against during evolution by founder effects and according to selection pressures such as climate, nutrition, and infectious disease, some of which may no longer be relevant in the modern world. Accordingly, while it is important to recognize the limits of human understanding and proceed with all due care, this does not necessarily mean that society should forswear any human intervention at all.

Overall, from a scientific viewpoint, some conclusions about likely benefits and risks of germline genome editing can be supported with a fair degree of certainty, while others remain uncertain and in need of further investigation and societal debate, calling for humility with respect to those conclusions. Assessing the probabilities of efficacy and risk is the focus of clinical trials, which can be viewed as a manifestation of the recognition of the limits of human knowledge.

Beyond the scientific assessment of risks and uncertainties, the question of the proper extent of human intervention in nature has long been discussed in spiritual and religious terms. In the contemporary West, where Christian traditions have had the most influence on what is today a more religiously diverse and often secular culture, these ideas are expressed in the debate about which tasks in improving or stewarding nature are the domain or obligation of humans and which are to be left to God (Cole-

人类尊严和对人类优生的恐惧

禁止修改生殖细胞的国际盟约、协定、国家宪法，包括欧盟条约，通常会援引"尊严"这一概念（Hennette-Vauchez，2011）。虽然这个术语有着多重含义，但是在有关生殖系基因组编辑的争论中最经常被援引的是，人是有价值的，仅仅是作为人的价值而不是靠能力大小或者是否能够服务他人的意志来决定的（Andorno，2005；Sulmasy，2008）。Emmanuel Kant 将人类能动性和自由意志视为人类尊严的本质部分。这一术语也能凸显出对人类有别于其他物种的特殊敬意、对人类智慧的欣赏，以及对促进人类自主和繁荣的保证。既然权利和其他个人主义的论点不能轻易用于解决对子孙后代或人性的担忧，"尊严"一词便成为"防止日新月异的生物技术被滥用的最终学说依据"（Andorno，2005）。即使仅限于预防严重的疾病或残疾，使用可遗传的生殖系基因组编辑的前景依旧引发关注，纯粹自愿的个人决定的累积也可改变对于接受较不严重的残疾的社会规范（Sandel，2004）。

残疾人权利社群也并非一个整体，它对遗传学技术如产前筛查的态度从支持到怀疑都存在（Chen and Schiffman，2000；Saxton，2000）。残疾人社群对于是否使用筛选技术也在进行长期和持续的辩论。这种紧张关系是真实、持续，且不可能完全和解的。然而，残疾社群活动家一直是对于使用技术筛选或评判儿童遗传优劣最活跃的批评者之一。Jackie Leach Scully 写出了自己的恐惧：自愿产前诊断及基因组编辑技术，将把我们置于一个对残疾零容忍的"滑坡"之中（下面进一步讨论的一个概念），甚至有回到过去被强制性实行的风险（Scully，2009）。

还有人写道，基因筛选（及基因组编辑）的预防政策"似乎反映出残疾本身对于残疾的孩子、其家庭及社会是极大的负担，因此在医疗保健中优先考虑避免残疾，而这个判断夸大了甚至错误地认为许多（甚至是绝大多数）困难是由于残疾所致"（Wasserman and Asch，2006）。同样的观察也见于残疾人士和医疗人员对特定疾病带来的压力的不同感受（Longmore，1995）。研究表明，"残疾儿童及其家庭在生活的满意度上与许多健康专

Turner, 1993; Vatican, 2015). This thinking reflects beliefs that are present across a variety of traditions and many centuries, including St. Francis' Canticle of Creation and the belief systems among some Native American nations.

In the Jewish tradition, on the other hand, there is an explicit obligation to build and develop the world in any way that is beneficial to people, and such improvements are viewed as a positive collaboration between God and humans, not as an interference with creation (Steinberg, 2006). Similarly, many Muslims and Buddhists view genetic engineering as just one of many welcome interventions to reduce suffering from disease (HDC, 2016; Inhorn, 2012; Pfleiderer et al., 2010). The question will always be how much human-directed intervention in nature and in humans themselves is appropriate or even permissible. This is a spiritual and practical question asked by both religious and nonreligious people, although somewhat more often by the former (Akin et al., 2017). Even among the religious, adherents of different faiths will have varying degrees of interest or concern (Evans, 2010).

Human Dignity and the Fear of Eugenics

International covenants, treaties, and national constitutions, including European treaties focused on banning germline modification, typically invoke the concept of dignity (Hennette-Vauchez, 2011). While this term has many meanings, it is most often invoked in the debate about germline genome editing to affirm that humans have value simply by virtue of being human and not because of their capacities, and thus cannot be treated as instruments of another's will (Andorno, 2005; Sulmasy, 2008). Emmanuel Kant viewed human agency and free will as essential aspects of human dignity. The term also can signal a special regard for humans as opposed to other species, an appreciation of the intellectual capacities of humans, and a commitment to promoting autonomy and human flourishing. Since rights and other individualistic arguments cannot easily be used to address concerns about future generations or humanity, "dignity" has been invoked to "provide an ultimate theoretical reason to prevent a misuse of emerging biotechnological powers" (Andorno, 2005). Even if limited to preventing serious disease or disability, the prospect of using heritable germline genome editing triggers concern that purely voluntary, individual decisions can collectively change social norms regarding the acceptance of less serious disabilities (Sandel, 2004).

The disability rights community is not monolithic, and its attitudes toward genetic technologies such as prenatal screening can vary from supportive to skeptical (Chen and Schiffman, 2000; Saxton, 2000). There has been a long and ongoing debate between different parts of the disability community with regard to use of screening technologies. These tensions are real, continuing, and unlikely to be

家所认为的儿童残疾对儿童及其家庭主要起负面影响的观点是相反的"（Parens and Asch，2000，p. 20）。

确实，有人担心这种技术的可用性可能会导致放弃生殖系基因组编辑的父母是有过失的判断。随着基因筛选的常规化，这个观点被严肃地讨论过（虽然最终被放弃了）（Malek and Daar, 2012; Wasserman and Asch, 2012; Sayres and Magnus, 2012）。其他人则担忧，随着直接受影响人群数量的下降，通过艰难奋斗才争取到的惠及残疾人的法律和政策会失去支持。

也有人可能会说这些关注反映出一种错误的二分法，对已出生的残疾儿童无条件的爱，以及对所有先天或后天残疾者的尊重，与在出生或受孕之前进行干预以避免疾病和残疾儿出生并不矛盾。产前诊断（伴随选择性让受影响的胎儿流产）和植入前诊断（伴随选择性植入未受影响的胚胎）爆炸性增长的几十年里，公众对残疾的接受度也同样大为增长（Hernandez et al., 2004; Makas, 1988; Steinbach et al., 2016）。这样看来"鼓励降低遗传性疾病发病率的尝试与继续尊重那些患有该疾病并为其特殊需要提供支持的做法可以齐头并进"（Kitcher，1997，p. 85）。

残疾人社群对使用筛选技术保持着长期和持续的紧张气氛。文献资料公开承认这种紧张气氛是真实、持续，且不可能完全消除的，通向利用基因组编辑来消除缺陷的每一步都必须经过公开讨论和谨慎执行，委员会也支持公众继续审议（Kitcher, 1997）（见第 7 章）。

公共政策已转向消除在就业和公共服务方面的歧视，为实现这一目标，政府在改变社交、物质和就业环境方面增加了公共投资，具体措施从无障碍建筑到手语演示再到十字路口的听觉信号都有。然而，各类措施包含的范围仍显不足，无法预判若人们在得知如果基因筛查和堕胎的法律并没有使得减少出生缺陷变得更为容易，则是否会有态度上显著的转变。尽管如此，这种转变在一定程度上缓解了降低残疾发病率必然会降低人们对残疾人的同情、接受或一体化进度的担忧。

经济和社会的公正

我们意识到生殖系基因组编辑技术在近期内难

resolved entirely. Still, disability activists have been among the most visible critics of using technology to screen for or determine the genetic qualities of children. Jackie Leach Scully writes of a fear that voluntary prenatal diagnostic techniques, which would also apply to genome editing, set us on a "slippery slope" (a concept discussed further below) toward intolerance of disability and even the risk of a return to the coercive practices of the past (Scully, 2009).

Others write that a policy of prevention by genetic screening (and by extension genome editing) "appears to reflect the judgment that lives with disabilities are so burdensome to the disabled child, her family, and society that their avoidance is a health care priority—a judgment that exaggerates and misattributes many or most of the difficulties associated with disability" (Wasserman and Asch, 2006). The same observation has been made concerning the differing perceptions of disabled persons and medical professionals about the degree of distress caused by a particular condition (Longmore, 1995). Studies suggest that "many members of the health professions view childhood disability as predominantly negative for children and their families, in contrast to what research on the life satisfaction of people with disabilities and their families has actually shown." (Parens and Asch, 2000, p. 20)

Indeed, there is concern that the availability of the technology might actually lead to a judgment that parents who forego germline genome editing are negligent, a theory seriously discussed (although ultimately abandoned) with respect to genetic screening when it became a common practice (Malek and Daar, 2012; Sayres and Magnus, 2012; Wasserman and Asch, 2012). Others have cited fears that hard-won successes at developing laws and policies that make the world accessible to those with disabilities will lose support when the number of persons directly affected declines.

One can argue that these concerns reflect a false dichotomy, and that unconditional love for a disabled child once born and respect for all people who are born with or who develop disabilities are not incompatible with intervening to avert disease and disability prior to birth or conception. And the decades that saw the explosion of prenatal diagnosis (accompanied by selective abortion of affected fetuses) and preimplantation diagnostics (accompanied by selective implantation of nonaffected embryos) are the same as those in which public attitudes toward disability became far more accepting (Hernandez et al., 2004; Makas, 1988; Steinbach et al., 2016). It would seem that "encouraging attempts to reduce the incidence of a genetic disease is compatible with continuing respect for those born with the disease and providing support for their distinctive needs" (Kitcher, 1997, p. 85).

The disability community is characterized by a long and ongoing tension with regard to use of screening

以被广泛应用，物种不会有剧烈转变，文化规范也不会立即改变，一些社会公正的争论点集中在该技术只能影响到少数人上。在这种框架下，这一技术不过是又一次体现了社会将大量资源用于开发一种只有相对少数的富人能够受益的技术，而这笔钱完全可以被投入用已有的技术来减轻千百万贫困人口的困苦（Cahill，2008）。反驳观点认为，在研究阶段也能包括一些不那么富裕的人，或者即使对罕有发生但迫切的疾病的治疗通常从富人开始，治疗费最终会变得不那么昂贵而使穷人也能承担。此外，实现生殖系基因组编辑的研究对其他疾病的保健干预也能提供参考。或者，更确切的说，事实上至少在美国，保健预算不是全局规划的，因此，在一个领域减少开支并不能确保省下来的资金能被投入到其他有需要的领域。

另一种担忧认为，如果可遗传的基因组编辑在那些富有或有更好保险的人群中变得普遍，它可以改变优势和弱势群体之间可避免疾病的发病率，并可以永久地建立哈里斯（2007）所谓的"平行种群"。虽然巨大的不平等现象已经存在，但争论仍在持续，而生殖系基因组编辑将在生物学的不平等中引入文化上的不平等。虽然因为更充足的营养带来的长期效果和在世界优势人群中使用疫苗，这种现象已然存在，但一些批评者仍担忧增加一种更持久获得更好健康的捷径所带来的影响（Center for Genetics and Society，2015）。这些担忧适用于一系列的医疗进步，并不仅限于基因组编辑。

滑坡学说

支持（至少不反对）生殖系修饰的许多学者将基因组编辑的可能用途按照接受度进行了排序。这种排序总是从最容易接受的、将单基因疾病基因转变为常见而无害的序列这一端开始，逐渐移向最难以接受的、与疾病无关的基因增强。滑坡学说断言，一旦迈出了从单基因疾病改变的第一步，在多年后很可能会转向多数人反对的非疾病性的基因增强。一个实施体细胞修饰的团体在《自然》杂志上写道，"许多人反对生殖系修饰，理由是允许哪怕明确无误的治疗干预，也可能让我们走上一条通往非治疗性基因强化的道路"（Lanphier et al., 2015）。

大多数评论家的"滑坡论"并不断言必然性，

technologies. The literature appears to support openly acknowledging that this tension is real, continuing, and unlikely to be resolved entirely, and that any step toward the use of genome editing to eliminate disabilities must be carried out with care and open discussion, and the committee supports this call for continued public deliberation (Kitcher, 1997) (see Chapter 7).

Public policy has shifted toward eliminating discrimination in employment or public services, and public investment in changing the social, physical, and employment environment to achieve this goal has increased, with measures ranging from accessible buildings to sign language presentations to aural signals for street crossings. The range of measures remains insufficient, however, and one cannot know whether this shift in attitude would have been even more dramatic if genetic screening and abortion laws had not made it easier to reduce the prevalence of birth defects. Nonetheless, this progress does to some extent address the concern that reducing the prevalence of disabilities will necessarily decrease empathy, acceptance, or integration of those who have them.

Economic and Social Justice

Recognizing that germline genome-editing technology is unlikely to be used widely in the near future and that drastic transformation of the species or immediate changes in cultural norms are unlikely, some social justice arguments focus on the effects of the technology's being accessible only to a few rather than to too many. In this framing, the technology is another example of a society's allocating considerable resources to develop a technology that will benefit only a relatively few wealthy people when this money could be used to relieve the suffering of millions of poor people through already existing technologies (Cahill, 2008). One counterargument is that the research phase may include those less well-off, or that even if treatments for rare but compelling diseases often start with the wealthy, they eventually become more affordable and available for the poor. Moreover, the research that will make germline genome editing possible will likely provide insights that will lead to health care interventions for other disorders as well. More to the point, perhaps, is the reality that—at least in the United States—health care budgets are not set globally, and therefore the decision to refrain from spending in one area will not necessarily result in the funds being redirected to another area of need.

Another concern is that if heritable genome editing were to become prevalent among those who are wealthier or better insured, it could change the prevalence of avoidable diseases between advantaged and disadvantaged populations and could permanently establish what Harris (2007) calls "parallel populations." While great inequality already exists,

而是声明概然性。他们的学说基于描绘社会实际如何运作的预测社会学，反对有关滑坡上放置栅栏和减速路障可以阻止非预期应用的说法（Volokh, 2003）。过去的许多科学上的进步——从重构手术（导致了美容整形手术）到产前致死性疾病的筛查（导致了疾病基因携带者的筛查，以及植入前筛选非致死乃至迟发性疾病基因）——也引发了类似的、对于滑向非必要甚至是反社会的应用的担忧。

反对生殖系编辑的人原则上不一定反对用对应的、常见而非疾病基因替代突变的疾病基因，因为这种替代不会给后代带来社会优势，而只是一种对儿童未来行为的修正，是时下现代医学的一部分。然而，许多反对者不相信基因组编辑会暂停在这里，他们观察到一些社会进程将使得滑坡更加倾斜。提供这项服务的部分医疗行业、寻求服务的患者团体都可能会投资，从而创造出强大的维护甚至扩张该服务的利益集团。例如，IVF 最初被开发用于规避输卵管阻塞，然而，它很快就被拓展到规避自然发生的大龄相关的生育率下降，甚至是绝经后不育，后来成为 PGD 的前提。同样，PGD 最初是为筛选掉有严重有害突变的胚胎而生，但后来扩展到不是所有人都认同的疾病或缺陷的情况，以及性别选择。

另一方面，滑坡学说也有他们的批评者。批评者指出了该学说固有的不确定性和许多类似的声明并没有发生的事实。事实上，尽管有着相反的预测，无论是 IVF 还是 PGD 都没有被用于为了方便或选择平常性状上。即便是提供了一种廉价的方式来"优化"男性遗传贡献的人工授精，也没有被广泛应用，除非男性伴侣缺席、不育，或有传递严重有害突变的风险。此外，虽然有证据表明当涉及捐献卵子时人们更倾向于"优化"（Klitzman, 2016），在已经需要使用捐献精子的那些妇女中，几乎没有人利用该机会从所谓的"诺贝尔精子库"（Plotz, 2006）获取精子。那些摒弃滑坡学说的人对可能被视为滑坡底部的情况往往不如支持者那般担忧。

许多试图在人类遗传修饰进化的滑坡上引入减速路障或增强摩擦的尝试集中在容易理解的个人/个体和后代/群体之间语言/认知的差异，从而建立起体细胞修饰和生殖系修饰的区别。批评者声称，目前关于跨越认知屏障（即跨越生殖系修饰）的争the argument continues, germline genome editing would make a culturally determined inequality into one that is biological. While such a phenomenon already exists in the form of durable effects of better nutrition and use of vaccines among the advantaged populations of the world, some critics are concerned about adding yet another, more durable form of superior access to better health (Center for Genetics and Society, 2015). These concerns apply to a range of health advances, and are not limited to genome editing.

The Slippery Slope

Many scholars who support (or at least are not opposed to) germline modifications align the possible uses of genome editing along a continuum of acceptability. This continuum almost always starts with converting single-gene disorders to a common, nondeleterious sequence at the most-acceptable end, and moves toward enhancements that are unrelated to disease on the least-acceptable end. The slippery slope claim is that taking the first step with single-gene disorders is likely to lead, in some number of years, to the conduct of nondisease enhancements that many would rather see prohibited. As one group involved with somatic modification wrote in the journal *Nature*, "many oppose germline modification on the grounds that permitting even unambiguously therapeutic interventions could start us down a path towards non-therapeutic genetic enhancement" (Lanphier et al., 2015).

The slippery slope arguments of most critics do not claim inevitability but are instead probabilistic. They are based on what could be described as predictive sociology about how societies actually function and rejection of the notion that placing barriers and speed bumps on the slippery slope will be a sufficient deterrent to less desirable uses (Volokh, 2003). Many scientific advances in the past—ranging from reconstructive surgery (which has led to plastic surgery for aesthetics) to prenatal screening for lethal disorders (which has led to screening of carriers for disease genes and preimplantation screening for nonlethal, even late-onset disorders) —have raised similar concerns about a slippery slope toward less compelling or even antisocial uses.

An opponent of editing the germline would not necessarily oppose on principle replacing a disease gene variant with a corresponding, common, non-disease gene, as such a change would give offspring no social advantage and is the type of instrumental action directed at future children that is currently part of modern medicine. Many opponents, however, do not believe genome editing would stop there and observe that a number of social processes make the slope more slippery. Parts of the medical profession might become invested in providing the service, or patient groups in seeking the service, creating powerful interest groups favoring its maintenance or even expansion. IVF, for example, was

论证明了滑坡的存在。

总的来说，滑坡学说并不是反对可遗传基因组编辑最初的、最迫切的应用。但是，虽然许多人认为法规可以建立有效的减速路障，滑坡学说的支持者提出了这样一个问题：社会是否以及如何能够制定足够强大的法规来平息对其逐渐走向不迫切和更有争议的应用的恐惧。事实上，他们会说，法规几乎无法阻碍下滑，因为法规是人文观点的体现，而人文观点的根本变化恰恰就是滑坡。

规 则 制 定

美国的规范

在美国，可遗传的基因组编辑受复杂的州和联邦法律及规范的管理（见第2章）。不同的州，由于对胎儿及胚胎研究的管理法律法规不同，因此，造成了这些研究在各州之间的合法性甚至临床应用有很大差异。由于目前对涉及人类胚胎研究资助的立法限制，研究可能无法获得美国联邦基金资助。如果可遗传的生殖系基因组编辑进入临床试验，美国食品药品监督管理局（FDA）将具有监管管辖权。改变了的细胞——不管是配子还是胚胎——都需要植入妊娠，这将触发FDA行使其曾经在2001年行使过的用来确保生殖性克隆不会在没有授权的情况下进行的同样的管辖权（FDA，2009）。FDA在这方面管辖权来源于其规范组织移植的权力。然而，由于体外受精IVF甚至PGD在FDA的管理规定出台及完善前早已问世，因此并没有像一些近年才有的新技术一样受到了严格的管理，而基因编辑技术则会被FDA完全监控管理。

在任何临床试验获得批准前，国立卫生研究院（NIH）重组DNA咨询委员会（RAC）（有公众意见和透明审查）、地方机构审查委员会（IRB）、地方机构生物安全委员会（IBC）和FDA必须逐步谨慎地对其审查（详见第2章）。如果生殖系基因组编辑在临床试验中获得成功并被批准上市，则也要经过监督机制规范批准后才能应用。

由于生殖系基因编辑涉及其他辅助生殖技术的使用，所以其应用的监督管理应与IVF和PGD相同。其中有些管理规范集中于供体材料的安全性、透明度及报告要求，如IVF，或者关注于PGD实验室

originally developed to circumvent fallopian tube blockage. It soon was extended, however, to circumventing naturally age-related decline in fertility and even postmenopausal infertility, and later became an enabling technology for PGD. Likewise, PGD was originally designed to select against embryos with serious deleterious mutations but later was expanded to conditions that not all agree are diseases or disabilities, as well as to sex selection.

On the other hand, slippery slope arguments have their critics, who point to their inherent uncertainty and the fact that many such claims do not come to pass. Indeed, despite predictions to the contrary, neither IVF nor PGD has come to be used for convenience or for selection of trivial traits. Even artificial insemination, which offers an inexpensive way to "optimize" the male genetic contribution, has not become a widespread practice except when the male partner is absent, infertile, or at risk of passing on a seriously deleterious mutation. Furthermore, among those women who already needed to use donor gametes, almost none took advantage of the opportunity to obtain semen from the so-called "Nobel sperm bank" (Plotz, 2006), although there has been evidence of a stronger tendency to "optimize" when it comes to egg donation (Klitzman, 2016). Those who reject slippery slope arguments often are less concerned than proponents about situations which might be viewed as the bottom of the slope.

Many of the attempts to introduce speed bumps or friction on the slippery slope in the evolution of genetic modification of humans have focused on the easily grasped linguistic/cognitive difference between a body/individual and offspring/society, thereby establishing the distinction between editing of somatic and germline cells. Critics would claim that the current debate about crossing the cognitive barrier (i.e., crossing the germline) is proof of the existence of a slippery slope.

Overall, slippery slope arguments do not depend on universal condemnation of the initial, most compelling applications of heritable genome editing. But while many think that regulation could establish effective speed bumps, proponents of slippery slope arguments raise the question of whether and how society can develop regulations that are sufficiently robust to quell the fear of a progressive move toward less compelling and more controversial applications. Indeed, they would say that regulations would do little to stop the progression down a slippery slope because regulations are based on cultural views, and it is the underlying change in cultural views that is precisely the slippery slope.

REGULATION

Regulation in the United States

In the United States, heritable genome editing would

的质控情况（尽管并不是实际诊断）。可遗传生殖系编辑可以与 IVF 和 PGD 结合应用，所以对可遗传基因编辑的管理也将会涉及对 IVF 和 PGD 的管理条例。例如，IVF 的管理规定就要求必须要有注册登记的场所，筛查供体配子的传染性疾病，符合组织操作规范（FDA，2001）。涉及 IVF 应用的项目必须向美国疾病控制与预防中心（CDC）汇报其成功受孕率[55]。

NASEM 发表了线粒体置换技术，该技术获得可遗传性的卵细胞少量线粒体（如非核的）DNA 变化，中国学者发表了利用非存活人胚胎进行基因编辑的成果，随后 NIH 发表声明不会资助任何涉及人类胚胎编辑的研究[56]。国立卫生研究院主任 Francis Collins 声明"国立卫生研究院不会资助任何在人胚胎上进行基因编辑的工作"，他指出"通过胚胎进行生殖系改变以达到某些临床目的，长期以来从多方面进行了争论，大家普遍认为是不能逾越的红线"（NIH，2015b）。并且，由于 Dickey-Wicker 修正法案（该法案禁止 HHS 资助类似的工作）及 RAC 的政策（拒绝审核此类研究）（见第 2 章），美国卫生研究院早已经不能资助此类项目。

NIH 的声明中也强调了任何临床上涉及经过编辑的胚胎移植和妊娠的临床试验申请（IND）都需要得到 FDA 的许可。FDA 从来没有收到或支持过任何生殖系编辑的方案，但显然是受到了目前研究方向的警告，美国国会在 2015 年 6 月召开了"人类基因编辑科学与伦理"的听证会[57]。紧随其后的就是综合性支出法案规定（见第 2 章），该法案禁止 FDA 利用其资源考虑任何生殖系编辑在临床上的尝试[58]。该禁令至少会持续到 2017 年 4 月，之后，该禁令或许会被延续，也或许被废除，这都要取决于下一个财政预算案。如果该禁令被解除，FDA 将再一次允许资助临床上类似工作的尝试，尽管如此，仍然会限制联邦基金在这方面的使用。

be subject to a complex landscape of state and federal laws and regulations (see Chapter 2). The legality of research, and perhaps even clinical application, would vary from state to state as a result of differing laws on fetal and embryo research. Federal funding for the research would likely be unavailable because of current legislative restrictions on funding research involving human embryos. Should heritable germline genome editing move into clinical investigations, the U.S. Food and Drug Administration (FDA) would have regulatory jurisdiction. The altered cells—whether gametes or the embryo—would need to be implanted for gestation, and this transfer would trigger the same FDA jurisdiction as that used by the agency in 2001 (FDA, 2009) when it determined that reproductive cloning could not proceed without authorization. This jurisdiction derives from the agency's power to regulate tissue transplantation. While IVF and even PGD were developed before the FDA policy in this area was fully developed and therefore have not been regulated as closely as more recent products, genome editing would fall squarely within FDA jurisdiction.

A careful stepwise process (outlined in more detail in Chapter 2) would include consideration by the National Institutes of Health's (NIH's) Recombinant DNA Advisory Committee (RAC) (with public comment and transparent review), local institutional review board (IRB) and local institutional biosafety committee (IBC) review, and FDA review before any decision about whether to permit clinical trials could be made. If germline genome editing succeeded in research trials and was approved for marketing, there would also be mechanisms for oversight in the postapproval context.

Because germline genome editing would involve the use of other assisted reproductive technologies, oversight of its use would likely involve the same statutes and regulations that apply to IVF and PGD. Some of these regulations focus on donor material safety, transparency and reporting requirements, as is the case with IVF, or on quality control of the laboratories (though not necessarily the actual diagnostics) used for PGD. Heritable germline editing would be performed in conjunction with IVF and PGD, thus could involve statutes and regulations that apply to those technologies. For example, IVF itself is subject to rules that require registration of facilities, screening of donor gametes for communicable diseases, and compliance with good tissue-handling standards (FDA, 2001). Programs using IVF

---

[55] *Assisted Reproductive Technology Programs*, 42 U.S.C. § 263a-1 (current through Public Law 114-38).

[56] NIH Statement can be accessed at https://www.nih.gov/about-nih/who-we-are/nih-director/statements/statement-nih-funding-research-using-gene-editing-technologies-human-embryos (accessed January 30, 2017).

[57] "The Science and Ethics of Engineered Human DNA." Hearing before the Subcommittee on Research and Technology, of the House Committee on Science, Space and Technology, June 16, 2015. https://science.house.gov/legislation/hearings/subcommittee-research-and-technology-hearing-science-and-ethics-genetically (accessed January 30, 2017).

[58] Consolidated Appropriations Act of 2016, HR 2029, 114 Cong., 1st sess. (January 6, 2015) (https://www.congress.gov/114/bills/hr2029/BILLS-114hr2029enr.pdf [accessed January 4, 2017]).

其他国家的立场及观点

遗传基因工程已经成为数十年来公众和学术界讨论的主题。其中突出的具有法律效力的还包括欧洲奥维耶多协议（European Oviedo convention），该协议仅允许基因工程用于以预防、诊断或治疗为目的，而不是人为地去创造下一代，因此，生殖系编辑也是被禁止的。尽管有35个国家签订了该协议，但只在29个批准该协议的国家中具有约束力（其中6个国家没有完全批准该协议），而且该协议还需通过各自国内的立法才能被实施（COE，2016）。

最近，如第1章中讨论的，由美国医学科学院（science and medicine academies of the United States）、英国和中国召集举办的2015年12月国际高峰论坛（International Summit）号召暂停一切可遗传的生殖系编辑。其声明如下：

在达到以下条件前，任何临床上尝试生殖系编辑都是不负责任的：

（1）基于对风险，潜在收益和备选方案的适当理解，相关的安全性和效率问题已经得到解决。

（2）公众对基因编辑还存在一定的担忧，因此，临床上此类的尝试必须要有健全的监管体系。

截至目前，以上这些标准在临床上都还没有达到：首先，该技术的安全性尚未完全清楚；其次，就目前而言，该技术能解决的问题还非常有限，并且很多国家已有明确的法律禁止进行生殖系编辑。但是，随着科学知识的不断完善及社会公众的意识不断变化，临床上是否应该进行生殖系编辑，今后还应该定期不间断地进行讨论（NASEM，2016d）。

同样，国际干细胞研究协会（the Interna-tional Society for Stem Cell Research）在2016年发表了再生医学研究指南，该指南指出，"在尚未完全解决该技术可能带来的科学及伦理道德问题之前，任何尝试进行人类胚胎基因编辑的工作都为时过早并且应该受到禁止"（ISSCR，2015）。

2005年，由来自多国专家自发组成的欣克斯顿小组（Hinxton group）发布声明称可遗传的基因编辑也许是可以被接受的，尽管他们在声明中做了很多说明（Hinxton Group，2015）。根据他们的声明，在走向人类生殖应用之前，许多重要的科学挑战和

also must report pregnancy success rates to the Centers for Disease Control and Prevention (CDC).[55]

Following the publication of the National Academies report on mitochondrial replacement techniques, which can result in heritable changes in small amounts of mitochondrial (i.e., non-nuclear) DNA present in the egg, and the publication of genome editing research in China using non-viable human embryos, NIH made a statement to the effect that it would not fund research involving genome editing of human embryos.[56] Francis Collins, Director of NIH, stated that NIH "will not fund any use of gene-editing technologies in human embryos." He noted that the "concept of altering the human germline in embryos for clinical purposes has been debated over many years from many different perspectives, and has been viewed almost universally as a line that should not be crossed" (NIH, 2015b). But this already something NIH could not fund, due to legal obstacles created by the Dickey-Wicker Amendment (forbidding HHS funding of such work) and the RAC policy (declining to review such work, as per requirement for NIH-funded projects) (see Chapter 2).

The NIH statement also highlighted the requirement for FDA approval of IND application for anything involving clinical trials for transfer and gestation of an edited embryo. The FDA had never received or approved a proposal to modify the germline, but apparently alarmed by the direction of research, the U.S. Congress held hearings in June 2015 on "The Science and Ethics of Engineered Human DNA."[57] This was followed by the omnibus spending bill provision, noted in Chapter 2, which prevents the FDA from using any of its resources to even consider an application to proceed with clinical trials.[58] This limitation will last at least through the end of April 2017. Beyond that date, the prohibition may be extended or deleted, depending upon the details of the next budget exercise. If the prohibition is lifted, the FDA will once again be permitted to entertain requests to being clinical trials in this area, though the restrictions on use of federal funds will remain.

Statements and Views from Other Bodies

Heritable genetic engineering has been the subject of public and academic discussion for decades. Salient instruments that have legal effect include the European Oviedo convention, which allows genetic engineering only for preventive, diagnostic, or therapeutic purposes and only where it is not aimed at changing the genetic makeup of a person's descendants, thus precluding germline genome editing. Although signed by 35 nations, it is binding only on those 29 that ratified it (6 nations did not ratify it in full), and even then requires implementation through domestic legislation (COE, 2016).

More recently, as discussed in Chapter 1, the organizers

问题必须要首先得到解决。该声明还列出了一系列与安全性和效率相关的技术上的问题，并且强调人们需要探索应该以一种什么样的态度去看待这个问题，以及在具体应用时又用什么样的法律法规进行管控。

法国科学院（The French National Academy）也认为尽管目前可遗传的基因编辑还没有得到批准，但人们可以静下心来思考什么情况下或将被允许，他们指出"包括生殖系细胞和胚胎的相关研究，只有在受到科学界和医学界的认可时才能进行"（ANM，2016）。

所有的这些声明都认为可遗传基因编辑在安全性和效率上还有许多问题没有解决，因此，目前在应用上进行可遗基因传编辑应被禁止。但他们同样也都认为，目前科技在迅猛进步，人们也不应该永久性地去禁止它。正如 Hinxton 小组建议的那样，"我们需要制定详细而又灵活的准则，并在此准则的基础上订立关于评价安全性和效率的标准"（Hinxton Group，2015）。

总结和建议

在有些情况下，对某些父母而言，他们若想降低自己孩子重大疾病和缺陷的发病风险，可遗传基因编辑是他们唯一或者最佳的选择。尽管可遗传基因编辑对遗传疾病有一定的治愈作用，但这仍难以消除公众对它的担心，尤其体现在对那些并不是非常严重或者还有其他替代治疗方案的遗传疾病上。这种担忧从认为人们不应该去干扰生物进化的过程，到对参与的个人还是社会可能造成的难以预料的后果。

在可遗传基因编辑被授权走向临床应用之前，还需要做大量的科学研究来评价其利弊风险。一旦卵子和精子等前体细胞基因编辑技术上的障碍被克服，今后利用基因编辑预防遗传疾病也许会成为现实。

美国食品药品监督管理局（FDA）作为美国生殖系基因编辑的主要监管部门，他们在做任何判断决定之前都会先去评价其利弊关系。广泛的公众讨论也认为首先应该权衡该技术所能带来的利与弊，之后再讨论是否应该授权在临床上尝试进行类似的工作。但 FDA 在法律职责上并没有被要求

of the December 2015 International Summit convened by the science and medicine academies of the United States, the United Kingdom, and China called for a pause of some undefined duration in any attempt at heritable germline editing. Their statement read:

> It would be irresponsible to proceed with any clinical use of germline editing unless and until
> (i) the relevant safety and efficacy issues have been resolved, based on appropriate understanding and balancing of risks, potential benefits, and alternatives, and
> (ii) there is broad societal consensus about the appropriateness of the proposed application. Moreover, any clinical use should proceed only under appropriate regulatory oversight.

At present, these criteria have not been met for any proposed clinical use: the safety issues have not yet been adequately explored; the cases of most compelling benefit are limited; and many nations have legislative or regulatory bans on germline modification. However, as scientific knowledge advances and societal views evolve, the clinical use of germline editing should be revisited on a regular basis" (NASEM, 2016d).

Similarly, the International Society for Stem Cell Research, when releasing its 2016 professional guidelines for regenerative medicine research, included the following: "Until further clarity emerges on both scientific and ethical fronts, the ISSCR holds that any attempt to modify the nuclear genome of human embryos for the purpose of human reproduction is premature and should be prohibited at this time" (ISSCR, 2015).

In 2015, a self-organized group of multinational experts called the Hinxton group published a statement exploring the possibility that heritable germline genome editing might be acceptable, albeit with many caveats (Hinxton Group, 2015). According to that statement "[p]rior to any movement toward human reproductive applications, a number of crucial scientific challenges and questions must be addressed." The statement proceeds to list a number of technical questions related to safety and efficacy and stresses the need to explore cultural attitudes and whether and how legal limits might be placed on particular uses.

The French National Academy also appears to have taken the position that while heritable germline genome editing is unacceptable now, one can contemplate a time when it might be permitted, stating that "this research, including that on germline cells and human embryos, should be carried out provided that it is scientifically and medically justified" (ANM, 2016).

These statements all recognize that issues of safety

在决定是否授权临床尝试之前，需将公众的道德价值观念纳入其考量范围。所以公众意见的讨论只会在RAC、立法及其他公众所能参与的场合进行（见第7章）。

我们应该谨慎对待可遗传基因编辑，但这也并不意味着它必须被禁止。当技术上的挑战逐渐被克服，并且其所能带来的利远大于弊时，也许可以考虑开始在临床上做类似的尝试。前提是只能限定在最迫切的情况下，并加之以完善的监管体系保护研究对象和他们的后代，同时还有足够的保障措施以防止一些尚未完全理解或者不很迫切的滥用。

建议5-1：临床上可遗传生殖系编辑应该只有在包含以下几点的强有力的监管体系下才能被授权进行：

- 没有其他的可替代方案；
- 仅限用于预防重大疾病或类似状况；
- 仅限编辑被明确证明或有很强证据支持导致疾病的致病基因；
- 仅限将致病基因转换为普通人群所对应的基因，并且证明该转换后几乎或完全没有其他副作用；
- 有足够的临床前或临床数据来评价其风险和潜在的健康福祉；
- 在试验过程中，须严格不间断监控该试验对参与者健康及安全造成的影响；
- 制订全面的计划来对接受基因编辑治疗者进行长期的、多代的后续跟踪，但同时也要尊重患者的自主选择权；
- 最大限度地透明化，但同时也要保证患者的隐私；
- 让公众广泛参与进来，长期不断地评估其风险，以及健康和社会福祉；
- 要有可靠的监管机制以防止重大疾病或状况以外的滥用。

考虑到生殖系基因编辑长期以来就是人们伦理道德讨论的焦点，同时考虑到当今社会的多元性，如果所有人都认同以上这些建议，反而是一件非常奇怪的事。甚至对同意以上建议的人而言，如

and efficacy associated with heritable germline genome editing are far from resolved and that attempts to apply this form of genome editing should not be made at this time. They all note, however, that the science is continuing to progress rapidly, and they avoid calling for permanent prohibitions. Indeed, the Hinxton group recommends that "a detailed but flexible roadmap [be] produced to guide the development of standards for safety and efficacy" (Hinxton Group, 2015).

CONCLUSIONS AND RECOMMENDATION

In some situations heritable germline editing would provide the only or the most acceptable option for parents who desire to have genetically related children while minimizing the risk of serious disease or disability in a prospective child. Yet while relief from inherited diseases could accrue from its use, there is significant public discomfort about germline genome editing, particularly for less serious conditions and for situations in which alternatives exist. These concerns range from a view that it is inappropriate for humans to intervene in their own evolution to anxiety about unintended consequences for the individuals affected and for society as a whole.

More research is needed before any germline intervention could meet the risk/benefit standard for authorizing clinical trials. But as the technical hurdles facing genome editing of progenitors of eggs and sperm are overcome, editing to prevent transmission of genetically inherited diseases may become a realistic possibility.

The primary U.S. entity with authority for the regulation of germline genome editing—the FDA—does incorporate value judgments about risks and benefits in its decision making. A robust public discussion about the values to be placed on the benefits and risks of heritable germline editing is needed now so that these values can be incorporated as appropriate into the risk/benefit assessments that will precede any decision about whether to authorize clinical trials. But the FDA does not have a statutory mandate to consider public views about the intrinsic morality of a technology when deciding whether to authorize clinical trials. That level of discussion takes place at the RAC, in legislatures and at other venues for public engagement, discussed in Chapter 7.

Heritable germline genome editing trials must be approached with caution, but caution does not mean they must be prohibited.

If the technical challenges are overcome and potential benefits are reasonable in light of the risks, clinical trials could be initiated, if limited to only the most compelling circumstances, subject to a comprehensive oversight framework that would protect the research subjects and their descendants; and have sufficient safeguards to

果他们认同的原因都一样的话，同样也是非常让人吃惊的。有些人是因为尊重父母想得到自己的孩子的那种情感，有些人是想孩子生下来尽可能健康。但我们在这一章的前面也提到过，有些人并不认为可遗传基因编辑对生出来的孩子有益，因为这些孩子可能就不会被怀上。还有其他一些人认为，尽管那些因为携带有遗传疾病的父母希望通过这个技术拥有自己亲生的孩子，但更应该考虑的是社会公众对此技术引发的担忧。也存在一些人认为我们上述建议 5-1 提到的标准根本不可能达到，而且一旦生殖系编辑被允许，那些监管机制也不能有效限制其完全按照建议中的内容去执行。如果上述建议的标准确实达不到的话，那么委员会认为生殖系基因编辑就不应被允许。委员会呼吁公众应该持续地参与进来并提出意见（见第 7 章），随着科学技术的不断进步，规章制度应不断完善以达到上述建议中的标准。

人们担忧可遗传基因编辑技术不成熟，以及未经批准而被使用。我们在本章中提出的标准可能在一些而并非所有管辖领域中可以实现。而正因为这种可能，人们担忧可能出现一些监管宽松或甚至完全无监管的地方诱使该技术提供者和使用者前往（Charo，2016a）。这将鼓励那些想通过提供"医疗旅游"以增加税收的国家放宽标准，毕竟这在干细胞治疗和线粒体置换技术上都已经发生过，导致滑向滑坡学说底线的竞赛（Abbott et al.，2010；Charo，2016a；Turner and Knoepfler，2016；Zhang et al.，2016）。"医疗旅游"就是以试图寻找快速、廉价，并且新的或少受监管的治疗手段为目的的，如果新技术一旦出现在宽松的管辖下，这种现象很难完全受控（Cohen，2015；Lyon，2017）。因此，建立全面完善的监管方法尤为重要。

截止到 2015 年年底，不管上述标准有没有达到，美国都没有考虑过尝试生殖系基因编辑。前面说过，美国国会通过的预算条款（该条款至少在 2017 年 4 月前有效）包含以下字句：

法案不允许相关基金用于此类研究。按联邦食品、药品及化妆品法案（21 U.S.C. 355（i））第 505 节（i）及公共健康法案中第 551 节（a）(3) 规定的能够豁免的用于研究目的的药品和生物制品，当它们涉及人为制造人类胚胎和带有可遗传修饰的情况时则不能被豁免，任何提交类似研究的申请将不会被

protect against inappropriate expansion to uses that are less compelling or less well understood.

RECOMMENDATION 5-1. Clinical trials using heritable germline genome editing should be permitted only within a robust and effective regulatory framework that encompasses

- the absence of reasonable alternatives;
- restriction to preventing a serious disease or condition;
- restriction to editing genes that have been convincingly demonstrated to cause or to strongly predispose to that disease or condition;
- restriction to converting such genes to versions that are prevalent in the population and are known to be associated with ordinary health with little or no evidence of adverse effects;
- the availability of credible pre-clinical and/or clinical data on risks and potential health benefits of the procedures;
- during the trial, ongoing, rigorous oversight of the effects of the procedure on the health and safety of the research participants;
- **comprehensive plans for long-term, multigenerational follow-up that still respect personal autonomy;**
- **maximum transparency consistent with patient privacy;**
- **continued reassessment of both health and societal benefits and risks, with broad on-going participation and input by the public; and**
- **reliable oversight mechanisms to prevent extension to uses other than preventing a serious disease or condition.**

Given how long modifying the germline has been at the center of debates about moral boundaries, as well as the pluralism of values in society, it would be surprising if everyone were to agree with this recommendation. Even for those who do agree, it would be surprising if they all shared identical reasoning for doing so. For some, it is about respecting parental desires for genetically related children. For others, it is primarily about allowing children to be born as healthy as possible. But as noted earlier in this chapter, some do not view heritable germline genome editing as a benefit to the resulting child, who otherwise might never have been conceived at all. And for others, the desire of parents who carry genetic disease to have a genetically related child through this technology, instead of having a genetically unrelated child, is not sufficient to outweigh the social concerns that have been raised. There are also those who think the final criterion of Recommendation 5-1 cannot be met, and

审核，并且豁免将失效。

这一规定使当前美国当局无法授权在临床上进行可遗传基因编辑的尝试，这也驱使了该技术的开发转向其他监管之下，有些有规范，而有些则没有。

that once germline modification had begun, the regulatory mechanisms instituted could not limit the technology to the uses identified in the recommendation. If it is indeed not possible to satisfy the criteria in the recommendation, the committee's view is that germline genome editing would not be permissible. The committee calls for continued public engagement and input (see Chapter 7) while the basic science evolves and regulatory safeguards are developed to satisfy the criteria set forth here.

Heritable germline genome editing raises concerns about premature or unproven uses of the technology, and it is possible that the criteria outlined here for responsible oversight would be achievable in some but not all jurisdictions. This possibility raises the concern that "regulatory havens" could emerge that would tempt providers or consumers to travel to jurisdictions with more lenient or nonexistent regulations to undergo the restricted procedures (Charo, 2016a). The result could be a "race to the bottom" that would encourage laxer standards in nations seeking revenues from medical tourism, as has happened with both stem cell therapy and mitochondrial replacement techniques (Abbott et al., 2010; Charo, 2016a; Turner and Knoepfler, 2016; Zhang et al., 2016). The phenomenon of medical tourism, which encompasses the search for faster and cheaper therapeutic options, as well as newer or less regulated interventions, will be impossible to control completely if the technical capabilities exist in more permissive jurisdictions (Cohen, 2015; Lyon, 2017). Thus, it is important to highlight the need for comprehensive regulation.

As of late 2015, the United States is unable to consider whether to begin germline genome editing trials, regardless of whether the criteria laid out above could be met. As noted above, a provision (in effect until at least April 2017) was passed in a budget bill,[59] in which Congress included the following language:

None of the funds made available by this Act may be used to notify a sponsor or otherwise acknowledge receipt of a submission for an exemption for investigational use of a drug or biological product under section 505(i) of the Federal Food, Drug, and Cosmetic Act (21 U.S.C. 355(i)) or section 351(a)(3) of the Public Health Service Act (42 U.S.C. 262(a)(3)) in research in which a human embryo is intentionally created or modified to include a heritable genetic modification. Any such submission shall be deemed to have not been received by the Secretary, and the exemption may not go into effect.

The current effect of this provision is to make it impossible for U.S. authorities to review proposals for clinical trials of germline genome editing, and therefore to drive development of this technology to other jurisdictions, some regulated and others not.

---

[59] Consolidated Appropriations Act of 2016, HR 2029, 114 Cong., 1st sess. (January 6, 2015) (https://www.congress.gov/114/bills/hr2029/BILLS-114hr2029enr.pdf [accessed January 4, 2017]).

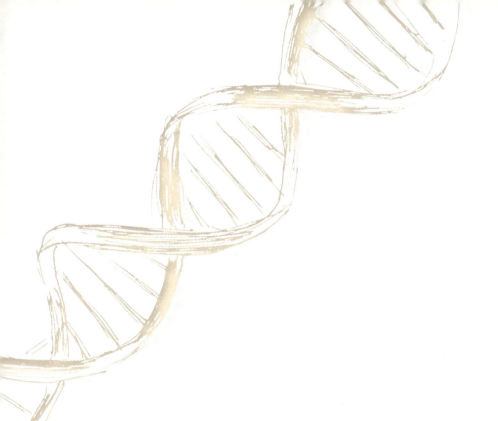

6

基因强化
Enhancement

普遍认为体细胞基因和细胞疗法在道德上是可被接受的。事实上，骨髓移植技术，即不同基因组成的细胞被引入患者，已经使用了几十年，并且使用基因疗法来治疗具有严重联合免疫缺陷的儿童，即所谓的"气泡男孩病"，也已进入临床治疗阶段（De Ravin et al., 2016）。除了安全性、有效性和知情同意等问题，赞同现代医学的人从不担心体细胞和基因治疗是否合法。基因组编辑技术在体细胞基因治疗和预防疾病中发挥着越来越大的作用（参见第 4 章）。

然而，最近的进展使得基因组编辑也可用于超出上述基因治疗和其他医学干预目的，因此重新提出了基因强化是否应该被规范或禁止的问题，以及体细胞强化或可遗传强化是否有显著差别的问题。

本章探讨了基因组编辑被用来达到通常所说的"基因强化"目的，这个术语本身是有问题的，它是指对现有条件的改变，确切地说是改进。强化的范围可以从很普通的（如染发）到进一步物理干预（如选择性的整容手术），甚至到更危险和被人质疑的（如竞技赛事中运动员使用一些类固醇等其他药物）。

强化通常被理解为是改变那些被认为"正常"的事物，包括强化之前作为整体的人类或是特定的个体。随之我们要问什么才是"正常"的？是平均值？是早已被自然天定的？还是运气使然？鉴于人

Somatic gene and cell therapies are widely seen as morally acceptable. Indeed, bone marrow transplantation, in which cells of differing genetic composition are introduced into patients, has been used for decades, and the use of gene therapy to treat children with severe combined immune deficiency, the so-called "bubble boy disease" is now a part of medical treatment (De Ravin et al., 2016). Beyond issues of safety, efficacy, and informed consent, there has been no concern about the legitimacy of somatic cell and gene therapies among those who generally endorse modern medicine. Genome editing is playing an increasing part in somatic gene therapy to treat and prevent disease (see Chapter 4).

Recent advances, however, have increased the possibility that genome editing could also be used for purposes that go beyond the kinds of gene therapy and other medical interventions discussed above, and therefore have raised anew the question of whether enhancement should be regulated or prohibited, and whether there are important differences depending upon whether the enhancements are somatic or heritable.

This chapter explores the possible applications of genome editing to achieve what is commonly described as "enhancement," a term that itself is problematic; it implies a change from—indeed an improvement upon—an existing condition. Enhancements may range from the mundane, such as cosmetic changes in hair color; to the more physically interventionist, such as elective cosmetic surgery; to the more dangerous and problematic, such as the use of some steroids and other drugs among athletes in competitive settings.

Enhancement is commonly understood to refer to changes that alter what is "normal," whether for humans as a whole or for a particular individual prior to enhancement. The question of what is meant by normal is then arises. Is it

类对于任何特定性状表现差异大，在确定平均值时，其实并没有任何基础来认定任何一种条件是正常的或有意义的。尽管如此，已经有一些尝试来描述正常范围，即在良好的环境下，能够享受生活和参与世界的能力。

本章首先审视了用于强化目的基因组编辑中的几个关键问题；然后分别讲述了用于强化目的的体细胞（不可遗传）和生殖系（可遗传）基因组编辑；最后给出总结和建议。

人类遗传多样性及"正常"和"自然"的定义

在开始讨论所谓的"基因强化"之前，有必要探索人类基因治疗和基因组编辑相关的术语是怎样无意识地引入有偏倚的判断。许多疾病与DNA变异相关，并且用于描述变异的术语"突变"经常有负面含义。在"正常"和"突变"或"致病"基因之间进行区分，后者被视为负面。"正常"这个术语也被用以形容表型，或由基因和环境的相互作用后产生的个体性状。值得注意的是，任何给定的性状（如身高、体重、力量、听觉或视力），"正常"可以指很广的范围，并且受许多因素影响，包括但不限于基因变异，因为它们经常相互和与环境因素作用。因此，"正常"这个术语表示一种范围，而不是某些理想状态。

尽管有证据表明"自然"的东西可以是安全的或本质上是危险的，但是"自然"一词同样具有积极的含义，反映了一种共同的观点，即自然产生更健康、通常比任何人为的东西更好。在本文中，存在于自然界中的基因变异可以支持健康或引起疾病，并且人类群体包含多数基因的多个变异（参见第4章）。因此，没有单一的"正常"人类基因组序列；相反，却存在多种变异人的基因组序列（IGSR，2016），所有这些变异都发生在世界范围的人类基因库中，并且在这个意义上是"自然的"，所有这些也都将是有利的或不利的。

在基因组中的任何给定位置，一些变异比其他变异更常见。一些是有益的而一些有害的，它们的效果有时取决于以下的因素，如变异是一个拷贝（杂合）还是两个拷贝（纯合），或该基因是否是性染色体偶联的（若基因处于Y染色体上是半合子，

average? Is it whatever nature has prescribed? Is it whatever luck has wrought? Given the wide range of capabilities exhibited by humans for any particular trait, there is little basis for deeming any one condition normal or any meaningful value in determining an average. Nonetheless, there have been some attempts to describe the range of conditions that, given the right environment, are consistent with an ability to appreciate life and participate in the world.

This chapter begins by reviewing several key issues in genome editing for enhancement. It then addresses in turn somatic (nonheritable) and germline (heritable) genome editing for enhancement purposes. The chapter ends with conclusions and recommendations.

HUMAN GENETIC VARIATION AND DEFINING "NORMAL" AND "NATURAL"

Before beginning to discuss so-called "enhancement," it is important to explore how terminology related to human gene therapy and genome editing has the potential to bias judgments unconsciously. Many diseases are associated with DNA variants, and the common term used to describe a variant—"mutation"—has therefore taken on a negative connotation in common parlance. A distinction often is made between "normal" and "mutant" or "disease (-causing)" genes, with the latter being viewed negatively. The term "normal" also is applied to phenotype, or the individual traits that result from interaction between the genotype and the environment. Here it is important to note that the "normal" distribution of any given trait (e.g., height, weight, strength, aural or visual acuity) covers a wide spectrum and can be affected by many factors, including but by no means limited to, gene variants, which often interact both with each other and with environmental factors. Hence, the term "normal" denotes a range or spectrum, not some ideal state.

The word "natural" has similarly taken on a positive connotation reflecting a common view that nature produces things that are healthier and generally better than anything artificial—this despite evidence demonstrating that "natural" things can be either safe or intrinsically dangerous. In the present context, genetic variants that exist in nature may either support health or cause disease, and the human population contains multiple variants of most genes (see Chapter 4). Thus, there is no single "normal" human genome sequence; rather, there are multiple variant human genomic sequences (IGSR, 2016), all of which occur in the worldwide human gene pool and, in that sense, are "natural," and all of which can be either advantageous or disadvantageous.

At any given position in the genome, some variants are more common than others. Some are beneficial and some detrimental, their effects at times depending on such factors as whether a person has one copy (heterozygosity) or two copies (homozygosity) of the variant, or whether the gene

而处于X染色体上是杂合子或纯合子）。另一个因素可能是个人生活的特定环境。举一个众所周知的例子——镰状细胞贫血。血红蛋白是红细胞携带氧的蛋白质。最常见的变体编码完全功能的蛋白质，但镰状细胞变异可以使蛋白质聚集并使红细胞扭曲成镰刀形状，如果基因的两个拷贝都是镰状细胞变异，则会导致镰状细胞性贫血。然而，如第5章所述，该变异是杂合子时赋予对疟疾的一些抗性，因此，镰状细胞变体通过自然选择在疟疾易感地区被保留下来（例如，非洲、印度和地中海地区），人群中镰状细胞疾病的缺点与抗疟疾的优点相平衡。因此，在这种情况下，哪种自然变体是有利的取决于环境。

在本报告中，委员会使用术语"变异"并尽可能避免使用"突变"或"正常"来指基因变异。然而，需要记住，许多变异是"自然的"，将与疾病相关的基因变异（如血红蛋白的镰状细胞变异）改变成为在群体中普遍存在的不是致病的拷贝（即"自然的"），这种改变被视为治疗或预防性改变。然而，还可以设想是否还有以下可能，即将基因改变为在人基因库中不存在的（或罕见的）、具有某些可以被视为"强化"性质的变异形式，并可能带来好处。这种改变是比用已知不引起疾病的共同人类拷贝来替代致病变异更激进的做法。

了解公众对基因强化的态度

个体强化有多种形式。它们可能需要大量的个人努力，如上钢琴课；或它们可以在很大程度上不需要个人努力，如佩戴牙齿矫正支架。它们可以是暂时的，如获益于早晨咖啡中的咖啡因；或者是持久的，如针对疾病的免疫接种。它们可能是容易逆转的，如染发；或者是难以逆转的，如整容手术；也可以是矫正性干预，如在去除白内障之后植入比人的自身敏感度更高的透镜。所有的这些因素都会影响这种改进在公平性和公众接受性等方面的评价。

尽管调查表明很多人支持利用基因治疗和基因工程改善现有个体和未出生孩子的健康（表6-1），但是这些新的、多种多样的"强化"方式在给人们带来兴奋情绪的同时也会引起焦虑。在2016年，皮尤（Pew）研究中心对超过4000个人的调查研究

is sex-chromosome–linked (hemizygous for a gene on the Y chromosome and either heterozygous or homozygous on the X chromosome). Another factor may be the particular environment in which a person lives. A well-known example is the case of sickle-cell anemia. Hemoglobin is the protein that carries oxygen in red blood cells. The most widespread variant encodes a fully functional protein, whereas the sickle-cell variant can cause the protein to aggregate and distort the red cells into a sickle shape if *both* copies of the gene are this variant (the homozygous state), which in turn causes the symptoms of sickle-cell disease. As noted in Chapter 5, however, being heterozygous for this variant confers some resistance to malaria, and for this reason, the sickle-cell variant has been maintained by natural selection in populations from malaria-prone areas (e.g., Africa, India, and the Mediterranean), the disadvantages of sickle-cell disease at the population level being balanced against the population-level advantages of resistance to malaria. So, in this case, which natural variant is advantageous depends on the environment.

In this report, the committee uses the term "variant" and eschews to the extent possible the use of "mutant" or "normal" in referring to gene variants. There is, however, a distinction worth keeping in mind. Many variants are "natural," and changing a gene variant that is associated with disease, such as the sickle-cell variant of hemoglobin, to a variant that is prevalent in the population (i.e., "natural") but not disease-causing can be viewed as a therapeutic or preventive change. It is also possible, however, to envision the possibility of changing a gene to a variant form that does not exist (or is rare) in the human gene pool but has some property that could be viewed as an "enhancement" since it is predicted to have a beneficial effect. Such a change is a more radical step than that of replacing a disease-causing variant with a common human variant known not to cause disease.

UNDERSTANDING PUBLIC ATTITUDES TOWARD ENHANCEMENT

Personal improvements take many forms. They can require significant personal effort, as in taking piano lessons, or they can be largely independent of personal effort, as in wearing teeth-straightening braces. They can be temporary, as in benefiting from the caffeine in morning coffee, or long-lasting, as in immunization against disease. They can be easily reversible, as in hair coloring, or reversible only with difficulty, as in cosmetic surgery. And they can be provided in connection with a corrective intervention, as in removing cataracts and inserting lenses that provide greater acuity than the person ever had naturally. All of these factors influence how improvements are evaluated in terms of fairness and public acceptability.

Although surveys indicate significant support for gene therapy and genetic engineering to improve health of both existing individuals and unborn children (see Table 6-1),

表明，焦虑超过了兴奋，不仅显现在体细胞基因组编辑方面，而且在通过机械和移植的强化方式方面（Pew research center，2016）。仅有一项研究不是决定性的，但是随着时间的推移和更多成功的例子，公众对一些其他有争议的领域（如体外受精）的新型干预措施变得可以接受。但皮尤研究中心和许多其他人认为，这一领域的政策制定需要充分注意公众的态度和理解。

the possibility of "enhancement" in new and potentially more wide-ranging ways can engender anxiety as well as enthusiasm. In 2016, a Pew study of surveys of more than 4,000 individuals revealed that anxiety outpaced enthusiasm, not only for enhancement through somatic genome editing but also for mechanical and transplant-related enhancement (Pew Research Center, 2016). A single study is not definitive, and public opinion on novel interventions in some other controversial areas (such as *in vitro* fertilization) has become more favorable over

表 6-1 关于公众对于基因治疗和基因编辑态度的调查结果总结

问题	投票（年份）	赞成 /%
对成人及孩子基因治疗和基因编辑的态度		
用于提高被治疗患者的健康		
赞成基因工程治疗疾病	Times-CNN-Yankelovich (1993)	79
如果能够通过改变基因来治愈患有致命疾病的人，你觉得那些人应该允许这样做吗？	Troika-Lifetime-PSRA（1991）	65
赞成科学家改变人类细胞的组成来治疗通常致命的遗传疾病	March of Dimes-Harris（1992）	87
赞成科学家改变人类细胞的组成，以减少在以后的生命中发生致命疾病的风险	March of Dimes-Harris (1992)	78
政府资助及对基因治疗的管制	STAT-HSPH-SSRS (2016)	64
联邦政府应该资助科学研究，开发新的基因治疗方法	STAT-HSPH-SSRS (2016)	59
FDA 应该批准基因治疗在美国使用	Hopkins-PSRA (2002)	59
改善儿童遗传健康	March of Dimes-Harris（1992）	84
赞成父母改变自己的基因，以防止他们的孩子有遗传病	OTA-Harris (1986)	84
赞成科学家改变人类细胞的组成，阻止儿童遗传通常致命的遗传病	March of Dimes-Harris（1992）	66
赞成科学家改变人类细胞的组成，阻止儿童遗传非致死性出生缺陷	OTA-Harris (1986)	77
如果你的孩子带有致命的遗传性疾病，愿意让孩子接受治疗，纠正那些基因吗？	March of Dimes-Harris（1992）	88
改善智力、身体特征或外貌	OTA-Harris (1986)	86
赞成基因工程改善人的智力	Times-CNN-Yankelovich (1993)	34
赞成科学家改变人类细胞的组成以提高儿童将遗传的智力水平	March of Dimes-Harris（1992）	42
赞成通过基因工程的办法提高个人的身体特征	OTA-Harris (1986)	44
赞成科学家改变人类细胞的组成以提高儿童将遗传的身体特征	Times-CNN-Yankelovich (1993)	25
赞成父母通过改变他们自己的基因，从而让孩子拥有更好的相貌、更强壮的身体、更聪明的头脑	March of Dimes-Harris（1992）	43
赞成科学家提供方法用以改变父母的基因，从而让孩子变得更聪明、更好看	OTA-Harris (1986)	44
关于出生前改变人基因组信息的态度	Hopkins-PSRA (2002)	20
为了以后的健康改变未出生婴儿的基因以减少他们发展为某些严重疾病的风险应该是合法的	Family Circle-PSRA (1994)	10
	STAT-HSPH-SSRS (2016)	26
改变婴儿的遗传特征以减少严重疾病的风险是医学进步的一种适当使用	Pew (2014)	46
如果未来的科学研发了通过改变孩子在子宫中的遗传结构来改变孩子的遗传特征的方法，并且你需要做出决定，那么你是否会考虑这样做以改善他/她的一般身体健康改变未出生婴儿的基因以改善他们的智力或身体特征应该是合法的	VCU (2003)	41
改变婴儿的遗传特征使婴儿更聪明是一种适当使用医学进展的方式	ABC (1990)	49
如果未来的科学研发了通过改变孩子在子宫中的遗传结构来改变孩子的遗传特征的方法，并且你需要做出决定，那么你是否会考虑这样做以改善他/她的智力	TAT-HSPH-SSRS (2016)	11
如果未来的科学研发了通过改变孩子在子宫中的遗传结构来改变孩子的遗传特征的方法，并且你需要做出决定，那么你是否会考虑这样做以改善他/她的身体特征，比如身高、体重	Pew (2014)	15
如果未来的科学研发了通过改变孩子在子宫中的遗传结构来改变孩子的遗传特征的方法，并且你需要做出决定，那么你是否会考虑这样做以改善他/她的头发、眼睛颜色和脸部特征	ABC (1990)	28
政府资助研究	ABC (1990)	13
联邦政府应该资助那些改变未出生婴儿的基因以减少他们发展为某些疾病的风险的科学研究	ABC (1990)	8
联邦政府应该资助那些改变未出生婴儿的基因以改善他们的特征（比如智力）、身体特征（比如运动能力）和外貌的科学研究	TAT-HSPH-SSRS (2016)	44
	TAT-HSPH-SSRS (2016)	14

TABLE 6-1 Summary of Public Attitudes toward Aspects of Gene Therapy or Genome Editing as Revealed by a Selection of Surveys (emphasis added)

Question	Poll (Year)	Percent Affirmative
Attitudes about gene therapy and gene editing in adults and children		
To improve health of the person being treated		
Approve of genetic engineering to cure a disease	Times-CNN-Yankelovich (1993)	79
If it is possible to cure people with fatal diseases by altering their genes, do you feel those people ought to be allowed to do this?	Troika-Lifetime-PSRA (1991)	65
Approve of scientists changing the makeup of human cells to cure a usually fatal genetic disease	March of Dimes-Harris (1992) OTA-Harris (1986)	87 83
Approve of scientists changing the makeup of human cells to reduce the risk of developing a fatal disease later in life	March of Dimes-Harris (1992) OTA-Harris (1986)	78 77
Government funding and regulation of gene therapy		
Federal government should fund scientific research on developing new gene therapy treatments	STAT-HSPH-SSRS (2016)	64
FDA should approve gene therapy treatments for use in the United States	STAT-HSPH-SSRS (2016)	59
To improve health inherited by child		
Approve of parents being offered a way to change their own genes to prevent their children from having a genetic disease	Hopkins-PSRA (2002)	59
Approve of scientists changing the makeup of human cells to stop children from inheriting a usually fatal genetic disease	March of Dimes-Harris (1992) OTA-Harris (1986)	84 84
Approve of scientists changing the makeup of human cells to stop children from inheriting a nonfatal birth defect	March of Dimes-Harris (1992) OTA-Harris (1986)	66 77
If you had a child with a usually fatal genetic disease, willing to have child undergo therapy to have those genes corrected	March of Dimes-Harris (1992) OTA-Harris (1986)	88 86
To improve intelligence, physical traits, or appearance		
Approve of genetic engineering to improve a person's intelligence	Times-CNN-Yankelovich (1993)	34
Approve of scientists changing the makeup of human cells to improve the intelligence level that children would inherit	March of Dimes-Harris (1992) OTA-Harris (1986)	42 44
Approve of genetic engineering to improve a person's physical appearance	Times-CNN-Yankelovich (1993)	25
Approve of scientists changing the makeup of human cells to improve the physical characteristics that children would inherit	March of Dimes-Harris (1992) OTA-Harris (1986)	43 44
Approve of parents being offered a way to change their own genes to have children who would be smarter, stronger, or better looking	Hopkins-PSRA (2002)	20

Question	Poll (Year)	Percent Affirmative
Approve if scientists offered parents a way to change their genes in order to have smarter or better-looking children	Family Circle-PSRA (1994)	10
Attitudes about changing human genomes before birth		
To improve future health		
Changing the genes of unborn babies to reduce their risk of developing certain serious diseases should be legal	STAT-HSPH-SSRS (2016)	26
Changing a baby's genetic characteristics to reduce the risk of serious diseases is an appropriate use of medical advances	Pew (2014)	46
	VCU (2003)	41
If future science developed the ability to change a child's inherited characteristics by changing the child's genetic structure in the womb and you were making the decision, would consider doing so to improve his/her general physical health	ABC (1990)	49
To improve intelligence, physical traits, or appearance		
Changing the genes of unborn babies to improve their intelligence or physical characteristics should be legal	STAT-HSPH-SSRS (2016)	11
Changing a baby's genetic characteristics to make the baby more intelligent is an appropriate use of medical advances	Pew (2014)	15
If future science developed the ability to change a child's inherited characteristics by changing the child's genetic structure in the womb and you were making the decision, would consider doing so to improve his/her intelligence	ABC (1990)	28
If future science developed the ability to change a child's inherited characteristics by changing the child's genetic structure in the womb and you were making the decision, would consider doing so to improve his/her body characteristics such as height and weight	ABC (1990)	13
If future science developed the ability to change a child's inherited characteristics by changing the child's genetic structure in the womb and you were making the decision, would consider doing so to improve his/her hair or eye color or facial characteristics	ABC (1990)	8
Government funding of research		
Federal government should fund scientific research on changing the genes of unborn babies to reduce their risk of developing certain diseases	STAT-HSPH-SSRS (2016)	44
Federal government should fund scientific research on changing the genes of unborn babies that aims to improve their characteristics, such as intelligence, or physical traits, such as athletic ability or appearance	STAT-HSPH-SSRS (2016)	14

NOTE: ABC = American Broadcasting Company; CNN = Cable News Network; FDA = U.S. Food and Drug Administration; HSPS = Harvard T.H. Chan School of Public Health; OTA = Office of Technology Assessment; PSRA = Princeton Survey Research Associates; VCU = Virginia Commonwealth University.
SOURCE: Blendon et al., 2016.

有时，治疗、预防和强化之间的界限模糊，甚至要治愈或预防的"疾病"的定义也是值得争论的。因此，预防或治疗疾病和缺陷（即"治疗"）与"强化"概念之间的区别可能不能完全反映公众态度或公共政策选择。对于使用基因治疗来预防疾病而不是治疗而言，公众不安的程度，似乎与"干预自然"和"越过不该越过的界限"相关联（Macer et al.，1995）（见第5章）。这些公众焦虑的产生对于那些与疾病治疗和预防无关的干预来说更是如此。如上所述，美国人对于基因强化的想法基本没有兴趣（Blendon et a.，2016；Pew Research Center，2016；见表6-1）。

这种缺乏热情的可能来自于对创新的犹豫，不论是从当代已经认为相当普通的事物，如咖啡和制冷，或是到至今仍然处在激烈辩论当中的事情，如转基因作物（Juma，2016），这已被证明是历史上的一个普遍现象。由于人们担心创新可能会影响文化特征，或者改变社会经济模式，这对于一部分人来说是有害无益的，因此产生了抵抗和怀疑的情绪。如果这些问题通过补救措施得以解决或被证明是无根据的，那么必要的创新就应该能被广泛接受。

目前还不清楚的是，通过基因组编辑来进行基因强化的方式是否会遵循这样的模式，或是因其颠覆性的应用被一直抵制，又或者随着技术的进步和新应用的出现而引发新的忧虑。"现状偏差"是指熟悉的偏好可以影响人们对一项创新的优点做出判断。对现状的倾向可能来自对过渡成本（即人们如何适应创新引起的情况）、对风险（假设创新具有的风险经不起现状衡量）、对关于自然的偏离（人们认为过去的自然进化过程已经为当前环境优化了人类），以及对个人的影响（科技将降低人与人之间关系的质量）的担心。

有人提出了一种测试现状偏差是否影响新技术评价的方法。在这种"逆转测试"中，可能有人会问，那些认为人们不应该具有比自身能力更大影响力的人，是否也会认为人们的影响力小更好（Bostrom and Ord，2006）。该测试旨在将对创新本身的关注与对任何离开现状的担忧区分开来。当它与关于"滑坡"理论（见第5章）放在一起时非常有用，因为它有助于区分是对今天使用的技术的担忧，还是对未来可能不必要的技术扩大化的担忧。

time and with evidence of successes. But the Pew study and many others suggest that policy in this area needs to be developed with full attention to public attitudes and understandings.

Sometimes, the lines between therapy, prevention, and enhancement are blurred, and even the definition of a "disease" that is to be cured or prevented can be open to debate. For this reason, the distinctions between preventing or treating disease and disability (i.e., "therapy") and the notion of "enhancement" may not fully capture either public attitudes or public policy options. To the extent that there is any public disquiet about the use of gene therapy for disease prevention as opposed to treatment, it appears to be linked to more generalized concerns about "meddling with nature" or "crossing a line we should not cross" (Macer et al., 1995) (see Chapter 5). This is even more true for interventions that appear unrelated to either disease treatment or prevention. As noted above, Americans appear to be largely unenthusiastic about the idea of "enhancement" (Blendon et al., 2016; Pew Research Center, 2016; see Table 6-1).

It is possible that this lack of enthusiasm is due in part to hesitation concerning innovation, which has been shown to be a common phenomenon throughout history, ranging from things now considered quite ordinary, such as coffee and refrigeration, to things still hotly debated, such as transgenic crops (Juma, 2016). Resistance or skepticism may be an outgrowth of concerns about the degree to which an innovation affects cultural identity or may distort socioeconomic patterns in a fashion that is harmful to at least some part of the population. If and when these concerns are either addressed through remedial measures or shown to be unwarranted, innovations that are needed or perceived as desirable become widely accepted.

What is unclear is whether genome editing for enhancement would follow such a pattern or would be such a disruptive application of a new technology that the resistance would persist over time or new concerns emerge as the technology progresses and new applications emerge. "Status quo bias," is a phenomenon in which the preference for what is familiar can affect the way people form judgments about the merits of an innovation. The predisposition toward the status quo may arise from concerns about transition costs (i.e., how people adapt to the circumstances arising from an innovation), about risk (with innovations assumed to have risks that are less amenable to measurement relative to the status quo), about deviation from what is natural (people holding the unwarranted belief that the past processes of natural evolution have optimized humans for the current environment), and about effects on individuals (concern that technology will diminish the quality of relationships between people).

如何区分基因治疗和基因强化

鉴于人们有多种干预方法来改变身体和个人环境,对所谓的基因强化的讨论就必须从工作定义开始。基因强化有若干不同的定义,例如,"提高一个人的能力到一个物种代表水平或者统计学正常水平之上","非治疗性介入以提高或扩展人类的特征",或者是"现有个体或者后代能力的提高"(Daniels,2000;NSF,2010;President's Council on Bioethics,2003)。还有一个定义特指给身体状态或功能带来超出恢复或者维持健康所需要的干预行为(Parens,1998)。这个定义将技术干预与目的并列阐述,因为大多数干预手段既可用来强化又可用来恢复。在这个定义下,例如,提高肌肉萎缩症患者的肌肉组织是恢复性的,而如果用于没有已知病变属于正常水平的个人就是强化。能够意识到基因组编辑的一个目的是基因强化是非常有用的,因为在美国大多数生物医学干预手段都受到FDA的监管,其法定职责会明确将产品授权的初始目的用途与基本的风险-收益平衡联系在一起,尽管授权后的应用可能会超出初始的目的。

另一个重要的定义是"基因治疗",通常被理解为包括对疾病的诊治(疾病本身的定义也值得探讨,后面讨论)。但是疾病的预防也经常被认为是包含在治疗范围内。例如,减少普通健康人和带有非恶性变异的人患乳腺癌的风险,这不是治疗疾病,只是预防可能发生的疾病,这种加强的抵抗力可以被认为是治疗性预防,类似抗感染免疫。另外,类似的例子有用降低胆固醇水平来降低人患心脏病的风险,还有他汀类药物和阿司匹林被广泛用来增强对心脏病的抵抗力,甚至用在那些患病风险不高的人上,这在美国社会已经普遍应用。用于类似目的的基因组编辑也会有相同的增进健康的作用。

A means of testing for whether status quo bias is affecting the evaluation of new technology has been suggested. In this "reversal test," one asks, for example, whether those who think people should not have more influence over their traits would also think it would be good if people had less influence (Bostrom and Ord, 2006). The test is intended to distinguish concerns about an innovation itself from concerns about any move away from the status quo. It can be useful when juxtaposed with arguments about a "slippery slope" (see Chapter 5) because it helps distinguish concerns about the technology as it is used today from concerns about future unwanted extensions of the technology.

DRAWING LINES: THERAPY VERSUS ENHANCEMENT

Given the wide range of other interventions people make to alter their bodies and their personal circumstances, any discussion of so-called "enhancement" must begin with a working definition. Enhancement has been variously defined as "boosting one's capabilities beyond the species-typical level or statistically normal range of function," "a nontherapeutic intervention intended to improve or extend a human trait," or "improvements in the capacities of existing individuals or future generations" (Daniels, 2000; NSF, 2010; President's Council on Bioethics, 2003). One definition focuses on interventions that improve bodily condition or function beyond what is needed to restore or sustain health (Parens, 1998). This is a definition that addresses intent as much as the technical intervention, as most interventions can be used either to "enhance" or to restore. For example, under this definition, improving musculature for patients with muscular dystrophy would be restorative whereas doing so for individuals with no known pathology and average capability would be considered enhancement. Recognizing the importance of intent as an aspect of "enhancement" is helpful, as most biomedical interventions will be subject (in the United States) to regulation by the FDA, whose statutory authority explicitly links the initial risk-benefit balance needed for approval to the "intended" use of the product, even though postapproval uses can range beyond that original intended purpose.

延伸内容 6-1

定义上的区别
Making Distinctions

20世纪70年代，一些核心的区别被加以诠释(Juengst, 1997; Walters and Palmer, 1997)。首先是体细胞和生殖系细胞基因组修饰的区别：体细胞强化只影响一个人，而可遗传性强化则可以传递给后代。可遗传性强化通常对包括基因库可能的影响和回到某种形式的优生学担忧进行讨论。其次是治疗和预防疾病（即基因治疗）与基因强化的区别。关于基因强化的讨论则聚焦在安全性和不平等优势方面（特别是在竞争环境下，如体育运动），当然"不平等"的定义也取决于环境。

科技的进步使得科学家期待减轻社会对基因工程的忧虑，特别是在1975年科学家组织的Asilomar会议讨论了重组DNA技术的风险和益处之后，进一步将争论推向了"医疗"，确切地说是"基因治疗"与"基因强化"的区别。重要的是，每一位负责任的科学家都想将当今的科技应用于体细胞诊治重疾上面。

下表是20世纪70年代初制定的一个有影响力的框表，它将细胞1和细胞2分别定义为"医疗"和"诊治遗传性疾病"(Anderson and Kulhavy, 1972, p. 109)。相反，强化（体细胞强化和可遗传性强化）分别定义为细胞3和细胞4。

治疗与增强比较

目的	体细胞	生殖系
治疗性疾病诊治	体细胞治疗（细胞1）	生殖系治疗（细胞2）
功能强化	体细胞强化（细胞3）	生殖系强化（细胞4）

到20世纪80年代中，科学家和生物伦理学家开始呼吁建立疾病和强化之间的道德界线，而不是体细胞和生殖系细胞之间的。美国国立卫生研究院(NIH)生物伦理主任John Fletcher写道，"最相关的道德区别在于使用目的是为了减少真正的痛苦还是为了改变与疾病相关性小甚至不相关的部分"(Fletcher et al., 1985)。基因治疗先驱Theodore Friedmann写道，"进行有效的疾病控制以及预防

Another definitional matter concerns the meaning of "therapy." It is understood to encompass treatment of a disease (the definition of which is itself subject to debate, as discussed in section on fairness and enhancement, below). But prevention of disease also often is viewed as being encompassed within therapy. To reduce risk of breast cancer in a person of average health and with nondeleterious variants, for example, is not to cure a disease or even to prevent one that is likely to occur, but this boosted resistance is often viewed as therapeutic prevention, akin to immunization against infectious diseases. A similar point pertains with respect to reducing cholesterol in persons at no more than average risk for heart disease; this is already a widespread practice in American society, and statins and aspirin are widely used to enhance resistance to heart disease pharmacologically even for those not at high risk. Genome edits for such purposes could be similarly health-promoting.

In the 1970s, certain central distinctions were drawn (Juengst, 1997; Walters and Palmer, 1997). First, the distinction between somatic and germline genome modifications was established: somatic enhancements affect only a single individual but heritable enhancements can be passed down through the generations. Discussions of heritable enhancements included concerns about possible effects on the gene pool and fears of a return to some form of eugenics. Second, a distinction was drawn between treating or preventing disease (therapy) and enhancement. Discussions of enhancement focused on issues such as safety and (especially in competitive environments such as sports) unfair advantage, with the definition of "unfair" highly dependent on context.

Available technology and the desire of prominent scientists to mitigate social concerns surrounding genetics—particularly after the Asilomar conference organized by scientists in 1975 to discuss the risks and benefits of recombinant DNA techniques—pushed the debate even further in the direction of the distinction between "medicine" or "therapy" versus "enhancement." It is important to note, as well, that somatic therapy to treat very severe diseases was all that any responsible scientist could imagine actually doing with the technology of the time.

An influential schematic was developed in the early 1970s (and depicted in table below), which defined cells 1 and 2 as "medicine" and as the "treatment of hereditary diseases,"

对发育早期和难治的细胞损害也许是生殖系治疗的最终作用"（Friedmann，1989）。1991年，当体细胞基因治疗最早的3个临床试验执行之际，NIH重组DNA顾问委员会（RAC）的人类基因治疗小组委员会主席就呼吁进行"一个详细的关于人类生殖系细胞基因干预治疗的公众讨论"（Walters，1991）。疾病的修复被定义为恢复到正常水平，而不是走向"优生学"。

respectively (Anderson and Kulhavy, 1972, p. 109), as opposed to enhancements (somatic or heritable) in cells 3 and 4.

TABLE　Schematic for Therapy versus Enhancement

Purpose	Somatic	Germline
Therapeutic Treatment of Disease	Somatic Therapy (Cell 1)	Germline Therapy (Cell 2)
Enhancement of Capabilities	Somatic Enhancement (Cell 3)	Germline Enhancement (Cell 4)

（非遗传）体细胞基因组编辑的公平性与强化性

考虑到上述区别，经过数十年基因治疗的研究和临床试验，国际上基本达成共识，即只要证明安全而且有效，则利用体细胞基因修饰治疗疾病，不但不应该被禁止，反而应该被鼓励。

在现有的DNA修饰工具出现之前，政府资助的研究集中在替代患病受损器官的实体器官移植，以及治疗白血病及其他威胁生命疾病的骨髓移植，尽管这些治疗需要在实体器官或者血液干/祖细胞中将患者本身的部分DNA序列进行替换。这些案例被明确地归类为传统的医学护理范畴。目前，很多国家政府资助的研究，也在支持基因治疗、再生医学以及最近开发的人类基因编辑等领域的发展，并应用于对各类疾病的造血干/祖细胞进行修饰，如镰状细胞贫血症和其他血液疾病，或者某些类型癌症等。这些研究先例，和许多其他类似的先例一样，已经建立了有力的科学、管理及伦理方面的监督结构（见第4章）。

如上所述，许多关于基因强化的伦理讨论，基于"治疗"和"强化"这两个概念的对比。但是，考虑到过去几十年医生的角色从患者的治疗师转变为疾病预防的健康促进者，"治疗"与"强化"二者的

By the mid-1980s, scientists and bioethicists had begun to call for the morally relevant line to be between disease and enhancement rather than somatic and germline. John Fletcher, then head of bioethics at the National Institutes of Health (NIH), wrote that "the most relevant moral distinction is between uses that may relieve real suffering and those that alter characteristics that have little or nothing to do with disease" (Fletcher et al., 1985). Gene therapy pioneer Theodore Friedmann wrote that "the need for efficient disease control or the need to prevent damage early in development or in inaccessible cells may eventually justify germline therapy" (Friedmann, 1989). In 1991, as the first three somatic gene therapy trials were under way, the chair of the Human Gene Therapy Subcommittee of NIH's Recombinant DNA Advisory Committee (RAC) called for "a detailed public discussion of the ethical issues surrounding germline genetic intervention in humans" (Walters, 1991). Disease correction was defined as returning to "normal functioning," but to go beyond that was labeled "eugenics."

Somatic (Nonheritable) Genome Editing, Fairness, and Enhancement

With these distinctions in mind, there appears to be broad international consensus, derived from decades of research and clinical trials for gene therapy, that a somatic intervention undertaken to modify a person's genetic makeup for purposes of treating disease is not only permissible but encouraged, provided it proves to be safe and effective.

Before the modern tools needed to modify DNA were developed, government-supported research was focused on developing solid-organ transplantation to replace damaged or diseased organs and on bone marrow transplantation and reconstitution to cure leukemia and other life-threatening disorders, even though these treatments required substituting donor DNA for the patient's DNA in the solid-organ or blood-forming cells. Those cases fell clearly under what is typically considered medical care. Government-supported research also has been conducted in many countries to advance the fields of gene therapy and regenerative medicine, and, more recently, human genome editing, to modify the DNA in the blood-forming cells of patients with sickle-cell disease and other blood disorders or some forms of cancer. These precedents, and many others like them, have built robust scientific, regulatory, and ethical oversight structures (see Chapter 4).

As noted above, many discussions of the ethics of enhancement have been based on contrasting the concepts of "therapy" and "enhancement." However, given the

界限需要加以明确，以包含范围广泛的预防性干预。例如，疫苗，它既不是治疗也不是强化，而是与两者擦边的混合；又如，当基因组编辑用于降低患有严重冠心病的患者的胆固醇水平时，这很可能被视为治疗；而基因组编辑用于患者的患有高胆固醇、面临冠心病风险因素的兄弟姐妹时，则可能会被视为是预防性措施；基因组编辑用来降低患者健康的21岁孩子的胆固醇，将疾病风险降低至普通人群平均正常水平以下，则可能被视为擦着预防和强化之间的界限了。因此需要介入"治疗—预防—强化"的范围定义，不论这三个界限范围是否还在讨论之中，或者这个界限范围还有可能随着介入的不同而变化。

随着人类对基因组及其序列变异体与其所在环境相关性认识的提高，能通过基因组编辑解决的性状越来越多，这个趋势增强了对"正常"的定义，以及与正常有偏差时是否就是疾病问题的疑虑。每个人可能都同意Tay-Sachs症的表现是不正常的，是一种疾病，但遗传性耳聋是否算是一种疾病就见仁见智了。从是否符合与人类普遍性状的角度来看，耳聋是不正常的特性，但确实有相当一部分人群拥有类似的症状，他们当中很多人反对聋人需要被"治愈"，并且不认同需要消除或者避免他们听力丧失的建议。

评论人已经注意到疾病的概念与其说总是客观的，不如说是受权利和偏见影响的社会认同的结果。例如，在20世纪50年代同性恋被认为是一种疾病，而现在已经极少有针对"治愈"同性恋的"治疗性"干预的课题了。30年代，"犯罪行为"被认为是一种遗传病。一些残疾人活动家开始质疑类似侏儒症或者耳聋这些性状应该被认为是疾病还是增加人类多样性的变异体。在正常和病态之间的界限应该在哪里划分，以及由谁来划分这些界限；划分界限的人应该具有何种权威；哪些社会层面应该包含在内或者应该排除；需要制定什么条例进行抗辩，这些问题都需要讨论。

一些变异体只是增加了发展一种疾病的概率，而其他与某些晚发型疾病相关，这些发现使得先前清晰的"疾病"界定变得模糊。早期伦理学的辩论是建立在人类遗传学先驱所用的语言上，称之为"天生的错误"，注意到大多数靶向修饰的性状都被称为"错误"，是意外发生的错误。对于正常的标准最大的挑战来自一些研究者认为减轻疾病的目的应该被称为强化。这种强化与其说改正错误，不如说

evolution of the role of the physician over the past several decades from a healer of the sick to a promoter of health through preventive measures, the therapy–enhancement duality needs to be modified to accommodate a wide range of preventive interventions, such as vaccines, that are neither therapy nor enhancement but blend into each at the edges. For example, while genome editing to lower the cholesterol level of a patient with severe coronary artery disease would likely be viewed as a therapy, and genome editing of a sibling of the patient with high cholesterol who also had other risk factors for coronary artery disease might be viewed as a preventive measure, genome editing to lower the cholesterol of a healthy 21-year-old child of the patient to reduce disease risk below what is average or "normal" in the general population might be viewed as approaching the line between prevention and enhancement. Interventions thus can be viewed as falling on a therapy–prevention–enhancement spectrum, although the boundaries between the three categories are still open to debate and will likely vary with the specifics of the intervention.

With the growth in understanding of the human genome and of which sequence variants are associated with which conditions, the number of traits that could be addressed by genome editing continues to grow. This growing potential again raises the question of what it means to be "normal" and whether deviations from "normality" are really a disease. Everyone would agree that the manifestation of Tay-Sachs disease is not normal and constitutes a disease, but opinions differ as to whether genetically caused deafness should be considered a disease. It is not normal in the sense of being typical or being consistent with the range of capabilities typically associated with the human species, but it can also be associated with membership in a community of persons sharing this characteristic, many of whom reject the notion that deaf people need to be "cured" or otherwise treated to eliminate or circumvent their lack of hearing.

Commentators have noted that the concept of disease is not always objective, but rather can be the result of social agreement influenced by power and prejudice. In the 1950s, for example, homosexuality was considered a disease, and even today it is occasionally the subject of "therapeutic" interventions aimed at "curing" it. In the 1930s, "criminality" was considered a genetic disorder. Some disability activists began to question whether such traits as dwarfism or deafness should be considered diseases instead of variants that enrich human diversity. It is a question that led to discussions about where the line is drawn between normal and pathological, as well as questions about who gets to draw these lines; what authority do they have to draw them; which social dimensions are included or excluded; and what provision is made to contest the decisions.

赋予少数幸运儿所具有的性状，如强化免疫功能或者强化捕获胆固醇的细胞受体（Juengst，1997；Parens，1998；Walters and Palmer，1997）。这些改变应该被视为强化，或者为那些没有足够幸运在出生就获得这些性状的人创造公平竞争环境，这也使治疗和强化之间的区别变得复杂，除非将预防置于二者之中。

不公平社会优势划分的演化

虽然体细胞/生殖细胞、疾病/强化之间的区别很有用，但它们（像大多数分类一样）并不完美。一些评论人已经转而关注一种干预的影响及其影响是否公平。不是通过个人努力（如锻炼或者音乐训练），而是通过外部因素（如染发或者整容手术）达到的改变，并以此获得能产生社会优势的能力，这是可以理解的，但这种优势是在物种特有属性的范围之内。对于一些人，这样的改变纯粹为了愉悦；对于另一些人，改变是一种为与那些有着最让人喜欢外表的人有"正常"或者"平等机会"的努力。外部诱发的改变可能提供显著或不同寻常的优势，比如更高的肌肉质量或者更敏锐的视觉，甚至不需要睡眠，这些改变引出对如何获得能力真实性的质疑，以及某些个体获得的能力会不会因为不是通过努力获得的而在某种程度上被轻视的问题。但是人们与生俱来的能力千差万别，一些明显优于正常人，这就提出是否以及什么时候这种优势会引发变得不公平的问题。

这是一个难以准确回答的问题，因为在人类种群中各种能力分布高度不均匀。除非认为天生很重要，否则很难从"非自然"、"不正常"或者"过度"的强化中梳理出使个体与其他人能力公平匹配的强化。此外，任何"正常"或者"平均"相关的强化的尝试也可能被归类为一种对抗普遍的"正常"却是生命中不好的方面（例如，与年龄相关的视力、听力和运动能力的下降）的"强化"，这个词隐含贬义。

社会已经能容忍在很多情况下使用非基因组编辑的手段（如白内障手术、髋关节置换）。一些社会团体对使用权不平等的回应是通过支持能为更多人提供更多保健的干预，规避对高成本的创新领域的科研投资和保险覆盖来解决。这可能是一种在经济条件下的回应，或者是一种哲学观点，疾病是生

The discovery of variants that simply increase the odds of developing a disease and others that are associated with diseases whose onset is in later life also has blurred the previously bright line demarcating "disease." Early ethical debate built on language used by the pioneers in human genetics, who referred to "inborn errors," noting that most traits targeted for modification were called "errors," as in mistakes from what was supposed to be. The greatest challenge for the normality standard came from some researchers considering what might best be called enhancements for the purpose of relieving disease. This enhancement would not correct errors, but rather instill traits that some lucky minority of humans already have, such as by enhancing immune function or adding cellular receptors to capture cholesterol (Juengst, 1997; Parens, 1998; Walters and Palmer, 1997). Such alterations could be viewed as "enhancements" or as leveling the playing field for those not lucky enough to have these traits at birth, and they also complicate the distinction between therapy and enhancement unless one includes prevention as an intermediate.

Evolution of the Unfair Social Advantage Demarcation

While both the somatic/germline and disease/enhancement distinctions have been useful, they (like most categories) are imperfect. Some commentators have focused instead on the effect of an intervention and whether that effect is "fair." Changes not made by personal effort (such as exercise or music practice) but by external forces (such as hair coloring and cosmetic surgery) are understood to have the capacity to generate a social advantage, but it is an advantage within the realm of species-typical attributes. For some, such changes are made purely for pleasure; for others they represent an effort to "normalize" or "even the odds" with those who have the most favored appearances. Externally induced changes that offer more significant or unusual advantages, such as those providing greater muscle mass or more acute vision or obviating the need for sleep, raise questions about the authenticity of the resulting capacity and whether the individual newly endowed with these capabilities is somehow diminished by having failed to earn them. Yet people are born with unearned varying capacities, some markedly superior to the norm, which raises the question of whether and when an advantage becomes "unfair."

This is a difficult question to answer precisely because of the highly uneven distribution of abilities in the human population. Unless one assigns great importance to fate, it is difficult to tease out enhancements that allow individuals to fairly match the capacities of others from those that are "unnatural," "abnormal," or "excessive." Furthermore, any attempt to relate enhancement to what is "normal" or "average" risks categorizing efforts to combat

命自然的一部分，不是所有治疗和预防的措施都是必需的。

另一些社会团体回应称，由医疗创新的使用权差异引起的不平等，应该通过尝试增加使用和保险覆盖，而不是通过限制新产品或新技术的研究或者市场来解决。解开这些差异需要区别研究本身的限制与保险可不可以覆盖治疗的决定，这个话题反过来需要询问保险首先是一种公益事业还是私人购买服务。

即使是那些倾向于扩展使用权以回应不平等的社会团体，对于某些人来说核心的问题是强化只是一种主要由个体获得的利益，而不是整个种群的利益。John Rawls 很有影响力的公正理论强调公平的思想。他观察到一些人生来健康、有才华，或者在一个良好的社会环境，这种运气既不是挣来的，也不是应得的。从这一点他总结不是所有人在所有结果中必须平等，但是社会资源的再次分配应该解决这种起始的不公平。这个概念导致所谓的基于平等的互惠，以至于只有不平等是种群优势，尤其是最不富有的人的优势时，才能被接受 (Rawls, 1999)。

一些人可能因此得出结论，个体不通过命运或者个人努力而获得以外的社会优势的强化是有问题的，而这种优势在任何方面并不对社会其他人有利，甚至会破坏竞争的意义。利用机会平等和对社会有利的不平等作为指导，可能有助于将那些一般能够忍受（假设风险与利益成正比）的强化方式与那些更具争议性的强化方式区分开来。当然，通过体细胞或生殖细胞基因组编辑的强化在任何情况下都不大可能是最大不平等的来源。但无论其贡献大小，那些最让人感觉不舒服的通过基因组编辑的强化仍可能引起忧虑。

看完这些主题，有人可能会得出结论，强化本身不是关注的焦点，其潜在意图和后续影响才是。有人回应这种关注应该集中在技术和应用上，限制那些最有可能用于加剧不可接受的不公平性的技术和应用。另一种不同的回应是应该坚持社会团体和政府工作，制造有利的、更通用的强化作用，集中在降低让人不满的不平等。这些回应也包含了监管政策的选择。

用于个体增强的非遗传性体细胞编辑的管理

针对人类强化的监控和伦理标准长期以来一直

widespread "normal" but undesirable aspects of life (e.g., age-related declining eyesight, hearing, and mobility) as a form of "enhancement," with all the pejorative connotations implied by the word.

Society already condones such efforts for many conditions (cataract surgery, hip replacement) using methods other than genome editing. Some respond to inequality in access by favoring interventions that provide more care to more people, and eschew research investment and insurance coverage in high-cost innovations. This may be a response to economic conditions, or a philosophical view of infirmity as a natural part of life, not necessarily in need of every possible measure for treatment and prevention.

Other societies have responded to inequalities that arise from differential access to medical innovations by trying to increase access and insurance coverage, rather than by restricting research or the marketing of new products or technologies. Unpacking such differences requires distinguishing between restrictions on the research itself and decisions about insurance to cover treatments, a topic that in turn requires inquiry into whether insurance is primarily a public good or a privately purchased service.

Even for societies that tend toward expanding access to respond to inequality, a core concern for some is that enhancements are yet another benefit that would accrue primarily to the individual, without a benefit to the population as a whole. John Rawls' influential theory of justice emphasizes the idea of equality. He observes that the luck with which someone is born healthy, talented, or in favored social circumstances is neither earned nor deserved. From this he concludes not that all people must be equalized in outcomes but that further distribution of social goods should be designed to account for this initial inequality. This notion leads to so-called equality-based reciprocity, such that inequalities should be tolerated only when they somehow accrue to the population's general advantage, in particular to the advantage of those least well-off (Rawls, 1999).

Some might conclude, therefore, that a problematic enhancement is one that confers a social advantage beyond that which an individual possesses by fate or through personal effort, and that does not benefit the rest of society in any way or undermines the implicit goals of a competition. Using equality of opportunity and societally useful inequality as guides may help distinguish those forms of enhancement that might generally be tolerated (assuming the risks are proportional to the benefits) from those that would be more controversial. Of course, somatic or germline genome editing for enhancement is very unlikely to be the most profound source of inequality in any setting. But those most uncomfortable with using genome editing for enhancement will likely still be concerned regardless of the size of its contribution.

Looking across these themes, one might conclude

是政策报告的主题。最近,美国生物伦理问题研究总统委员会重点关注使用影响神经功能药物的人类强化(Bioethics Commission,2015),欧洲开放文化遗产优秀项目(EPOCH)概括了最普遍的监管模式、学术界和生物伦理学家在这个议题中的角色,以及未有确凿证据但需要确定某种可能性的领域中应发挥的作用(European Commission,2012)。早期的努力包括美国国家科学基金会(NSF)在2009年的报告(2010年)和美国总统生物伦理委员会在2003年的报告(2003年)。在英国,医学科学院、英国科学院、皇家工程学院和皇家学会共同于2012年召开了一次关于新兴技术改进的政策性研讨会,这种新兴技术的改进可能给工作场所带来影响(AMS et al.,2012)。法国国家伦理与生命科学咨询委员会和新加坡国家生物伦理委员会于2013年发表了专门关于神经增强(NCECHLS,2013)和神经科学研究(BAC Singapore,2013)的报告。所有这些报告都反映了来自医学、生物伦理和学术界的广泛意见,并为已经提出的对于强化的关注提供了丰富的信息来源,以及在清晰地描绘治疗、预防和强化之间的差异时遭遇的深刻挑战。

在美国,与其他基因治疗一样,基因组编辑的强化应用治疗将会受到FDA、RAC、IBC和IRB及立法机关的监控(见第2章)。RAC为体细胞强化提议的讨论提供了场所[60]。IRB和FDA会根据强化对个人、公众健康和环境安全的风险考虑强化带给个体、科学和社会的利益是否合理。虽然对文化或社会道德的关注很重要,但通常不在IRB的职权范围之内;该条例规定,IRB"不应该将应用研究中获得的知识而可能产生的长期影响(例如,研究对公共政策可能产生的影响)认为是属于其责任范围内的研究风险"(45 CFR Sec. 46.111(a)(2))。

因此,如果协议具有对个人大有益处的潜力,并且那些人愿意接受更大的风险,则监管机构和IRB可能同意这符合风险与可能利益匹配的标准。然而,如果监管机构和IRB决定,强化对于个人或科学没有真正的利处,即使最小的风险也不合理。随着人类基因组编辑技术的改进,有充分的理由相信,对于个人的健康和安全风险会减少。如果这些风险变得微不足道,人们可能会认为,证明风险合

---

[60] 目前,RAC不接受生殖系编辑方法的审查(见第2章和第5章)。

that enhancement per se is not the focus of concern, but rather the underlying intent and subsequent effect. One response to this concern is to focus on the technologies and applications and to restrict those most likely to be used to unacceptably exacerbate inequalities. A different response is to insist that communities and governments work to make advantageous enhancements available more generally and focus on reducing undesirable inequalities. Within this range of responses lies the choice of governance policy.

Governance of Nonheritable Somatic Editing for Enhancement of the Individual

The governance and ethics of human enhancement have long been the subject of policy reports. Most recently, the U.S. Presidential Commission for the Study of Bioethical Issues focused on human enhancement related to the use of drugs that affect neurological function (Bioethics Commission, 2015), and the European Excellence in Processing Open Cultural Heritage (EPOCH) project summarized the prevailing modes of governance and the roles of academics and bioethicists in these debates, as well as areas of missing evidence needed to identify real possibilities (European Commission, 2012). Earlier efforts include a 2009 report by the U.S. National Science Foundation (NSF) (2010) and a 2003 report by the U.S. President's Council on Bioethics (2003). In the United Kingdom, the Academy of Medical Sciences, British Academy, Royal Academy of Engineering, and The Royal Society came together in 2012 for a policy-focused workshop on emerging technological enhancements that could affect the workplace (AMS et al., 2012). And the French National Advisory Committee on Ethics and the Life Sciences and the Singapore National Bioethics Commission both produced reports in 2013 focused specifically on neuroenhancement (NCECHLS, 2013) and neuroscience research (BAC Singapore, 2013). All of these reports reflect broad input from the medical, bioethical, and academic communities and provide a rich source of information on the concerns that have been raised about enhancements, as well as the profound challenges entailed in clearly delineating the differences among therapy, prevention, and enhancement.

In the United States, governance of enhancement applications of genome editing would fall, as with other gene therapy, to the FDA, the RAC, the IBCs and the IRBs, and the legislature (see Chapter 2). The RAC can provide a venue for discussion of somatic enhancement proposals.[60] IRBs and the FDA look at whether the benefits the enhancement might provide to the individual, to science,

---

[60] At the moment, the RAC is not accepting germline editing protocols for review (see Chapters 2 and 5).

理的潜在利益也将下降。因此，随着技术的改进，其应用可以从严重疾病，延伸到不太严重的疾病、预防，以及更长远的强化，但这需要明确。

在美国，同样重要的是要记住，一旦医疗产品被批准用于特定目的和人群，赞助商仅限于为这些有适应症状的进行营销，但医生可以自由地自己做出判断，并为其他用途和其他人群开处方[61]（见第4章）。这种"非适应证"的使用使监管问题在美国和其他具有类似规则的司法管辖区（如欧洲联盟）复杂化，因为这会很难限制新医疗产品仅在那些具有最佳风险/收益比率和公众普遍支持的情况下使用。

关于强化，这种规范计划引起了一些关注，一些产品将被批准用于治疗或预防疾病，但随后被用于风险更高或不太合理的适应证外的用途。然而，如第4章所述，与许多药物不同，这些被编辑细胞的特异性将会限制其在适应证以外的应用。虽然可以想象希望变得更强壮的具有健康肌肉组织的那些人，对于用于肌营养不良的基因组编辑的细胞治疗的兴趣，但例子不多。被编辑细胞的特异性将使得滥用在可预见的未来不太可能。

此外，FDA有一些权限来限制适应证外的应用，如通过特殊患者测试或不良事件报告的要求，同时国会也可以通过立法来特别地禁止某些用途，如人类生长激素的情况（见延伸内容6-2）。其他司法管辖区有类似的权力和选择。然而，关注适应证外使用的可能的范围也是必要的，并且对于适应证外的使用是需要控制的。

结合正式的规范程序，监管的许多其他方面将影响基因组编辑是否及如何被用来强化个体。其中包括专业指南，其直接影响医生的行为，设立标准防止医生行为不当（Campbell and Glass, 2000; Mello, 2001）。提供不当治疗保险的保险公司也可以影响医生提供某些服务的意愿（Kessler, 2011）。在另一种情况下，保险公司通过选择支付使用认可的技术而发挥作用，而技术被认可部分取决于它们的使用目的，以及使用是否必要或是否有其他选择。

---

[61] 59联邦公报59,820,59,821（1994年11月18日）。"一旦[药物]产品被批准用于市场，医生可以用不包括在批准适应证内的药物为患者群体的使用或治疗方案来开处方。"通知继续说明"未经批准的"，或更准确地说，"适应证外"的使用在某些情况下可能是适当和合理的，并且可能从实际反映出医学文献中广泛报道过的药物治疗的方法。

and to society are reasonable in light of the risks to the individual, to public health, and to environmental safety. But concerns about culture or societal morals, while important, are generally not within the IRB's remit; the regulations state that the IRB "should not consider possible long-range effects of applying knowledge gained in the research (for example, the possible effects of the research on public policy) as among those research risks that fall within the purview of its responsibility" (45 CFR Sec. 46.111(a)(2)).

Thus if a protocol holds the potential for great benefit to individuals, and those individuals are willing to accept greater risk, the regulator and IRB may agree that this meets the standard of the possible benefits being reasonable in relation to the risks. If the regulator and IRB decide, however, that there are no real benefits of an enhancement—either to the individual or to science—then even a minimal risk is unjustified. As human genome editing improves technologically, there is every reason to believe that the health and safety risks to individuals will diminish. If these risks become de minimis, one might assume that the potential benefits required to justify the risks also will decline. Thus as the technology improves, its application could extend from serious illnesses, to less serious illnesses, to prevention, and in the long-term, to enhancement, however defined.

In the United States, it is also important to keep in mind that once a medical product has been approved for a particular purpose and population, the sponsor is limited to marketing it for these "labeled" indications, but individual physicians are free to use their own judgment and prescribe the product for other uses and other populations[61] (see Chapter 4). This "off-label" use complicates the question of governance in the United States and in other jurisdictions with similar rules, such as the European Union, because it makes it more difficult to restrict the use of new medical products to those situations that have the best risk/benefit ratios and the general support of the public.

With regard to enhancements, this regulatory scheme has raised concern that some products will be approved for treatment or prevention of disease but then be used off-label for riskier or less well-justified uses. As noted in Chapter 4, however, the specificity of these edited cells will limit the range of off-label uses for unrelated indications far more than is the case with many drugs. While one might imagine a genome-edited cell therapy for muscular dystrophy being of interest to those with healthy muscle tissue who wish to become even stronger,

---

[61] 59 *Federal Register* 59, 820, 59, 821 (November 18, 1994). "Once a [drug] product has been approved for marketing, a physician may prescribe it for uses or in treatment regimens of patient populations that are not included in approved labeling." The notice goes on to state that "'unapproved' or, more precisely, 'unlabeled' uses may be appropriate and rational in certain circumstances, and may in fact reflect approaches to drug therapy that have been extensively reported in medical literature."

延伸内容 6-2

人类生长激素
Human Growth Hormone

长期使用人类生长激素的经验，说明了在治疗和预防以及治疗和强化之间很难找到明确的界限。这也是一个教训，技术或干预的可获性的突然变化如何使得这些以前不重要的边界突然变成专业准则、舆论观点及法律监管的主题。

人类生长激素（hGH）一直以来是一种稀缺的商品，它仅用于那些缺乏正常水平激素的患者的治疗（Ayyar，2011）。自 1985 年，随着合成人类生长激素的发展，大众可负担的人类生长激素供应量持续增加，引发了是否应限制其使用的激烈辩论。人类生长激素现在主要给予那些不明原因的比同龄人身材矮小但健康状况良好的儿童或成年人，这种身材矮小是指身高小于人群的 1%（人群中 99% 的人都高于此高度）。一些人尽管具有正常的身高和生长速度，仍致力于寻找生长激素，实现高于平均身高或强度的水平。那么在这些案例中，人类生长激素的使用是否反映了良好的风险收益比，是否达到治疗、预防或强化的目的，以及哪些可能是合适的，这些都需要认真地进行科学和伦理调查。事实上，激素将被使用在不能为自己做决定的儿童身上，也增加了问题的复杂性。

在 20 世纪 80 年代和 90 年代，使用生长激素的风险和可能的益处是不确定的，并且该药物仅被批准用于严重缺乏生长激素的儿科病例。但是随着时间的推移，生长激素治疗被证明对于血液中具有低激素水平的患者，或者具有正常水平的激素但生长严重受阻的患者是相当安全和有效的。研究显示，根据治疗的剂量和时间的不同，儿童身高可增长 10%~25%（Allen and Cuttler, 2013; Maiorana and Cianfarani, 2009）。但治疗不是完全无风险的，患者往往伴随相对轻微或中度不良症状，如呼吸道阻塞和头痛（Bell et al., 2010; Cohen et al., 2002; Kemp et al., 2005; Lindgren and Ritzen, 1999; Willemsen et al., 2007）。此外，长期风险的研究相当少，一些证据表明，使用生长激素的儿童，成人阶段中风的风险也会增加（Ichord, 2014）。

The experience with human growth hormone illustrates the difficulty of finding clear boundaries between therapy and prevention, and between treatment and enhancement. It is also a lesson in how a sudden change in the ease or availability of a technology or intervention can make those boundaries—previously of little importance—suddenly become the subject of professional norms, public opinion and legal controls.

Human growth hormone (hGH) was a scarce commodity for many years, and its use was largely limited to 'treatment' for those who lacked normal levels of the hormone (Ayyar, 2011). With the development of synthetic human growth hormone in 1985, the newly increased supply of affordable hGH triggered a lively debate about whether there ought to be constraints on its use. For example, hGH could now be given to children or adults in good health but with unexplained short stature, ranging from <1st percentile of height for their age to being only slightly shorter than their peers. Some others sought hGH despite being of normal height and growth rate, with an interest in attaining above- average height or strength. Whether or not administration of hGH in these and many other cases reflected a favorable risk -benefit ratio, whether they qualified as treatment, prevention, or enhancement, and which of those uses might be appropriate, required careful scientific and ethical investigation. The fact that the hormone would be administered to children too young to make decisions for themselves added to the complexity of the conversation.

The risks and possible benefits of hGH administration were uncertain in the 80's and 90's, and the drug was approved only for severe cases of growth hormone deficiency in children. But over time growth hormone therapy was shown to be reasonably safe and effective for patients with low levels of the hormone in their blood or who experienced severely stunted growth despite normal levels of hormone; studies showed it bringing children up to the 10th to 25th percentile, depending on dosage and timing of treatment (Allen and Cuttler, 2013; Maiorana and Cianfarani, 2009). But the treatment is not entirely risk- free; children can experience relatively minor or moderate adverse events such as respiratory congestion and headache (Bell et al., 2010; Cohen et al., 2002; Kemp et al., 2005; Lindgren and Ritzen, 1999; Willemsen et al., 2007). In addition, long-term risks have had considerably less study, and there is some evidence to suggest that children who receive hGH may have a slightly increased risk of stroke as adults (Ichord, 2014).

FDA批准儿童和成人仅在很窄的适应症列表中使用人类生长激素，包括生长障碍的儿童（如慢性肾功能不全、特纳综合征、普拉德-威利综合征，Noonan综合征）或严重不明原因的身材矮小的儿童。对于严重缺乏生长激素的成人，治疗方案可能适度提高身体组分、运动能力和骨骼的完整性（Molitch et al.，2011）。与艾滋病相关的消瘦综合征或短肠综合征的成年人也有资格使用生长激素进行治疗（Ayyar, 2011; Cook and Rose, 2012; Cuttler and Silvers, 2010）。

尽管没有证据表明补充生长激素增加了健康个体的肌肉力量或有氧运动能力（Liu et al., 2007），但一些健康成年人使用生长激素，希望它能作为提高机体性能的药物，或将减慢衰老过程。但生长激素可能增加高风险的严重不良反应，包括糖尿病、癌症、高血压、肌肉疼痛、关节疼痛、软组织的肿胀和炎症、腕管综合征和男性乳腺组织扩大（Liu et al., 2008; Perls and Handelsman, 2015）。此外，一些证据表明，对生长激素具有遗传抗性的人享有更长的生命期，这表明给予正常衰老的成年人生长激素，实际上可能缩短他们的生命（Suh et al., 2008）。尽管存在这些危险，生长激素仍被一些运动员和老年人使用（DEA, 2013）。

FDA负责处方药物的监管，联邦法律禁止赞助商对超出其批准适应证的用途进行广告和营销药物。但是医生通常可以根据自己的专业知识和经验进行判断，自由地为未经FDA评估的其他适应证（称为"非"处标签方）开药。这是一个常见的现象，在不断变化的信息和医生经验的情况下，确实带来了公认的益处，但生长激素"强化"的非适应证使用的滑坡趋势，导致国会采取不寻常的步骤进行立法限制。根据1990年的《犯罪控制法》[62]第61条，

The FDA approved human growth hormone for use in children and adults for a limited and narrowly defined list of indications, including children with growth disorders (e.g. chronic renal insufficiency, Turner's syndrome, Prader-Willi syndrome, Noonan syndrome) or severe unexplained short stature. For adults with severe growth hormone deficiency, treatment can result in modest gains in body composition, exercise capacity, and skeletal integrity (Molitch et al., 2011). They, as well as adults with AIDS-associated wasting syndrome, or short bowel syndrome also qualified for treatment (Ayyar, 2011; Cook and Rose, 2012; Cuttler and Silvers, 2010).

Despite the absence of evidence that hGH supplementation increases muscle strength or aerobic exercise capacity in healthy individuals (Liu et al., 2007), some healthy adults have been drawn to hGH in hopes it would serve as a performance-enhancing drug or would slow the normal aging process. But this has been linked to a high risk of serious adverse effects, including diabetes, cancer, hypertension, muscle pain, joint pain, swelling and inflammation of soft tissue, carpal tunnel syndrome, and enlarged breast tissue in men (Liu et al., 2008; Perls and Handelsman, 2015). Additionally, there is some evidence that people with genetic resistance to growth hormone enjoy longer life spans, suggesting that giving hGH to otherwise normally aging adults might actually shorten their lives (Suh et al., 2008). Despite these dangers, hGH was used by some athletes and aging adults (DEA, 2013).

The FDA is responsible for regulation of prescription drugs, and federal law prohibits sponsors from advertising and marketing drugs for uses that go beyond their approved indications. But physicians are generally free to prescribe drugs for other indications not evaluated by the FDA (known as 'off-label' prescription), based on their own professional judgment and expertise. This is a common phenomenon with widely acknowledged benefits in cases of evolving information and physician experience, but the slippery slope trend toward off-label so-called "enhancement" uses of hGH led Congress to take the unusual step of enacting legislative restrictions. According to the Crime Control Act of 1990[62], the

---

[62] 该美国法典标题21§333（e）禁止分配人类生长激素。
(1) 除第（2）条另有规定外，任何人故意或意图分发人类生长激素，除用于人类使用，可处以不超过5年的监禁，第18条或两者都授权了刑罚。其中将人类生长激素用于治疗疾病或其他已认可的医疗状况不会受到刑罚，根据本标题第355条，这种使用是已获卫生和公共服务部部长的授权并且要按照医生的指示进行。
(2) 任何人犯第（1）条所列的任何罪行，而该罪行涉及未满18岁的人，可处以不超过10年的监禁，第18条或两者都授权了刑罚。
(3) 任何违反本第（1）和（2）条的定罪，均视为是违反"管制物质法案"的重罪[21 U.S.C. 801等]，以便根据该法令[21 U.S.C. 853]第413条进行相应没收。
(4) 如本小部分中所使用的，术语"人生长激素"是指人蛋氨生长素，生长激素或其任一种的类似物。
(5) 授权药品监管局调查本部分所规定的可处罚的罪行。

[62] US Code Title 21 § 333(e) Prohibited distribution of human growth hormone
(1) Except as provided in paragraph (2), whoever knowingly distributes, or possesses with intent to distribute, human growth hormone for any use in humans other than the treatment of a disease or other recognized medical condition, where such use has been authorized by the Secretary of Health and Human Services under section 355 of this title and pursuant to the order of a physician, is guilty of an offense punishable by not more than 5 years in prison, such fines as are authorized by title 18, or both.
(2) Whoever commits any offense set forth in paragraph (1) and such offense involves an individual under 18 years of age is punishable by not more than 10 years imprisonment, such fines as are authorized by title 18, or both.
(3) Any conviction for a violation of paragraphs (1) and (2) of this subsection shall be considered a felony violation of the Controlled Substances Act [21 U.S.C. 801 et seq.] for the purposes of forfeiture under section 413 of such Act [21 U.S.C. 853].
(4) As used in this subsection the term "human growth hormone" means somatrem, somatropin, or an analogue of either of them.
(5) The Drug Enforcement Administration is authorized to investigate offenses punishable by this subsection.

对于未经批准的适应证，如抗衰老、与年龄有关的情况，或提高运动成绩，人类生长激素的持有和使用是一种重罪，可处以罚款和监禁。FDA 和药品执行机构都对联邦食品、药品和化妆品法案的此修正案进行解释，意味着人类生长激素的非处方使用目前来看是非法的（Cronin，2008；FDA，2012a）。

种系（遗传）基因组的编辑和强化

正如第 5 章所述，生殖系基因组编辑展现出诱导可遗传性变化的前景，可能会影响几代人，而不仅仅是由基因组编辑的胚胎或配子所长成的孩子。在围绕基因组编辑改良的讨论中，这种前景可能会加深一些人的不安，担心那些与疾病治疗、疾病预防及修复重要的身体或社会缺陷完全无关的应用。这种担心不仅来自于上述对于体细胞基因组编辑的关注，还受到优生学那悠久而令人不安的历史所影响，那是一段包含强迫措施甚至种族屠杀的历史。这段历史充斥着各种根据种族、宗教信仰、国家来源和经济状况来建立人类素质等级制度的教条，它充分证实了科学概念（如自然选择）和公众福利措施（如公共卫生）是如何被用于残忍且毁灭性的社会政策的目的。这些考虑引发了以下疑问，即可遗传性生殖系编辑的"改良"应用是否应该被完全禁止，或者用不同于那些纯粹的非遗传性体细胞编辑的措施进行严格的限制？

优生学

"优生学"这一名词最初使用于 19 世纪末期，其定义是通过给予"更合适的种群或血统以更好的机会来迅速取得对劣等种群或血统的压倒性优势（Kevles，1985），从而达到改良人种的目的。"它的中心思想就是为了达到改良人种的目的，鼓励拥有"优秀"血统的人类繁育更多子孙后代，而让那些"劣等"血统的人类少生或者不生。当时人们对于哪些是真正可遗传的品质还所知甚少，因而许多优生学家都以现在看来无法接受的方式来推广他们的社会偏见。在英国，优生学家们认为"好的"品

distribution of human growth hormone for non-approved indications, such as for anti-aging, age-related conditions, or enhancing athletic performance, is a felony punishable by fines and imprisonment. Both the FDA and the Drug Enforcement Agency interpret this amendment to the Federal Food, Drug, and Cosmetic Act strictly to mean that off-label prescription of hGH is now illegal (Cronin, 2008; FDA, 2012a).

other examples are more difficult to envision. The specificity of edited cells will make such applications less likely for the foreseeable future.

In addition, the FDA has some authority to restrict off-label uses—for example, through requirements for special patient testing or adverse event reporting—and the U.S. Congress can pass legislation to specifically prohibit certain uses, as has been the case for human growth hormone (see Box 6-2). Other jurisdictions have similar powers and choices. Nonetheless, attention to the possible range of off-label uses is necessary, and the need for some control over off-label use can be anticipated.

In conjunction with formal regulatory processes, a number of other aspects of governance will affect whether and how genome editing is used for enhancement. These include professional guidelines, which influence physician behavior directly and set standards against which physician behavior is judged in cases of possible malpractice (Campbell and Glass, 2000; Mello, 2001). Insurers that offer malpractice coverage also can influence the willingness of physicians to offer certain services (Kessler, 2011). In another capacity, insurers play a role by choosing to cover the cost of using approved technologies based in part on the purpose for which they are going to be used and whether the use is necessary or elective.

GERMLINE (HERITABLE) GENOME EDITING AND ENHANCEMENT

As noted in Chapter 5, germline genome editing presents the prospect of inducing heritable changes that could affect multiple generations, not just the child who developed from a genome-edited embryo or gametes. In the context of the discussions around enhancement, this prospect may deepen some of the disquiet concerning those applications that are most distant from disease treatment, disease prevention, and correction of significant physical or social disadvantages relative to the norm. This disquiet is influenced not only by the concerns outlined above with respect to somatic genome editing but also by the long and troubling history of eugenics, a history that included coercive measures and even genocide. This

质存在于贵族上层阶级中。他们推测上层社会的优良品质是可以遗传的，因此穷人就应该生育更少的孩子。在美国，最初的优生学动机含有种族或者种族特点。优生学的一个目的就是阻止具有"不良"品质的种族移民进入美国。这种思想的巅峰之作是1924年颁布的《移民控制法案》，该法案限制来自东欧和南欧的人移民美国，由时任总统的卡尔文·柯立芝签署，他之前就宣称"美国必须是美国人的美国"，因为"生物法则显示日耳曼民族和其他种族混合时会遭到弱化"（Kevles，1985）。

直到20世纪20年代，很多国家都将人们的个人品质和其种族与阶层分开研究，试图鉴别出那些可能具有诸如"智力低下"和"犯罪行为"遗传性状的人群，并竭力阻止他们繁衍后代。这些优生学计划并不是个人自愿的，许多重罪犯和妇女被强制性绝育。其中最著名的一个例子是美国最高法院的Oliver Wendell Holmes大法官准许对一名"智力低下"的妇女Carrie Buck进行绝育手术，他认为这样的绝育手术是公正的，因为"这对整个世界都有好处。如果需要采取行动，而不仅仅是等着对罪犯的后代犯罪时进行惩罚，或者让他们因愚钝而挨饿，社会就应该阻止那些明显不合适的人群继续繁衍自己的后代。强制接种疫苗的原则应被推广到切断妇女们的输卵管上。连续三代都生出低能儿那就禁止再生育。"[63] 优生学计划曾是进步社会变革的一部分，并且曾经妄想通过改良遗传品质来增长人口数量（Lombardo，2008）。

优生学的逻辑思想在纳粹德国被引申到极端，在那里，那些被认为有遗传缺陷的人首先会被实施绝育，之后又惨遭杀害。优生学纯粹性的逻辑思想就是种族大屠杀的一部分，它导致了上百万被描述成劣等的人类的死亡，其中基本是犹太人、吉普赛人和残疾人。根据历史学家丹尼尔·凯夫利斯的描述，在对种族大屠杀进行揭露之后，人们意识到"一条鲜血汇成的大河最终从1933年的德国绝育法案流淌至奥斯维辛和布痕瓦尔德集中营"（Kevles，1985，p.118）。

革新优生学

对种族大屠杀的揭露并不意味着优生学的结

---

[63] *Buck v. Bell*, 274 U.S. 200 (1927).

history is replete with dogma that creates hierarchies of human quality based on race, religion, national origin, and economic status, and it demonstrates how scientific concepts, such as natural selection, and public welfare measures, such as public hygiene, can be subverted for purposes of cruel and destructive social policies. These considerations lead to the question of whether "enhancement" applications of heritable germline editing should be prohibited entirely or significantly restricted in ways measurably different than those for purely non-heritable somatic editing?

Eugenics

The term "eugenics" was first used in the late 19th century to define the goal of improving the human species by giving "the more suitable races or strains of blood a better chance of prevailing speedily over the less suitable" (Kevles, 1985). The general idea was to create schemes for encouraging people with "good" bloodlines to have more children and those with "bad" bloodlines to have fewer or no children in order to improve the human species. Given their extremely limited understanding of what traits were truly heritable, eugenicists in various societies applied their social biases in ways now deemed unacceptable. In Britain, eugenicists assumed that the "good" traits were those found among the upper classes. They inferred that the fine qualities of the aristocracy were heritable, so the poor should simply produce fewer children. In the United States, the original eugenic impulse involved race or ethnicity. One eugenic goal was to keep races with "bad" traits from immigrating to America. The peak effort in meeting this goal was the 1924 immigration control act, which limited immigrants from Eastern and southern Europe. It was signed by President Coolidge, who had earlier claimed that "America must be kept American" because "biological laws show that Nordics deteriorate when mixed with other races" (Kevles, 1985).

By the 1920s, countries increasingly looked at people's individual qualities independent of their race and class, trying to identify those with supposedly genetic traits such as "feeblemindedness" and "criminality" and discouraging them from reproducing. These eugenics programs were not necessarily voluntary, and many felons and women were forcibly sterilized. Most famously, Justice Oliver Wendell Holmes of the U.S. Supreme Court wrote the opinion that allowed the sterilization of Carrie Buck, a "feebleminded" woman, concluding that sterilization was justified because "[i]t is better for all the world, if instead of waiting to execute degenerate offspring for crime, or to let them starve for their imbecility, society can prevent those who are manifestly unfit from continuing their kind. The principle that sustains compulsory vaccination is broad enough to cover cutting the Fallopian tubes. Three generations of

束，而仅是以种族论为基础的、强制性的、带有国家意志的优生学的结束。许多科学家摒弃了主流的优生学观点，赫尔曼·穆勒在1935年写道，优生学已经"绝望地堕落"成为伪科学中的一员，它正在被"种族主义和阶级偏见的拥护者、教会和国家既得利益的捍卫者、法西斯主义者、纳粹主义者以及极右反动派们"所鼓吹（Kevles，1985，p. 164）。穆勒（Muller）和其他著名的科学家如朱利安·赫胥黎（Julian Huxley）一起创立了"革新"优生学，致力于阻止带有遗传疾病的人生育和鼓励更多具有"优越"基因的人生育，而不管这些人的种族和阶层。人们被鼓励自愿地为种群利益来改变他们的生殖行为。最值得争议的部分在于，这些思想家希望人类控制他们自己的进化，提高种群的质量，如使人类变得更聪明。从20世纪50年代到70年代早期，有关的伦理学争论非常广泛，它们经常集中在对一个物种进行遗传学改造的目标是什么。这个20世纪早期的主题又被重新提起，一些革新优生学的批评者提出人类应该满足于他们目前所存在的方式。一般来说，伦理争议是物种改造的目标，或者究竟是否有这样的目标（Evans，2002）。

在1953年，沃森和克里克诠释了DNA自我复制的结构基础（Watson and Crick，1953），让大家明白基因中的DNA是如何编码信息的。这个发现改变了有关优生学的伦理学争论，正如人们意识到的，如果基因实际上是结构可以被描绘的化学物质，社会就不再需要依靠"谁和谁结婚"，人类可以被化学性地改造成为具有"更多的""好"基因的物种。当代著名科学家罗伯特·辛斯海姆（Robert Sinsheimer）的反应就非常典型，他在1969年写道，"旧的优生学需要进行连续的筛选，留下有用的，淘汰无用的。而新的优生学原则上允许将所有无用的基因转换至最高的遗传学水平…因为我们有这样的潜力在人类身上去创造出新基因和新品质"（Sinsheimer，1969，p. 13）。神学家保罗·拉姆塞（Paul Ramsey）在1970年写道，这样的方法将会使"人"成为"他自身的创造者"，从而产生一门全新的神学（Ramsey，1970，p. 144）。

imbeciles are enough."[63] Eugenics programs were part of progressive social reforms and were thought to uplift the population by improving genetic qualities (Lombardo, 2008).

The logic of eugenics was taken to its extreme conclusion in Nazi Germany, where those perceived to have genetically derived limitations were first sterilized and in later years killed. The logic of eugenic purity was a part of the Holocaust, which resulted in the deaths of millions of people portrayed as genetic inferiors, primarily Jews, Roma, and those with disabilities. According to historian Daniel Kevles, after revelation of the Holocaust, people realized that "a river of blood would eventually run from the German sterilization law of 1933 to Auschwitz and Buchenwald" (Kevles, 1985, p. 118).

Reform Eugenics

Revelation of the Holocaust was not the end of eugenics, only the end of race-based, coercive, state-mandated eugenics. Many scientists had already rejected the mainstream eugenic view, with Hermann Muller writing in 1935 that eugenics had become "hopelessly perverted" into a pseudoscientific facade for "advocates of race and class prejudice, defenders of vested interests of church and state, Fascists, Hitlerites, and reactionaries generally" (Kevles, 1985, p. 164). Muller and other prominent scientists, such as Julian Huxley, would create a "reform" eugenics that sought to stop the reproduction of people with genetic disease and encourage more reproduction by people with "superior" genes—from whatever race and class. People would be encouraged to change their reproductive practices voluntarily for the good of the species. Most notably for future debates, these thinkers wanted humanity to seize control of its own evolution and improve the species in various ways, such as by making humans more intelligent. The ethical debate of the 1950s through the early 1970s was quite broad, often focused on what the goals for genetic modification as a species should be. In a theme that would recur from this era forward, some critics of reform eugenics averred that humans should be satisfied with the way they are. In general, the ethical debate was about the genetic goals of the species—or whether to have such goals at all (Evans, 2002).

In 1953, Crick and Watson described the structural basis of how DNA duplicates itself (Watson and Crick, 1953), leading to understanding of how the DNA of genes encodes information. This discovery changed the ethical debate concerning eugenics as people realized that if genes were actually chemicals whose structure could be characterized, society no longer would have to rely on "who mates with whom"; rather, people could be chemically modified to have "more" of the "good" genes. Robert Sinsheimer, a prominent scientist of the time,

物种是否需要被改良；如果是，采用何种方式，这是直到20世纪70年代早期伦理争论的核心。其中的一个论题（直到今天仍存在于某些超人论者中）是人类进化是否应该留待自然选择来进行，因为这是一个随机且非常缓慢的过程。例如，罗马尼亚化学家 Corneliu Giurgea 在1964年合成了吡拉西坦，该药物能提高认知能力，他写道，"人类将不会被动地等待数百万年让进化来赋予人更好的大脑"（Giurgea, 1973）。事实上，随着对地球上气候变化的担忧和想象中的火星殖民计划，一些超人论者正在讨论人类是否需要干预他们自身的进化以适应他们正在创造的未来（Bostrom, 2005; Rosen, 2014）。但是在20世纪70年代，做出改变的复杂程度和决定何种改变合适，都会让很多人重新思考著名生物学家 Bernard Davis 提出的"创造性预测的无法控制"，他提醒读者在现在看来显而易见的事实，诸如"强化特征"是针对多种基因，因此很难或者不可能被改造（Davis, 1970, p. 1279）。

那个时代的技术限制也助长了争论的形成。当体细胞治疗都还不是那么成功的时候是无法去想象体细胞强化的，所以，任何对于体细胞强化的想法都会被认为过于冒险而且收效甚微。同样的，如果通过病毒载体对体细胞进行改造的早期尝试仅仅在很少一部分细胞中取得成功，又如何使精子、卵子或受精卵获得改变呢？但是，在体细胞强化和可遗传性改变同时到来之时，尽管没有技术上的可行性，公众的担忧却是最强的。优生学家们的目标——使物种变得更好——就被放在这个范畴之内，因而，优生学被摒弃之后，随之而来的是生殖系编辑的可能性。

对于生殖系强化带来的滑坡效应的担忧

革新优生学时期之后生殖系强化的反对者们担忧长期的文化或社会变革带来的社会滑坡效应（见第5章）。体细胞治疗最终会引发通过生殖系基因工程对物种进行完善的行为，这将是一个随着体细胞治疗的技巧和精度逐步提高使得其他相关应用变得更加有益和安全的过程吗？这些反对者们愿意支持体细胞治疗是因为他们认为体细胞和生殖系之间有着显著的差距，而公众也会在改造的个体和改造的后代之间做出明确的区分（Burgess and Prentice,

was typical in his response, writing in 1969 that "the old eugenics would have required a continual selection for breeding of the fit, and a culling of the unfit. The new eugenics would permit in principle the conversion of all of the unfit to the highest genetic level ... for we should have the potential to create new genes and new qualities yet undreamed in the human species" (Sinsheimer, 1969, p. 13). Theologian Paul Ramsey wrote in 1970 that such proposals would make "man" "his own self-creator" and lead to a new theology of science (Ramsey, 1970, p. 144).

Whether to improve the species and if so, in what way was the core of the ethical debate until the early 1970s. A discussion then—one continuing today among some transhumanists—is whether human evolution should be left to processes of natural selection, which are random and occur very slowly. For example, Corneliu Giurgea, the Romanian chemist who synthesized Piracetam in 1964 and showed that it might act in cognitive enhancement, said, "Man is not going to wait passively for millions of years before evolution offers him a better brain" (Giurgea, 1973). Indeed, with the specter of climate change on earth and the imagined colonization of Mars, some transhumanists today discuss whether humans need to intervene in their own evolution to cope with the future they are creating (Bostrom, 2005; Rosen, 2014). But in the 1970s, the evident complexity of making changes, let alone determining which changes are desirable, led many to rethink what prominent biologist Bernard Davis would dub "Promethean predictions of unlimited control." He reminded readers of facts now considered obvious, such as that most "enhanced characteristics" are polygenic and thus difficult or impossible to modify (Davis, 1970, p. 1279).

Technological limits of that era also helped shape the debate. It was impossible to imagine somatic enhancements when somatic therapy was not yet successful, so any claim of a somatic enhancement would have been considered too risky, with very little benefit. Similarly, if early attempts at modifying somatic cells through viral vectors were successful in only a small number of the cells, how could sperm, eggs, or zygotes be changed? But where enhancement and heritable change came together, even though not yet technically feasible, public concern was greatest. And the goal of the eugenicists—to make the species better—was placed in this category, thus comingling the rejection of eugenics with the possibilities for germline editing.

Slippery Slope Concerns about Germline Enhancement

Opponents of germline enhancement from the reform eugenics era forward were concerned largely with long-term cultural or social changes that could occur as the result of a sociological slippery slope process (see also Chapter 5). That is, would somatic therapy eventually

2016)。

这种滑坡效应的争论出现在 1981 年，当时美国最高法院做出了一项裁决，允许对基因工程改造的生命形式授予专利（Evans，2002），这引起了有关"人类生命的基本属性和人类个体的尊严及价值"的关注（President's Commission，1982，p. 95）。当时的总统生物伦理委员会撰写了一份名为《剪接生命》的报告，报告中重新表述了伦理学的争论，因而这份报告将会"对公共政策考虑大有裨益"（President's Commission，1982，p. 20）。为了使伦理诉求成为可执行的法律，就必须抛弃关于未来文化伤害的争论或主张人性的作用不是用来改造自身，并且需要更加具体且即时的结论，而不是推测性的。在这份报告中，委员会陈述了"没有根据可以推断任何目前的或者计划中的基因工程形式，无论其使用人类或者非人类物质，存在着内在的错误或者本身是反宗教的"（President's Commission，1982，p. 77）。这份报告建立了风险、利益及个人权利的框架，它可以作为政府管理这门新兴科学的框架，如通过 RAC 的人类基因治疗小组委员会。这种方法并没有特别地去考虑更广泛或者更长期的社会效应，如滑坡效应，是因为它更关注于对个体人类的即时效应。

20 世纪 80 年代之前的伦理学争论在 2001 年又重新回归，乔治 W. 布什总统任命了一个联邦生物伦理学委员会，该委员会宣称"避免成本和利益的功利主义的核算或者是基于个人'权利'的狭隘分析"。与之相对的，它宣称要将它的考虑建立在"对人类繁衍和人类治疗这个广泛的平面上，挖掘其中更深刻的意义"（President's Council on Bioethics，2003）。该委员会主要担忧不平等关系的升级和父母对子女至高无上的权力（President's Council on Bioethics，2003，p. 44）。报告没有对有关人类基因改造的政策产生重大影响，但它明确了美国社会利用伦理学透镜来审视生殖系基因改造。

这个布什总统时期的委员会的担忧是生殖系强化可能促使人们将孩子视为可以设计和操纵的对象，这也是长久以来一些社会科学家和人类学家所忧虑的。政治理论家 Michael Sandel 写道，"将孩子视为珍贵礼物的正确方法是接受他们到来时的模样，而不是把其当作我们设计的对象、我们意愿

lead to efforts to enhance the species through germline engineering, a process that might evolve as skill and familiarity with somatic therapy made it easier to imagine other applications as helpful and safe? These opponents were willing to endorse somatic therapy because they thought the somatic/germline distinction was culturally strong, and the public would make a clear distinction between modifying individuals and modifying their offspring (Burgess and Prentice, 2016).

This sort of slippery slope argument emerged in 1981 after a U.S. Supreme Court decision that allowed the patenting of genetically engineered life forms (Evans, 2002). Concerns were raised about "the fundamental nature of human life and the dignity and worth of the individual human being" (President's Commission, 1982, p. 95). The presidential bioethics commission of that era wrote a report entitled *Splicing Life*, in which it reformulated the ethical debate so that the report would be "meaningful to public policy consideration" (President's Commission, 1982, p. 20). To make ethical claims legally actionable meant moving away from arguments about future cultural harms or claims that it is not the role of humanity to modify itself. Consequences needed to be more concrete and near-term, not speculative. In the report, the commission stated that it "could find no ground for concluding that any current or planned forms of genetic engineering, whether using human or nonhuman material, are intrinsically wrong or irreligious *per se*" (President's Commission, 1982, p. 77). The report established a framework of risks and benefits and the rights of individuals that would serve as a framework for government regulation of this new science, such as through the human gene therapy subcommittee of the RAC. It was an approach not particularly amenable to consideration of broader and longer-term social effects, for instance slippery slopes, due to its focus on more immediate effects on identifiable persons.

The pre-1980s ethics debate returned in 2001 with the appointment of a federal bioethics commission by President George W. Bush. This commission claimed that it "eschewed a thin utilitarian calculus of costs and benefits, or a narrow analysis based only on individual 'rights.'" Instead, it claimed to ground its reflections "on the broader plane of human procreation and human healing, with their deeper meanings" (President's Council on Bioethics, 2003). Most notably, the commission was concerned about the promotion of inequality; about parents having the ultimate power over their children (Council on Bioethics, 2003, p. 44). The report did not have a strong impact on policy regarding human genetic modification, but made it clear that some in U.S. society viewed germline genetic modification through this particular ethical lens. The concern of the Bush-era commission that germline enhancement might encourage people to view children as something to be designed

的产物或是实现我们野心的工具。父母的爱不取决于其子女天生所具有的才能和素质"（Sandel，2013，p. 349）。其含义之一就是潜在的父母应避免直接进行有利于他们未来子女的改造，不仅是因为他们可以看到自己的孩子与众不同，也是因为点滴而又不直接的文化积累使他们可以看到所有孩子的不同之处。但是，对于这种观点的批评者们可以质疑危害只是推测性的，因为目前个人和社会关系的自由程度在自由民主社会所允许的范围之内是比较好的。

另外提出的一个疑虑主要是围绕父母是否应该对于其后代的品质负有越来越大的责任。Sandel写道，"我们正在减少偶然性，更关注选择性。父母对为他们的子女选择或不选择正确的品质负上责任"（Sandel，2004）。

同样的想法是用更为宗教化的语言"制造"和"产生"来进行表达的。神学家Gilbert Meilaender描述设计孩子的遗传品质就像是"制造"他们，而不设计就是"产生"他们。更为重要的是，和Sande一样，他声明"我们所产生的就如同我们一样，而我们所制造的则不是。它是我们自由决定的产物，它的命运由我们来决定"（Meilaender，1996）。在这个观点中，产生（非制造）对人类尊严和人类权利是很重要的，因为"我们生而平等，无论我们富贵贫贱、卓越与否，只因为没有任何人是其他人的'制造者'"（Meilaender，2008，p. 264）。这些对于客观化的担忧包括生殖系转变到与普通健康有关的基因，更有可能的是由健康有关的甚或超出该范畴的生殖系强化所引出。

当然还存在其他观点。其中之一认为无论人类对于遗传的未来进行何种选择——不管是否人类更像物体的观念——人类终究还是人类。正如生物伦理学的奠基人之一Joseph Fletcher在1971年写道，"人类既是制造者，也是选择者，更是设计者，事物能被更加理性地人工获取或者更刻意的话，人类也会如此……真正的区别只存在于偶然或随机的繁衍还是理性或选择性的"（Fletcher，1971，pp. 780-781）。

其他人已经讨论过，父母的裁量权允许大范围的尝试，只要它们对于孩子的身体或生理发展没有显著性的危害（Robertson，2008）。正如第5章中所讨论的，这需要纯粹的生殖权利框架，该框架

and manipulated has long been a concern of some social scientists and humanists. Political theorist Michael Sandel wrote that "to appreciate children as gifts is to accept them as they come, not as objects of our design or products of our will or instruments of our ambition. Parental love is not contingent on the talents and attributes a child happens to have" (Sandel, 2013, p. 349). One implication is that potential parents should refrain from making modifications that would directly benefit their future child, not necessarily because they would see their own child differently, but rather because it might in a tiny, indirect yet cumulative way promote a culture that comes to see all children differently. But critics of this view would argue the harm is speculative, and that this level of freedom in the relationship of the individual to society is well within the range of what we allow in liberal democratic societies.

Another concern that has been raised revolves around whether parents might become increasingly viewed as responsible for the qualities of their offspring. According to Sandel, "we attribute less to chance and more to choice. Parents become responsible for choosing, or failing to choose, the right traits for their children" (Sandel, 2004).

A similar idea is expressed in the more religious language of making versus begetting. Theologian Gilbert Meilaender describes designing the genetic qualities of one's children as akin to "making" them, whereas nondesign is "begetting" them. More important, and like Sandel, he states that "what we beget is like ourselves. What we make is not; it is the product of our free decision, and its destiny is ours to determine" (Meilaender, 1996). By this view, begetting (i.e., nondesign) is critical to human dignity and human rights because "we are equal to each other, whatever our distinctions in excellence of various sorts, precisely because none of us is the 'maker' of another one of us" (Meilaender, 2008, p. 264). These concerns about objectification might possibly apply to germline conversion to genes associated with ordinary health, but would more likely be raised by enhancements, health-related or beyond.

There are other views, of course. One might say that making choices about our genetic future—whether they increase the perception that humans are more like objects or not—is precisely human. As Joseph Fletcher, one of the founders of bioethics, wrote in 1971: "Man is a maker and a selecter and a designer, and the more rationally contrived and deliberate anything is, the more human it is.... [T]he real difference is between accidental or random reproduction and rationally willed or chosen reproduction" (Fletcher, 1971, pp. 780-781).

Others have argued that parental discretion allows for a wide range of practices, provided they are not significantly harmful to the physical or psychological development of a child (Robertson, 2008). As discussed in Chapter 5, this would require a pure reproductive rights framework that

必须扩展它的限制以包含加强或者减弱人类特征的权利（Robertson，2004）。这也是父母的自由，包括对婴儿和儿童的大范围的改善，包括一些生物医学措施，如整容手术、使用生长激素治疗未知原因的身材矮小和使用一些性能增强药物。引申开来，我们可以说这种自由应该包括具有相似风险和利益平衡的生殖系强化。因此，有很少理由去认为，在美国，有关生育的法律案例能阻止政府来监管生殖系基因组编辑，即使它有一个理性的基础。

学术性超人学说的出现也对这些争论起到了推波助澜的作用。超人学家们不仅讨论某种形式的、以改良或强化为导向的生殖系编辑的伦理学合法性，他们甚至会讨论父母的责任及伦理义务，即利用强化可能性使自己的孩子受益（Persson and Savelescu，2012）。其中一位哲学家提出，我们有义务为那些还不能为自己做决定的人做出最好的决定，这些人包括未来的孩子及他们的后代（Harris，2007）。这些是有关道德义务的观点，然而，在美国法规或法院判决（或者其他国家的）中，没有把他们视作法律问题。

总体来说，在过去的半个世纪里出现了两种截然不同的方法来评估生殖系强化的伦理学基础。其中一种方法的焦点是更加社会性的和哲学性的。它不仅包含通常所有关于生殖系编辑的担忧，正如第5章中所述，也包含改变我们如何看待自己孩子的担忧，还包含由于强化的可遗传性多代累积创造和增加社会不平等的担忧。尽管个体强化的好处能被视作评判个体干预的理由，但这些分析常常还有滑坡效应的担忧和以往优生学运动的回响。

另一种方法认为疾病和强化之间的区别还是大有用处的，因为它可以很好地追踪对于个人风险和利益的评估。这种评估是管理机构如FDA的关注焦点，他们负责审查批准新的医学产品，研究监管机构负责监督对于临床试验参与者及其他一些可能在试验中遭遇健康风险者的保护。当治疗或预防疾病时，若被治疗个体的功能特征能够被提高到生命必需水平之上，则临床试验的收益就能更高。反过来，收益的高低也取决于子孙后代可能遭遇的健康风险和疾病预防的能力。

考虑到人类生殖系基因组编辑还没有进行治疗

must be stretched to its limits in order to encompass the right to enhance or diminish traits (Robertson, 2004). It is a vision of parental liberty that already encompasses a wide range of enhancements of infants and children, including biomedical measures such as cosmetic surgeries, the use of growth hormone for short stature of unknown cause, and the use of some performance-enhancing drugs. By extension, it could be argued that this liberty encompasses germline enhancements with similar risk/benefit balances. Here again, though, there is very little reason to think that in the United States, the constitutional cases on parenting would prevent the government from banning germline genome editing if it had a rational basis for doing so.

There has also been the emergence of academic transhumanism as a contributor to these debates. Transhumanists argue not only for the ethical legitimacy of some forms of enhancement-oriented germline editing, but perhaps even for parental responsibility and an ethical obligation to take advantage of such enhancement possibilities for the benefit of one's children (Persson and Savelescu, 2012). One philosopher has argued that we are obligated to make the best possible decisions for those who cannot decide for themselves, and this would include our future children and their descendants (Harris, 2007). These are arguments about moral obligations, however, as nothing in U.S. statutes or judicial decisions (or those of other countries) imposes them as a matter of law.

Overall, two distinct approaches to evaluating the ethics of germline enhancement have emerged over the last half century. In one, the focus is more societal and philosophical. It encompasses not only the concerns raised about germline editing in general, as described in Chapter 5, but also concerns about altering how we view our children, and about creating or increasing social inequities in a multi-generational fashion due to the heritability of the enhancement. Even where the benefits of an individual enhancement might be regarded as justification for an individual intervention, these analyses often feature a concern about slippery slopes and an echo of eugenics movements of the past.

In another approach, the disease/enhancement distinction remains largely useful, as it tracks well to the evaluation of individual risks and benefits. This evaluation is the focus of the regulatory bodies, such as the FDA, that review new medical products for approval and the research oversight bodies that oversee protection of clinical trial participants and others who might be put at physical risk from the trials. When diseases are cured or prevented, the benefit of trials is seen as greater relative to when functional traits are improved beyond what is necessary for a typical life. In turn, this gradation of benefits is balanced against health risks for offspring and future generations, including the potential for disease prevention.

Given that human germline genome editing has not yet been tested clinically for therapeutic or preventive purposes, it seems clear that germline genome editing for

或预防目的的临床试验，我们很清楚，以强化而非出于治愈或战胜疾病与缺陷为目的的生殖系基因组编辑，在目前是不大可能满足潜在收益和可容忍的风险匹配的标准来开始临床试验的。即使风险程度随着更多的经验和信息而减小，哪怕只有很小的风险，都不会允许任意可选的生殖系编辑拥有过大的利益。

总结和建议

在进行任何基因组编辑干预之前还需要取得非常巨大的科学进步，除了治疗或预防疾病和缺陷，还能满足进行临床研究的风险／收益标准。这个结论对体细胞和可遗传性生殖系干预都适用。公众对于应用基因组编辑对人体特征和能力进行超越人类正常健康状态的所谓"强化"是非常不安的。因此，需要对除疾病和缺陷的治疗或预防之外目的的基因组编辑所产生的个人和社会利益与风险进行更多和更广泛的公众讨论。这些讨论应该包含对于引入或加重社会不平等的潜在担忧，因此在做出是否授权临床研究的决定之前，应将这些有价值的信息适当地整合进入风险／收益评估之中。

建议 6-1：监管机构不应该在目前授权进行除疾病或残疾的治疗或预防之外的体细胞或生殖系基因组编辑的临床试验。

建议 6-2：政府部门应该鼓励对于除疾病或残疾的治疗或预防之外目的的人类体细胞基因组编辑监管的公众讨论和政策辩论。

purposes of enhancement, that is, for reasons not clearly intended to cure or combat disease and disability, is very unlikely at this time to meet the standard of possible benefit and tolerable risk needed to initiate clinical trials. Even as risks recede with greater experience and information, truly discretionary and elective germline edits would be unlikely to have benefits outweighing even minor health risks.

CONCLUSIONS AND RECOMMENDATIONS

Significant scientific progress will be necessary before any genome-editing intervention for indications other than the treatment or prevention of disease or disability can satisfy the risk/benefit standards for initiating a clinical trial. This conclusion holds for both somatic and heritable germline interventions. There is significant public discomfort with the use of genome editing for so-called "enhancement" of human traits and capacities beyond those typical of adequate health. Therefore, a robust public discussion is needed concerning the values to be placed upon the individual and societal benefits and risks of genome editing for purposes other than treatment or prevention of disease or disability. These discussions would include consideration of the potential for introducing or exacerbating societal inequities, so that these values can be incorporated as appropriate into the risk/benefit assessments that will precede any decision about whether to authorize clinical trials.

RECOMMENDATION 6-1. Regulatory agencies should not at this time authorize clinical trials of somatic or germline genome editing for purposes other than treatment or prevention of disease or disability.

RECOMMENDATION 6-2. Government bodies should encourage public discussion and policy debate regarding governance of somatic human genome editing for purposes other than treatment or prevention of disease or disability.

7

公 众 参 与
Public Engagement

CRISPR-Cas9 是一种用于人类基因组编辑领域的研究工具，由于这一技术的出现，使得针对这类技术和应用进行广泛公开对话的呼吁变得更为迫切。这些呼吁不仅来自于伦理学家和社会科学家（如 Jasanoff et al.，2015），也来自于生物医学科学家（Bosley et al.，2015；Doudna，2015），以及多个智囊团、生物伦理团体和科学或专业协会，如 Hinxton 集团（Chan et al.，2015）、Nuffield 理事会（2006b）和遗传与社会中心（2015）。

此呼吁本身并不是新近才提出的，早在 1975 年 2 月的 Asilomar 会议上，一个国际科学家小组就讨论了重组 DNA 的使用，并认为对其使用应该进行严格控制（Berg et al.，1975），此建议被涵盖在一份美国参议院人力资源小组委员会的健康与科学研究报告中，该报告指出："我们应准确定义在科学与社会交界面上出现的严肃问题，并设立机制和模式解决这些问题，这对社会愈发重要"（Powledge and Dach，1977，第 1 页）。

这些早期的工作已演变成一种"日益增长的最高等级的政治承诺，使公民在影响他们生活的决策中拥有更多的发言权，而公民的参与会使政府更加积极和负责"（Cornwall，2008，p. 11）。英国上议院在 2000 年的一份报告中，建议将公众对话作为政策进程的一个强制性组成部分，包括将公开会议作为公民的正规参与途径（U.K. House of Lords，2000）。而美国《纳米技术研究和开发

The emergence of CRISPR/Cas9 as a research tool in the area of human genome editing has lent new urgency to calls for a broad public dialogue about these technologies and their applications. These calls have come from ethicists and social scientists (e.g., Jasanoff et al., 2015), as well as biomedical scientists (Bosley et al., 2015; Doudna, 2015) and multiple think tanks, bioethics groups, and scientific/professional societies, such as The Hinxton Group (Chan et al., 2015), the Nuffield Council (2006b), and the Center for Genetics and Society (2015).

The idea itself is not new. At the Asilomar Conference in February 1975, an international group of scientists discussed the use of recombinant DNA and decided that strict controls should be placed on its use (Berg et al., 1975). The concerns expressed by that group are reflected in a report to the U.S. Senate Committee on Human Resources Subcommittee on Health and Scientific Research. That report argued that it was "increasingly important to society that the serious problems which arise at the interface between science and society be carefully identified, and that mechanisms and models be devised, for the solution of these problems" (Powledge and Dach, 1977, p. 1).

These early efforts have evolved into a "growing political commitment at the highest levels to giving citizens more of a voice in the decisions that affect their lives, and to engaging citizens in making government more responsive and accountable" (Cornwall, 2008, p. 11). In a 2000 report, the UK House of Lords recommended that dialogue with the public be a mandatory and integral part of policy processes, including the use of public meetings as a tool for formal citizen engagement (UK House of Lords, 2000). Likewise, the 2003 U.S. Nanotechnology Research and Development

法案》也在2003年授权"通过公民小组、协商会议和教育活动等参与机制召开定期和持续的公共讨论"。[64]

这些工作的部分起因是希望预判公众对具有潜在争议技术的反应,以便"避免不合理地抑制创新、诋毁新技术或创造贸易壁垒"(Holdren et al.,2011,p. 1)。有研究表明,"在透明地融合不同观点的政策进程中",决策者和公共利益相关者之间有意义的互动,可以提高公众对围绕新兴技术监管或政策决策合理性的认识(Posner et al., 2016, p. 1760)。2008年,美国国家研究委员会(NRC)的一份共识报告中也有相似的结论,报告认为公众参与(环境)评估和决策不仅可以增进对合法性的认知,也可以改善决策的质量(Holdren et al., 2011, p. 1;NRC,2008)。

一个典型的例子就是,在转基因生物技术(GMO)出现期间,新兴的基因工程领域由于缺乏与不同公众群体之间有意义的沟通互动,遭受了不可挽回的损害。例如,有关Bt玉米对帝王蝶幼虫有害影响的公开辩论"不仅导致了孟山都股票(Monsanto stock)价值下降近10%,也导致了日本对进口Bt玉米的限制,还导致了欧洲委员会(布鲁塞尔)对Bt转基因玉米批准过程的冻结及要求暂停在美国进一步种植Bt玉米的呼吁"等一系列后果(Shelton and Roush, 1999)。类似这些辩论最终减慢和停止了一些具有生物强化性能的新型食品的开发和推广,例如,黄金大米,其被认为可以降低缺乏维生素A(VAD)相关疾病的发生率。据世界卫生组织估计,全球约2.5亿人患有VAD(其中将近一半儿童患者来自于发展中国家),VAD是儿童致盲的主要原因,全球有25万~50万例儿童致盲来源于VAD(Achenbach, 2016)。美国国家科学院2016年5月的一份报告显示,在常规和转基因作物中适当增加微量营养素"对数百万人的健康有良好的影响……"(NASEM,2016c)。

一些学者指出,人类基因组编辑技术的出现,引出了日益增多的伦理、监管和社会政治等方面的问题,这些问题远远超出了生物学家之间关于技术风险和利益的讨论(Jasanoff et al., 2015),甚至超出了社会科学家和伦理学家提出的哲学和社

Act mandated "convening of regular and ongoing public discussions, through mechanisms such as citizens' panels, consensus conferences, and educational events."[64]

These efforts have been motivated in part by the desire to anticipate the reactions of members of the public to potentially controversial technologies and to "avoid unjustifiably inhibiting innovation, stigmatizing new technologies, or creating trade barriers" (Holdren et al., 2011, p. 1). Research also has shown that engaging meaningfully with decision makers and public stakeholders "in processes … that incorporate diverse perspectives transparently" can increase public perceptions of the legitimacy of regulatory or policy decisions surrounding emerging technologies (Posner et al., 2016, p. 1760). These findings echo the conclusion of a 2008 National Research Council (NRC) consensus report that public participation in (environmental) assessment and decision making has the potential to improve not just perceptions of legitimacy but also the quality of decisions (Holdren et al., 2011, p. 1; NRC, 2008).

One could argue, for example, that a lack of meaningful engagement with different publics during the emergence of genetically modified organisms (GMOs) did irreparable damage to the emerging scientific field of genetic engineering. Public debates about harmful effects of Bt corn on larvae of monarch butterflies, for instance, led to "a nearly 10% drop in the value of Monsanto stock, possible trade restrictions by Japan, freezes on the approval process for Bt-transgenic corn by the European Commission (Brussels), and calls for a moratorium on further planting of Bt-corn in the United States" (Shelton and Roush, 1999). These debates have also slowed or halted the development and introduction of biofortification of some foods, such as golden rice, which has the potential to reduce disease caused by a vitamin A deficiency (VAD). The World Health Organization estimates that 250 million people suffer from VAD (nearly half of them children in the developing world), and it is a leading cause of childhood blindness, affecting between a quarter and a half million children worldwide (Achenbach, 2016). A May 2016 report from the National Academies found that increasing micronutrients at appropriate levels in both conventional and genetically engineered crops "could have favorable effects on the health of millions of people" (NASEM, 2016c).

Some scholars have argued that human genome editing has raised, and will continue to raise, ethical, regulatory, and sociopolitical questions that go well beyond discussions of technical risks and benefits identified by biologists (Jasanoff et al., 2015) or even philosophical and sociopolitical concerns raised by social scientists and ethicists (Sarewitz, 2015). These scholars argue that the risks and benefits associated with human genome editing should not be defined solely by the scientific community,

---

[64] 21st Century Nanotechnology Research and Development Act, Public Law 108-153 (December 3, 2003).

会政治关注的范畴（Sarewitz，2015）。他们认为，与人类基因组编辑相关的风险和利益不应仅由科学界来定义，对这些风险和利益的全面理解需要广泛的公众辩论，应该广泛吸取不同群体的意见来确定如何定义相关概念。这就意味着随着基因组编辑技术及其应用的发展，监管机构在区分治疗和保健或缺陷和疾病等概念时也需要通过长期的公众讨论来确定。

公众舆论、世界观和宗教信仰在公共政策形成方面的作用，在不同国家之间，甚至在同一个国家不同时代之间，都存在相当大的差异。在神权政治的国家，关于胚胎研究、种系基因组编辑甚至体细胞治疗的公共政策可以通过明确的宗教教义来支撑。即使在非神权的政治国家，宗教信仰也可能强烈影响一个人的道德观，从而影响个人观点和政治倾向。某些国家可能在宪法上对政府规划或政策与宗教机构之间不同程度的分离存在明确要求。因此，公众参与将在全球范围内通过多种方式成为公共政策形成的一部分（Pew Research Center，2008）。

在美国，诸如食品药品监督管理局（FDA）的管理机构具有法定权限，可以基于风险和利益的技术考量来准许或禁止医疗产品的销售，但没有权力拒绝批准用于（或可能用于）通常视为不道德目的的特定产品，这样的禁令如果要被颁布，必须由立法机关出台，同时该禁令不能与宪法保护的权利相违背。在这些事项上，公众意见将成为健全决策的重要组成部分。

本章首先将论述公共参与所涵盖的广义概念，然后会依次综述美国国内和国际社会公众参与实践的方式，接下来将讨论从过去的公众参与事件中吸取的经验教训，最后一节将给出结论和建议。

公众参与：广义概念

美国科学促进会荣誉首席执行官 Alan Leshner 先生很好地概述了指导公众参与的广义概念：

我们需要让公众参与一个关于科学技术及其产品的双向对话，这个对话将更开放和诚实，不仅包括其利益，而且包括其限制、危险和误区。我们需

and that a comprehensive understanding of risks and benefits will require broad public debates that are highly inclusive with respect to the range of voices and how relevant concepts are defined. This argument suggests, as genome editing technologies and applications develop, the need for ongoing public discussion about how regulatory bodies should draw distinctions between such things as therapy and enhancement or disability and disease.

There is considerable variation among countries, or even over time in the same country, with respect to the role of public opinion, world view, and religious affiliation in the formation of public policy. In theocracies, public policy with respect to embryo research, germline genome editing, and even somatic therapy may be shaped by explicit reference to religious doctrine. Even in formally secular countries, religious belief may strongly influence individual morality, which in turn is expressed in personal opinions and political preferences. Other countries may have constitutional requirements for various degrees of separation between government programs or policies and religious institutions. For this reason, public engagement will be part of public policy formation in a variety of ways across the globe (Pew Research Center, 2008).

In the United States, regulatory agencies such as the U.S. Food and Drug Administration (FDA) have statutory authority to permit or prohibit the marketing of medical products based on technical considerations of risk and benefit. There is no authority to refuse approval for a particular product that is intended for (or might be used for) a purpose that many view as immoral. If such a prohibition were to be enacted, it would emerge from the legislature, subject to limitation where the prohibition would abridge constitutionally protected rights. In matters such as these, public input is an important component of sound decision making.

This chapter begins by delineating the broad concepts encompassed by public engagement. It then reviews in turn U.S. and international public engagement practices. The chapter turns next to lessons learned from past public engagement efforts. The final section presents conclusions and recommendations.

PUBLIC ENGAGEMENT: BROAD CONCEPTS

The broad concepts that guide public engagement have been outlined nicely by Alan Leshner, CEO Emeritus of the American Association for the Advancement of Science:

We need to engage the public in a more open and honest bidirectional dialogue about science and technology and their products, including not only their benefits but also their limits, perils, and pitfalls. We need to respect the public's perspective and concerns even when we do not fully share

要尊重公众的视角和关注点，即使我们不能全盘接受它们，我们也需要建立一种伙伴关系，以便能够回应公众的需求 (Leshner, 2003, p. 977)。

同时，《自然》杂志上刊登了一篇社论，呼吁科学家参与这些公开讨论，将专业知识引入更广泛的对话中，即使"这样的公开讨论可能会让许多研究人员觉得不自在"(Nature, 2017)。

在实践中，公众参与有许多不同形式，讨论所有不同形式的优缺点超出了本报告的范围 (Rowe and Frewer, 2005; NRC, 2008; Scheufele, 2011)。但是，我们可以尽量阐明三种广泛的参与原则，用于指导拓宽人类基因组编辑话语的行为，以期囊括最大数量的相关观点和利益相关者。

公众参与的第一个原则是结果的质量。美国国家科学院之前的报告中已经确定了诸多有助于高质量监管或政策决定的公众参与行为要素 (NRC, 1996, 2008; NASEM, 2016a)。而对于人类基因组编辑来说，其中有四个因素需要特别注意。首先，系统地考虑和衡量公众参与的最大影响范围，以及各种不确定性，包括对技术、医疗和科学问题之外的风险及利益的考量，也涵盖了"所有感兴趣和受影响的各方"的观点和知识 (NRC, 1996, p. 3)；第二个因素是要确定所有潜在的政策或监管选择的范围；第三个要素是优质的公共参与机制，因为这会"综合考虑现实和价值观，特别是世界的预期变化对人们价值观的影响"(NRC, 2008, p. 235)；最后，本行业之外的公众成员能发现和提出甚至连监管机构或专家都没有想到的问题与解决方案。

公众参与的第二个原则是结果的合法性。合法性往往与一些相关因素有关。首先，公众参与的过程必须是透明的，并且被所有参与者认为是公平的和可行的 (Hadden, 1995)；其次，如前文中提到的，公众参与需要考虑到所有感兴趣或潜在受影响方的价值观、利益和关注点；最后，公众参与须以符合相关法律法规的方式进行。

此外，前两个参与原则需要与第三个原则——行政效率——相平衡。"全面参与的目标需要考虑到行政效率的因素，以确保及时作出决策"(NASEM, 2016c, p. 56)，同时还要防范有实力雄厚或组织有序的选区主导公共话语，淹没

them, and we need to develop a partnership that can respond to them. (Leshner, 2003, p. 977)

In the same vein, an editorial in Nature called on scientists to participate in these public discussions and bring their expertise to the wider conversation, even when "such public discussions may take many researchers outside their comfort zone" (*Nature*, 2017).

In practice, public engagement takes many different forms, and it is beyond the scope of this report to discuss the advantages and disadvantages of different modalities (NRC, 2008; Rowe and Frewer, 2005; Scheufele, 2011). Nonetheless, it is possible to articulate at least three broad principles of engagement that can be applied to guide any effort to broaden the discourse on human genome editing to include the maximum number of relevant viewpoints and stakeholders.

The first principle of public engagement relates to *quality of outcomes*. Previous National Academies reports have identified a host of factors that contribute to high-quality regulatory or policy decisions informed by engagement efforts (NASEM, 2016a; NRC, 1996, 2008). Four such factors are particularly noteworthy for engagement efforts surrounding human-genome editing. First, is considering and weighing systematically the widest possible range of effects, as well as the uncertainties surrounding them. This includes consideration of risks and benefits that go beyond technical, medical, or scientific questions and encompass "perspectives and knowledge of [all] interested and affected parties" (NRC, 1996, p. 3). Second, is identifying the full range of potential policy or regulatory options. Third, quality public engagement mechanisms "deal with both facts and values and in particular with how anticipated changes in the world will affect the things people value" (NRC, 2008, p. 235). And finally, members of the lay public are able to ask questions and suggest solutions that may not have been imagined by regulators or experts.

A second principle of public engagement is the *legitimacy of outcomes*. Legitimacy tends to be connected to a number of related factors. First, processes for public engagement are transparent and perceived by all participants as fair and competent (Hadden, 1995). Second, echoing some of the considerations outlined earlier, public engagement identifies values, interests, and concerns of all interested or potentially affected parties. Finally, engagement is pursued in a manner consistent with relevant laws and regulations.

These first two engagement principles, however, need to be balanced with a third— *administrative efficiency*. "The goal of full participation needs to be considered in light of the need for administrative efficiency to ensure that decisions are made in a timely manner" (NASEM, 2016c, p. 56), while also guarding against the risk that well-resourced or well-organized constituencies will

其他声音。

基于这些广泛的原则，公众参与行为通常围绕一个或多个以下过程进行（Rowe and Frewer，2005）：

- 交流／信息——确保决策相关信息（包括伦理、监管和政治考量）有效地达到社会的最大横截面。换句话说，不论形式如何，公共参与行为不能局限于易于接触的受众（例如，受教育程度较高的人群）或参与程度高的群体（例如，为患者辩护团体、宗教团体、与妇女权利／性别问题相关团体，以及环境活动家）。
- 反馈——最大限度地由相关和受影响的公众向倡议的发起者（例如，联邦或州级政府的监管机构、政策制定者）传达决策相关信息。构建和征求反馈的过程通常由发起人开始，而一个更有效的反馈则需要与利益和倡导团体之间进行互动。这些团体可以作为一个工具，通过它们可以聚集大量拥有相同地位和目标的人群，然而这些团体的民主程度或者反映其成员各种意见的准确程度依然存在较大差别（Seifter，2015）。
- 参与——最大限度地在相关公众和政策参与者之间交换所有决策相关信息及价值考量。对话和协商能够转变观点，增加发起者和公众参与者的信息与认识。丹麦共识会议（下文进一步讨论）作为政策制定过程的正式部分，之后参与性活动的许多实例都基于或至少部分基于（政策）决策者和公众成员之间的正式磋商（Danish Board of Technology，2006，2010a，b）。

不管过程如何，公众参与的目的不是诱导或促进公众对新兴技术的接受。从这个意义来说，公众参与是对所谓的"知识赤字模型"的直接反应（Brossard and Lewenstein，2009）——通过关闭知识赤字并在非专业受众中建立相关的科学素养，从而促进公众对新技术的接受，这一观点是可能的（或可取的）。公众参与模式以两种方式偏离知识赤字模型。首先，他们假定了解越多的公众就越能接受新的技术，但这一假设缺乏实证数据的支持

dominate the public discourse and drown out other voices.

Based on these broad principles, public engagement efforts typically are built around one or more of the following processes (Rowe and Frewer, 2005):

- *Communication/information*—ensuring that decision-relevant information (including ethical, regulatory, and political considerations) effectively reaches maximum cross sections of society. In other words, public engagement efforts—regardless of format—need to focus not just on easy-to-reach audiences (e.g., more educated segments of the population) or highly invested groups (e.g., patient advocacy groups, religious communities, groups concerned with women's rights/gender issues, environmental activists).
- *Consultation*—conveying decision-relevant information from the maximum number of relevant and affected public(s) to the sponsors of the initiative (e.g., regulatory agencies, policy makers at the federal and potentially even state levels). The process of structuring and soliciting this consultation typically is initiated by the sponsor. One possibility for more efficient consultation is to interact with interest and advocacy groups. These groups can be a vehicle for reaching large numbers of people who share a position or goal, although with the caveat that these groups vary widely in how democratically they arrive at their positions or how accurately they reflect the range of views of their members (Seifter, 2015).
- *Participation*—exchanging all decision-relevant information and value considerations among the maximum number of relevant public(s) and policy actors. Dialogue and negotiation transform opinions and increase information and awareness among both sponsors and public participants. Many examples of participatory activities are based at least partially on formal deliberations among (policy) decision makers and members of the public, along the lines of Danish consensus conferences (discussed further below) that provide formal input into policy-making processes (Danish Board of Technology, 2006, 2010a,b).

Regardless of the process, the purpose of engagement efforts is not to create or increase public acceptance of emerging technologies. In this sense, public engagement is a direct response to what is known as the "knowledge deficit model" (Brossard and Lewenstein, 2009)—the notion that it is possible (or desirable) to increase public acceptance of new technologies by closing knowledge deficits and building relevant scientific literacy among nonexpert audiences. Public engagement models deviate from knowledge deficit models in two ways. First, they acknowledge that very limited empirical data support the assumption that more informed

(Scheufele, 2013);其次,这种模式误用了这两个观点:①建立公众对科学的支持在所有情况下都是可取的;②公开辩论甚至争论总是不可取的。相反,公众参与旨在促进"在专业知识、力量和价值观上存在差异的群体之间的知识、观点和偏好的共享和交流"(NASEM,2016b,p. 22)。因此,公众参与"使所有利益相关者有机会在一个技术开发或部署之前讨论其潜在的风险、利益和影响;可以激发对公众利益重要问题的关注;鼓励所有涉及同一问题的各种群体的公民参与和表达意见"(NASEM,2016a,p. 1~7)。

不同公众参与模式在实现这些目标上的效果同样缺乏实证数据,现有文献主要是地方和区域层面的公众参与实例。因此,很难得到适用于不同情境的具体形式或高效参与流程,也很难预测如何针对广泛多样性的参与者、监管问题和主题进行定制(NASEM,2016a)。

美国的实践

对不同的议题,不同的国家和行政辖区,将上述原则转化为实践的方式差异很大。在美国,奥巴马政府将关于公众参与的理念纳入负责任发展的政策下:"关于新兴技术(如纳米技术、合成生物学和基因工程等)的创新不仅需要协调研究和开发,还需要适当和平衡的监督"(Holdren et al.,2011,p. 1)。在实践中,这一概念包括两个关键过程,即上一节中提到的交流和反馈:

- 沟通:"联邦政府应积极向公众传达与新技术相关的潜在利益和风险的信息"(Holdren et al.,2011,p. 2)。
- 反馈:"在可行并受到有效约束的情况下(如涉及国家安全和机密商业信息),应推进相关信息,为利益相关者介入和公众参与提供充足的机会,这对于促进问责制、改善决策、增加信任和确保官员获得范围足够的广泛信息反馈很重要"(Holdren et al.,2011,p. 2)。

在美国,国会颁布了管理科学及卫生保健等相关法律,所有这些法律都受到民主程序的影响,正

citizens will be more accepting of emerging technologies (Scheufele, 2013). Second, such models abandon the notions (1) that building public support for science is, in all cases, desirable; and (2) that public debate or even controversy is always undesirable. Instead, public engagement is designed to facilitate "the sharing and exchange of knowledge, perspectives, and preferences between or among groups who often have differences in expertise, power, and values" (NASEM, 2016b, p. 22). As a result, it "gives all stakeholders opportunities to discuss the potential risks, benefits, and consequences of a technology before it is developed or deployed; can motivate attention to issues important to the public good; and ideally encourages civic participation and expression of views by all the diverse groups that are concerned with an issue" (NASEM, 2016a, pp. 1-7).

Empirical evidence also is limited on how effective different models of public engagement are in achieving these goals, and the existing literature draws mainly on examples of engagement activities from the local and regional levels. As a result, it is difficult either to point to specific structures or processes for effective engagement that are applicable across contexts or to predict how they can be tailored to fit the wide diversity of participants, regulatory questions, and topics (NASEM, 2016a).

U.S. PRACTICES

Across different issues, countries, and jurisdictions, efforts to translate the principles discussed above into practice have varied widely. In the United States, the Obama Administration codified the idea of public engagement under the label *responsible development*: "Innovation with respect to emerging technologies—such as nanotechnology, synthetic biology, and genetic engineering, among others—requires not only coordinated research and development but also appropriate and balanced oversight" (Holdren et al., 2011, p. 1). In practice, this concept involves *communication* and *consultation*, as described in the previous section, as two key processes:

- *Communication*: "The Federal Government should actively communicate information to the public regarding the potential benefits and risks associated with new technologies" (Holdren et al., 2011, p. 2).
- *Consultation*: "To the extent feasible and subject to valid constraints (involving, for example, national security and confidential business information), relevant information should be developed with ample opportunities for stakeholder involvement and public participation. [This is] important for promoting accountability, for improving decisions, for increasing trust, and for ensuring that officials have access to widely dispersed information" (Holdren et al., 2011, p. 2).

是这些民主程序才使得公众舆论和选民利益受到当选官员的关注。专门机构通常拥有执法权（在大多数情况下，是行政服务部门的总体政策方向），执行根据公众意见制定出的相关法规。

从某种程度而言，这种公共参与的性质和范围与监督机构有所不同，但所有机构都受一系列法律条文的约束，这些条文在《行政程序法》中加以概述并由法院进行解释。参与程序包括：提前得知所要拟立的法规，寻求发表意见的机会，要求相关机构解释规则的合理性，以及对所提意见采纳或否决的理由。这些正式规则制定过程包括有关公众参与的相对简单的规则。通常，在高度复杂或快速演变的领域（如生物技术或生命科学）中，实行的指导或代理行为是缺乏监管和没有相关法律约束的，只有部分行为接受正式的公开评议。最后，对于诸如基因组编辑的新研究，可以由国家卫生研究院（NIH）重组 DNA 咨询委员会（RAC）组织，或由食品药物监督管理局（FDA）召集专家咨询委员会进行额外的不具约束力的审查。

公众参与科学政策是一项全球性的事务，其中一个例子就是欧盟委员会在"地平线 2020"资助计划中引入了负责任研究和创新（Responsible Research and Innovation，RRI）的概念，以增加利益相关者在研究方向上的参与度和影响力。它提倡"在科学和技术问题上，由公民和民间社会组织共同创造未来，并且引入通常情况下不会相互影响的尽可能多样化的参与者"（European Commission，2016a）。它包括绘制与创新相关的伦理图谱，并在公众的帮助下，在正式法律及自愿标准和实践中，确定最佳的管理方式（European Commission，2016b）。它还包括"欧洲公民在科学上的观点、意见和想法"项目（"Views，Opinions，and Ideas of Citizens in Europe on Science，VOICE"），以及其成千上万的参与者组成的焦点小组和其他实践活动（European Commission，2016c）。

现有的用于公众参与基因组编辑策略的基础架构

除了公众参与政策的大体框架之外，目前体细胞基因治疗研究的批准程序还有一些需要公众参与的特殊场合。由于有些政策在各州级别就可以制定，如某些州制定的关于胚胎、克隆和干细胞研究集资的相关政策，在这种情况下，各个州本身就是

In the United States, Congress enacts laws governing science, health care, and the like, all of which are influenced by democratic processes that bring public opinion and constituent interests to the attention of elected officials. Specialized agencies generally are empowered to implement law (subject in most cases to overall policy direction from the executive branch) through the use of regulations that are developed with public input.

The nature and extent of this public engagement vary to some degree with the overseeing agency, but all agencies are governed by a set of legal rules that have been outlined in the Administrative Procedures Act and subsequently interpreted by the courts. The engagement techniques include advance notice of proposed rulemaking, an opportunity to comment, and a requirement that the agency explain its rationale for a rule and why it adopted or rejected comments. These formal rulemaking processes include relatively straightforward rules regarding public engagement. Often in the case of a highly complex or rapidly evolving area (such as biotechnology or life sciences), subregulatory, legally nonbinding guidances or agency practices apply, some of which also are subject to formal public comment. Lastly, for new research such as genome editing, there can be additional nonbinding review by the National Institutes of Health's (NIH's) Recombinant DNA Advisory Committee (RAC) or by an expert advisory committee convened by the FDA.

Public engagement in science policy is a global affair. One example is the introduction of the concept of RRI, Responsible Research and Innovation, by the European Commission in the Horizon 2020 funding program, to increase stakeholder participation and influence in research directions. It is about "co-creating the future with citizens and civil society organisations, and also bringing on board the widest possible diversity of actors that would not normally interact with each other, on matters of science and technology" (European Commission 2016a) . It includes an effort to map the ethical issues relevant to innovation and, with the help of the public, to identify the best options for governance, whether in formal law, or in voluntary standards and practices (European Commission, 2016b). It also includes the VOICES (Views, Opinions, and Ideas of Citizens in Europe on Science) project, with its thousands of participants in focus groups and other exercises (European Commission, 2016c).

Existing Infrastructure for Public Involvement in Genome Editing Policy

In addition to the general framework for public engagement in policy, there are some particularly relevant opportunities for public engagement in the current approval process for somatic gene therapy research. To the extent that some policy can be formulated at the state level, as has happened in some states with respect to embryo research, cloning, and funding for stem cell research, the states

与联邦机构交流的利益相关者（尽管这涉及更复杂的因素，包括对州权力和独立性的考量）(Seifter, 2014b)。在联邦制体系中，如美国、欧洲和澳大利亚，必须考虑公众参与和政策制定的区域因素，这样做除了考虑到政策制定在地方一级发生的可能性外，还可以增加对行政规则采纳的集中式联邦政策合法性的理解 (Seifter, 2014a)。

在美国，存在联邦级别的公众参与平台，但这些平台都是有限和被动的，不够完善，并不适用于如人类基因组编辑之类重要的新技术，尤其是可能应用到可遗传生殖系的研究。相关平台如下。

国立卫生研究院重组 DNA 咨询委员会 (NIH, RAC)

目前，RAC 是公众参与人类转基因或基因组编辑协议的监督的最大平台。根据 2016 年 4 月颁布的修订规程，RAC 只对引出新的或重要的科学、社会或伦理问题的人类转基因方案进行审查。RAC 依据这些标准在 2016 年 6 月审查了美国第一个人类 CRISPR / Cas9 基因组编辑试验。更详细的描述见第 2 章，RAC 的审查过程是透明的，为公众参与提供了平台。RAC 会议依据《联邦咨询委员会法》开展，该法要求事先公布对公众开放的会议（除非有特殊例外情况），预留充足时间征询公众意见。此外，近来 RAC 会议已进行网络直播，以便在会议后实时查看或存档。NIH 的生物技术活动办公室还为希望了解新的 RAC 会议和活动的个人创建了电子邮件列表。

虽然这些条款允许感兴趣的公众有限度地参与，但他们是被动的，因为他们主要适用于有利益相关而寻求会议信息的那一部分公众。在现有的组织形式下，RAC 不善于吸引舆论和提高公众参与度，因此不能很好地开展从不同群体的广大群众（大部分人只对手头上问题感兴趣，这就是所谓"公众"）中收集建议并进行对话的工作。

美国食品药品监督管理局 (FDA)

FDA 是美国第二大参与基因治疗审批的机构，由于新药临床 (IND) 申请包含机密商业信息（相关内容详见第 2 章），因此提交的申请内容需要保密，不能供公众审核或评论。然而，如前所述，clinicaltrials.gov 网站现在已经完成改进，公众可以更方便地获取有关试验及其结果的数据，产品赞

themselves can be stakeholders that engage with the federal agencies (albeit with complex goals that include concerns about state power and independence) (Seifter, 2014b). In federalist systems, such as those not only of the United States but also of Europe and Australia, it is important to include consideration of regional opportunities for public engagement and policy making as well. Beyond allowing for the possibility that policy making may well occur at the local level, doing so has the potential to increase the perceived legitimacy of centralized, federal policies adopted by administrative rule (Seifter, 2014a).

At the federal level in the United States, opportunities for public engagement exist but tend to be limited and passive, and not the type of fuller public engagement that may be appropriate for an important new technology such as human genome editing, especially as potentially applied to the germline. These opportunities are described below.

National Institutes of Health's Recombinant DNA Advisory Committee

At present, the RAC provides the greatest opportunity for public involvement in the oversight of human gene-transfer or genome-editing protocols. Under the modified procedures enacted in April 2016, the RAC reviews only human gene-transfer protocols that present novel or significant scientific, societal, or ethical concerns. The RAC reviewed the first human CRISPR/Cas9 genome-editing trial in the United States pursuant to these criteria in June 2016. As described in greater detail in Chapter 2, the RAC's process is transparent and provides opportunities for public involvement. RAC meetings are conducted in accordance with the Federal Advisory Committee Act, which requires public advance notice of meetings that are open to the public (unless certain exceptions apply) and where time is made available for public comment. In addition, recent RAC meetings have been webcast for viewing in real time or from the archives after the meeting, and NIH's Office of Biotechnology Activities has created an email list for individuals who wish to be informed of new RAC meetings and activities.

Although these provisions allow for limited participation by interested members of the public, they are passive in that they apply primarily to a subset of the public that has an existing interest and that seeks out the meeting information. In its current form, the RAC lacks scholarly expertise in public opinion or public engagement research, and is therefore not as well positioned to spearhead efforts to seek input from, or dialogues with, different communities of people at large who have an interest in the issue at hand, often referred to as "publics."

U.S. Food and Drug Administration

The FDA is the second major institutional player

助商可以选择公开部分或全部 IND 信息。此外，当 FDA 科学咨询委员会审查基因治疗或基因组编辑方案时，若会议要求对公众开放，那么公众代表也必须列入咨询委员会名册。一旦生物许可证被批准，相关信息也必须公布出来。但总体而言，FDA 审查过程缺乏与咨询机构（如 RAC）类似的透明度。

FDA 会定期召开信息公开会议，如转基因鲑鱼案例（FDA，2010），或举办更广泛的公众讨论，如现代化生物技术类产品监管系统的案例（FDA，2015c）。FDA 的咨询委员会会议以及更广泛的会议机制可适用于讨论依赖于基因组编辑的疗法，对于讨论针对或可能用于未被临床试验认可的"增强"疗法的产品尤为重要。

国家生物伦理委员会

许多国家都有向其政府提供建议和为公共对话提供场所的专门机构，这样的机构存在于除南极洲以外的每个大陆。它们通常由行政或立法部门任命，并就生物伦理学的一系列主题提供分析和政策建议。1996 年 11 月，应美国"国家生物伦理咨询委员会"和法国"全国民主科学技术协会"邀请，几个国家的生物伦理委员会举行了一次高峰会议[65]，并从这次后举办了一系列全球高峰会议，最近的一次是在德国举办，而下一次会议将于 2018 年在塞内加尔举办。

像许多其他国家一样，美国一直以生物伦理委员会作为公众参与和向政府提出政策建议的机构（见延伸内容 7-1）。这项传统始于 20 世纪 70 年代国家生物医学和行为研究人类受试者保护委员会的创立，该委员会的工作引起了临床试验管理的重大变革。美国联邦政府随后陆续组织了许多委员会来解决生物伦理问题，这些委员会的共同点包括公开见证的平台、开放性会议、委员会讨论记录，以及公众的参与和关注，这些都能显著影响到会议结论和建议的执行情况。尽管各机构在决策方面仍受到法律的限制，但这些平台为广纳言论提供了重要的渠道，甚至在必要的情况下可以发起机构授权或监管方面的立法改革。

in gene-therapy approvals in the United States, but as described in more detail in Chapter 2, the Investigational New Drug (IND) application contains confidential business information. Therefore, submissions are proprietary and not available for public review or comment. As noted earlier, however, the website ClinicalTrials.gov has now been amended to increase public access to data about trials and their results and a product sponsor may elect to make some or all of the IND information publicly available. Furthermore, if a gene-therapy or genome-editing protocol is reviewed by an FDA scientific advisory committee, that meeting is open to the public, and a public representative must be included on the advisory committee roster. Once a biological license is approved, additional information is available for public posting. Nonetheless, overall the FDA review process lacks the transparency associated with an advisory body such as the RAC.

The FDA periodically calls informational public meetings, as it did for genetically engineered salmon (FDA, 2010), or hosts a more general public discussion, such as that on how the regulatory system for biotechnology products should be modernized (FDA, 2015c). The FDA's advisory committee meetings, as well as its more generalized meeting mechanisms, would be available for discussion of therapies that depend on genome editing, and they might be particularly useful for discussion of products aimed at or likely to be used off-label for "enhancement."

National Bioethics Commissions

Many nations have bodies to provide advice to their governments or to provide venues for public conversation and such entities exist on every continent except Antarctica. They are usually appointed by the executive or legislative branches, and offer analyses and policy recommendations on a range of topics in bioethics. In November 1996, a group of national bioethics commissions met for a summit, at the invitation of the American "National Bioethics Advisory Commission" and the French "Comité Consultatif National d'Ethique pour les Sciences de la Vie et de la Santé."[65] Since then, there have been a series of global summits of these bodies, the most recent in Germany and the next to take place in 2018 in Senegal.

Like many other countries, the United States has a long tradition of using bioethics commissions as a venue for both public participation and advice to the government on policy options (see Box 7-1). The tradition began in the 1970s with the creation of the National Commission for the Protection of Human Subjects of Biomedical and Behavioral Research, whose work led to substantial changes in the management of clinical

---

[65] See http://www.who.int/ethics/partnerships/globalsummit/en (accessed January 30, 2017).

延伸内容 7-1

美国国家级别的生物伦理委员会
U.S. National-Level Bioethics Commissions

生物伦理问题研究总统委员会，2009—2016

总统生物伦理委员会，2001—2009

国家生物伦理咨询委员会，1996—2001

人体辐射实验咨询委员会，1994—1995

生物医学伦理咨询委员会，1988—1990

生物医学和行为研究中的医学伦理问题总统委员会，1978—1983

国家生物医学和行为研究人类受试者保护委员会，1974—1978

Presidential Commission for the Study of Bioethical Issues, 2009–2016 President's Council on Bioethics, 2001–2009

National Bioethics Advisory Commission, 1996–2001

Advisory Committee on Human Radiation Experiments, 1994–1995 Biomedical Ethical Advisory Committee, 1988–1990

President's Commission for the Study of Ethical Problems in Medicine and in Biomedical and Behavioral Research, 1978–1983

National Commission for the Protection of Human Subjects of Biomedical and Behavioral Research, 1974-1978

机构监督

所有基因治疗和基因组编辑的研究必须在当地由研究机构的机构审查委员会（IRB）和机构生物安全委员会（IBC）批准，这两个委员会的成员中都必须包括公众代表。根据现行的美国法规，IRB 必须至少有一个"不隶属于该机构"的成员，并且其成员所具备的专业类型中必须有"社会态度"的敏感性[66]。同样，IBC 必须包括至少两个不隶属于该机构的"代表社会态度"的成员。

国际上的实践

如上所述，在欧洲，许多围绕公众参与科学的工作都是以负责任创新理念为指导的，即一个"透明的、互动的过程，社会活动者和创新者就创新过程及其适销产品的（道德）可接受性、可持续性和社会需要性进行相互回应"（Schomberg，2012）。这一理念包括如上所述的信息和咨询的要素，是许多国家政策制定和决策过程中公众参与的有力保障。

美国之外的各种公共利益相关者参与形式概述详见表 7-1。

trials. The U.S. federal government has assembled many subsequent commissions to address bioethical issues. Common features of these commissions include opportunities for public testimony, open meetings, the availability of transcripts of commission discussions, and the evident effect of this public participation and observation on both the decisions made at the meetings and the implementation of the resulting recommendations. Although agencies often are restricted by law with regard to what they can consider in their decisions, these venues provide an important outlet for broader considerations that, if necessary, can lead to legislative changes in agency mandates or regulatory approaches.

Institutional Oversight

All gene-therapy and genome-editing studies must be approved at the local level by a research institution's institutional review board (IRB) and institutional biosafety committee (IBC), both of which are required to include public representatives in their membership. Under current U.S. regulations, an IRB must have at least one member "who is not otherwise affiliated with the institution," and among the types of expertise its membership must represent is sensitivity to "community attitudes."[66] Likewise, an IBC must include at least two members not affiliated with the institution that "represent community attitudes."

INTERNATIONAL PRACTICES

As noted above, in Europe, many of the efforts

---

[66] S21 CFR § 56.107.

表 7-1 一些国家有关公众参与的例子

英国	丹麦	法国
• 招募政府以外的人员，特别是在通信和网络资源方面的专家，建立起公众咨询体系 • 针对单一的问题进行磋商 • 公民可以以广泛多样的方式提供信息，包括研讨会、会议在线调查问卷和互动网站论坛	• 长期的公众咨询经验 • 政策制定者没有考虑到公民团体提出的紧急伦理问题 • 通过一个独立的组织来为政府提供建议和指导，使得报告的内容被政策的决定者考虑到	• 公众注意到了新的社会需求，并且需要新的法律途径来规范新的技术 • 公众的讨论有广泛的传播途径 • 增加了其他新的论坛，供公众协商与讨论

TABLE 7-1 Attributes of Public Engagement: Selected Examples

United Kingdom	Denmark	France
• Solicits entities outside of the government, especially specialists in communication and web resources, to create its public consultation structure • Single-issue focus in its consultation • Wide variety of ways citizens provide input, including workshops, meetings, online questionnaires, interactive website forums	• Long-standing experience in public consultation • Emergence of ethical issues raised by citizen groups that policy makers have not considered • Report content taken into consideration by policy decision makers via an independent agency that informs and advises the government	• Citizen panels bring attention to new social demands and needs for new legal approaches to novel technologies • Widespread media dissemination of panel discussions • Proliferation of other forums for discussion subsequent to the public consultation

例如，丹麦长期以来都有举行公民共识会议的传统，这被认为具有广泛的代表性，其会议结果在接下来的政策制定中会得到认真谨慎的对待。然而，在某种程度上，强调共识限制了公民提出不同意见的自由（表 7-2）。虽然在人类基因组编辑领域中没有一个国家可以全部实现公众参与的所有目标，但是一个共同的目标是：通过传达公民对有争议的技术的观点和态度，以便尽可能地丰富和扩大专家、政治家及有关各方之间的辩论的范围（Scheufele, 2011）。

surrounding public engagement in science have been guided by the idea of *responsible innovation*, i.e., a "transparent, interactive process by which societal actors and innovators become mutually responsive to each other with a view to the (ethical) acceptability, sustainability and societal desirability of the innovation process and its marketable products" (Schomberg, 2012). This includes elements of *information* and *consultation*, as described above, but also a strong commitment to formal *public participation* in policy formation and decision making in many countries.

An overview of different non-U.S. efforts to engage with public stakeholders is provided in Table 7-1.

表 7-2 丹麦的公众参与：公民共识会议

形式	优点	缺点
• 这些代表了最早和最多参考的共识会议之一（20 世纪 80 年代末） • 招募具有代表性样本；约 2000 名随机选择的公民被邀请参加 • 丹麦技术委员会（DBoT，由丹麦议会创建的一个独立机构）选择 14~16 名小组成员参加会议 • 由一名专家记者向参会的公民介绍他们参与的议题的基本情况 • 在他们之间进行广泛的讨论 • 让参会公民起草一份报告以表明他们对某一问题的立场	• 最终的报告将由 DBoT 认真处理，并且用它来为丹麦议会提供建议。因此，公民的声音将响彻政府的辩论厅 • 在任何形式的立法辩论之前，先让丹麦公民参与讨论，使决策者能够发现他们没有考虑到的伦理问题，并给予他们时间，根据公民报告来调整政策提案 • 这增加了公民对政府的信任水平 • 许多学者和其他人也认为丹麦共识会议本身就是一个典范	• 通过这种自上而下的制度选择出来的公民小组成员并不能真正代表整个国家 • 对公民小组成员的筛选进行严格控制，以尽量减少因年龄、地理位置、性别、社会经济地位、认知及性格上的差异，以及对话题的兴趣水平而造成的不平衡，这往往会形成一个"理想化"的公民小组，用这个小组来代表实际的情况并不科学 • 这个高度选择性的过程意味着那些被选择的公民已经对当前的议题有显著的兴趣。因此，忽略那些需要更多了解和学习当前议题的公民的意见 • 最终公民的报告必须是一个共识，因此在文件中不允许包含不一致的意见和看法

TABLE 7-2 Public Engagement in Denmark: Consensus Conferences

Modalities	Strengths	Weaknesses
• These represent one of the earliest and most referenced of consensus conferences (late 1980s). • A representative sample is recruited; around 2,000 randomly selected citizens are invited to apply. • The Danish Board of Technology (DBoT, an independent body created by the Danish Parliament) then selects 14-16 panel members to participate in the conference. • These citizens are introduced to and briefed on the topic(s) at hand by an expert journalist. • They then meet for extensive discussion among themselves. • Finally, they draft a report that expresses their stance on a given issue.	• The final report is taken seriously by the DBoT, which uses it to advise the Danish Parliament. Hence citizens' voices are fully integrated into the debating chambers of the government. • This solicitation of input from selected Danish citizens prior to any form of legislative debate allows policy makers to identify ethical issues they may not even have considered, as well as gives them time to adjust policy proposals in light of the citizen report. • This in turn increases citizens' levels of trust in government. • Many academics, and others as well, consider the Danish consensus conference to be a model unto itself.	• The members of a citizen panel selected through such a topdown system can never truly be representative of the population at large. • The careful monitoring of citizen panel members to minimize imbalances due to age, geographic location, gender, socioeconomic status, and cognitive and personality differences, as well as level of interest in the topic(s), tends to create an "idealized" group of people meant to represent the real world, when such is not the case. • Hence, this highly selective procedure means that those selected already have a pronounced interest in the topic at hand. Thus inclusion of those who need more exposure to and education on the topic under consideration is neglected. • The final citizen's report must be a consensus; hence no dissension is either allowed or included in the document.

在设计公众参与活动时，决策部门可从有先例的国家中吸取教训，这既需要避免小范围的公民样本和观点，也需要避免过度集中且自上而下的（在大的集权国家系统中可能难以避免这样的结构）决策。解决这些潜在问题的方法之一是学习英国，其将许多公共参与的活动"外包"给独立的非政府实体（Sciencewise，2016；Wilsdon，2015），但用别国数据来进行对本国适用性的评估非常有限，并且，当涉及由国家来制定未来的公共政策时，有一个中心组织者在协商和讨论方面可能是具有优势的。

公众参与性研讨活动的经验

公众参与对有关实施人类基因组编辑所带来的社会、伦理、法律和政治方面的影响的讨论至关重要。考虑到已有的吸引公众参与的基础架构，以及本章讨论的一般参与原则，委员会意识到使用不同的程序来处理围绕着基因组编辑的不同类型的问题具有特别重要的意义。

同时，所有这些工作都必须遵守早先概述的关于有效公众参与的所有原则，并且还需要两个额外的参与原则来避免潜在的危险。

第一个原则，任何涉及公众参与的工作都必须将公众舆情调研与公众参与性研讨活动区分开来。前者使用了社会科学的方法来收集公众的意见，把

For example, Denmark has a longstanding tradition of consensus conferences for which broad representation is sought, and whose results are taken seriously in the policy-making process. To some extent, however, the emphasis on consensus constrains their ability to present dissenting views (see Table 7-2). Although no country can fully satisfy all the objectives of public engagement in the realm of human germline genome editing, one widely shared objective is to strive to "enrich and expand the scope of traditional debate between experts, politicians, and interested parties by communicating citizens' views and attitudes on controversial technologies" (Scheufele, 2011).

In designing public engagement activities, there are lessons to be learned from previous national efforts. These include the need to avoid having a small or overly selective sample of citizens and viewpoints and having a structure that is overly centralized and controlled in a top-down fashion (a structure that may be difficult to avoid in large, centralized national systems). One way to address these potential pitfalls is to follow the UK approach of "outsourcing" many public engagement activities to independent and nongovernmental entities (Sciencewise, 2016; Wilsdon, 2015). However, independent data with which to evaluate the efficacy of these efforts in other countries and their applicability to the U.S. context are limited. Moreover, when the process involves formulating future public policy for a country at large, it may be advantageous to have a central organizer of these types of consultations.

LESSONS LEARNED FROM PUBLIC ENGAGEMENT ACTIVITIES

Public engagement efforts are crucially important for

代表性样本中得到的结论推广到更大的群体中。这样的工作包括一些可量化的指标,如在样本中观察到的某些特征在一般群体中出现的可能性有多大(例如,Dillman et al.,2014)。公众舆情调研在了解公众对信息的需求、对风险和收益的认知,以及对问题的不同态度方面是非常有帮助的(Scheufele,2010)。相比之下,公众参与性研讨活动(如合议会议或公开会议)通常由专业性或相关性更强的代表参与,这些团体可以帮助决策者和科学界在决策过程中更早地意识到道德、法律或社会方面存在的问题。通常情况下,公众参与性研讨活动所吸引的特殊人群和推动双边对话的社会动力(即使是专业的)也都缺少将研究结果推广到公众的能力(Merkle,1996;Scheufele,2011),这也就需要调研对象既包括专业的群体,也要有广泛的群众基础。

第二个重要原则涉及决策者及参与研讨的召集人沟通交流的科学家类型。有关人类的新信息的解释表述有标准的参考信息和架构(Goffman,1974;Kahneman and Tversky,1984)。因此,在会议材料,以及被选作特定应用的案例中关于专业技术术语的表述差别很小,不足以显著改变参与者的初始态度和讨论的整体性质(Anderson et al.,2013)。因此,在公众参与性研讨活动中,有关专家撰写书面会议材料和专题介绍时不能仅凭自己的看法和观点,而更需要充分借鉴经验社会学进行预估,以确保最小化先验性偏见,进而促进包容性和广泛性的讨论,而不是人为地局限在专业性的技术和科学层次。

更 进 一 步

美国当前的基础架构充分融汇了当前基因治疗模式的公众意见,包括商业和公共资助的涉及人类基因组编辑的基础研究。正如之前所述,美国的科学受制于卓有成效的质量控制、监督机制和伦理要求。而其中许多机制,如IRBs,已经涉及了公众意见。与此同时,公众通过在联邦、州和地方各级的投票,也可以提出关于优先资助的系统性的意见、规章及基础研究的其他方面。

在美国,当前监管基础架构中建立的参与机制足以解决人类基因组编辑技术在体细胞上的应用要求,但这并不意味着它们不需要改进。由诸如RAC等团体采用的参与过程是传达相关信息并

guiding societal and political debates about the social, ethical, legal, and political aspects of applications of human genome editing. Given the infrastructures already in place to engage the public, as well as the general principles for engagement discussed in this chapter, the committee sees particular value in an approach that uses different processes for engagement for different types of questions surrounding genome editing.

At the same time, it is essential that all such efforts adhere to all the principles for effective public engagement outlined earlier and that they apply two additional engagement principles to help avoid potential pitfalls.

First, any effort to engage the public broadly needs to distinguish between systematic public opinion research and public engagement exercises. The former uses social scientific methods to measure public opinion in ways that allow for generalization from representative samples to larger populations. Such efforts include quantifiable indicators of how likely it is that certain characteristics observed in a sample occur in the general population (e.g., Dillman et al., 2014). Public opinion research is particularly useful for identifying informational needs, perceptions of risk and benefit, or other attitudinal variables among different publics (Scheufele, 2010). By contrast, public engagement exercises such as consensus conferences or public meetings typically rely on representatives of highly interested and knowledgeable groups that can help policy makers or the scientific community identify ethical, legal, or societal considerations early in the policy-making process. The specific populations on which most engagement exercises draw and the social dynamics that drive conversational settings—even with professional moderators—often limit the ability to generalize findings from such exercises to broader public opinion (Merkle, 1996; Scheufele, 2011). Parallel efforts with focus groups and broad public surveys of more randomly selected samples may be needed.

A second important principle relates to communicating the specific types of science on which policy makers or other conveners of engagement exercises seek input. Human beings interpret new information by using existing frames of reference (Goffman, 1974; Kahneman and Tversky, 1984). Fairly minor differences in how scientific techniques are described in meeting materials or the examples that are chosen for particular applications can significantly alter initial attitudes among participants, as well as the overall nature of discussions (Anderson et al., 2013). Thus it is important that written meeting materials or presentations by experts during public engagement exercises not be developed only by technical experts based on their perceptions of relevance or appropriateness. Instead, they need to be systematically pretested using empirical social science to ensure that they minimize a priori biases and allow for inclusive, broad discussions

与相关方协商，在极少数情况下，他们还需要为公众真正参与制定监管规则而努力。理想情况下，参与人类基因组编辑的所有监督机构将扩充其参与工作的形式，以开发出更系统化和可持续的公众参与模式。需要强调的是，现有的公众参与模式需要扩大，以帮助监管机构界定"治疗"、"增强"、"疾病"与"缺陷"等术语的定义和范围。要做到这些，需要监管机构向具体委员会输送具有相关专业知识的成员（如 FDA 咨询委员会），用来评估结果的好坏及新编辑的基于细胞或组织的产品的达标程度。

对于任何人类生殖系基因组编辑的应用，符合本章概述的参与原则的广泛的、包容的、有意义的公共意见将是推动进展的必要条件。为此，持续关注公众的态度、信息缺陷及出现的新问题至关重要。这些公众参与成果有助于政策机构实现既能够通过向不同的公众群体通报信息和提供政策相关的科学信息来进行有效地沟通，也能够确定需要系统性的工作来创造早期公众参与基础建设的领域（NASEM 2016a）。

这些鼓励公众参与的不断的尝试需要直接与政策制定过程联系起来（NRC，2008，p. 19）。此外，围绕不断变得复杂的问题我们还需坚持公开讨论，以便告知监管机构和决策者可否为将个人和社会的价值置于这种"增强"干预措施的临床试验之前的利益和风险上的做法授权。

为了促进和监控这种公众参与的有效性，联邦机构需要考虑为研究提供资助计划，以便：第一，促进对人类基因组编辑在社会政治、伦理和法律方面所带来的的短期和长期影响的理解；第二，评估在监管或决策基础架构中建立公众参与（沟通、协商和参与）方面所做的各种努力的效力；第三，评估公众参与如何能够且应该影响的不同的政策制定领域。基因组计划的经验包括将"伦理、法律和社会问题"作为其科学研究总体的一部分，以及国家科学基金会资助的社会纳米技术中心的经验可能为公众参与基因组编辑提供有用的参考来安排研究议程或资助计划。[67]

总结和建议

通过由技术专家和社会科学家进行的公众参

that are not constrained artificially to the technical or scientific aspects of the subject.

MOVING FORWARD

Current infrastructure in the United States adequately includes public input for current modes of gene therapy, including both commercially and publicly funded basic research involving human genome editing. As discussed earlier, science in the United States is subject to well-functioning quality controls, oversight mechanisms, and ethical controls. Many of these mechanisms, such as IRBs, already involve public input. Similarly, the public has a means of providing systematic input on funding priorities, regulations, and other aspects of basic research through electoral choices at the federal, state and local levels.

Engagement mechanisms built into current regulatory infrastructures in the United States are sufficient address somatic applications of human genome-editing techniques, but this does not mean they cannot be improved. The engagement processes employed by groups such as the RAC are communication of relevant information and consultation with affected or interested parties. In rarer cases, efforts have also been undertaken to provide for true public *participation* in regulatory rulemaking. Ideally, all oversight bodies involved in human genome editing would expand their portfolio of engagement efforts to develop more systematic and sustainable modes of public participation. In particular, an expansion of current modes of public engagement will be necessary to help regulatory bodies define the definitions of and boundaries between such terms as therapy and enhancement or disease and disability. These efforts might be aided by regulatory agencies' adding members with relevant expertise to specific committees, such as FDA advisory committees that need to evaluate the benefits of an indication and the degree of unmet need for a new edited cell- or tissue-based product.

For any consideration of applications of genome editing of the human germline, extensive, inclusive, and meaningful public input consistent with the principles of engagement outlined in this chapter would be a necessary condition for moving forward. To this end, ongoing monitoring of public attitudes, information deficits, and emerging concerns would be essential. These public engagement efforts would allow agencies and other policy bodies to (1) communicate effectively by informing different publics and providing policy-relevant scientific information, and (2) identify areas requiring systematic efforts to create infrastructures for public engagement early in the process (NASEM, 2016a).

The complex issues surrounding enhancement will also require an ongoing public debate to inform regulators and policy makers about the individual and societal values

与，将加强通过基因组编辑来促进人类医学的进步，社会科学家应通过进行系统的公众舆情分析，形成有针对性的宣传材料，并尽量减少可能妨碍讨论和辩论的人为偏见或约束。

美国现有的公众参与基础架构足以满足对基础科学和人类基因组编辑实验室研究的监督。同样，作为当前美国监管基础架构的一部分，用于公众交流和协商的机制也可用于解决围绕体细胞基因组编辑的开发的公众交流问题。

衡量对生殖系编辑未来应用而带来的技术，以及社会的利益和风险需要更加规范化的工作，广泛征求公众意见并鼓励公开辩论，而不仅仅是停留在目前的状态。此外，围绕不断增加的复杂问题将需要持续的公开辩论，以便告知监管机构和决策者，能否为将个人和社会价值的利益和风险置于这种增强干预措施的临床试验之前的做法授权。

在其他新兴科学和技术领域发展出的为实现有效和包容性的公众参与而探索出的做法和原则，也为基因组编辑领域的公众参与提供了有益的借鉴。

建议 7-1：将人类基因组编辑技术用于任何超出治疗或预防疾病及缺陷的临床试验之前，应该允许广泛的和包容性的公众参与。

建议 7-2：在公众广泛、持续的参与和投入下，正在进行的关于健康、社会的益处和风险评估，应该优先考虑可遗传的生殖系基因组编辑的临床试验。

建议 7-3：公众参与应纳入到人类基因组编辑的政策制定过程，并且应持续监测公众的态度、信息缺失和对基因强化问题的急切关注。

建议 7-4：在资助人类基因组编辑的研究时，联邦机构应考虑资金支持以下近期研究和战略：

- 明确需要系统地和早期努力来征求公众参与的领域；
- 发展必要的内容并进行有效地沟通；
- 在现有的基础架构的背景下改善公众参与。

建议 7-5：当资助人类基因组编辑的研究时，

to be placed on the benefits and risks before clinical trials for such enhancement interventions could be authorized.

These ongoing efforts to encourage public engagement would need to be tied directly to the policy-making process (NRC, 2008, p. 19). Furthermore, the complex issues surrounding enhancement will require ongoing public debate to inform regulators and policy makers about the individual and societal values to be placed on the benefits and risks before clinical trials for such interventions could be authorized.

To facilitate and monitor the effectiveness of such engagement efforts, federal agencies would need to consider funding programs for research to (1) promote understanding of the long-and short-term sociopolitical, ethical, and legal aspects of human genome editing; (2) evaluate the efficacy of various efforts to build public engagement (communication, consultation, and participation) into regulatory or policy-making infrastructures; and (3) assess how public engagement can and should influence different areas of policy making. Experiences with the genome initiative's program to include consideration of "ethical, legal and social issues" as part of its overall funding of scientific research, and experiences with the Centers for Nanotechnology in Society funded by the National Science Foundation might provide useful frameworks for structuring similar research agendas or funding programs for public engagement for genome editing.[67]

CONCLUSIONS AND RECOMMENDATIONS

Efforts to advance human medicine through genome editing will be strengthened by public engagement informed by technical experts and by social scientists who undertake systematic public opinion research, develop appropriate communication materials, and minimize artificial biases or constraints that would hinder discussion and debate.

Existing public communication and engagement infrastructures in the United States are sufficient to address oversight of basic science and laboratory research on human genome editing. Similarly, mechanisms for public communication and consultation that are part of the current U.S. regulatory infrastructures are also available to address public communication around development of human somatic-cell genome editing.

Weighing the technical and societal benefits and risks of applications of future uses of germline editing will require more formalized efforts to solicit broad public input and encourage public debate than are currently in place. Furthermore, the complex issues surrounding

---

[67] See https://www.genome.gov/elsi/ (accessed January 30, 2017); https://www.nsf.gov/news/news_summ.jsp?cntn_id=117862 (accessed January 30, 2017).

联邦机构应考虑包括资助有以下目标的研究：

- 研究编辑人类生殖系的社会政治、伦理和法律方面的；
- 研究超出疾病及缺陷的治疗或预防的基因组编辑用途的社会政治、伦理和法律方面的；
- 评估建立公众沟通和参与纳入监管或政策制定基础架构的努力的有效性的。

enhancement will require an ongoing public debate to inform regulators and policy makers about the individual and societal values to be placed on the benefits and risks before clinical trials for such enhancement interventions could be authorized.

The practices and principles developed for effective and inclusive public engagement on other emerging areas of science and technology provide a valuable base to inform public engagement on genome editing.

RECOMMENDATION 7-1. Extensive and inclusive public participation should precede clinical trials for any extension of human genome editing beyond treatment or prevention of disease or disability.

RECOMMENDATION 7-2. Ongoing reassessment of both health and societal benefits and risks, with broad ongoing participation and input by the public, should precede consideration of any clinical trials of heritable germline genome editing.

RECOMMENDATION 7-3. Public participation should be incorporated into the policy-making process for human genome editing and should include ongoing monitoring of public attitudes, informational deficits, and emerging concerns about issues surrounding "enhancement."

RECOMMENDATION 7-4. When funding human genome-editing research, federal agencies should consider including funding to support near-term research and strategies for

- identifying areas that require systematic and early efforts to solicit public participation,
- developing the necessary content and communicating it effectively, and
- improving public engagement within the context of existing infrastructure.

RECOMMENDATION 7-5. When funding human genome-editing research, federal agencies should consider including funding for research aimed at

- understanding the sociopolitical, ethical, and legal aspects of editing the human germline;
- understanding the sociopolitical, ethical, and legal aspects of uses for genome editing that go beyond treatment or prevention of disease or disability; and
- evaluating the efficacy of efforts to build public communication and engagement on these issues into regulatory or policy-making infrastructures.

8

原则和建议概要
Summary of Principles and Recommendations

基因组编辑在推进基础科学和治疗应用的进步方面拥有巨大的潜力。将基因组编辑方法应用于人类细胞、组织、生殖系细胞和胚胎的基础实验研究有助于提高我们对一般人体生物学的理解，包括进一步增加我们对人类生育能力、繁殖和发育的知识，以及加深对疾病的理解乃至建立新的治疗方法。这种研究正在现有的监督系统内迅速进行。基因组编辑早已用于某些特定遗传疾病的体细胞治疗的临床试验，并在体细胞基因治疗研究的监管系统的管理下进行。此外，新方法的出现提供了未来在国内和跨国法律的限度内编辑生殖系细胞以预防遗传性疾病的遗传性传播的可能性。与此同时，基因组编辑技术在评估遗传改变的正当性、危险性及反社会性等方面对现有的监管机构和公众评估管理系统提出了新的挑战。本章总结了委员会关于总体原则和具体的结论，以及关于对这一快速发展的研究和应用领域执行与监管的具体建议。

人类基因组编辑监管原则

基因组编辑很有希望用于预防、改善或消除许多人类疾病和缺陷，但这需要进行伦理研究和临床运用。

Genome editing offers great potential to advance both fundamental science and therapeutic applications. Basic laboratory research applying genome-editing methods to human cells, tissues, germline cells and embryos has great promise for improving our understanding of normal human biology, including furthering our knowledge of human fertility, reproduction, and development, as well as providing deeper understanding of disease and establishing new approaches to treatment. Such research is proceeding rapidly within existing oversight systems. Genome editing is already entering clinical testing for somatic treatment of certain genetic diseases, subject to regulatory systems designed to oversee human somatic-cell gene-therapy research. Furthermore, recently developed methods offer the future possibility of editing germline cells to prevent heritable transmission of genetic disease, within the limits of domestic and transnational law. At the same time, genome-editing technologies challenge regulators and the public to evaluate existing governance systems to determine whether there are genetic alterations that are insufficiently justified, too risky, or too socially disruptive to be pursued at this time. This chapter summarizes the conclusions of the committee relating overarching principles and specific conclusions to its recommendations for the conduct and oversight of this burgeoning area of research and applications.

OVERARCHING PRINCIPLES FOR GOVERNANCE OF HUMAN GENOME EDITING

Genome editing has great promise for preventing, ameliorating or eliminating many human diseases and conditions. Along with this promise comes the need for ethically responsible research and clinical use.

建议 2-1：以下原则应该作为支持人类基因组编辑的监督系统、研究和临床应用的底线：
- 福利提升
- 透明度
- 慎重
- 负责任的科学
- 尊重人权
- 公平
- 跨国合作

反过来，基于这些原则，在设计基因组编辑的管理系统时需要兑现的几项基本责任：

福利提升：福利提升的原则是向受影响者提供福利以及预防其受到伤害，这在生物伦理学文献中经常被称为有益和无害的原则。

遵守这一原则的责任包括：①促进个人健康和幸福的人类基因组编辑的应用，如治疗或预防疾病，同时使具有高度不确定性的针对个体的早期应用中的风险最小化；②确保任何人类基因组编辑应用的风险和利益合理的平衡。

透明度：透明度原则要求以开放和共享的方式为利益相关者提供信息。

遵守这一原则的责任包括：①承诺及时地披露尽可能全面的信息；②要将有意义的公众意见和其他新型和颠覆性的技术投入到与人类基因组编辑相关的政策制定过程中。

慎重：慎重原则是指对参加研究或接受临床护理的患者的治疗要在有足够和强有力的证据支持下，仔细和审慎地进行。

遵守这一原则的责任包括在适当的监督下谨慎和渐进地进行，并根据未来进步和文化观点而多次进行重新评估。

负责任的科学：负责任科学的原则是研究的最高标准，从实验室到临床，符合国际和专业规范。

遵守这一原则的责任包括：①承诺高质量的实验设计和分析；②对协议和结果数据的适当地审查和评估；③透明度；④纠正假的或误导性的数据和分析。

尊重人权：尊重人权的原则需要认识到个人的尊严，承认个人选择的中心性和尊重个人的决定。所有人都有平等的道德价值，不论他们的遗传素质如何。

RECOMMENDATION 2-1. The following principles should undergird the oversight systems, the research on, and the clinical uses of human genome editing:
- **Promoting well-being**
- **Transparency**
- **Due care**
- **Responsible science**
- **Respect for persons**
- **Fairness**
- **Transnational cooperation**

In turn, these principles result in a number of responsibilities when devising a governance system for genome editing:

Promoting well-being: *The principle of promoting well-being supports providing benefit and preventing harm to those affected, often referred to in the bioethics literature as the principles of beneficence and nonmaleficence.*

Responsibilities that flow from adherence to this principle include (1) pursuing applications of human genome editing that promote the health and well-being of individuals, such as treating or preventing disease, while minimizing risk to individuals in early applications with a high degree of uncertainty; and (2) ensuring a reasonable balance of risk and benefit for any application of human genome editing.

Transparency: *The principle of transparency requires openness and sharing of information in ways that are accessible and understandable to stakeholders.*

Responsibilities that flow from adherence to this principle include (1) a commitment to disclosure of information to the fullest extent possible and in a timely manner, and (2) meaningful public input into the policy-making process related to human genome editing, as well as other novel and disruptive technologies.

Due care: *The principle of due care for patients enrolled in research studies or receiving clinical care requires proceeding carefully and deliberately, and only when supported by sufficient and robust evidence.*

Responsibilities that flow from adherence to this principle include proceeding cautiously and incrementally, under appropriate supervision and in ways that allow for frequent reassessment in light of future advances and cultural opinions.

Responsible science: *The principle of responsible science underpins adherence to the highest standards of research, from bench to bedside, in accordance with international and professional norms.*

Responsibilities that flow from adherence to this principle include a commitment to (1) high-quality experimental design and analysis, (2) appropriate review and evaluation of protocols and resulting data, (3) transparency, and (4) correction of false or misleading data or analysis.

Respect for persons: *The principle of respect for persons requires recognition of the personal dignity of all individuals, acknowledgment of the centrality of personal*

遵守这一原则的责任包括：①对所有个人的同等价值的承诺；②尊重和促进个人决策；③承诺防止过去被滥用的优生学行为再次出现；④致力于消除对残疾的偏见。

公平：公平原则要求对于类似情况一视同仁，而风险和收益是公平合理的分配的（分配公正）。

遵守这一原则的责任包括：①公平分配研究的负担和利益；②广泛和公平地获得人类基因组编辑的临床应用的利益。

跨国合作：跨国合作的原则是承诺以合作的方式去研究和管理，同时尊重不同的文化背景。

遵守这一原则的责任包括：①尊重不同国家的政策；②尽可能协调监管标准和程序；③不同的科学团体和负责的监管机构之间的跨国合作和数据共享。

这些原则和责任可以以管理基因组编辑的具体建议的形式来实现，如下所述。

美国监督人类基因组编辑的现有机制

在美国，州和联邦层面的现有法律和资助政策将控制从实验室研究、临床前测试和临床试验到临床应用的所有阶段的人类基因组编辑。现有的系统，虽然有改进的余地，可以用来去管理当前可预期的人类基因组编辑的使用问题，但一些未来的使用问题仍需要严格的标准和进一步的公开讨论。

使用基因组编辑的实验研究

基因组编辑作为开展人体细胞和组织相关研究的工具，在很大程度上与其他类型的实验研究以相同的方式接受监管，包括生物安全审查和实验实践的一般标准等。其他的政策也已到位，比如用以管理人类细胞、组织或胚胎研究的捐赠和使用规范等。这些考虑了一些因素，如组织是否是临床程序遗留下来的，或是为了研究而专门通过干预才获得的。如果所涉及的人体组织通过所携带的信息或与其相关联的信息可轻易确认捐赠者的身份信息，则需要一些额外的受试者保护措施，包括某种形式的同意和机构审查委员会（IRB）的审查。

其他考虑也同样适用于使用人类胚胎进行实验

choice, and respect for individual decisions. All people have equal moral value, regardless of their genetic qualities.

Responsibilities that flow from adherence to this principle include (1) a commitment to the equal value of all individuals, (2) respect for and promotion of individual decision making, (3) a commitment to preventing recurrence of the abusive forms of eugenics practiced in the past, and (4) a commitment to destigmatizing disability.

Fairness: *The principle of fairness requires that like cases be treated alike, and that risks and benefits be equitably distributed (distributive justice).*

Responsibilities that flow from adherence to this principle include (1) equitable distribution of the burdens and benefits of research and (2) broad and equitable access to the benefits of resulting clinical applications of human genome editing.

Transnational cooperation: *The principle of transnational cooperation supports a commitment to collaborative approaches to research and governance while respecting different cultural contexts.*

Responsibilities that flow from adherence to this principle include (1) respect for differing national policies, (2) coordination of regulatory standards and procedures whenever possible, and (3) transnational collaboration and data sharing among different scientific communities and responsible regulatory authorities.

These principles and responsibilities can be fulfilled in the form of specific recommendations for regulation of genome editing, as presented below.

EXISTING U.S. OVERSIGHT MECHANISMS FOR HUMAN GENOME EDITING

In the United States, existing laws and funding policies at the state and federal levels will govern human genome editing at all stages from laboratory research through preclinical testing and clinical trials to clinical application. The existing systems, while always having room for improvement, can be deployed to manage currently anticipated uses of human genome editing, but some future uses will require stringent criteria and further public debate.

Laboratory Research Using Genome Editing Methods

The use of genome editing as a laboratory research tool in human somatic cells and tissues would largely be governed in the same way as other types of laboratory research, which are subject to institutional biosafety review and general standards of laboratory practice. Additional policies are also in place to govern the donation and use of human cells, tissues or embryos for research. These take account of factors such as whether the tissue is left over from a clinical procedure or is obtained through intervention specifically for research,. If tissue has information within or linked to it that makes the donor's identity readily ascertainable, then

室研究的基因组编辑（非怀孕目的）。在美国，联邦资助使用胚胎的研究一般会受到 Dickey-Wicker 修正案禁止，但一些州和私人来源的研究是允许的。这种用途将受到管制人类生殖和受孕产品的一些法律制度的约束。NIH 人类胚胎研究小组（1994）、国家科学院人类胚胎干细胞研究指南和国际干细胞研究学会指南的建议继续影响着这一领域的研究实践的发展。

过去曾经探索过关于在实验室中使用人类胚胎进行基因组编辑研究产生的伦理和监管等方面的问题，例如：胚胎在伦理学中的状态；研究用胚胎或使用将被丢弃的胚胎的可接受性；在研究中使用胚胎的法律或自愿等因素的限制。在其他国家也产生了同样的关于伦理的考虑。即使意识到在研究中使用人类胚胎的科学价值，这种做法在许多司法管辖范围内也是有限的、受阻的，甚至被禁止的。纯粹用于非生殖研究目的的人胚胎的基因组编辑将受这些伦理规范和政策的约束。然而，如果允许，已经在为其他形式的胚胎研究实施监督的程序应该可以保证研究的必要性和质量。

对使用人类细胞和组织的实验研究的监督是负责任科学原则的体现，其包括高质量的实验设计和方案审查。科学通过严格的同行评审和结果发表而进步，也从共享和获得可支持该领域持续发展的数据而受益。透明度原则支持依照相关法律最大限度地共享信息。在使用人类胚胎进行的研究中，尊重不同国家政策的多样性不应成为跨国合作的障碍，包括数据共享、监管当局的合作，以及在可能的情况下统一标准。

总结和建议：基础实验研究

涉及人类基因组编辑的实验研究，即不与患者接触的研究，与其他人类组织体外基础研究遵循相同的管理方式，并且根据现有的道德规范和管理制度制定了相关政策。

该政策不仅包括体细胞的研究，而且包括被允许用于研究目的的接受捐赠，以及使用人类配子和胚胎的研究，尽管有人对这项政策中的个别条款有异议，但它仍然是有效的。与人类生育力和繁殖相关的重要学科和临床问题需要对人类配子及其祖细胞、胚胎和多能干细胞的持续

additional human subjects protections such as the need for some form of consent and an institutional review board (IRB) review generally will also apply.

Additional considerations apply to the use of genome editing for laboratory research using human embryos (with no aim of establishing a pregnancy). In the United States, federal funding for research using embryos generally is prohibited by the Dickey-Wicker amendment, but some state and private sources for such research are available. Such uses would be subject to some of the legal regimes governing human reproduction and products of conception. Recommendations of the 1994 National Institutes of Health (NIH) Human Embryo Research Panel, the National Academies of Sciences, Engineering, and Medicine's Guidelines for Human Embryonic Stem Cell Research, and the International Society for Stem Cell Research guidelines continue to shape research practices in this area.

The ethical and regulatory considerations posed by genome editing research using human embryos in the laboratory have been explored in the past: the moral status of the embryo; the acceptability of making embryos for research or using embryos that would otherwise be discarded, and legal or voluntary limits that apply to the use of embryos in research. These same ethical considerations are raised in other countries. Even with recognition of the scientific value of using human embryos in research, the practice is limited, discouraged or even prohibited in many jurisdictions. Genome editing of human embryos purely for non-reproductive research purposes will be subject to those same ethical norms and policies. Where permitted, however, oversight procedures already in place for other forms of embryo research should provide assurance of the necessity and quality of the research.

Oversight of laboratory research using human cells and tissues is an expression of the principle of *Responsible Science*, which includes high-quality experimental design and protocol review. Science proceeds by rigorous peer review and publication of results, and also benefits from sharing and access to data that can support continued development of the field. The principle of *Transparency* supports sharing information to the fullest extent possible consistent with applicable law. Respect for diversity among nations in domestic policy on research using human embryos should not be an obstacle to *Transnational Cooperation*, including data sharing, collaboration by regulatory authorities and, where possible, harmonization of standards.

Conclusions and Recommendation: Fundamental Laboratory Research

Laboratory research involving human genome editing—that is, research that does not involve contact with patients—follows regulatory pathways that are the same as those for other basic laboratory *in vitro* research

实验研究，这对非直接遗传的基因编辑的医学和科学研究目的是必要的，如果将来试图进行可遗传的基因编辑，该研究也会提供有价值的信息和技术。

建议 3-1：现有的用于审查和评估人类细胞和组织基因编辑基础实验的监管基础架构和方法理应用于评估将来基本的人类基因组编辑实验研究。

基于治疗和预防疾病和残疾的体细胞基因编辑

基因编辑最直接的临床应用是利用人类体细胞预防或治疗疾病和残疾。事实上，此类研究已在临床试验当中了。在美国，使用体细胞基因编辑技术的临床应用归食品药品监督管理局（FDA）管辖，该机构管控基于人体组织和细胞的治疗活动。启动任何基因编辑的临床试验都需经食品药品监督管理局事先批准，而机构审查委员会（IRB）将监督试验参与者的招募、咨询和不良事件监测等方面。体细胞基因编辑的临床试验相关的监管评估将以类似于其他医疗治疗相关的监管评估模式进行，包括试验风险最小化、分析参与者的潜在收益与风险是否合理，以及参与者是否在自愿和知情的前提下被招募和登记。在美国，额外的监管包括生物安全委员会的地方级安全审查和在美国国立卫生研究院（NIH）重组DNA咨询委员会主持下的国家级审查，这些额外监管是通过特殊和新颖方案的通用方法。

有些伦理规范和监管制度已经适用于其他形式的基因治疗开发，那么这些规范和制度足以用于管理以预防和治疗疾病和残疾为目的的有关体细胞基因组编辑的新应用，而监管监督的重点是防范未经授权或未成熟的基因组编辑技术的应用。

在一些情况下，对子宫内的胎儿体细胞进行基因组编辑的实验也需要考虑进来。例如，对于发育早期的一些可能产生严重影响的遗传疾病，在胎儿时基因编辑会比产后干预有更显著的效果，这将对产后孩子的健康起着关键作用。但子宫内的基因组编辑需要特别注意是否知情同意，以及胎儿生殖细胞和生殖祖细胞基因编辑的中靶与脱靶风险。

体细胞基因组编辑的规范建议是根据总体原则制定得出的。体细胞基因组编辑在研究和临

with human tissues, and raises issues already managed under existing ethical norms and regulatory regimes.

This includes not only work with somatic cells, but also the donation and use of human gametes and embryos for research purposes, where this research is permitted. While there are those who disagree with the policies embodied in some of those rules, the rules continue to be in effect. Important scientific and clinical issues relevant to human fertility and reproduction require continued laboratory research on human gametes and their progenitors, human embryos and pluripotent stem cells. This research is necessary for medical and scientific purposes that are not directed at heritable genome editing, though it will also provide valuable information and techniques that could be applied if heritable genome editing were to be attempted in the future.

RECOMMENDATION 3-1. Existing regulatory infrastructure and processes for reviewing and evaluating basic laboratory genome-editing research with human cells and tissues should be used to evaluate future basic laboratory research on human genome editing.

Somatic Cell Genome Editing for Treatment or Prevention of Disease and Disability

The most immediate clinical applications of genome editing will be in human somatic cells for the treatment or prevention of disease and disability. Indeed such research is already in clinical trials. In the United States, clinical applications that use somatic cell genome editing fall under jurisdiction of the U.S. Food and Drug Administration (FDA), which regulates human tissue and cell-based therapies. Initiation of any genome-editing clinical trial requires prior approval by the FDA, and IRBs will also oversee aspects such as recruitment, counseling and adverse-event monitoring for trial participants. Regulatory assessments associated with clinical trials of somatic-cell genome editing will be similar to those associated with other medical therapies, including minimization of risk, analysis of whether risks to participants are reasonable in light of potential benefits, and whether participants are recruited and enrolled with appropriate voluntary and informed consent. Additional oversight in the United States includes local safety reviews by institutional biosafety committees and national-level review opportunities under the auspices of the NIH Recombinant DNA Advisory Committee (RAC), for both specific, novel protocols and for general approaches.

The ethical norms and regulatory regimes already developed for other forms of gene therapy are adequate for managing new applications involving somatic genome editing with the purpose of treating or preventing disease and disability. But regulatory oversight should also emphasize prevention of unauthorized or premature applications of genome editing.

In some circumstances, it may also be desirable to consider undertaking genome editing in the somatic

床上应用的一个重要目标是促进健康，透明和负责任的科学研究对于提高工作质量是必需的，而适当的关心确保了规范在兼顾风险和成果的情况下的逐步应用，最后需要反复评估来及时响应不断变化的科学和临床信息。随着治疗和预防医疗技术的发展，需要平等地对待和尊重个人公平获得这些进步的要求，保护个人对于允许或者拒绝使用这些治疗的选择权，尊重做出选择的生命尊严。

总结和建议：体细胞治疗

总体而言，大部分公众支持使用基因治疗手段（甚至包括进行基因编辑）来预防或治疗疾病和残疾。

体细胞中进行人类基因组编辑对于预防或治疗多种疾病，以及改善现有基因治疗技术在临床试验中的安全性、有效性和高效性有巨大前景。虽然基因编辑技术仍在持续优化中，但当下的目标仅只适用于预防或治疗疾病和残疾，并不能兼顾其他不太紧要的用途。

而针对基因治疗的伦理规范和监管机制可以适用于这些用途。相关体细胞基因组编辑的临床试验监管评估类似于其他医疗治疗相关的监管评估，包括试验风险最小化、分析参与者的潜在收益与风险是否合理、参与者是否在自愿和知情的前提下被招募和登记。监管监督还需要法律权威和执法能力来防止未经授权或过早地应用基因组编辑，监管机构需要不断更新具体技术方面的知识。他们的评估不仅需要考虑基因组编辑系统的技术背景，还需要考虑适合的临床应用，以便可以权衡预期的风险和收益。基因编辑的脱靶率将随着技术平台、细胞类型、靶基因组序列和其他因素而变化，所以暂时可不规范用于体细胞基因组编辑特异性（如可接受脱靶事件的效率）的单一标准。

建议 4-1：现有用于审查及评估基于治疗或预防疾病和残疾的体细胞基因治疗的监管基础架构和流程也应该用于评估利用基因组编辑技术来进行的体细胞基因治疗。

建议 4-2：此时，监管机构应授权仅用于预防或治疗疾病和残疾的临床试验或细胞疗法。

cells of a fetus *in utero*, for example where fetal editing could be significantly more effective than post-natal intervention for genetic diseases with devastating effects early in development. The potential benefit to the resulting child would be key. But *in utero* genome editing would also require special attention to issues surrounding consent and to any increased risk of on-target or off-target modifications to fetal germ cells or germ cell progenitors.

Recommendations for regulating somatic cell genome editing are informed by several of the overarching principles. An important goal in both the research and clinical uses of somatic genome editing is *Promoting Well-being*. *Transparency* and *Responsible Science* are necessary for advancing the research with confidence in the quality of the work, while *Due Care* ensures that the applications proceed incrementally with careful attention to risks and benefits, as well as reassessments that allow timely response to changing scientific and clinical information. As therapeutic and preventive medical technologies are developed, *Fairness* and *Respect for Persons* call for attention to equitable access to the benefits of these advances, protection of individual choice to pursue or decline use of these therapies, and respect for the dignity of all persons regardless of that choice.

Conclusions and Recommendations: Somatic Therapy

In general, there is substantial public support for the use of gene therapy (and by extension, gene therapy that uses genome editing) for the treatment and prevention of disease and disability. Human genome editing in somatic cells holds great promise for treating or preventing many diseases and for improving the safety, effectiveness, and efficiency of existing gene therapy techniques now in use or in clinical trials. While genome-editing techniques continue to be optimized, however, they are best suited only to treatment or prevention of disease and disability and not to other less pressing purposes.

The ethical norms and regulatory regimes already developed for gene therapy can be applied for these applications. Regulatory assessments associated with clinical trials of somatic cell genome editing will be similar to those associated with other medical therapies, encompassing minimization of risk, analysis of whether risks to participants are reasonable in light of potential benefits, and determining whether participants are recruited and enrolled with appropriate voluntary and informed consent. Regulatory oversight also will need to include legal authority and enforcement capacity to prevent unauthorized or premature applications of genome editing, and regulatory authorities will need to continually update their knowledge of specific technical aspects of the technologies being applied. At a minimum, their assessments will need to consider not only the technical context of the genome-editing system but also the proposed clinical application so that anticipated risks and benefits can be weighed. Because off-target events will vary with the platform technology, cell type, target genome

建议 4-3：当对人类体细胞基因组编辑应用潜在的风险和益处进行评估时，监管机构应评估该应用的安全性和有效性，意识到脱靶率可能随技术平台、细胞类型、靶基因组位置以及其他因素而发生变化。

建议 4-4：在考虑是否允许超出预防或治疗疾病和残疾范围的体细胞基因组编辑临床试验之前，监管机构应该进行透明和包容的公开政策辩论。

遗传基因组编辑

在子代中产生遗传变异的基因组编辑（可遗传的生殖系编辑）有减轻由遗传疾病所造成痛苦的潜力。然而，它也引起了超出个人风险和益处考虑以外的关注。考虑到安全性和功效的不确定性，人类生殖系基因组编辑目前不可被批准，但该技术正在迅速发展，可遗传的生殖系编辑在经过认真考虑后可能在不久的将来经过认真考虑后有变成现实的可能。在某些情况下，生殖细胞或者胚胎中的基因组编辑可能是一些希望在基因上将遗传疾病和残疾传递风险最小化的父母唯一或最可接受的选择。

历史上，对人类基因组进行可遗传变异的可能性存在着争论。因为这种变异的影响可能是多代的，潜在的益处和潜在的危害都可能加倍。这种编辑对于未来出生的减轻遗传疾病负担的儿童，以及那些希望拥有一个在基因上关联却不传递缺陷基因的孩子的人们都有益处。另一方面，人们对于这种人类干预形式的明智性和适当性给予了关注。预期的基因组编辑本身可能有意想不到的后果，如果遗传，也会影响到后代。与其他形式的先进医疗技术一样，此类基因组编辑也会出现使用平等的问题。可遗传生殖系编辑的前景也引发类似于胚胎植入前和产前遗传筛选的关注，例如，纯粹自愿的个体决定可能共同改变包容残疾的社会规范。

可遗传生殖系的编辑

在一些情况下，可遗传生殖系编辑将成为希望未来拥有基因上有关联的孩子，同时将传递严重疾病或残疾的风险最小化的父母的唯一或最可接受的选择。然而，尽管应用可遗传生殖系的编辑会减轻遗传疾病，但对于生殖系基因组编辑存在着显著的

sequence, and other factors, no single standard for somatic genome-editing specificity (e.g., acceptable off-target event rate) can be set at this time.

RECOMMENDATION 4-1. Existing regulatory infrastructure and processes for reviewing and evaluating somatic gene therapy to treat or prevent disease and disability should be used to evaluate somatic gene therapy that uses genome editing.

RECOMMENDATION 4-2. At this time, regulatory authorities should authorize clinical trials or approve cell therapies only for indications related to the treatment or prevention of disease or disability.

RECOMMENDATION 4-3. Oversight authorities should evaluate the safety and efficacy of proposed human somatic cell genome-editing applications in the context of the risks and benefits of intended use, recognizing that off-target events may vary with the platform technology, cell type, target genomic location, and other factors.

RECOMMENDATION 4-4. Transparent and inclusive public policy debates should precede any consideration of whether to authorize clinical trials of somatic cell genome editing for indications that go beyond treatment or prevention of disease or disability.

Heritable Genome Editing

Genome editing that creates genetic changes heritable by future generations (heritable germline editing) has the potential to alleviate the suffering caused by genetically inherited diseases. However, it also raises concerns that extend beyond consideration of individual risks and benefits. Although human germline genome editing would not currently be approvable given the uncertainty about safety and efficacy, the technology is advancing rapidly, and heritable germline editing may, in the not-so-distant future, become a realistic possibility that needs serious consideration. There are circumstances in which genome editing in germline cells or embryos might be the only or most acceptable option for prospective parents who wish to have a genetically related child while minimizing the risk of transmitting a serious disease or disability.

There is a history of debate around the possibility of making heritable changes to the human genome. Because the effects of such changes could be multi-generational, both the potential benefits and the potential harms could be multiplied. Benefits from such editing would accrue to any future child born with reduced burden from genetically inherited disease, and to the prospective parents seeking to have a genetically related child without fear of passing along a disease.. On the other hand, concerns have been raised about the wisdom and appropriateness of this form of human intervention. The intended genome edits themselves might have unintended consequences which, if inherited, would also affect descendants. As with other forms of advanced medical technologies, questions of equality of

公众不安，特别是对于不太严重的情况和存在可替换方案的情况。这些关注的范围包括从认为人类不宜干预他们自身的演化的观点到对于影响个人和整个社会的非预期后果的焦虑。

在任何生殖系干预可以满足授权临床试验的风险／收益标准之前，需要进行更多的研究。但是随着对于卵子和精子的祖细胞进行基因组编辑的技术障碍被攻克，用基因组编辑来防止传递遗传性疾病的可能成为现实。

授权管理生殖系基因组编辑的主要美国实体机构——食品和药品监督管理局（FDA），的确在其决策中纳入了关于风险和收益的价值判断。现在需要对关于可遗传生殖系编辑的收益和风险的价值进行强有力的公开讨论，以便在任何关于是否授权临床试验的决定之前，这些价值可以适当地纳入风险／收益评估。但是，食品和药品监督管理局（FDA）在决定是否授权临床试验时，没有法定的授权来考虑公众对于一个技术的道德本质的看法。讨论细节在重组DNA咨询委员会（RAC）部分，相关的立法机构和其他公众参与场所层次的讨论详见第7章。

进行可遗传的生殖系基因组编辑实验必须谨慎，但谨慎并不意味着它们必须被禁止。

在有风险的情况下，如果技术挑战被克服并且潜在的收益是合理的，则可以开始临床试验；如果仅限于最有说服力的情况，则综合的监督将用于保护研究对象及其后代，并具有足够的保护措施，以防止其不适当地应用到人们察觉不到或者理解甚浅的地方。

建议5-1：可遗传生殖系基因组编辑相关的临床试验必须被限制在如下的健康和有效的管理框架内：

- 缺乏其他可行治疗办法；
- 仅限于预防某种严重疾病；
- 仅限于编辑已证明会导致或与疾病和残疾密切相关的基因；
- 仅限于把基因编辑成在群体中普遍存在且很少或没有副作用的健康版本；
- 就这些程序的风险和对于健康的潜在收益而言，可靠的临床前和／或临床数据的实用性；
- 在试验期间，持续严格地监督可能影响研究

access arise. The prospect of heritable germline editing also triggers concerns similar to those earlier raised by preimplantation and prenatal genetic screening, i.e., that purely voluntary, individual decisions might collectively change social norms about the acceptance of disabilities.

Conclusions and Recommendations: Heritable Germline Editing

In some situations heritable germline editing would provide the only or the most acceptable option for parents who desire to have genetically related children while minimizing the risk of serious disease or disability in a prospective child. Yet while relief from inherited diseases could accrue from its use, there is significant public discomfort about germline genome editing, particularly for less serious conditions and for situations in which alternatives exist. These concerns range from a view that it is inappropriate for humans to intervene in their own evolution to anxiety about unintended consequences for the individuals affected and for society as a whole.

More research is needed before any germline intervention could meet the risk/benefit standard for authorizing clinical trials. But as the technical hurdles facing genome editing of progenitors of eggs and sperm are overcome, editing to prevent transmission of genetically inherited diseases may become a realistic possibility.

The primary U.S. entity with authority for the regulation of germline genome editing—the FDA—does incorporate value judgments about risks and benefits in its decision making. A robust public discussion about the values to be placed on the benefits and risks of heritable germline editing is needed now so that these values can be incorporated as appropriate into the risk/benefit assessments that will precede any decision about whether to authorize clinical trials. But the FDA does not have a statutory mandate to consider public views about the intrinsic morality of a technology when deciding whether to authorize clinical trials. That level of discussion takes place at the RAC, in legislatures and at other venues for public engagement, discussed in Chapter 7.

Heritable germline genome editing trials must be approached with caution, but caution does not mean they must be prohibited.

If the technical challenges are overcome and potential benefits are reasonable in light of the risks, clinical trials could be initiated, if limited to only the most compelling circumstances, are subject to a comprehensive oversight framework that would protect the research subjects and their descendants; and have sufficient safeguards to protect against inappropriate expansion to uses that are less compelling or less well understood.

RECOMMENDATION 5-1. Clinical trials using heritable germline genome editing should be permitted only within a robust and effective regulatory framework that encompasses

- 参与者的健康和安全的方面；
- 尊重人们长期的、延续后代的自主权；
- 与患者隐私相符合的最大程度透明化；
- 在公众广泛参与和投入情况下，不间断地评估基因编辑技术带来的健康和社会福利及风险；
- 利用可靠的监督机制来防止基因编辑技术扩展到预防严重疾病和残疾以外的用途。

考虑到种系改变已经成为很多人关于道德底线的讨论重点，以及考虑到社会的多元化，所以如果每个人都同意这一意见，这将令人惊讶。即使那些确实同意这一建议的人，如果他们的理由是相同的，这也是令人惊讶的。还有一些人认为建议 5-1 的标准不能达到，因为一旦开始生殖系修饰，所建立的监督机制就不能将技术限制在特定的用途。委员会的观点是，如果确实不可能满足建议中的标准，生殖系基因组编辑是不能被批准的。委员会为此呼吁进行持续的公众参与和投入，同时发展基础科学和制定监管保障措施以满足上述标准。

可遗传的生殖系基因组编辑也引起人们对该技术早期或未被论证的使用的担忧，并且在这里概括出来的可靠的监督标准可能只能在一些有监管的地方实现。这些引起了人们担心可能出现"监管避风港"，诱使提供者或消费者前往具有宽松法规或不存在法规的灰色地带去违规进行生殖系基因编辑。这种医疗旅游现象包括寻找更快和更便宜的治疗选择，同时存在较新或较少管制措施区域，如果该技术在缺少必要的管辖的地区存在，则将不可能完全被管控。因此，强调对可遗传的生殖系基因组编辑的全面监督是十分重要的。

截止到 2015 年年末，美国监管部门不论能否达到上述标准，均不批准生殖系基因组编辑的相关实验，并在一项最早要到 2017 年 4 月才能够生效的预算法案中通过了一项条款[68]，写明：

根据联邦食品药品化妆品法案（21 U.S.C. 355(i)）第 505(i) 条以及公共保健服务法（42 U.S.C 262(a)(3)）第 351(a) 条规定，本法案所提供的资金不得用于通知监管机构，也不得用于提交关于研

- the absence of reasonable alternatives;
- **restriction to preventing a serious disease or condition;**
- **restriction to editing genes that have been convincingly demonstrated to cause or to strongly predispose to that disease or condition;**
- restriction to converting such genes to versions that are prevalent in the population and are known to be associated with ordinary health with little or no evidence of adverse effects;
- the availability of credible pre-clinical and/or clinical data on risks and potential health benefits of the procedures;
- during the trial, ongoing, rigorous oversight of the effects of the procedure on the health and safety of the research participants;
- comprehensive plans for long-term, multigenerational follow-up that still respect personal autonomy;
- maximum transparency consistent with patient privacy;
- **continued reassessment of both health and societal benefits and risks, with broad on-going participation and input by the public; and**
- reliable oversight mechanisms to prevent extension to uses other than preventing a serious disease or condition.

Given how long modifying the germline has been at the center of debates about moral boundaries, as well as the pluralism of values in society, it would be surprising if everyone were to agree with this recommendation. Even for those who do agree, it would be surprising if they all shared identical reasoning for doing so. There are also those who think the final criterion of Recommendation 5-1 cannot be met, and that once germline modification had begun, the regulatory mechanisms instituted could not limit the technology to the uses identified in the recommendation. If, indeed, it is not possible to satisfy the criteria in the recommendation, the committee's view is that germline genome editing would not be permissible. The committee calls for continued public engagement and input while the basic science evolves and regulatory safeguards are developed to satisfy the criteria set forth here.

Heritable germline genome editing also raises concerns about premature or unproven uses of the technology, and it is possible that the criteria outlined here for responsible oversight would be achievable in some but not all jurisdictions. This possibility raises the concern that "regulatory havens" could emerge that would tempt providers or consumers to travel to jurisdictions with more lenient or nonexistent regulations to undergo the restricted procedures. The phenomenon of medical tourism, which encompasses the search for faster and cheaper therapeutic options, as well as newer or less regulated interventions, will be impossible to control completely if the technical capabilities exist in more

---

[68] *Consolidated Appropriations Act of 2016*, HR 2029, 114 Cong., 1st sess. (January 6, 2015) (https://www.congress.gov/114/bills/hr2029/BILLS-114hr2029enr.pdf [accessed January 4, 2017]).

究药物或生物制品用来故意制造人类胚胎或进行可遗传的生殖系修饰。

目前，这一规定的结果是美国当局不可能审查关于生殖系基因组编辑的临床试验的建议，因此促使该技术向其他司法管辖区发展，某些会进行监管，而另一些则不设监管。

不以治疗或预防疾病和残疾为目的的基因组编辑

体细胞基因组编辑在治疗用途上的发展和生殖系基因组编辑在未来治疗用途上的发展引起了大众对于如何定义疾病和残疾的讨论，以及对用于预防和治疗所进行的基因编辑应该在哪里设置怎样合适的界限。和其他技术一样，人类基因组编辑技术可以应用于更广泛的目的，包括提高人类能力到超出正常范围。要定义基因强化的概念很困难。治疗、预防和基因强化，这三者的定义之间的界限不明确或者说就一般情况而言不容易辨别，甚至于"疾病"的定义也可以进行开放地讨论。因此，如何区分治疗或预防疾病和残疾与基因强化的概念是具有挑战性的。因此，基因组编辑潜在用途的可接受性在持续下降。通过将致病性遗传变异编辑为非有害性变异来解决严重遗传疾病通常是在其范围内最能够接受的，而基因编辑产生的与疾病无关的基因强化通常是最不可能被接受的。本报告描述了用于治疗或预防疾病和残疾的基因组编辑与用于治疗或预防疾病和残疾之外目的的基因组编辑之间的区别，但没有得出如何定义基因强化的结论。

体细胞基因组编辑和可遗传的生殖系基因编辑原则上都可以产生基因强化。与其他基因组编辑的潜在应用一样，个体风险和收益将与评估这类编辑技术相关联。但是遗传改良的可能性引起了一些额外的伦理和社会问题，对于这一问题没有简单的答案，却有不同的意见。

总结和建议：基因组编辑在防治疾病之外的目的

除非治疗或预防疾病和残疾的方法能够满足开启临床试验的风险和收益标准，否则我们需要有明确的科学进展后才能开展基因组编辑干预试验。这个结论适用于体细胞和可遗传的生殖系干预试验。为了所谓的强化人类特性及超出正常健康水平而使用基因编辑技术已经遭到公众的不满。因此，需

permissive jurisdictions. Thus, it is important to highlight the need for comprehensive regulation.

As of late 2015, the United States is unable to consider whether to begin germline genome editing trials, regardless of whether the criteria laid out above could be met. A provision (in effect until at least April 2017) was passed in a budget bill,[68] in which Congress included the following language:

None of the funds made available by this Act may be used to notify a sponsor or otherwise acknowledge receipt of a submission for an exemption for investigational use of a drug or biological product under section 505(i) of the Federal Food, Drug, and Cosmetic Act (21 U.S.C. 355(i)) or section 351(a)(3) of the Public Health Service Act (42 U.S.C. 262(a)(3)) in research in which a human embryo is intentionally created or modified to include a heritable genetic modification. Any such submission shall be deemed to have not been received by the Secretary, and the exemption may not go into effect.

The current effect of this provision is to make it impossible for U.S. authorities to review proposals for clinical trials of germline genome editing, and therefore to drive development of this technology to other jurisdictions, some regulated and others not.

Genome Editing for Purposes Other Than Treating or Preventing Disease or Disability

Both the ongoing development of therapeutic uses of somatic genome editing, and possible future development of therapeutic uses of germline genome editing, raise the issue of defining disease and disability, and the question of how and where to set appropriate boundaries for treatment and prevention of these conditions. Like other technologies, human genome editing methods may be applied for a wider range of purposes, including to enhance human capacities beyond the normal range. It is difficult to define the concept of enhancement. The lines between what are considered therapy, prevention, and enhancement are not rigid or easily discernible in all cases, and even the definition of what is considered a "disease" can be open to debate. For this reason, distinguishing between treating or preventing disease and disability on the one hand, and a notion of enhancement on the other, is challenging. Possible uses of genome editing thus fall along a continuum of acceptability. Addressing serious genetic disorders by converting causative genetic variants to non-deleterious variants generally fall at the most acceptable end of this spectrum, while editing to produce enhancements unrelated to disease typically fall at the least acceptable end. The report draws a distinction between genome editing for the purpose of treating or preventing disease or disability and genome editing for purposes other than treating or preventing disease or disability, without concluding that there is, as yet, any general consensus as to how to define the blurry boundaries of enhancement more clearly.

Editing to create a genetic enhancement could, in

要进行强有力的公众讨论，来关注基因组编辑有关个人和社会利益的风险和价值，而不是单纯局限在治疗或预防疾病和残疾的目的。这些讨论将包括考虑引入或加剧社会不平等的可能性，以便这些价值观可以纳入到任何关于是否授权临床试验决定的风险／收益评估之前。

建议 6-1：监管机构目前不应授权体细胞或生殖系基因组编辑用于治疗或预防疾病和残疾以外目的的临床试验。

建议 6-2：政府机构应鼓励关于治疗体细胞人类基因组编辑用于治疗或预防疾病和残疾之外目的的公众讨论和政策辩论。

公众参与管理人类基因组编辑的作用

人类通过基因组编辑来促进人类医学上的努力将由于公众参与而得到加强，这种参与对于目前监管框架下无法有效利用的潜在用途特别重要。特别是在美国，监管机构倾向于关注个人和公众的健康及安全，而不是关于对社会习俗和文化可能产生影响的问题。在其他论坛中也经常有人强调这些关注，虽然在咨询委员会中有讨论，但是缺乏法律效力，除非可以反映在政府已授权的以有限权利为基础的立法中。就决定是否以及如何允许开发新技术而言，其他国家的系统更明确地考虑到公众的态度，各国对于政府权力的法律约束各不相同。

公众参与

由技术专家和社会科学家进行的公众参与，将加强通过基因组编辑促进人类医学的努力，并且经由社会科学家进行系统的舆情分析，形成适当的宣传材料，可以尽量减少可能妨碍讨论及辩论的人为偏见或约束。

美国现有的公众交流和参与的基础架构足以监管人类基因组编辑的基础科学实验研究。同样，用于公众传播和协商的途径作为当前美国监管基础架构的一部分也可以用于解决围绕人体细胞基因组编辑的公众传播。

衡量未来运用生殖系编辑在技术、社会上的利益和风险需要有比目前更有效的努力，来征求公众的意见并鼓励公开辩论。此外，围绕基因强化的复杂问题将需要持续的公开辩论，以便向监管者和

principle, be undertaken in the context of somatic-cell genome editing or in the context of heritable germline editing. As with other potential applications of genome editing, individual risks and benefits would be associated with the assessment of such editing. But the possibility of genetic enhancement raises a number of additional ethical and social concerns, for which easy answers are not available and differences of opinion are likely.

Conclusions and Recommendations: Genome Editing for Purposes Other Than Treatment or Prevention of Disease

Significant scientific progress will be necessary before any genome-editing intervention for indications other than the treatment or prevention of disease or disability can satisfy the risk/benefit standards for initiating a clinical trial. This conclusion holds for both somatic and heritable germline interventions. There is significant public discomfort with the use of genome editing for so-called "enhancement" of human traits and capacities beyond those typical of adequate health. Therefore, a robust public discussion is needed concerning the values to be placed upon the individual and societal benefits and risks of genome editing for purposes other than treatment or prevention of disease or disability. These discussions would include consideration of the potential for introducing or exacerbating societal inequities, so that these values can be incorporated as appropriate into the risk/benefit assessments that will precede any decision about whether to authorize clinical trials.

RECOMMENDATION 6-1. Regulatory agencies should not at this time authorize clinical trials of somatic or germline genome editing for purposes other than treatment or prevention of disease or disability.

RECOMMENDATION 6-2. Government bodies should encourage public discussion and policy debate regarding governance of somatic human genome editing for purposes other than treatment or prevention of disease or disability.

The Role of Public Engagement in Governance of Human Genome Editing

Efforts to advance human medicine through genome editing will be strengthened by public engagement, and this engagement will be particularly critical for the potential uses that are not captured effectively by current regulatory frameworks. In the United States in particular, regulatory authority tends to focus primarily on health and safety of individuals and the public, and not on issues surrounding possible effects on social mores and culture. These latter concerns are regularly addressed in other fora, such as advisory committees, but lack legal force unless reflected in legislation that is grounded in the limited powers granted to government. Other countries have systems that more explicitly account for public attitudes

决策者明确在授权这种基因强化干预措施的临床试验之前，需要将个人和社会价值考虑在收益和风险中。

为了公众参与的有效性和包容性而在其他新兴科学和技术领域开发的做法及原则为公众参与基因组编辑提供了宝贵的基础。

建议 7-1：将人类基因组编辑技术用于任何超出治疗或预防疾病和残疾的临床试验之前，应该允许广泛的和包容性的公众参与。

建议 7-2：在公众广泛、持续参与和投入下，正在进行的关于健康、社会的益处和风险评估，应该优先考虑这种可遗传的生殖系基因组编辑的临床试验。

建议 7-3：公众参与应纳入人类基因组编辑的政策制定过程，并且应包括持续监测公众态度、信息缺失和对基因强化问题的急切关注。

建议 7-4：在资助人类基因组编辑的研究时，联邦机构应考虑资金支持以下近期研究和战略：
- 明确需要系统地和早期努力来征求公众参与的领域；
- 发展必要的内容并进行有效的沟通；
- 在现有的基础架构的背景下改善公众参与。

建议 7-5：当资助人类基因组编辑的研究时，联邦机构应考虑包括资助有以下目标的研究：
- 研究编辑人类生殖系的社会政治、伦理和法律方面；
- 研究超出疾病及缺陷的治疗或预防的基因组编辑用途的社会政治、伦理和法律方面；
- 评估建立公众沟通和参与纳入监管或政策制定基础架构的努力的有效性。

in deciding whether and how to permit new technologies to be developed, with widely varying degrees of legal constraint on governmental authority.

Conclusions and Recommendations: Public Engagement

Efforts to advance human medicine through genome editing will be strengthened by public engagement informed by technical experts and by social scientists who undertake systematic public opinion research, develop appropriate communication materials, and minimize artificial biases or constraints that would hinder discussion and debate.

Existing public communication and engagement infrastructures in the United States are sufficient to address oversight of basic science and laboratory research on human genome editing. Similarly, mechanisms for public communication and consultation that are part of the current U.S. regulatory infrastructures are also available to address public communication around development of human somatic-cell genome editing.

Weighing the technical and societal benefits and risks of applications of future uses of germline editing will require more formalized efforts to solicit broad public input and encourage public debate than are currently in place. Furthermore, the complex issues surrounding enhancement will require an ongoing public debate to inform regulators and policy makers about the individual and societal values to be placed on the benefits and risks before clinical trials for such enhancement interventions could be authorized.

The practices and principles developed for effective and inclusive public engagement on other emerging areas of science and technology provide a valuable base to inform public engagement on genome editing.

RECOMMENDATION 7-1. Extensive and inclusive public participation should precede clinical trials for any extension of human genome editing beyond treatment or prevention of disease or disability.

RECOMMENDATION 7-2. Ongoing reassessment of both health and societal benefits and risks, with broad ongoing participation and input by the public, should precede consideration of any clinical trials of heritable germline genome editing.

RECOMMENDATION 7-3. Public participation should be incorporated into the policy-making process for human genome editing and should include ongoing monitoring of public attitudes, informational deficits, and emerging concerns about issues surrounding "enhancement."

RECOMMENDATION 7-4. When funding human genome-editing research, federal agencies should consider including funding to support near-term research and strategies for

identifying areas that require systematic and early efforts to solicit public participation,

developing the necessary content and communicating it effectively, and improving public engagement within the context of existing

infrastructure.

RECOMMENDATION 7-5. When funding human genome-editing research, federal agencies should consider including funding for research aimed at

understanding the sociopolitical, ethical, and legal aspects of editing the human germline;

understanding the sociopolitical, ethical, and legal aspects of uses for genome editing that go beyond treatment or prevention of disease or disability; and

evaluating the efficacy of efforts to build public communication and engagement on these issues into regulatory or policy-making infrastructures.

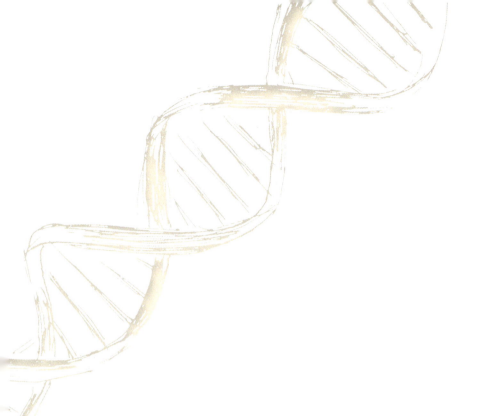

参考文献
References

AAP (American Academy of Pediatrics). 2014. Off-label use of drugs in children. *Pediatrics* 133(3): 563-567.

Abbott, K. W. A., D. J. Sylvester, and G. E. Marchant. 2010. Transnational regulation: Reality or romanticism? In *International handbook on regulating nanotechnologies,* edited by G. Hodge, D. Bowman, and A. Maynard. Cheltenham, UK: Edward Elgar Publishing. Pp. 525-544.

Abudayyeh, O. O., J. S. Gootenberg, S. Konermann, J. Joung, I. M. Slaymaker, D. B. Cox, S. Shmakov, K. S. Makarova, E. Semenova, L. Minakhin, K. Severinov, A. Regev, E. S. Lander, E. V. Koonin, and F. Zhang. 2016. C2c2 is a single-component programmable RNA-guided RNA-targeting CRISPR effector. *Science* 353(6299): aaf5573.

Achenbach, J.. 2016. 107 Nobel laureates sign letter blasting Greenpeace over GMOs. *The Washington Post*, June 29. https://www.washingtonpost.com/news/speaking-of-science/wp/2016/06/29/more-than-100-nobel-laureates-take-on-greenpeace-over-gmo-stance/?utm_term=.feb89580ad48 (accessed October 18, 2016).

Akin, H., K. M. Rose, D. A. Scheufele, M. Simis-Wilkinson, D. Brossard, M. A. Xenos, and E. A. Corley. 2017 *(forthcoming)*. Mapping the landscape of public attitudes on synthetic biology. *BioScience*. doi: 10.1093/biosci/biw171.

Alghrani, A., and M. Brazier. 2011. What is it? Whose it? Re-positioning the fetus in the context of research. *The Cambridge Law Journal* 70(1): 51-82.

American Cancer Society. 2015. *Off-label drug use: What is off-label drug use?* http://www.cancer.org/treatment/treatmentsandsideeffects/treatmenttypes/chemotherapy/off-label-drug-use (accessed January 5, 2017).

AMS (Academy of Medical Sciences), AMRC (Association of Medical Research Charities), BBSRC (Biotechnology and Biological Sciences Research Council), MRC (Medical Research Center), and Wellcome Trust. 2015. *Genome editing in human cells—initial joint statement.* https://wellcome.ac.uk/sites/default/files/wtp059707.pdf (accessed January 4, 2017).

AMS, British Academy, Royal Academy of Engineering, and The Royal Society. 2012. *Human enhancement and the future of work.* https://www.acmedsci.ac.uk/viewFile/publicationDownloads/135228646747.pdf (accessed January 5, 2017).

Anderson, A. A., J. Kim, D. A. Scheufele, D. Brossard, and M. A. Xenos. 2013. What's in a name? How we define nanotech shapes public reactions. *Journal of Nanoparticle Research* 15(2): 1-5.

Anderson, R. C., and R. W. Kulhavy. 1972. Learning concepts from definitions. *American Educational Research Journal* 9(3): 385-390.

Andorno, R. 2005. The Oviedo convention: A European legal framework at the intersection of human rights and health law. *Journal of International Biotechnology Law* 2(4): 133-143.

Anguela, X. M., R. Sharma, Y. Doyon, J. C. Miller, H. Li, V. Haurigot, M. E. Rohde, S. Y. Wong, R. J. Davidson, and S. Zhou. 2013. Robust ZFN-mediated genome editing in adult hemophilic mice. *Blood* 122(19): 3283-3287.

ANM (Académie Nationale de Médecine). 2016. *Genome editing of human germline cells and embryos.* http://www.academie-medecine.fr/wp-content/uploads/2016/05/report-genome-editing-ANM-2.pdf (accessed January 4, 2017).

Arras, J. 2016. Theory and bioethics. In *The Stanford encyclopedia of philosophy*, edited by E. N. Zalta. http://plato.stanford.edu/archives/sum2016/entries/theory-bioethics (accessed January 4, 2017).

Au, P., D. A. Hursh, A. Lim, M. C. Moos, S. S. Oh, B. S. Schneider, and C. M. Witten. 2012. FDA oversight of cell therapy clinical trials. *Science Translational Medicine* 4(149): 149fs31.

Ayyar, V. S. 2011. History of growth hormone therapy. *Indian Journal of Endocrinology and Metabolism* 15(Suppl. 3): S162-S165.

BAC (Bioethics Advisory Committee) Singapore. 2013. *Ethical, legal and social issues in neuroscience research: A consultation paper*. http://www.bioethics-singapore.org/index/publications/consultation-papers.html (accessed November 4, 2016).

Baltimore, D., P. Berg, M. Botchan, D. Carroll, R. A. Charo, G. Church, J. E. Corn, G. Q. Daley, J. A. Doudna, M. Fenner, H. T. Greely, M. Jinek, G. S. Martin, E. Penhoet, J. Puck, S. H. Sternberg, J. S. Weissman, and K. R. Yamamoto. 2015. A prudent path forward for genomic engineering and germline gene modification. *Science* 348(6230): 36-38.

Bamford, K. B., S. Wood, and R. J. Shaw. 2005. Standards for gene therapy clinical trials based on pro-active risk assessment in a London NHS Teaching Hospital Trust. *QJM: Monthly Journal of the Association of Physicians* 98(2): 75-86.

Barrangou, R., and E. G. Dudley. 2016. CRISPR-based typing and next-generation tracking technologies. *Annual Review of Food Science and Technology* 7: 395-411.

BBAW (Berlin-Brandenburg Academy of Sciences and Humanities). 2015. *Human genome surgery— towards a responsible evaluation of a new technology*. http://www.gentechnologiebericht.de/bilder/BBAW_Human-Genome-Surgery_PDF-A1b-1.pdf (accessed January 4, 2017).

Bell, J., K. L. Parker, R. D. Swinford, A. R. Hoffman, T. Maneatis, and B. Lippe. 2010. Long-term safety of recombinant human growth hormone in children. *Journal of Clinical Endocrinology & Metabolism* 95(1): 167-177.

Berg, P., and J. E. Mertz. 2010. Personal reflections on the origins and emergence of recombinant DNA technology. *Genetics* 184(1): 9-17.

Berg, P., D. Baltimore, S. Brenner, R. O. Roblin, and M. F. Singer. 1975. Asilomar conference on DNA recombinant molecules. *Science* 188(4192): 991-994.

Bertero, A., M. Pawlowski, D. Ortmann, K. Snijders, L. Yiangou, M. C. de Brito, S. Brown, W. G. Bernard, J. D. Cooper, E. Giacomelli, L. Gambardella, N. R. F. Hannan, D. Iyer, F. Sampaziotis, F. Serrano, M. C. F. Zonneveld, S. Sinha, M. Kotter, and L. Vallier. 2016. Optimized inducible shRNA and CRISPR/Cas9 platforms for *in vitro* studies of human development using hPSCs. *Development* 143: 4405-4418.

Bioethics Commission. 2015. *Gray matters: Topics at the intersection of neuroscience, ethics, and society*(Vol. 2) . Washington, DC: Presidential Commission for the Study of Bioethical Issues.

Blakeley, P., N. M. Fogarty, I. del Valle, S. E. Wamaitha, T. X. Hu, K. Elder, P. Snell, L. Christie, P. Robson, and K. K. Niakan. 2015. Defining the three cell lineages of the human blastocyst by single-cell RNA-seq. *Development* 142(18): 3151-3165.

Blendon, R.J., M. T. Gorski, and J. M. Benson, M.A. 2016. The public and the gene-editing revolution. *New England Journal of Medicine* 374(15): 1406-1411.

Boggio, A. 2005. Italy enacts new law on medically assisted reproduction. *Human Reproduction* 20(5): 1153-1157.

Booth, C., H. B. Gaspar, and A. J. Thrasher. 2016. Treating immunodeficiency through HSC gene therapy. *Trends in Molecular Medicine* 22(4): 317-327.

Borg, J. J., G. Aislaitner, M. Pirozynski, and S. Mifsud. 2011. Strengthening and rationalizing pharmacovigilance in the EU: Where is Europe heading to? A review of the new EU legislation on pharmacovigilance. *Drug Safety* 34(3): 187-197.

Bosley, K. S., M. Botchan, A. L. Bredenoord, D. Carroll, R. A. Charo, E. Charpentier, R. Cohen, J. Corn, J. Doudna, G. Feng, H. T. Greely, R. Isasi, W. Ji, J. -S. Kim, B. Knoppers, E. Lanphier, J. Li, R. Lovell- Badge, G. S. Martin, J. Moreno, L. Naldini, M. Pera, A. C. F. Perry, J. C. Venter, F. Zhang, and Q. Zhou. 2015. CRISPR germline engineering: The community speaks. *Nature Biotechnology* 33(5): 478-486.

Bostrom, N. 2005. In defense of posthuman dignity. *Bioethics* 19(3): 202-214.

Bostrom, N., and T. Ord. 2006. The reversal test: Eliminating status quo bias in applied ethics. *Ethics* 116(4): 656-679.

Brossard, D., and B. Lewenstein. 2009. A critical appraisal of model of public understanding of science: Using practice to inform theory. In *Understanding science: New agendas in science sommunication*, edited by L. Kahlor and P. Stout. New York: Routledge. Pp. 11-39.

Brown, K. V. 2016. Inside the garage labs of DIY gene hackers, whose hobby may terrify you. *Fusion,* March 29. http://fusion.net/story/285454/diy-crispr-biohackers-garage-labs (accessed January 9, 2016).

Buchanan, A., D. W. Brock, N. Daniels, and D. Wikler. 2001. *From chance to choice: Genetics and justice*. New York: Cambridge University Press.

Burgess M., and D. Prentice. 2016. Let Congress know to take it slow on human gene editing. *Dallas News*, December 28.

Cahill, L. S. 2008. *Germline genetics, human nature, and social ethics*. Cambridge, MA: MIT Press.

Califf, R. M. 2017. Benefit-risk assessments at the U.S. Food and Drug Administration: Finding the balance. *Journal of the American Medical Association*. http://jamanetwork.com/journals/jama/fullarticle/2599251 (accessed February 3, 2017).

Califf, R. M., and R. Nalubola. 2017. FDA's science-based approach to genome edited products. *FDA Voice*, January 18. http://blogs.

fda.gov/fdavoice/index.php/2017/01/fdas-science-based-approach-to-genome-edited-products (accessed February 2, 2017).

Campbell, A., and K. C. Glass. 2000. Legal status of clinical and ethics policies, codes, and guidelines in medical practice and research. *McGill Law Journal* 46(2): 473-489.

Carroll, D. 2014. Genome engineering with targetable nucleases. *Annual Review of Biochemistry* 83: 409-439.

Center for Genetics and Society. 2015. *Extreme genetic engineering and the human future: Reclaiming emerging biotechnologies for the common good.* http: //www.geneticsandsociety.org/downloads/Human_Future_Exec_Sum.pdf (accessed January 6, 2017).

Chan, J. L., L. N. C. Johnson, M. D. Sammel, L. DiGiovanni, C. Voong, S. M. Domchek, and C. R. Gracia. 2016. Reproductive decision-making in women with BRCA1/2 mutations. *Journal of Genetic Counseling* 1-10.

Chan, S., P. J. Donovan, T. Douglas, C. Gyngell, J. Harris, R. Lovell-Badge, D. J. Mathews, and A. Regenberg. 2015. Genome editing technologies and human germline genetic modification: The Hinxton Group consensus statement. *The American Journal of Bioethics* 15(12): 42-47.

Chapman, K. M., G. A. Medrano, P. Jaichander, J. Chaudhary, A. E. Waits, M. A. Nobrega, J. M. Hotaling, C. Ober, and F. K. Hamra. 2015. Targeted Germline Modifications in Rats Using CRISPR/Cas9 and Spermatogonial Stem Cells. *Cell Reports* 10(11): 1828-1835.

Charo, R. A. 2016a. On the road (to a cure?)—stem-cell tourism and lessons for gene editing. *New England Journal of Medicine* 374(10): 901-903.

Charo, R. A. 2016b. The legal and regulatory context for human gene editing. *Issues in Science and Technology* 32(3): 39.

Charpentier, E., and J. A. Doudna. 2013. Biotechnology: Rewriting a genome. *Nature* 494(7439): 50-51.

Chen, E. A., and J. F. Schiffman. 2000. Attitudes toward genetic counseling and prenatal diagnosis among a group of individuals with physical disabilities. *Journal of Genetic Counseling* 9(2): 137-152.

Cho, S. W., S. Kim, J. M. Kim, and J.-S. Kim. 2013. Targeted genome engineering in human cells with the Cas9 RNA-guided endonuclease. *Nature Biotechnology* 31(3): 230-232.

Choulika, A., A. Perrin, B. Dujon, and J.-P. Nicolas. 1995. Induction of homologous recombination in mammalian chromosomes by using the I-SceI system of Saccharomyces cerevisiae. *Molecular and Cellular Biology* 15(4): 1968-1973.

CIOMS (Council for International Organizations of Medical Sciences). 2012. *Bioethics.* http: //www.cioms.ch/index.php/2012-06-07-19-16-08/about-us/bioethics (accessed January 4, 2017).

CIRM (California Institute for Regenerative Medicine). 2015. *Scientific and medical accountability standards.* https: //www.cirm.ca.gov/our-funding/chapter-2-scientific-and-medical-accountability-standards (accessed January 4, 2017).

Cockburn, K., and J. Rossant. 2010. Making the blastocyst: Lessons from the mouse. *Journal of Clinical Investigation* 120(4): 995-1003.

COE (Council of Europe). 2016. *Chart of signatures and ratifications of Treaty 164.* http: //www.coe.int/en/web/conventions/full-list/-/conventions/treaty/164/signatures (accessed November 3, 2016).

Coghlan, A. 2016. Exclusive: "3-parent" baby method already used for infertility. *New Scientist*, October 10. https: //www.newscientist.com/article/2108549-exclusive-3-parent-baby-method-already-used-for-infertility (accessed November 3, 2016).

Coghlan, A. 2017. First baby born using three-parent technque to treat infertility. *New Scientist*, January 18. https: //www.newscientist.com/article/2118334-first-baby-born-using-3-parent-technique-to-treat-infertility (accessed January 19, 2017).

Cohen, I. G. 2015. *Patients and passports: Medical tourism, law, and ethics* (1st Edition). New York: Oxford University Press.

Cohen, P., G. M. Bright, A. D. Rogol, A. M. Kappelgaard, and R. G. Rosenfeld. 2002. Effects of dose and gender on the growth and growth factor response to GH in GH-deficient children: Implications for efficacy and safety. *Journal of Clinical Endocrinology & Metabolism* 87(1): 90-98.

Cole-Turner, R. 1993. *The new genesis: Theology and the genetic revolution.* Louisville, KY: Westminster/John Knox Press.

Cong, L., F. A. Ran, D. Cox, S. Lin, R. Barretto, N. Habib, P. D. Hsu, X. Wu, W. Jiang, and L. A. Marraffini. 2013. Multiplex genome engineering using CRISPR/Cas systems. *Science* 339(6121): 819-823.

Cook, D. M., and S. R. Rose. 2012. A review of guidelines for use of growth hormone in pediatric and transition patients. *Pituitary* 15(3): 301-310.

Cornwall, A. 2008. *Democratising engagement: What the U.K. can learn from international experience.* London, U.K.: Demos.

Corrigan-Curay, J. 2013. *NIH Recombinant DNA Advisory Committee (RAC) and gene transfer research.* Presentation at the First Meeting on Independent Review and Assessment of the Activities of the NIH Recombinant DNA Advisory Committee, Washington, DC, June 4. https: //www.nationalacademies.org/hmd/Activities/Research/ReviewNIHRAC/2013-JUN-04.aspx (accessed November 4, 2016).

Corrigan-Curay, J., M. O'Reilly, D. B. Kohn, P. M. Cannon, G. Bao, F. D. Bushman, D. Carroll, T. Cathomen, J. K. Joung, and D. Roth. 2015. Genome editing technologies: Defining a path to clinic. *Molecular Therapy* 23(5): 796-806.

Costantini, F., and E. Lacy. 1981. Introduction of a rabbit beta-globin gene into the mouse germ line. *Nature* 294(5836): 92-94.

Council of Europe. 2015. *Statement on genome editing technologies*. http: //www.coe.int/en/web/bioethics/-/gene-editing (accessed October 21, 2016).

Couzin-Frankel, J. 2016. Ailing fetuses to be treated with stem cells. *Science*, April 14. http: //www.sciencemag.org/news/2016/04/ailing-fetuses-be-treated-stem-cells (accessed January 5, 2017).

Cox, D. B. T., R. J. Platt, and F. Zhang. 2015. Therapeutic genome editing: Prospects and challenges. *Nature Medicine* 21(2): 121-131.

Cronin, R. 2008. Bureaucrats vs. physicians: Have doctors been stripped of their power to determine the proper use of human growth hormone in treating adult disease. *Washington University Journal of Law & Policy* 27: 191. http: //openscholarship.wustl.edu/cgi/viewcontent.cgi?article=1141&context=law_journal_law_po licy (accessed January 6, 2017).

Cuttler, L., and J. Silvers. 2010. Growth hormone and health policy. *The Journal of Clinical Endocrinology & Metabolism* 95(7): 3149-3153.

Cyranoski, D. 2016. Chinese scientists to pioneer first human CRISPR trial. *Nature* 535: 476-477.

Daniels, N. 2000. Normal functioning and the treatment-enhancement distinction. *Cambridge Quarterly of Healthcare Ethics* 9(3): 309-322.

Danish Board of Technology. 2006. *The consensus conference*. http: //www.tekno.dk/subpage.php3?article=468&toppic=kategori12&language=uk (accessed December 10, 2016).

Danish Board of Technology. 2010a. *A clear message from world citizens to COP15 politicians*. http: //www.tekno.dk/article/offentliggrelse -af-policy-rapport-om-wwviews-resultater-19-november/ (accessed December 11, 2016).

Danish Board of Technology. 2010b. *Profile of the Danish Board of Technology*. http: //www.tekno.dk/about-dbt-foundation/?lang=en (accessed December 11, 2016).

Davis, B. D. 1970. Prospects for genetic intervention in man. *Science* 170(3964): 1279-1283.

de la Noval, B. D. 2016. Potential implications on female fertility and reproductive lifespan in BRCA germline mutation women. *Archives of Gynecology and Obstetrics* 294(5): 1099-1103.

de Melo-Martín, I. 2012 A parental duty to use PGD: More than we bargained for? *American Journal of Bioethics* 12(4): 14-15.

De Ravin, S. S., X. Wu, S. Moir, L. Kardava, S. Anaya-O'Brien, N. Kwatemaa, P. Littel, N. Theobald, U. Choi, and L. Su. 2016. Lentiviral hematopoietic stem cell gene therapy for x-linked severe combined immunodeficiency. *Science Translational Medicine* 8(335): 335ra57.

DEA (Drug Enforcement Administration). 2013. *Human growth hormone (trade names: Genotropin®, Humatrope®, Norditropin®, Nutropin®, Saizen®, Serostim®)*. http: //www.deadiversion.usdoj.gov/drug_chem_info/hgh.pdf (accessed January 4, 2017).

Decruyenaere, M., G. Evers-Kiebooms, A. Boogaerts, K. Philippe, K. Demyttenaere, R. Dom, W. Vandenberghe, and J. P. Fryns. 2007. The complexity of reproductive decision-making in asymptomatic carriers of the Huntington mutation. *European Journal of Human Genetics: EJHG* 15(4): 453-462.

Deglincerti, A., G. F. Croft, L. N. Pietila, M. Zernicka-Goetz, E. D. Siggia, and A. H. Brivanlou. 2016. Self-organization of the *in vitro* attached human embryo. *Nature* 533(7602): 251-254.

Delaney, J. J. 2012 Revisiting the non-identity problem and the virtues of parenthood. *American Journal of Bioethics* 12(4): 24-26.

Dever, D. P., R. O. Bak, A. Reinisch, J. Camarena, G. Washington, C. E. Nicolas, M. Pavel -Dinu, N. Saxena, A. B. Wilkens, S. Mantri, N. Uchida, A. Hendel, A. Narla, R. Majeti, K. I. Weinberg, and M. H. Porteus. 2016. CRISPR/Cas9 β-globin gene targeting in human haematopoietic stem cells. *Nature* 539: 384-389.

Devereaux, M., and M. Kalichman. 2013. ESCRO committees—not dead yet. *The American Journal of Bioethics* 13(1): 59-60.

DeWitt, M. A., W. Magis, N. L. Bray, T. Wang, J. R. Berman, F. Urbinati, S.-J. Heo, T. Mitros, D. P. Muñoz, and D. Boffelli. 2016. Selection-free genome editing of the sickle mutation in human adult hematopoietic stem/progenitor cells. *Science Translational Medicine* 8(360): 360ra134-360ra134.

Dillman, D. A., J. D. Smyth, and L. M. Christian. 2014. *Internet, phone, mail, and mixed-mode surveys: The tailored design method* (4th Edition). Hoboken, NJ: Wiley.

Ding, Y., H. Li, L.-L. Chen, and K. Xie. 2016. Recent advances in genome editing using CRISPR/Cas9. *Frontiers in Plant Science* 7: 703.

Doudna, J. 2015. Perspective: Embryo editing needs scrutiny. *Nature* 528(7580): S6-S6.

Doudna, J. A., and E. Charpentier. 2014. The new frontier of genome engineering with CRISPR-Cas9. *Science* 346(6213).

DRZE (Deutsche Referenzzentrum für Ethik in den Biowissenschaften). 2016. *Selected national and international laws and regulations*. http: //www.drze.de/in-focus/stem-cell-research/laws-and-regulations (accessed October 25, 2016).

Dudding, T., B. Wilcken, B. Burgess, J. Hambly, and G. Turner. 2000. Reproductive decisions after neonatal screening identifies cystic fibrosis. *ADC Fetal & Neonatal Edition* 82(2): F124-F127.

East-Seletsky, A., M. R. O'Connell, S. C. Knight, D. Burstein, J. H. Cate, R. Tjian, and J. A. Doudna. 2016. Two distinct RNase activities of CRISPR-C2c2 enable guide-RNA processing and RNA detection. *Nature* 538(7624): 270-273.

Editing humanity. 2015. *The Economist*, August 22.

EGE (European Group on Ethics in Science and New Technologies). 2016. *Statement on gene editing*. https: //ec.europa.eu/research/ege/pdf/gene_editing_ege_statement.pdf (accessed January 5, 2017).

EMA (European Medicines Agency). 2006. *Guideline on non-clinical testing for inadvertent germline transmission of gene transfer vectors*. http: //www.ema.europa.eu/docs/en_GB/document_library/Scientific_guideline/2009/10/WC5000 03982.pdf (accessed February 2, 2017).

Enserink, M. 2016. Swedish academy seeks to stem "crisis of confidence" in wake of Macchiarini scandal. *Science Magazine*, February 11. http: //www.sciencemag.org/news/2016/02/swedish-academy-seeks-stem-crisis-confidence-wake-macchiarini-scandal (accessed January 5, 2017).

European Commission. 2012. *Ethics in public policy making: The case of human enhancement*. http: //cordis.europa.eu/result/rcn/153896_en.html (accessed November 4, 2016).

European Commission. 2016a. *Horizon 2020 Work Programme 2016-2017*. http: //ec.europa.eu/research/participants/data/ref/h2020/wp/2016_2017/main/h2020-wp1617-swfs_en.pdf (accessed January 6, 2017).

European Commission. 2016b. *Public engagement in responsible research and innovation*. https: //ec.europa.eu/programmes/horizon2020/en/h2020-section/public-engagement-responsible-research-and-innovation (accessed January 6, 2017).

European Commission. 2016c. *Voices*. http: //www.ecsite.eu/activities-and-services/projects/voices (accessed January 6, 2017).

Evans, J. H. 2002. *Playing god?: Human genetic engineering and the rationalization of public bioethical debate*. Chicago, IL: University of Chicago Press.

Evans, J. H. 2010. *Contested reproduction: Genetic technologies, religion, and public debate*. Chicago, IL: University of Chicago Press.

Ezkurdia, I., D. Juan, J. M. Rodriguez, A. Frankish, M. Diekhans, J. Harrow, J. Vazquez, A. Valencia, and M. L. Tress. 2014. Multiple evidence strands suggest that there may be as few as 19 000 human protein-coding genes. *Human Molecular Genetics* 23(22): 5866-5878.

FDA (U.S. Food and Drug Administration). 1991. Points to consider in human somatic cell therapy and gene therapy. *Human Gene Therapy* 2(3): 251-256.

FDA. 1993. Application of current statutory authorities to human somatic cell therapy products and gene therapy products; notice. *Federal Register* 58(197): 53248-53251. http: //www.fda.gov/downloads/BiologicsBloodVaccines/SafetyAvailability/UCM148113.pdf (accessed January 5, 2017).

FDA. 2000. *Guidance for industry: Formal meetings with sponsors and applicants for PDUFA products*. Rockville, MD: FDA. http: //www.fda.gov/OHRMS/DOCKETS/98fr/990296g2.pdf (accessed November 4, 2016).

FDA. 2001. *IND meetings for human drugs and biologics: Chemistry, manufacturing, and controls information*. Rockville, MD: FDA. http: //www.fda.gov/downloads/Drugs/GuidanceComplianceRegulatoryInformation/Guidances/uc m070568.pdf (accessed September 1, 2013).

FDA. 2006. *Guidance for industry: Gene therapy clinical trials—observing subjects for delayed adverse events*. Rockville, MD: FDA. http: //www.fda.gov/downloads/BiologicsBloodVaccines/GuidanceComplianceRegulatoryInforma tion/Guidances/CellularandGeneTherapy/ucm078719.pdf (accessed September 1, 2013).

FDA. 2009. *Cloning*. http: //www.fda.gov/BiologicsBloodVaccines/CellularGeneTherapyProducts/Cloning/default.htm (accessed January 5, 2017).

FDA. 2010. *Background document: Public hearing on the labeling of food made from the AquAdvantage salmon*. http: //www.fda.gov/downloads/Food/GuidanceRegulation/GuidanceDocumentsRegulatoryInform ation/LabelingNutrition/UCM223913.pdf (accessed January 5, 2017).

FDA. 2012a. *Guidance for industry: Preclinical assessment of investigational cellular and gene therapy products*. http: //www.fda.gov/BiologicsBloodVaccines/GuidanceComplianceRegulatoryInformation/Guida nces/CellularandGeneTherapy/ucm376136.htm (accessed February 3, 2017).

FDA. 2012b. *Import alert #66-71: Detention without physical examination of human growth hormone (HGH), also known as Somatropin*. http: //www.accessdata.fda.gov/cms_ia/importalert_204.html (accessed January 6, 2017).

FDA. 2012c. *Vaccine, blood, and biologics: SOPP 8101.1: Scheduling and conduct of regulatory review meetings with sponsors and applicants*. Rockville, MD: FDA. http: //www.fda.gov/BiologicsBloodVaccines/GuidanceComplianceRegulatoryInformation/Proced uresSOPPs/ucm079448.htm (accessed September 1, 2013).

FDA. 2015a. *Fast track, breakthrough therapy, accelerated approval, priority review*. http: //www.fda.gov/forpatients/approvals/fast/ucm20041766.htm (accessed January 4, 2017).

FDA. 2015b. *Guidance for industry: Considerations for the design of early-phase clinical trials of cellular and gene therapy products*. Rockville, MD: FDA. http://www.fda.gov/downloads/Biologi.../UCM359073.pdf (accessed February 1, 2017).

FDA. 2015c. *Modernizing the regulatory system for biotechnology products: First public meeting*. http://www.fda.gov/NewsEvents/MeetingsConferencesWorkshops/ucm463783.htm (accessed October 19, 2016).

FDA. 2016a. *Manufacturer communications regarding unapproved uses of approved or cleared medical products*. http://www.fda.gov/NewsEvents/MeetingsConferencesWorkshops/ucm489499.htm (accessed October 25, 2016).

FDA. 2016b. *Public hearing; request for comments—draft guidances relating to the regulation of human cells, tissues or cellular or tissue-based products*. http://www.fda.gov/BiologicsBloodVaccines/NewsEvents/WorkshopsMeetingsConferences/ucm 462125.htm (accessed October 26, 2016).

FEAM (Federation of European Academies of Medicine) and UKAMS (United Kingdom Academy of Medical Sciences). 2016. *Human genome editing in the EU*. http://www.acmedsci.ac.uk/more/events/human-genome-editing-in-the-eu (accessed January 5, 2017).

Fletcher, J. 1971. Ethical aspects of genetic controls: Designed genetic changes in man. *New England Journal of Medicine* 285(14): 776-783.

Fletcher, J. C., K. Berg, and K. E. Tranøy. 1985. Ethical aspects of medical genetics. *Clinical Genetics* 27(2): 199-205.

Flicker, L. S. 2012 Acting in the best interest of a child does not mean choosing the "best" child. *American Journal of Bioethics* 12(4): 29-31.

Frankel, M. S., and A. R. Chapman. 2000. *Human inheritable genetic modifications: Assessing scientific, ethical, religious, and policy issues*. Washington, DC: American Association for the Advancement of Sciences. https://www.aaas.org/sites/default/files/migrate/uploads/germline.pdf (accessed January 6, 2017).

Franklin, S. 2013. *Biological relatives: IVF, stem cells, and the future of kinship*. www.oapen.org/download?type=document&docid=469257 (accessed January 4, 2017).

Friedmann, T. 1989. Progress toward human gene therapy. *Science* 244(4910): 1275-1281.

Friedmann, T., P. Noguchi, and C. Mickelson. 2001. The evolution of public review and oversight mechanisms in human gene transfer research: Joint roles of the FDA and NIH. *Current Opinion in Biotechnology* 12(3): 304-307.

Friedmann, T., E. C. Jonlin, N. M. P. King, B. E. Torbett, N. A. Wivel, Y. Kaneda, and M. Sadelain. 2015. ASGCT and JSGT joint position statement on human genomic editing. *Molecular Therapy* 23(8): 1282.

Frum, T., and A. Ralston. 2015. Cell signaling and transcription factors regulating cell fate during formation of the mouse blastocyst. *Trends in Genetics* 31(7): 402-410.

Fu, Q., M. Hajdinjak, O. T. Moldovan, S. Constantin, S. Mallick, P. Skoglund, N. Patterson, N. Rohland, I. Lazaridis, and B. Nickel. 2015. An early modern human from Romania with a recent Neanderthal ancestor. *Nature* 524(7564): 216-219.

Gaj, T., S. J. Sirk, S.-L. Shui, and J. Liu. 2016. Genome-editing technologies: Principles and applications. *Cold Spring Harbor Perspectives in Biology*.

Gardner, R. L., and J. Rossant. 1979. Investigation of the fate of 4-5 day post-coitum mouse inner cell mass cells by blastocyst injection. *Journal of Embryology & Experimental Morphology* 52: 141-152.

Genovese, P., G. Schiroli, G. Escobar, T. Di Tomaso, C. Firrito, A. Calabria, D. Moi, R. Mazzieri, C. Bonini, and M. C. Holmes. 2014. Targeted genome editing in human repopulating haematopoietic stem cells. *Nature* 510(7504): 235-240.

George, B. 2011. Regulations and guidelines governing stem cell based products: Clinical considerations. *Perspectives in Clinical Research* 2(3): 94-99.

Giurgea, C. 1973. The "nootropic" approach to the pharmacology of the integrative activity of the brain 1, 2. *Conditional Reflex* 8(2): 108-115.

Goldsammler, M., and A. Jotkowitz. 2012 The ethics of PGD: What about the physician? *American Journal of Bioethics* 12(4): 28-29.

Goffman, E. 1974. *Frame analysis: An essay on the organization of experience*. New York: Harper & Row.

Green, R. 1994. The ethics of embryo research. *The Washington Post*, October 18.

Hadden, S. G. 1995. Regulatory negotiation as citizen participation: A critique. In *Fairness and competence in citizen participation*, edited by O. Renn, T. Webler, and P. Wiedemann. Dordrecht, Netherlands: Springer Science. Pp. 239-252.

Halevy, T., J. C. Biancotti, O. Yanuka, T. Golan-Lev, and N. Benvenisty. 2016. Molecular characterization of down syndrome embryonic stem cells reveals a role for RUNX1 in neural differentiation. *Stem Cell Reports* 7(4): 777-786.

Hamzelou, J. 2016. Exclusive: World's first baby born with new "3 parent" technique. *New Scientist*, September 27. https://www.newscientist.com/article/2107219-exclusive-worlds-first-baby-born-with-new-3-parent-technique (accessed November 3, 2016).

Harris, J. 2007. *Enhancing evolution: The ethical case for making people better*. Princeton, NJ: Princeton University Press.

Hashimoto, M., Y. Yamashita, and T. Takemoto. 2016. Electroporation of Cas9 protein/sgRNA into early pronuclear zygotes generates non-mosaic mutants in the mouse. *Developmental Biology* 418(1): 1-9.

Hayakawa, K., E. Himeno, S. Tanaka, and T. Kunath. 2014. Isolation and manipulation of mouse trophoblast stem cells. *Current Protocols in Stem Cell Biology* 1: 1E.4.

Hayashi, K., S. Ogushi, K. Kurimoto, S. Shimamoto, H. Ohta, and M. Saitou. 2012. Offspring from oocytes derived from *in vitro* primordial germ cell-like cells in mice. *Science* 338(6109): 971-975.

HDC (Halal Industry Development Corporation). 2016. *How does Islam view genetic engineering?* http: //www.hdcglobal.com/publisher/pid/b368dc7b-039b-4335-9df3-8c015cbb33af/container/contentId/cc170e96-408d-485d-8ec3-f63644df412c (accessed November 3, 2016).

Health Canada. 2016. *News release—government of Canada plans to introduce regulations to support the Assisted Human Reproduction Act.* http: //news.gc.ca/web/article-en.do?nid=1131339&tp=1 (accessed January 5, 2017).

Hennette-Vauchez, S. 2011. A human dignitas? Remnants of the ancient legal concept in contemporary dignity jurisprudence. *International Journal of Constitutional Law* 9(1): 32-57.

Hermann, B. P., M. Sukhwani, F. Winkler, J. N. Pascarella, K. A. Peters, Y. Sheng, H. Valli, M. Rodriguez, M. Ezzelarab, G. Dargo, K. Peterson, K. Masterson, C. Ramsey, T. Ward, M. Lienesch, A. Volk, D. K. Cooper, A. W. Thomson, J. E. Kiss, M. C. Penedo, G. P. Schatten, S. Mitalipov, and K. E. Orwig. 2012. Spermatogonial stem cell transplantation into rhesus testes regenerates spermatogenesis producing functional sperm. *Cell Stem Cell* 11(5): 715-726.

Hernandez, B., C. B. Keys, and F. E. Balcazar. Disability rights: Attitudes of private and public sector representatives. *Journal of Rehabilitation* 70(1): 28-37.

HFEA (Human Fertilisation and Embryology Authority). 2014. *Third scientific review of the safety and efficacy of methods to avoid mitochondrial disease through assisted conception: Update.* http: //www.hfea.gov.uk/docs/Third_Mitochondrial_replacement_scientific_review.pdf (accessed January 5, 2017).

HFEA. 2016a. *Guidance: Mitochondrial donation.* http: //www.hfea.gov.uk/9931.html (accessed November 3, 2016).

HFEA. 2016b. *U.K.'s independent expert panel recommends "cautious adoption" of mitochondrial donation in treatment.* http: //www.hfea.gov.uk/10559.html (January 4, 2017).

HHS (U.S. Department of Health and Human Services). 1979. *The Belmont Report* . https: //www.hhs.gov/ohrp/regulations-and-policy/belmont-report (January 5, 2017).

Hikabe, O., N. Hamazaki, G. Nagamatsu, Y. Obata, Y. Hirao, N. Hamada, S. Shimamoto, T. Imamura, K. Nakashima, and M. Saitou. 2016. Reconstitution *in vitro* of the entire cycle of the mouse female germ line. *Nature* 539: 299-303.

Hinxton Group. 2015. *Statement on genome editing technologies and human germline genetic modification.* http: //www.hinxtongroup.org/hinxton2015_statement.pdf (accessed July 21, 2016).

Hirsch, F., Y. Levy, and H. Chneiweiss. 2017. Crispr–cas9: A european position on genome editing. *Nature* 541(7635): 30-30. http: //dx.doi.org/10.1038/541030c

Hoban, M. D., S. H. Orkin, and D. E. Bauer. 2016. Genetic treatment of a molecular disorder: Gene therapy approaches to sickle cell disease. *Blood* 127(7): 839-848.

Hockemeyer, D., and Jaenisch, R. 2016. Induced pluripotent stem cells meet genome editing. *Cell Stem Cell* 18(5): 573-586.

Holdren, J. P., C. R. Sunstein, and I. A. Siddiqui. 2011. *Memorandum: Principles for regulation and oversight of emerging technologies.* https: //www.whitehouse.gov/sites/default/files/omb/inforeg/for-agencies/Principles-for-Regulation-and-Oversight-of-Emerging-Technologies-new.pdf (accessed January 6, 2017).

Howden, S. E., B. McColl, A. Glaser, J. Vadolas, S. Petrou, M. H. Little, A. G. Elefanty, and E. G. Stanley. 2016. A Cas9 variant for efficient generation of indel-free knockin or gene-corrected human pluripotent stem cells. *Stem Cell Reports* 7(3): 508-517.

Hsu, P. D., E. S. Lander, and F. Zhang. 2014. Development and applications of CRISPR-Cas9 for genome engineering. *Cell* 157(6): 1262-1278.

Huang, K., T. Maruyama, and G. Fan. 2014. The naive state of human pluripotent stem cells: A synthesis of stem cell and preimplantation embryo transcriptome analyses. *Cell Stem Cell* 15(4): 410-415.

Hubbard, N., D. Hagin, K. Sommer, Y. Song, I. Khan, C. Clough, H. D. Ochs, D. J. Rawlings, A. M. Scharenberg, and T. R. Torgerson. 2016. Targeted gene editing restores regulated CD40L function in X-linked hyper-IgM syndrome. *Blood* 127(21): 2513-2522.

Hughes, J. 2004. *Citizen cyborg: Why democratic societies must respond to the redesigned human of the future.* Cambridge, MA: Westview Press.

ICH (International Council for Harmonisation of Technical Requirements for Pharmaceuticals for Human Use). 2006. *General principles to address the risk of inadvertent germline integration of gene therapy vectors.* http: //www.ich.org/fileadmin/Public_Web_Site/ICH_Products/Consideration_documents/GTDG _Considerations_Documents/ICH_Considerations_General_Principles_Risk_of_IGI_GT_Vectors. pdf (accessed February 2, 2017).

Ichord, R. N. 2014. Adult stroke risk after growth hormone treatment in childhood first do no harm. *Neurology* 83(9): 776-777.

IGSR (Institute for Governmental Service and Research). 2016. *About the IGSR and the 1000 Genomes Project*. http: //www.internationalgenome.org/about (accessed November 4, 2016).

Inhorn, M. C. 2012. *The new Arab man: Emergent masculinities, technologies, and Islam in the Middle East*. Princeton, NJ: Princeton University Press.

IOM (Institute of Medicine). 2005. *Guidelines for human embryonic stem cell research* (Vol. 23). Washington DC: The National Academies Press.

IOM. 2014. *Oversight and review of clinical gene transfer protocols: Assessing the role of the Recombinant DNA Advisory Committee*. Washington, DC: The National Academies Press.

IOM. 2016. *Mitochondrial replacement techniques: Ethical, social, and policy considerations*. Washington, DC: The National Academies Press.

Irie, N., L. Weinberger, W. W. Tang, T. Kobayashi, S. Viukov, Y. S. Manor, S. Dietmann, J. H. Hanna, and M. A. Surani. 2015. SOX17 is a critical specifier of human primordial germ cell fate. *Cell* 160(1-2): 253-268.

ISSCR (International Society for Stem Cell Research). 2015. *Statement on human germline genome modification*. http: //www.isscr.org/home/about-us/news-press-releases/2015/2015/03/19/statement-on-human-germline-genome-modification (accessed June 15, 2016).

ISSCR. 2016a. *Guidelines for stem cell research and clinical translation*. http: //www.isscr.org/docs/default-source/guidelines/isscr-guidelines-for-stem-cell-research-and-clinical-translation.pdf?sfvrsn=2 (accessed January 5, 2017).

ISSCR. 2016b. *Updated guidelines for stem cell research and clinical translation*. http: //www.isscr.org/home/about-us/news-press-releases/2016/2016/05/12/isscr-releases -updated-guidelines-for-stem-cell-research-and-clinical-translation (accessed November 4, 2016).

Jasanoff, S., J. B. Hurlbut, and K. Saha. 2015. CRISPR democracy: Gene editing and the need for inclusive deliberation. *Issues in Science and Technology* 32(1): 25-32.

Jasin, M. 1996. Genetic manipulation of genomes with rare-cutting endonucleases. *Trends in Genetics* 12(6): 224-228.

Jinek, M., K. Chylinski, I. Fonfara, M. Hauer, J. A. Doudna, and E. Charpentier. 2012. A programmable dual-RNA-guided DNA endonuclease in adaptive bacterial immunity. *Science* 337(6096): 816-821.

Jinek, M., A. East, A. Cheng, S. Lin, E. Ma, and J. Doudna. 2013. RNA-programmed genome editing in human cells. *eLife* 2: e00471.

Juengst, E. T. 1991. Germ-line gene therapy: Back to basics. *Journal of Medicine and Philosophy* 16(6): 587-592.

Juengst, E. T. 1997. Can enhancement be distinguished from prevention in genetic medicine? *Journal of Medicine and Philosophy* 22(2): 125-142.

Juma, C. 2016. *Innovation and its enemies: Why people resist new technologies*. New York: Oxford University Press.

Kahneman, D., and A. Tversky. 1984. Choices, values, and frames. *American Psychologist* 39(4): 341-350.

Kajaste-Rudnitski, A., and L. Naldini. 2015. Cellular innate immunity and restriction of viral infection: Implications for lentiviral gene therapy in human hematopoietic cells. *Human Gene Therapy* 26(4): 201-209.

Kasowski, M., F. Grubert, C. Heffelfinger, M. Hariharan, A. Asabere, S. M. Waszak, L. Habegger, J. Rozowsky, M. Shi, and A. E. Urban. 2010. Variation in transcription factor binding among humans. *Science* 328(5975): 232-235.

Kemp, S. F., J. Kuntze, K. M. Attie, T. Maneatis, S. Butler, J. Frane, and B. Lippe. 2005. Efficacy and safety results of long-term growth hormone treatment of idiopathic short stature. *Journal of Clinical Endocrinology & Metabolism* 90(9): 5247-5253.

Kessler, D. P. 2011. Evaluating the medical malpractice system and options for reform. *The Journal of Economic Perspectives* 25(2): 93-110.

Kessler, D. A., J. P. Siegel, P. D. Noguchi, K. C. Zoon, K. L. Feiden, and J. Woodcock. 1993. Regulation of somatic -cell therapy and gene therapy by the Food and Drug Administration. *New England Journal of Medicine* 329(16): 1169-1173.

Kevles, D. J. 1985. *In the name of eugenics: Genetics and the uses of human heredity*. Cambridge, MA: Harvard University Press.

Kitcher, P. 1997. *The lives to come: The genetic revolution and human possibilities*. New York: Simon & Shuster.

Klitzman, R. 2017. Buying and selling human eggs: Infertility providers' ethical and other concerns regarding egg donor agencies. *BMC Medical Ethics* 17(1): 71.

Kohn, D. B., M. H. Porteus, and A. M. Scharenberg. 2016. Ethical and regulatory aspects of genome editing. *Blood* 127(21): 2553-2560.

Konermann, S., M. D. Brigham, A. E. Trevino, J. Joung, O. O. Abudayyeh, C. Barcena, P. D. Hsu, N. Habib, J. S. Gootenberg, H. Nishimasu, O. Nureki, and F. Zhang. 2015. Genome-scale transcriptional activation by an engineered CRISPR-Cas9 complex. *Nature* 517: 583-588.

Krukenberg, R. C., D. L. Koller, D. D. Weaver, J. N. Dickerson, and K. A. Quaid. 2013. Two decades of Huntington disease testing: Patient's demographics and reproductive choices. *Journal of Genetic Counseling* 22(5): 643-653.

Kuhlmann, I., A. M. Minihane, P. Huebbe, A. Nebel, and G. Rimbach. 2010. Apolipoprotein E genotype and hepatitis C, HIV and herpes simplex disease risk: A literature review. *Lipids in Health and Disease* 9: 8.

Ladd, R., and E. Forman. 2012. A duty to use IVF? *American Journal of Bioethics* 12(4): 21-22. Lanphier, E., F. Urnov, S. E. Haecker, M. Werner, and J. Smolenski. 2015. Don't edit the human germ line. *Nature* 519(7544): 410-411.

Laventhal, N., and M. Constantine. 2012 The harms of a duty: Misapplication of the best interest standard. *American Journal of Bioethics* 12(4): 17-19.

Lazaraviciute, G., M. Kauser, S. Bhattacharya, P. Haggarty, and S. Bhattacharya. 2014. A systematic review and meta-analysis of DNA methylation levels and imprinting disorders in children conceived by IVF/ICSI compared with children conceived spontaneously. *Human Reproduction Update* 20(6): 840-852.

Le Page, M. 2016. Exclusive: Mexico clinic plans 20 "three-parent" babies in 2017. *New Scientist*, December 9. https: //www.newscientist.com/article/2115731-exclusive-mexico-clinic-plans-20-three-parent-babies-in-2017 (accessed January 4, 2017).

Ledford, H. 2015. Biohackers gear up for genome editing. *Nature* 524(7566): 398-399.

Leopoldina. 2015. *The opportunities and limits of genome editing*. http: //www.leopoldina.org/nc/en/publications/detailview/?publication%5bpublication%5d=699& cHash=4d49c84a36e655feacc1be6ce7f98626 (accessed January 6, 2017).

Leshner, A. I. 2003. Public engagement with science. *Science* 299(5609): 977. http: //www.sciencemag.org/content/299/5609/977.short (accessed January 6, 2017).

Lindgren, A. C., and E. M. Ritzen. 1999. Five years of growth hormone treatment in children with Prader-Willi syndrome. Swedish National Growth Hormone Advisory Group. *Acta Paediatrica Supplement* 88(433): 109-111.

Liu, H., D. M. Bravata, I. Olkin, S. Nayak, B. Roberts, A. M. Garber, and A. R. Hoffman. 2007. Systematic review: The safety and efficacy of growth hormone in the healthy elderly. *Annals of Internal Medicine* 146(2): 104-115.

Liu, H., D. M. Bravata, I. Olkin, A. Friedlander, V. Liu, B. Roberts, E. Bendavid, O. Saynina, S. R. Salpeter, and A. M. Garber. 2008. Systematic review: The effects of growth hormone on athletic performance. *Annals of Internal Medicine* 148(10): 747-758.

Lomax, G. P., and A. O. Trounson. 2013. Correcting misperceptions about cryopreserved embryos and stem cell research. *Nature Biotechnology* 31(4): 288-290.

Lombardo, P. A. 2008. *Three generations, no imbeciles: Eugenics, the Supreme Court, and Buck v. Bell*. Baltimore, MD: JHU Press.

Long, C., J. R. McAnally, J. M. Shelton, A. A. Mireault, R. Bassel-Duby, and E. N. Olson. 2014. Prevention of muscular dystrophy in mice by CRISPR/Cas9-mediated editing of germline DNA. *Science* 345(6201): 1184-1188.

Long, C., L. Amoasii, A. A. Mireault, J. R. McAnally, H. Li, E. Sanchez-Ortiz, S. Bhattacharyya, J. M. Shelton, R. Bassel-Duby, and E. N. Olson. 2016. Postnatal genome editing partially restores dystrophin expression in a mouse model of muscular dystrophy. *Science* 351(6271): 400-403.

Longmore, P. K. 1995. Medical decision making and people with disabilities: A clash of cultures. *Journal of Law, Medicine & Ethics* 23(1): 82-87.

Lu, Y.-H., N. Wang, and F. Jin. 2013. Long- term follow-up of children conceived through assisted reproductive technology. *Journal of Zhejiang University. Science B* 14(5): 359-371.

Lyon, J. 2017. Sanctioned U.K. trial of mitochondrial transfer nears. *Journal of the American Medical Association*. http: //jamanetwork.com/journals/jama/fullarticle/2599746 (accessed February 3, 2017).

Macer, D. R., S. Akiyama, A. T. Alora, Y. Asada, J. Azariah, H. Azariah, M. V. Boost, P. Chatwachirawong, Y. Kato, V. Kaushik, F. J. Leavitt, N. Y. Macer, C. C. Ong, P. Srinives, and M. Tsuzuki. 1995. International perceptions and approval of gene therapy. *Human Gene Therapy* 6(6): 791-803.

Macer, D. R. J. 2008. Public acceptance of human gene therapy and perceptions of human genetic manipulation. *Human Gene Therapy* 3(5): 511-518.

Machalek, A. Z. 2009. *Comparing genomes to find what makes us human*. https: //publications.nigms.nih.gov/computinglife/compare_genome.htm (accessed November 3, 2016).

Majumder, M. A. 2012. More mud, less crystal? Ambivalence, disability, and PGD. *American Journal of Bioethics* 12(4): 26-28.

Mak, T. W. 2007. Gene targeting in embryonic stem cells scores a knockout in Stockholm. *Cell* 131(6): 1027-1031.

Makas, E. 1988. Positive attitudes toward disabled people: Disabled and nondisabled persons' perspectives. *Journal of Social Issues* 44(1): 49-61.

Malek, J., and Daar, J. 2012. The case for a parental duty to use preimplantation genetic diagnosis for medical benefit. *American Journal of Bioethics* 12(4): 3-11.

Mali, P., L. Yang, K. M. Esvelt, J. Aach, M. Guell, J. E. DiCarlo, J. E. Norville, and G. M. Church. 2013. RNA-guided human genome engineering via Cas9. *Science* 339(6121): 823-826.

Malkki, H. 2016. Huntington disease: Selective deactivation of Huntington disease mutant allele by CRISPR-Cas9 gene editing. *Nature Reviews Neurology* 12(11): 614-615.

Malm, H. 2012. Moral duty in the use of preimplantation genetic diagnosis. *American Journal of Bioethics* 12(4): 19-21.

Maresca, M., V. G. Lin, N. Guo, and Y. Yi Yang. 2013. Obligate Ligation-Gated Recombination (ObLiGaRe): Custom-designed nuclease-mediated targeted integration through nonhomologous end joining. *Genome Research* 23(3): 539-546.

Margottini, L. 2014. Final chapter in Italian stem cell controversy? *Science*, October 7.

Martin, A. K., and B. Baertschi. 2012. In favor of PGD: The moral duty to avoid harm argument. *American Journal of Bioethics* 12(4): 12-13.

Mawer, S. 1998. *Mendel's dwarf*. New York: Harmony Books.

Maxmen, A. 2015. Easy DNA editing will remake the world. Buckle up. *Wired*, July 2015.

McClain, L. E., and A. W. Flake. 2016. *In utero* stem cell transplantation and gene therapy: Recent progress and the potential for clinical application. *Best Practice & Research Clinical Obstetrics & Gynaecology* 31: 88-98.

Meilaender, G. 1996. Begetting and cloning. *First Things (New York, NY)* 74: 41-43.

Meilaender, G. 2008. *Human dignity: Exploring and explicating the council's vision*. https://bioethicsarchive.georgetown.edu/pcbe/reports/human_dignity/chapter11.html (accessed January 6, 2017).

Mello, M. M. 2001. Of swords and shields: The role of clinical practice guidelines in medical malpractice litigation. *University of Pennsylvania Law Review* 149(3): 645-710.

Merkle, D. M. 1996. The polls—review—the National Issues Convention deliberative poll. *Public Opinion Quarterly* 60(4): 588-619.

Molitch, M. E., D. R. Clemmons, S. Malozowski, G. R. Merriam, and M. L. Vance. 2011. Evaluation and treatment of adult growth hormone deficiency: An Endocrine Society clinical practice guideline. *The Journal of Clinical Endocrinology & Metabolism* 96(6): 1587-1609.

More, M. 1990. Transhumanism: Towards a futurist philosophy. In *Extropy*, 6th ed. https://www.scribd.com/doc/257580713/Transhumanism-Toward-a-Futurist-Philosophy (accessed February 2, 2017).

Mullin, E. 2016. Despite the hype over gene therapy, few drugs are close to approval. *MIT Technology Review*, September 29. https://www.technologyreview.com/s/602467/despite-the-hype-over-gene-therapy-few-drugs-are-close-to-approval (accessed January 22, 2017).

Murray, M., and K. Luker. 2015. *Cases on reproductive rights and justice*. St. Paul, MN: West Academic Publishing.

Naldini L. 2015. Gene therapy returns to centre stage. *Nature* 526(7573): 351-360.

NAS (National Academy of Sciences). 2002. *Scientific and medical aspects of human reproductive cloning*. Washington, DC: The National Academies Press.

NASEM (National Academies of Sciences, Engineering, and Medicine). 2016a. *Communicating science effectively: A research agenda*. Washington, DC: The National Academies Press.

NASEM. 2016b. *Gene drives on the horizon: Advancing science, navigating uncertainty, and aligning research with public values*. Washington, DC: The National Academies Press.

NASEM. 2016c. *Genetically engineered crops: Experiences and prospects*. Washington, DC: The National Academies Press.

NASEM. 2016d. *International summit on human gene editing: A global discussion*. Washington, DC: The National Academies Press.

NASEM. 2016e. *Mitochondrial replacement techniques: Ethical, social, and policy considerations*. Washington, DC: The National Academies Press.

Nature. 2017. Why researchers should resolve to engage in 2017. Nature Editorial. *Nature* 541(5). http://www.nature.com/news/why-researchers-should-resolve-to-engage-in-2017-1.21236?WT.mc_id=FBK_NatureNews (accessed January 12, 2017).

NCECHLS (National Consultative Ethics Committee for Health and Life Sciences) . 2013. *Opinion no. 122: The use of biomedical techniques for "neuroenhancement" in healthy individuals: Ethical issues*. http://www.ccne-ethique.fr/sites/default/files/publications/ccne.avis_ndeg122eng.pdf (accessed January 5, 2017).

NCSL (National Conference of State Legislatures). 2016. *Embryonic and fetal research laws, January 1, 2016*. http://www.ncsl.org/research/health/embryonic-and-fetal-research-laws.aspx (accessed October 31, 2016).

Nelson, C. E., C. H. Hakim, D. G. Ousterout, P. I. Thakore, E. A. Moreb, R. M. C. Rivera, S. Madhavan, X. Pan, F. A. Ran, and W. X. Yan. 2016. *In vivo* genome editing improves muscle function in a mouse model of Duchenne muscular dystrophy. *Science* 351(6271): 403-407.

Nelson, E. 2013. *Law, policy and reproductive autonomy*. Portland, OR: Hart Publishing.

NIH (National Institutes of Health). 1994. *Report of the Human Embryo Research Panel*. https://repository.library.georgetown.edu/bitstream/handle/10822/559352/human_embryo_vol_1.pdf?sequence=1&isAllowed=y (accessed January 5, 2017).

NIH. 2004. NIH and FDA launch new human gene transfer research data system. *Journal of Investigative Medicine* 52(5): 286. http://jim.bmj.com/content/jim/52/5/286.1.full.pdf (accessed November 4, 2016).

NIH. 2011. *Charter: Recombinant DNA Advisory Committee*. Bethesda, MD: NIH. http://oba.od.nih.gov/oba/RAC/Signed_RAC_Charter_2011.pdf (accessed October 1, 2013).

NIH. 2013a. *Frequently asked questions about the NIH review process for human gene transfer trials*. Bethesda, MD: NIH. http://

oba.od.nih.gov/oba/ibc/FAQs/NIH_Review_Process_HGT.pdf (accessed September 1, 2013).

NIH. 2013b. *Frequently asked questions of interest to IBCs*. Bethesda, MD: NIH. http: //oba.od.nih.gov/oba/ibc/FAQs/IBC_Frequently_Asked_Questions7.24.09.pdf (accessed October 1, 2013).

NIH. 2013c. *NIH guidelines for research involving recombinant or synthetic nucleic acid molecules*. Bethesda, MD: NIH.

NIH. 2015a. *NIH research involving introduction of human pluripotent cells into non-human vertebrate animal pre-gastrulation embryos*. NOT-OD-15-158. https: //grants.nih.gov/grants/guide/notice-files/NOT-OD-15-158.html (accessed January 5, 2017).

NIH. 2015b. *Statement on NIH funding of research using gene-editing technologies in human embryos*. https: //www.nih.gov/about-nih/who-we-are/nih-director/statements/statement -nih-funding-research-using-gene-editing-technologies-human-embryos (accessed January 5, 2017).

NIH. 2016a. *NIH guidelines for research involving recombinant or synthetic nucleic acid molecules*. Washington, DC: HHS.

NIH. 2016b. *Request for public comment on the proposed changes to the NIH guidelines for human stem cell research and the proposed scope of an NIH steering committee's consideration of certain human-animal chimera research*. https: //grants.nih.gov/grants/guide/notice-files/NOT-OD-16-128.html (accessed November 4, 2016).

NIH. 2016c. *State initiatives for stem cell research*. https: //stemcells.nih.gov/research/state-research.htm (accessed October 25, 2016).

NIH. 2016d. *Stem cell policy*. https: //stemcells.nih.gov/policy.htm (accessed October 25, 2016). NRC (National Research Council). 1996. *Understanding risk: Informing decisions in a democratic society*. Washington, DC: National Academy Press.

NRC. 2008. *Public participation in environmental assessment and decision making*. Washington, DC: The National Academies Press.

NRC. 2010. *Final report of the National Academies' Human Embryonic Stem Cell Research Advisory Committee and 2010 amendments to the National Academies' guidelines for human embryonic stem cell research*. Washington, DC: The National Academies Press.

NRC and IOM. 2007. *2007 amendments to the National Academies' guidelines for human embryonic stem cell research*. Washington, DC: The National Academies Press.

NRC and IOM. 2008. *2008 amendments to the National Academies' guidelines for human embryonic stem cell research*. Washington, DC: The National Academies Press.

NRC and IOM. 2010. *Final report of the National Academies' Human Embryonic Stem Cell Research Advisory Committee and 2010 amendments to the National Academies' guidelines for human embryonic stem cell research*. Washington, DC: The National Academies Press.

NSF (National Science Foundation). 2010. Ethics of human enhancement: 25 questions & answers. *Studies in Ethics, Law, and Technology* 4(1): 1-49.

Nuffield Council. 2015. *Naturalness*. http: //nuffieldbioethics.org/project/naturalness (accessed November 3, 2016).

Nuffield Council. 2016a. *Genome editing: An ethical review*. http: //nuffieldbioethics.org/project/genome-editing/ethical-review-published-september-2016 (accessed January 6, 2017).

Nuffield Council. 2016b. *Public dialogue on genome editing: Why? When? Who?:* http: //nuffieldbioethics.org/wp- content/uploads/Public-Dialogue-on-Genome-Editing-workshop-report.pdf (accessed January 6, 2017).

OBA (Office of Biotechnology Activities). 2013. *Office of Biotechnology Activities welcome page*. http: //oba.od.nih.gov (accessed September 1, 2013).

O'Connor, K. 2012. Ethics of fetal surgery. In *The Embryo Project Encyclopedia*. Tempe, AZ: The Embryo Project at Arizona State University.

Oktay, K., V. Turan, S. Titus, R. Stobezki, and L. Liu. 2015. BRCA mutations, DNA repair deficiency, and ovarian aging. *Biology of Reproduction* 93(3): 67.

O'Reilly, M., A. Shipp, E. Rosenthal, R. Jambou, T. Shih, M. Montgomery, L. Gargiulo, A. Patterson, and J. Corrigan-Curay. 2012. NIH oversight of human gene transfer research involving retroviral, lentiviral, and adeno-associated virus vectors and the role of the NIH Recombinant DNA Advisory Committee. In *Gene Transfer Vectors for Clinical Application*, Vol. 507, edited by F. Theodore. Bethesda, MD: Academic Press. Pp. 313-335.

Orr-Weaver, T. L., J. W. Szostak, and R. J. Rothstein. 1981. Yeast transformation: A model system for the study of recombination. *Proceedings of the National Academy of Sciences of the United States of America* 78(10): 6354-6358.

Parens, E. 1995. Should we hold the (germ) line? *The Journal of Law, Medicine & Ethics* 23(2): 173-176.

Parens, E. 1998. *Enhancing human traits: Ethical and social implications*. Washington, DC: Georgetown University Press.

Parens, E., and A. Asch. 2000. *Prenatal testing and disability rights*. Washington, DC: Georgetown University Press.

Pera, M. F. 2014. In search of naivety. *Cell Stem Cell* 15(5): 543-545.

Perls, T., and D. J. Handelsman. 2015. Disease mongering of age-associated declines in testosterone and growth hormone levels. *Journal of the American Geriatrics Society* 63(4): 809-811.

Persson, I., and J. Savulescu. 2012. *Unfit for the future: The need for moral enhancement.* Oxford, U.K.: Oxford University Press.

Petropoulos, S., D. Edsgard, B. Reinius, Q. Deng, S. P. Panula, S. Codeluppi, A. Plaza Reyes, S. Linnarsson, R. Sandberg, and F. Lanner. 2016. Single-cell RNA-seq reveals lineage and X chromosome dynamics in human preimplantation embryos. *Cell* 165(4): 1012-1026.

Pew Research Center. 2008. Stem cell research at the crossroads of religion and politics. *Pew Forum on Religion & Public Life*, July 17. http://www.pewforum.org/2008/07/17/stem-cell-research-at-the-crossroads-of-religion-and-politics (accessed July 17, 2008).

Pew Research Center. 2016. *U.S. public wary of biomedical technologies to enhance human abilities.* http://www.pewinternet.org/2016/07/26/u-s-public-wary-of-biomedical-technologies-to-enhance-human-abilities (accessed January 5, 2015).

Pfleiderer, G., G. Brahier, and K. Lindpaintner. 2010. *Genethics and religion.* Basel, Switzerland: Karger Medical and Scientific Publishers.

Pickering, F. L., and A. Silvers. 2012. A wrongful case for parental tort liability. *American Journal of Bioethics* 12(4): 15-17.

Plotz, D. 2006. *The genius factory: The curious history of the Nobel Prize Sperm Bank.* New York: Random House, Inc.

Poirot, L., B. Philip, C. Schiffer-Mannioui, D. Le Clerre, I. Chion-Sotinel, S. Derniame, P. Potrel, C. Bas, L. Lemaire, R. Galetto, C. Lebuhotel, J. Eyquem, G. W.-K. Cheung, A. Duclert, A. Gouble, S. Arnould, K. Peggs, M. Pule, A. M. Scharenberg, and J. Smith. 2015. Multiplex Genome-Edited T-cell Manufacturing Platform for "Off-the-Shelf" Adoptive T-cell Immunotherapies. *Cancer Research* 75(18): 3853-3864.

Pollard, K. S. 2016. Decoding human accelerated regions. *The Scientist*, August 1. http://www.the-scientist.com/?articles.view/articleNo/46643/title/Decoding-Human-Accelerated-Regions (accessed November 3, 2016).

Poolman, E. M., and A. P. Galvani. 2007. Evaluating candidate agents of selective pressure for cystic fibrosis. *Journal of the Royal Society, Interface* 4(12): 91-98.

Porteus, M. 2016. Genome editing: A new approach to human therapeutics. *Annual Review of Pharmacology and Toxicology* 56: 163-190.

Posner, S. M., E. McKenzie, and T. H. Ricketts. 2016. Policy impacts of ecosystem services knowledge. *Proceedings of the National Academy of Sciences of the United States of America* 113(7): 1760-1765.

Powledge, T. M., and L. Dach, eds. 1977. *Biomedical research and the public: Report to the Subcommittee on Health and Scientific Research, Committee on Human Resources, U.S. Senate.* Washington, DC: U.S. Government Printing Office.

Präg, P., and M. C. Mills. 2015. Assisted reproductive technology in Europe. Usage and regulation in the context of cross-border reproductive care. *Families and Societies* 43(1-23). http://www.familiesandsocieties.eu/wp-content/uploads/2015/09/WP43PragMills2015.pdf (accessed January 6, 2017).

President's Commission. 1982. *Splicing life: A report on the social and ethical issues of genetic engineering with human beings.* Washington, DC: President's Commission for the Study of Ethical Problems in Medicine and Biomedical and Behavioral Research. https://bioethics.georgetown.edu/documents/pcemr/splicinglife.pdf (accessed January 6, 2017).

President's Commission. 1983. *Deciding to forego life-sustaining treatment: A report on the ethical, medical and legal issues in treatment decisions.* Washington, DC: U.S. Government Printing Office.

President's Council on Bioethics. 2003. *Beyond therapy.* Washington, DC: President's Council on Bioethics.

Qasim, W., H. Zhan, S. Samarasinghe, S. Adams, P. Amrolia, S. Stafford, K. Butler, C. Rivat, G. Wright, and K. Somana. 2017. Molecular remission of infant B-ALL after infusion of universal TALEN gene-edited CAR T cells. *Science Translational Medicine* 9(374): eaaj2013.

Qi, L. S., M. H. Larson, L. A. Gilbert, J. A. Doudna, J. S. Weissman, A. P. Arkin, and W. A. Lim. 2013. Repurposing CRISPR as an RNA-guided platform for sequence-specific control of gene expression. *Cell* 152(5): 1173-1183.

Qiao, J., and H. L. Feng. 2014. Assisted reproductive technology in China: Compliance and non-compliance. *Translational Pediatrics* 3(2): 91.

Quinn, G. P., S. T. Vadaparampil, S. Tollin, C. A. Miree, D. Murphy, B. Bower, and C. Silva. 2010. BRCA carriers' thoughts on risk management in relation to preimplantation genetic diagnosis and childbearing: When too many choices are just as difficult as none. *Fertility and Sterility* 94(6): 2473-2475.

Rainsbury, J. 2000. Biotechnology on the RAC-FDA/NIH regulation of human gene therapy. *Food and Drug Law Journal* 55: 575-600.

Ramsey, P. 1970. *Fabricated man: The ethics of genetic control* (Vol. 6). New Haven, CT: Yale University Press.

Rawls, J. 1999. *A Theory of Justice,* 2nd ed. Cambridge, MA: Belknap Press.

Reardon, S. 2016. First CRISPR clinical trial gets green light from U.S. panel. *Nature News*, June 22. http://www.nature.com/news/first-crispr-clinical-trial-gets-green-light-from-us-panel-1.20137 (accessed June 22, 2016).

Reeves, R. 2016. Second gene therapy wins approval in Europe. *BioNews*, Issue 854. http://www.bionews.org.uk/page_656625.asp (accessed February 2, 2017).

Regalado, A. 2015. Engineering the perfect baby. *MIT Technology Review*, March 5.

Rine, J., and A. P. Fagen. 2015. The state of federal research funding in genetics as reflected by members of the Genetics Society of America. *Genetics* 200(4): 1015-1019.

Robertson, J. A. 2004. Procreative liberty and harm to offspring in assisted reproduction. *American Journal of Law & Medicine* 30(1): 7-40.

Robertson, J. A. 2008. Assisting reproduction, choosing genes, and the scope of reproductive freedom. *George Washington Law Review* 76(6): 1490-1513.

Robillard, J. M., D. Roskams-Edris, B. Kuzeljevic, and J. Illes. 2014. Prevailing public perceptions of the ethics of gene therapy. *Human Gene Therapy* 25(8): 740-746.

Rosen, L. 2014. What Mars One needs is genetically altered human colonists. *h+ Magazine*, April 14. http://hplusmagazine.com/2014/04/14/what-mars-one-needs-is-genetically-altered-human-colonists (accessed January 4, 2017).

Rossant, J. 2015. Mouse and human blastocyst-derived stem cells: Vive les differences. *Development* 142(1): 9-12.

Rossetti, M., M. Cavarelli, S. Gregori, and G. Scarlatti. 2012. HIV-derived vectors for gene therapy targeting dendritic cells. In *HIV interactions with dendritic cells*. New York: Springer. Pp. 239-261.

Roux, P., F. Smih, and M. Jasin. 1994a. Expression of a site-specific endonuclease stimulates homologous recombination in mammalian cells. *Proceedings of the National Academy of Sciences of the United States of America* 91(13): 6064-6068.

Roux, P., F. Smih, and M. Jasin. 1994b. Introduction of double-strand breaks into the genome of mouse cells by expression of a rare-cutting endonuclease. *Molecular and Cellular Biology* 14(12): 8096-8106.

Rowe, G., and L. J. Frewer. 2005. A typology of public engagement mechanisms. *Science Technology & Human Values* 30(2): 251-290.

Rulli, T. 2014. Preferring a genetically-related child. *Journal of Moral Philosophy* 13(6): 669-698.

Saitou, M., and H. Miyauchi. 2016. Gametogenesis from pluripotent stem cells. *Cell Stem Cell* 18(6): 721-735.

Sandel, M. 2004. The case against perfection. *The Atlantic Monthly* 293(3): 51-62.

Sandel, M. 2013. The case against perfection. In *Society, ethics, and technology*, 5th ed., edited by M. Winston and R. Edelbach. Boston, MA: Wadsworth, Cengage Learning. Pp. 343-354.

Sander, J. D., and J. K. Joung. 2014. CRISPR-Cas systems for editing, regulating and targeting genomes. *Nature Biotechnology* 32(4): 347-350.

Sarewitz, D. 2015. Science can't solve it. *Nature* 522(7557): 413-414.

Sasaki, K., S. Yokobayashi, T. Nakamura, I. Okamoto, Y. Yabuta, K. Kurimoto, H. Ohta, Y. Moritoki, C. Iwatani, H. Tsuchiya, S. Nakamura, K. Sekiguchi, T. Sakuma, T. Yamamoto, T. Mori, K. Woltjen, M. Nakagawa, T. Yamamoto, K. Takahashi, S. Yamanaka, and M. Saitou. 2015. Robust *in vitro* induction of human germ cell fate from pluripotent stem cells. *Cell Stem Cell* 17(2): 178-194.

Saxton, M. 2000. Why members of the disability community oppose prenatal diagnosis and selective abortion. In *Prenatal testing and disability rights*, edited by E. Parenz and A. Ash. Washington, DC: Georgetown University Press. Pp. 147-165.

Sayres, L. C., and D. Magnus. 2012. Duty-free: The non-obligatory nature of preimplantation genetic diagnosis. *American Journal of Bioethics* 12(4): 1-2.

Scharschmidt, T., and B. Lo. 2006. Clinical trial design issues raised during Recombinant DNA Advisory Committee review of gene transfer protocols. *Human Gene Therapy* 17(4): 448-454.

Scheufele, D. A. 2010. Survey research. In *Encyclopedia of science and technology communication*, Vol. 2, edited by S. H. Priest. Thousand Oaks, CA: SAGE Publications. Pp. 853-856.

Scheufele, D. A. 2011. *Modern citizenship or policy dead end? Evaluating the need for public participation in science policy making, and why public meetings may not be the answer*. Joan Shorenstein Center on the Press, Politics and Public Policy Research Paper Series (#R-43). http://shorensteincenter.org/wp-content/uploads/2012/03/r34_scheufele.pdf (accessed January 6, 2017).

Scheufele, D. A. 2013. Communicating science in social settings. *Proceedings of the National Academy of Sciences of the United States of America* 110 (Suppl. 3): 14040-14047.

Schomberg, R. 2012. Prospects for technology assessment in a framework of responsible research and innovation. In *Technikfolgen abschätzen lehren: Bildungspotenziale transdisziplinärer methoden*, edited by M. Dusseldorp and R. Beecroft. Wiesbaden: VS Verlag für Sozialwissenschaften. Pp. 39-61.

Sciencewise. 2016. *Sciencewise—the U.K.'s national centre for public dialogue in policy making involving science and technology issues*. http://www.sciencewise-erc.org.uk/cms (accessed October 19, 2016).

Scully, J. L. 2009. Towards a bioethics of disability and impairment. In *The handbook of genetics and society: Mapping the new genomic era*, edited by P. Atkinson, P. Glasner, and M. Lock. London, U.K.: Routledge. Pp. 367-381.

Seifter, M. 2014a. States, agencies, and legitimacy. *Vanderbilt Law Review* 67(2): 443-504.

Seifter, M. 2014b. States as interest groups in the administrative process. *Virginia Law Review* 100: 953-1025.

Seifter, M. 2015. Second-order participation in administrative law. *UCLA Law Review* 1301-1363. http: //www.uclalawreview.org/second-order-participation-in-administrative-law (accessed January 4, 2017).

Shahbazi, M. N., A. Jedrusik, S. Vuoristo, G. Recher, A. Hupalowska, V. Bolton, N. M. Fogarty, A. Campbell, L. G. Devito, and D. Ilic. 2016. Self-organization of the human embryo in the absence of maternal tissues. *Nature Cell Biology* 18(6): 700-708.

Shalala, D. 2000. Protecting research aubjects—what must be done. *New England Journal of Medicine* 343(11): 808-810.

Sharma, R., X. M. Anguela, Y. Doyon, T. Wechsler, R. C. DeKelver, S. Sproul, D. E. Paschon, J. C. Miller, R. J. Davidson, and D. Shivak. 2015. *In vivo* genome editing of the albumin locus as a platform for protein replacement therapy. *Blood* 126(15): 1777-1784.

Shelton, A. M., and R. T. Roush. 1999. False reports and the ears of men. *Nature Biotechnology* 17(9): 832-832.

Shetty, G., R. K. Uthamanthil, W. Zhou, S. H. Shao, C. C. Weng, R. C. Tailor, B. P. Hermann, K. E. Orwig, and M. L. Meistrich. 2013. Hormone suppression with GnRH antagonist promotes spermatogenic recovery from transplanted spermatogonial stem cells in irradiated cynomolgus monkeys. *Andrology* 1(6): 886-898.

Sinsheimer, R. L. 1969. The prospect for designed genetic change. *American Scientist* 57(1): 134-142.

Skerrett, P. 2015. A debate: Should we edit the human germline? *STAT*, November 30. https: //www.statnews.com/2015/11/30/gene-editing-crispr-germline (accessed January 5, 2017).

Slaymaker, I. M., L. Gao, B. Zetsche, D. A. Scott, X. Winston, W. X. Yan, and F. Zhang. 2016. Rationally engineered Cas9 nucleases with improved specificity. *Science* 351(6268): 84-88.

Solter, D. 2006. From teratocarcinomas to embryonic stem cells and beyond: A history of embryonic stem cell research. *Nature Reviews Genetics* 7(4): 319-327.

Specter, M. 2015. The gene hackers. *The New Yorker*, November 16.

Steinbach, R. J., M. Allyse, M. Michie, E. Y. Liu, and M. K. Cho. 2016. "This lifetime commitment": Public conceptions of disability and noninvasive prenatal genetic screening. *American Journal of Medical Genetics* 170(2): 363-374.

Steinberg, A. 2006. Introduction, Ch. 3. In *Halakhic-Medical Encyclopedia* (Hebrew), 2nd ed., Vol. 1. Pp. 158-163.

Steinbrook, R. 2002. Improving protection for research subjects. *New England Journal of Medicine* 346(18): 1425-1430.

Steinbrook, R. 2004. Science, politics, and federal advisory committees. *New England Journal of Medicine* 350(14): 1454-1460.

Suh, Y., G. Atzmon, M.-O. Cho, D. Hwang, B. Liu, D. J. Leahy, N. Barzilai, and P. Cohen. 2008. Functionally significant insulin-like growth factor I receptor mutations in centenarians. *Proceedings of the National Academy of Sciences of the United States of America* 105(9): 3438-3442.

Sulmasy, D. P. 2008. *Dignity and bioethics: History, theory, and selected applications*. https: //bioethicsarchive.georgetown.edu/pcbe/reports/human_dignity/chapter18.html (accessed January 6, 2017).

Suzuki, K., Y. Tsunekawa, R. Hernandez-Benitez, J. Wu, J. Zhu, E. J. Kim, F. Hatanaka, M. Yamamoto, T. Araoka, Z. Li, M. Kurita, T Hishida, M. Li, E. Aizawa, S. Guo, S. Chen, A. Goebl, R. D. Soligalla, J. Qu, T. Jiang, X. Fu, M. Jafari, C. R. Esteban, W. T. Berggren, J. Lajara, E. Nuñez-Delicado, P. Guillen, J. M. Campistol, F. Matsuzaki, G. H. Liu, P. Magistretti, K. Zhang, E. M. Callaway, K. Zhang, and J. C. Belmonte. 2016. *In vivo* genome editing via CRISPR/Cas9 mediated homology-independent targeted integration. *Nature* 540: 144-149.

Tabebordbar, M., K. Zhu, J. K. Cheng, W. L. Chew, J. J. Widrick, W. X. Yan, C. Maesner, E. Y. Wu, R. Xiao, and F. A. Ran. 2016. *In vivo* gene editing in dystrophic mouse muscle and muscle stem cells. *Science* 351(6271): 407-411.

Takahashi, K., and S. Yamanaka. 2006. Induction of pluripotent stem cells from mouse embryonic and adult fibroblast cultures by defined factors. *Cell* 126(4): 663-676.

Takahashi, K., K. Tanabe, M. Ohnuki, M. Narita, T. Ichisaka, K. Tomoda, and S. Yamanaka. 2007. Induction of pluripotent stem cells from adult human fibroblasts by defined factors. *Cell* 131(5): 861-872.

Takefman, D. 2013. *The FDA review process*. Presentation at the first meeting on independent review and assessment of the activities of the NIH Recombinant DNA Advisory Committee, Washington, DC, June 4. https: //www.nationalacademies.org/hmd/Activities/Research/ReviewNIHRAC/2013-JUN-04.aspx (accessed November 4, 2016).

Takefman, D., and W. Bryan. 2012. The state of gene therapies: The FDA perspective. *Molecular Therapy* 20(5): 877-878.

Taylor, T. H., S. A. Gitlin, J. L. Patrick, J. L. Crain, J. M. Wilson, and D. K. Griffin. 2014. The origin, mechanisms, incidence and clinical consequences of chromosomal mosaicism in humans. *Human Reproduction Update* 20(4): 571-581.

Tebas, P., D. Stein, W. W. Tang, I. Frank, S. Q. Wang, G. Lee, S. K. Spratt, R. T. Surosky, M. A. Giedlin, and G. Nichol. 2014. Gene editing of CCR5 in autologous CD4 T cells of persons infected with HIV. *New England Journal of Medicine* 370(10): 901-910.

The 1000 Genomes Project Consortium. 2015. A global reference for human genetic variation. *Nature* 526(7571): 68-74.

The Washington Post. 1994. Embryos: Drawing the line. *The Washington Post*, October 2.

Turner, L., and P. Knoepfler. 2016. Selling stem cells in the USA: Assessing the direct-to-consumer industry. *Cell Stem Cell* 19(2):

154-157.

U. K. House of Lords. 2000. *Select committee on science and technology—third report.* http://www.publications.parliament.uk/pa/ld199900/ldselect/ldsctech/38/3801.htm (accessed December 14).

UN (United Nations). 1948. *Universal Declaration of Human Rights.* http://www.un.org/en/universal-declaration-human-rights/index.html (accessed January 5, 2017).

UN. 2006. *Convention on the Rights of Persons with Disabilities (CRPD).* https://www.un.org/development/desa/disabilities/convention-on-the-rights-of-persons-with-disabilities.html (accessed January 5, 2017).

UNESCO (UN Educational, Scientific and Cultural Organization). 2004a. International declaration on human genetic data. *European Journal of Health Law* 11: 93-107.

UNESCO. 2004b. *National legislation concerning human reproductive and therapeutic cloning.* http://unesdoc.unesco.org/images/0013/001342/134277e.pdf (accessed January 5, 2017).

UNESCO. 2005. *Universal declaration on bioethics and human rights.* Paris, France: UNESCO. http://www.unesco.org/new/en/social-and -human-sciences/themes/bioethics/bioethics-and-human-rights (accessed January 4, 2017).

UNESCO. 2015. *Report of the International Bioethics Committee on updating its reflection on the human genome and human rights.* http://www.coe.int/en/web/bioethics/-/gene-editing (accessed October 21, 2016).

UNICEF (UN Children's Fund). 1990. *Convention on the Rights of the Child.* https://www.unicef.org/crc (accessed January 5, 2017).

Urnov, F. D., E. J. Rebar, M. C. Holmes, H. S. Zhang, and P. D. Gregory. 2010. Genome editing with engineered zinc finger nucleases. *Nature Reviews Genetics* 11(9): 636-646.

van Delden, J. J. M. and van der Graaf, R. 2016. Revised CIOMS international ethical guidelines for health-related research involving humans. *The JAMA Network*, December 6. http://jamanetwork.com/journals/jama/fullarticle/2592245 (accessed January 4, 2017).

Vatican. 2015. *Encyclical letter laudato si' of the holy father Francis on care for our common home.* http://w2.vatican.va/content/francesco/en/encyclicals/documents/papa-francesco_20150524_enciclica-laudato-si.html (accessed January 4, 2017).

Vernot, B., S. Tucci, J. Kelso, J. G. Schraiber, A. B. Wolf, R. M. Gittelman, M. Dannemann, S. Grote, R. C. McCoy, and H. Norton. 2016. Excavating Neanderthal and Denisovan DNA from the genomes of Melanesian individuals. *Science* 352(6282): 235-239.

Volokh, E. 2003. The mechanisms of the slippery slope. *Harvard Law Review* 116(4): 1026 -1137. Waddington, S. N., M. G. Kramer, R. Hernandez-Alcoceba, S. M. Buckley, M. Themis, C. Coutelle, and J. Prieto. 2005. *In utero* gene therapy: Current challenges and perspectives. *Molecular Therapy* 11(5): 661-676.

Wailoo, K., A. Nelson, and C. Lee. 2012. *Genetics and the unsettled past: The collision of DNA, race, and history.* New Brunswick, NJ: Rutgers University Press.

Walters, L. 1991. Human gene therapy: Ethics and public policy. *Human Gene Therapy* 2(2): 115-122. Walters, L., and J. G. Palmer. 1997. *The ethics of human gene therapy.* New York: Oxford University Press.

Wasserman, D., and A. Asch. 2006. The uncertain rationale for prenatal disability screening. *The Virtual Mentor: VM* 8(1): 53-56.

Wasserman, D., and A. Asch. 2012. A duty to discriminate? *American Journal of Bioethics* 12(4): 22-24.

Watson, J. D., and F. H. Crick. 1953. Molecular structure of nucleic acids. *Nature* 171(4356): 737-738.

Werner, M. and A. Plant. 2016. Collaboration between National Institute of Standards and Technology and the Standards Coordinating Body for Regenerative Medicines. Presentation of the Alliance for Regenerative Medicine, NASEM Regenerative Medicine Forum, Washington, DC, October 14, 2016.

Willemsen, R. H., M. van Dijk, Y. B. de Rijke, A. W. van Toorenenbergen, P. G. Mulder, and A. C. Hokken- Koelega. 2007. Effect of growth hormone therapy on serum adiponectin and resistin levels in short, small-for-gestational-age children and associations with cardiovascular risk parameters. *Journal of Clinical Endocrinology & Metabolism* 92(1): 117-123.

Wilsdon, J. 2015. Let's keep talking: Why public dialogue on science and technology matters more than ever. *The Guardian*, March 20. https://www.theguardian.com/science/political-science/2015/mar/27/lets-keep-talking -why-public-dialogue-on-science-and-technology-matters-more-than-ever (accessed October 19, 2016).

WMA (World Medical Association). 2013. World Medical Association Declaration of Helsinki: Ethical principles for medical research involving human subjects. *Journal of the American Medical Association* 310(20): 2191-2194.

Wolf, S. M., R. Gupta, and P. Kohlhepp. 2009. Gene therapy oversight: Lessons for nanobiotechnology. *Journal of Law, Medicine and Ethics* 37(4): 659-684.

Wozniak, M. A., R. F. Itzhaki, E. B. Faragher, M. W. James, S. D. Ryder, and W. L. Irving. 2002. Apolipoprotein E-epsilon 4 protects against severe liver disease caused by hepatitis C virus. *Hepatology* 36(2): 456-463.

Wright, A. V., J. K. Nunez, and J. A. Doudna. 2016. Biology and applications of CRISPR systems: Harnessing nature's toolbox for genome engineering. *Cell* 164(1-2): 29-44.

Wu, Y., D. Liang, Y. Wang, M. Bai, W. Tang, S. Bao, Z. Yan, D. Li, and J. Li. 2013. Correction of a genetic disease in mouse via use of CRISPR-Cas9. *Cell Stem Cell* 13(6): 659-662.

Wu, Y., H. Zhou, X. Fan, Y. Zhang, M. Zhang, Y. Wang, Z. Xie, M. Bai, Q. Yin, D. Liang, W. Tang, J. Liao, C. Zhou, W. Liu, P. Zhu, H. Guo, H. Pan, C. Wu, H. Shi, L. Wu, F. Tang, and J. Li. 2015. Correction of a genetic disease by CRISPR-Cas9-mediated gene editing in mouse spermatogonial stem cells. *Cell Research* 25(1): 67-79.

Yin, H., W. Xue, S. Chen, R. L. Bogorad, E. Benedetti, M. Grompe, V. Koteliansky, P. A. Sharp, T. Jacks, and D. G. Anderson. 2014. Genome editing with Cas9 in adult mice corrects a disease mutation and phenotype. *Nature Biotechnology* 32(6): 551-553.

Zetsche, B., J. S. Gootenberg, O. O. Abudayyeh, I. M. Slaymaker, K. S. Makarova, P. Essletzbichler, S. E. Volz, J. Joung, J. van der Oost, A. Regev, E. V. Koonin, and F. Zhang. 2015. Cpf1 is a single RNA-guided endonuclease of a class 2 CRISPR-Cas system. *Cell* 163(3): 759-771.

Zhang, J., H. Liu, S. Luo, A. Chavez-Badiola, Z. Liu, S. Munne, M. Konstantinidis, D. Wells, and T. Huang. 2016. First live birth using human oocytes reconstituted by spindle nuclear transfer for mitochondrial DNA mutation causing leigh syndrome. *Fertility and Sterility* 106(3): e375-e376.

Zheng, W., H. Zhao, E. Mancera, L. M. Steinmetz, and M. Snyder. 2010. Genetic analysis of variation in transcription factor binding in yeast. *Nature* 464(7292): 1187-1191.

Zhou, Q., M. Wang, Y. Yuan, X. Wang, R. Fu, H. Wan, M. Xie, M. Liu, X. Guo, Y. Zheng, G. Feng, Q. Shi, X. Y. Zhao, J. Sha, and Q. Zhou. 2016. Complete meiosis from embryonic stem cell-derived germ cells *in vitro*. *Cell Stem Cell* 18(3): 330-340.

Zhu, Y.-Y., and R. Rees. 2012. *Vaccines, blood, & biologics: Clinical review, October 2, 2012—Ducord.* http://www.fda.gov/BiologicsBloodVaccines/CellularGeneTherapyProducts/ApprovedProducts/u cm326333.htm (accessed September 24, 2013).

附录 A 基因组编辑的基础科学
Appendix A The Basic Science of Genome Editing

本附录介绍了与基因治疗和基因编辑的基础科学相关的技术和历史背景的一些问题。为了尽可能提高这些资料的易懂性,在第 3 章和第 4 章中可以找到对这些资料更简单的概述。本附录包括关于以下主题的详细资料:

- DNA 的断裂和修复;
- 规律成簇间隔短回文重复 (CRISPR) 系统 /CRISPR 关联内切核酸酶 (Cas9) 基因编辑——巨大核酸酶、锌指核酸酶和转录激活因子样效应核酸酶 (TALENs);
- CRISPR / Cas9 系统的发展;
- 基因编辑的准确性;
- 提高 CRISPR / Cas9 的特异性;
- 基因编辑的质量控制和质量保证;
- 使用无 DNA 切割活性的 Cas9 蛋白 (dCas9) 调节转录或进行表观遗传修饰;
- 转基因动物中的基因靶向;
- 胚胎中的基因编辑;
- 可遗传生殖细胞基因编辑的替代路径;
- 线粒体基因组编辑。

基因治疗和基因组编辑

近年来我们明显地看到基因治疗解决人类疾病的潜力,并且在应用中取得了很大进展 (Cox et al., 2015; Naldini, 2015)。基因治疗是指替代有

This appendix provides technical and historical context for a number of issues related to the basic science of gene therapy and gene editing. Although an effort has been made to maximize the accessibility of this material, a simpler summary of this material can be found in Chapters 3 and 4. This appendix includes detailed material on the following topics:

- breakage and repair of DNA
- precursors of the Clustered Regularly Interspersed Short Palindromic Repeats (CRISPR) system/ CRISPR-associated endonuclease (Cas9) gene editing—meganucleases, zinc fingers, and transcription activator-like effector nucleases (TALENs)
- development of CRISPR/Cas9
- the accuracy of gene editing
- enhancing the specificity of CRISPR/Cas9
- quality control and quality assurance for gene editing
- use of dead Cas9 (dCas9) to regulate transcription or to make epigenetic modifications gene targeting in transgenic animals
- gene editing in embryos
- alternative routes to heritable germline
- editing editing the mitochondrial genome

GENE THERAPY AND GENOME EDITING

The potential for gene therapy to address human disease has been evident for some years, and much progress has been made in its applications (Cox et al., 2015; Naldini, 2015). Gene therapy refers to the replacement of faulty genes, or the addition of new genes as a means to cure disease or improve the ability to fight disease. Genome editing is one aspect of gene

缺陷的基因，或添加新的基因作为治疗疾病或提高抵抗疾病能力的手段。基因组编辑是基因治疗的一个方面。根据在单个细胞和非人体组织中进行的广泛实验室前期研究结果，已初步建立了基因治疗的方法，并正在建立在活体生物中进行基因添加、删除或修饰的方法。主要进展包括发展技术方法来开发在特定位置切割基因组 DNA 的分子工具，实现 DNA 序列中的靶向改变。近年来，几种这样的方法已经被引入并且有效地应用于临床。

过去 5 年中，基于细菌对病毒感染的免疫机制的基础研究——一种全新的基因编辑技术被发展起来。最先被开发用于人类细胞基因组编辑的此类系统，被称为 CRISPR /Cas9 系统，由 RNA 导向靶位点，比以前的方法更简单、更快捷，并且花费更少。CRISPR / Cas9 系统设计的简易性，以及显著的特异性和高效性已经革命性地改变了基因组编辑领域，并重新引起了人们对探索人类基因组编辑潜力的兴趣。作为基因组编辑工具的 CRISPR/Cas9 系统的开发是建立在早期研究的坚实基础上的。

基因组 DNA 的断裂和修复

基因组及其组成基因由双链 DNA、构成这些 DNA 可意外地（如受到辐射）受到破坏，有目的地被一种被称为限制性内切核酸酶（通常又称核酸酶）的蛋白质作用，产生双链断裂（DSBs）。

细胞具有修复 DNA 中断裂双链的机制，并且这些机制可用来引起 DNA 序列的改变。在细菌、酵母和哺乳动物系统中的突破性工作表明，DNA 双链断裂的缺口能通过非同源末端连接（NHEJ）显著提高 DNA 修复的速率，其中断裂的末端被重新连接（图 A-1）。这种非同源末端连接修复常常导致不同长度的 DNA 序列的缺失或插入，能引起基因功能的破坏（Rouet et al., 1994）。

然而，如果将同源 DNA 片段作为供体模板引入细胞中，同源定向修复（HDR）可以导致更准确的修复；或者如果同源延伸中包括特定的改变，它可以将特定的精确变化引入受体基因组 DNA（图 A-1）。这些细胞 DNA 修复机制已被用于开发能够以非常精确的方式编辑基因或基因组的几种方法。

therapy. Established approaches to gene therapy have been based on the results of extensive prior laboratory research on individual cells and on nonhuman organisms, establishing the means to add, delete, or modify genes in living organisms. Key advances include the development of techniques for generating molecular tools for cutting the DNA of genomes in specific places to allow targeted alterations in the DNA sequence. Over recent years, several such methods have been introduced and used effectively in clinical applications.

Within the past 5 years, a completely novel system has been developed based on fundamental research on bacterial systems of immunity to viral infections. The first such system to be developed for use in genome editing of human cells, known as CRISPR/Cas9, is based on RNA-guided targeting and is much simpler, faster, and cheaper than earlier methods. The ease of design, together with the remarkable specificity and efficiency of the CRISPR/Cas9 system has revolutionized the field of genome editing and reignited interest in the potential for editing of the human genome. The development of the CRISPR/Cas9 system as a programmable genome-editing tool was built on a firm foundation of earlier research.

BREAKAGE AND REPAIR OF GENOMIC DNA

Genomes and their constituent genes are made of double-stranded DNA; this DNA can be broken accidentally (e.g., by radiation) or purposefully, using proteins called endonucleases (often called nucleases) that can generate double-strand breaks (DSBs) in DNA.

Cells have mechanisms to repair DSBs in DNA, and these mechanisms can be used to generate alterations in the DNA sequence. Groundbreaking work in bacteria, yeast, and mammalian systems shows that DSBs dramatically stimulate the rate of DNA repair by nonhomologous end joining (NHEJ), in which the broken ends are reattached (see Figure A-1). Such NHEJ repair often results in the deletion or insertion of DNA sequences of varying length, which can disrupt gene function (Rouet et al., 1994).

However, if a homologous stretch of DNA is introduced into the cell as a donor template, homology-directed repair (HDR) can lead to more accurate repair or, if specific alterations are included in the homologous stretch, it can introduce specific precise changes into the recipient genomic DNA (see Figure A-1). These cellular DNA repair mechanisms have been used to develop several methods that allow genes or the genome to be edited in a very precise manner.

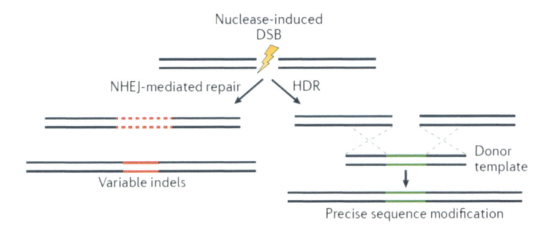

图 A-1 由 NHEJ 或 HDR 修复核酸酶诱导的 DNA 双链断裂介导的基因组编辑结果。
FIGURE A-1 Genome-editing outcomes are mediated by repair of nuclease-induced double-strand DNA breaks by NHEJ or HDR.
注：DSB = 双链断裂；HDR = 同源定向修复；NHEJ = 非同源末端连接。
来源：由 Sander 和 Joung 修订，2014。
NOTE: DSB = double-strand break; HDR = homology-directed repair; NHEJ = nonhomologous end joining.
SOURCE: Modified from Sander and Joung, 2014.

CRISPR / CAS9 基因编辑系统的先驱

在 CRISPR / Cas9 系统发展之前，三种基于核酸酶系统的不同基因编辑策略——巨大转移酶、锌指核酸酶（ZFN）和转录激活因子样效应（TALE）核酸酶，被用来实现 DNA 靶向切割（图 A-2）。

这三种方法都取得了重大进展，使利用靶向核酸酶消除致病基因及修复受损或突变的基因成为可能，开启了生物学和医学的新时代。

锌指核酸酶 (ZFNs)

锌指是已经进化以识别并结合特定 DNA 序列的蛋白质片段。对天然锌指的认识帮助我们开发了锌指核酸酶（ZFNs），这是一种经过设计的、能切割 DNA 的酶（图 A-2）。这些酶在基因组工程中的应用代表着开创性的、史诗性的蛋白质工程。蛋白质工程的两个主要进展使得锌指核酸酶的开发成为可能：① Berg（Desjarlais and Berg，1992）、Pabo（Rebar and Pabo，1994）和 Wells（Jamieson et al.，1994）开始通过对 DNA 结合特异性的个性化定制对锌指蛋白进行工程化改造；②这种个性化定制的锌指与 DNA 切割蛋白、Fok I 核酸酶（Kim et al.，1996）之间的融合产生的"人造限制酶"，可以通过在基因组中的特定位置对 DNA 进行切割，促进位点特异性基因组工程（Bibikova et al.，2001，2002，2003）。

PRECURSORS OF CRISPR/CAS9 GENE EDITING

Three distinct strategies based on nuclease systems for generating targeted cleavages in DNA preceded the development of CRISPR/Cas9—meganucleases, zinc finger nucleases (ZFNs), and transcription activator-like effector (TALE) nucleases (see Figure A-2). All three have already enabled major advances in establishing the feasibility of using such targeted nucleases both to eliminate disease-causing genes as well as to repair damaged or mutated genes, ushering in a new era in biology and medicine.

Zinc Finger Nucleases (ZFNs)

Zinc fingers are segments of protein that have evolved to recognize and bind to specific DNA sequences. Knowledge gained from natural zinc fingers led to the development of ZFNs as designer DNA-cutting enzymes (see Figure A-2), and their use in genome engineering represents pioneering, even heroic, protein engineering. Two major advances in protein engineering enabled the development of ZFNs: (1) the engineering of zinc finger proteins with designed DNA-binding specificity pioneered by Berg (Desjarlais and Berg, 1992), Pabo (Rebar and Pabo, 1994), and Wells (Jamieson et al., 1994); and (2) the generation of fusions between such designer zinc fingers and a DNA-cleaving protein, FokI nuclease (Kim et al., 1996), yielding "artificial restriction enzymes" that could be used to promote site-specific genome engineering by creating DSBs at defined locations in the genome (Bibikova et al., 2001, 2002, 2003).

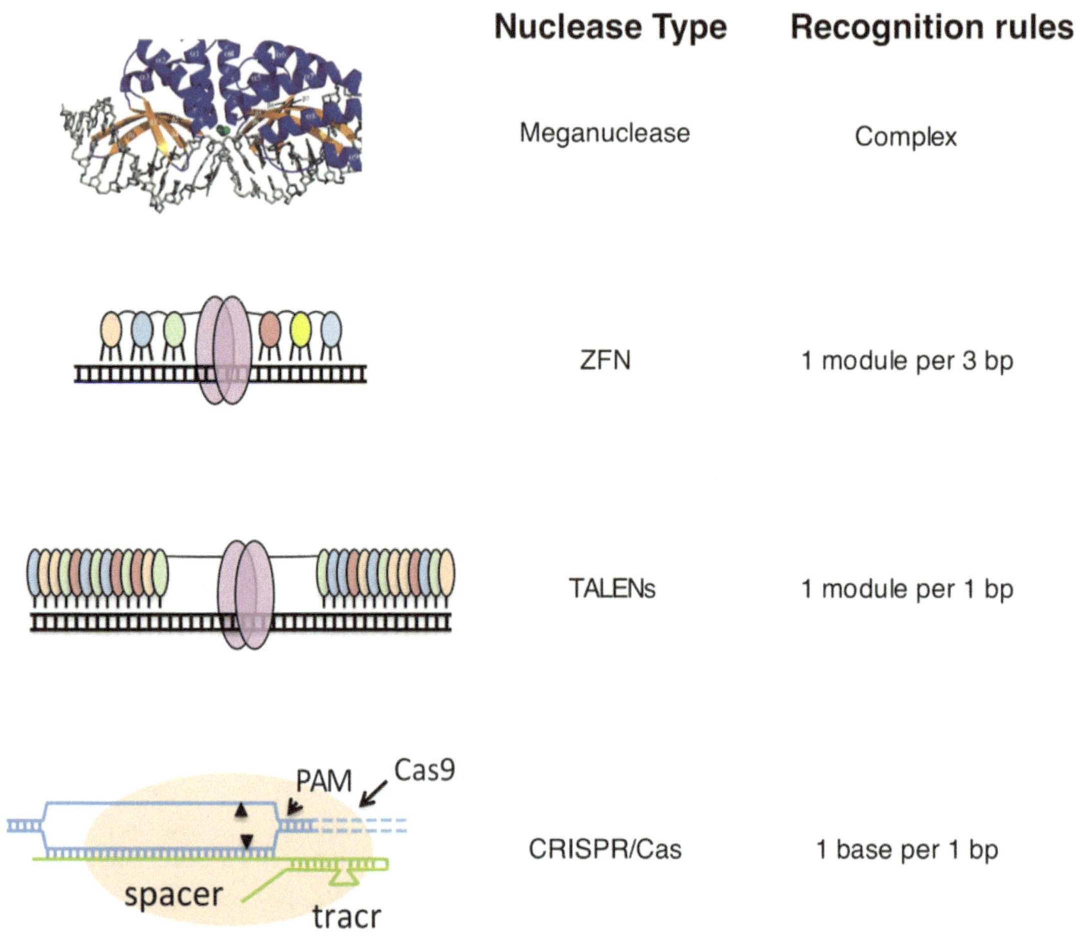

Nuclease Type	Recognition rules
Meganuclease	Complex
ZFN	1 module per 3 bp
TALENs	1 module per 1 bp
CRISPR/Cas	1 base per 1 bp

图 A-2 本附录中讨论的靶向核酸酶的工作原理图。巨大核酸酶显示为 E-DreI（工程化 I-Dmo Ⅰ / Cre Ⅰ）的晶体结构的示意图。对于锌指核酸酶（ZFNs）和转录激活因子样效应核酸酶（TALENs）的说明，黑色水平线代表 DNA，紫色椭圆形代表 Fok Ⅰ 核酸酶结构域，核酸酶的其他模块显示为多色，表示能识别 DNA 中的不同碱基对。对于 CRISPR / Cas 系统，Cas9 蛋白以橙色椭圆表示，DNA 显示为蓝色，向导 RNA（间隔区和反式激活嵌合 RNA 的嵌合体——参见文本）以绿色显示。箭头指向 DNA 切割位点。

FIGURE A-2 Schematic of targetable nucleases discussed in this appendix. The meganuclease is a schematic of the crystal structure of E-DreI (Engineered I-DmoI/CreI). For the zinc-finger nucleases (ZFNs) and transcription activator–like effector nucleases (TALENs) illustrations, DNA is represented horizontally in black, purple ovals represent the FokI nuclease domains, and other modules of the nucleases are multi-colored to indicate that they recognize different bases in the DNA. For the CRISPR/Cas system, the Cas9 protein is represented by an orange oval, DNA is in blue, and guide RNA (chimera of spacer and tracr RNAs—see text) is shown in green. Arrowheads point to the sites of DNA cleavage.

资料来源：Carroll，2014。（经生物化学年鉴许可修改，第 83 卷 ©2014 年鉴，http://www.annualreviews.org）；Chevalier 等，2002。

SOURCES: Carroll, 2014. (Modified with permission from the Annual Review of Biochemistry, Volume 83 © 2014 by Annual Reviews, http://www.annualreviews.org); Chevalier et al., 2002.

利用这些发现，Sangamo Therapeutics 公司将锌指核酸酶由实验室工具开发成可用于使致病基因失活（通过非同源末端连接修复）或修正现有基因中的错误（通过同源重组修复）的治疗药物。后一个目标是特别具有挑战性的，因为同源重组修复的效率通常比不精确的、易突变的非同源末端连接修复低得多。通过克服这些困难，Sangamo 公司现在已经有了几个进行中的人体临床试验[69]。目前进行

Using these findings, Sangamo Therapeutics developed ZFNs from laboratory tools into therapeutic agents that could be used to disable disease-causing genes (by NHEJ) or to correct errors in existing genes (by HDR). The latter goal is particularly challenging, as the efficiency of HDR is usually much lower than the imprecise and mutagenic repair by NHEJ. By overcoming some of these difficult challenges, Sangamo now has several ongoing human clinical trials,[69] the most advanced of which is in the treatment of HIV/

---

[69] See http://www.sangamo.com/pipeline/index.html (accessed January 7, 2017).

得最深入的一个项目是通过敲除 CCR5 HIV 受体进行针对艾滋病毒/艾滋病的治疗，使骨髓移植后的 HIV 病毒清除成为可能。锌指核酸酶的其他应用也已经或正在进行临床试验（NCT02695160 [血友病 B] 和 NCT02702115 [MPS1 酶的体内编辑]）。

转录激活因子样效应核酸酶 (TALENs)

像 ZFN 一样，TALENs 由能识别特定 DNA 序列的 DNA 结合蛋白和核酸酶效应结构域融合而成，能对 DNA 进行切割（Joung and Sander，2013）（见图 A-2）。TALE 是含有 DNA 结合结构域的细菌分泌蛋白，含有一系列长度为 32~34 个氨基酸残基的保守区段，每个残基有两个不同的氨基酸。对单一 DNA 碱基对的结合特异性主要由这两个不同的氨基酸决定，这使我们可以通过选择含有合适氨基酸的重复肽段组合来改造特定的 DNA 结合域。

生物技术公司 Cellectis 报道了在英国首次基于 TALEN 技术的基因编辑临床试验的成功案例，实验的对象是一名患有不可治愈的急性淋巴细胞白血病（ALL）的女孩（Qasim et al.，2017）。尽管 ZFNs 和 TALENs 的应用在很大程度上是重叠的，但由于 TALENs 具有更强的识别信号，因此 TALENs 具有相对容易设计的优点。

巨大核酸酶

巨大核酸酶具有非常长的 DNA 结合识别位点，高达 40 个核苷酸（Silva et al.，2011）（见图 A-2）。由于它们的长度，即使在复杂的人类基因组中，巨大核酸酶也很少存在天然位点。使用巨大核酸酶的挑战是很难随心所欲地设计新的核酸酶来靶向结合目标序列。但也有一些案例是通过设计改变 DNA 结合位点和将巨大核酸酶与 TALE DNA 结合元件组合而取得成功的。然而，由于找到替代方法相对容易，巨大核酸酶不太可能在人类基因组编辑中被大量使用。

CRISPR / CAS9 系统的发展

细菌适应性免疫系统 CRISPR

CRISPR 系统能够为细菌提供适应性免疫，该发现代表了一个概念上的重大进步，并且对 CRISPR/Cas9 基因组工程的发展非常关键。以下是对其发现

AIDS where deletion of the CCR5 HIV coreceptor has the potential to enable the elimination of HIV following bone-marrow transplantation. Other applications of ZFNs have also been, or are being, tested in clinical trials (NCT02695160 [hemophilia B] and NCT02702115 [*in vivo* editing MPS1]).

Transcription Activator-Like Effector Nucleases (TALENs)

Like ZFNs, TALENs are composed of a DNA-binding protein that recognizes particular DNA sequences fused to a nuclease effector domain to achieve the cleavage (Joung and Sander, 2013) (see Figure A-2). TALEs are secreted bacterial proteins with DNA-binding domains that contain a series of conserved blocks of sequence 32-34 residues long, each with two divergent amino acids. These divergent amino acids are largely responsible for determining the DNA-binding specificity for a single DNA base pair, which allows engineering of specific DNA-binding domains by choosing combinations of repeat segments containing appropriate amino acids. The biotech firm Cellectis reported the successful conduct of the first-ever TALEN-based gene-editing clinical trial in the United Kingdom on a girl with incurable acute lymphoblastic leukemia (ALL) (Qasim et al, 2017). Although the applications of ZFNs and TALENs are largely overlapping, TALENs have the advantage of relative ease of design because of the robust recognition code.

Meganucleases

Meganucleases are nucleases with very long DNA-binding recognition sites, up to 40 nucleotides (Silva et al., 2011) (see Figure A-2). As a result of their length, it is exceedingly unlikely that natural sites would be present by chance even in complex human genomes. The challenge with meganucleases is the difficulty in designing new nucleases to target a sequence of interest at will. Some success has been achieved by designed changes to DNA-binding sites and combining meganucleases with TALE DNA-binding elements. However, meganucleases are unlikely to be much used in human genome editing given the relative simplicity of the alternative methods.

DEVELOPMENT OF CRISPR/CAS9

CRISPR as a Bacterial Adaptive Immunity System

The discovery that CRISPR systems provide adaptive immunity to bacteria represents a major conceptual advance in its own right. This discovery was also critical to the development of CRISPR/Cas9 genome engineering.

过程中关键事件的简要概述（对于更完整的综述，参见 Doudna and Charpentier，2014）。

CRISPR（即规律成簇间隔短回文重复序列）位点是在对细菌基因组分析的基础上首次被鉴定出来的。从这些研究中，人们推测 CRISPR 位点的间隔区（非重复区）来自于噬菌体（感染细菌的病毒）基因组 DNA，并且提出了 CRISPR 能够为抵抗外来基因（Makarova et al.，2006；Mojica et al.，2005；Pourcel et al.，2005）提供防御机制的假说。研究中的重大实验突破表明，通过将噬菌体基因组的元件整合到 CRISPR 位点，CRISPR 可以使细菌获得对噬菌体的抵抗力，这证明 CRISPR 是一种新的适应性免疫形式（Barrangou et al.，2007）。2010 年，II 型 CRISPR / Cas 系统被证明能够介导对入侵噬菌体 DNA 的切割（Garneau et al.，2010）。而在 2011 年，一种 CRISPR 相关的 RNA，即反式激活嵌合 RNA（tracrRNA）被发现（Deltcheva et al.，2011），CRISPR 相关基因 Cas9 被证明是 II 型 Cas 基因中唯一的蛋白质编码基因，是提供防御功能所必需的基因（Sapranauskas et al.，2011）。

可编程内切核酸酶 Cas9 的开发

在 CRISPR / Cas9 的基因组编辑方法的开发过程中，关键性的进步来自于 2012 年 Doudna 和 Charpentier 的实验室所开展的工作。他们将构建的 CRISPR 相关蛋白 Cas9 和两种小 RNA——从 CRISPR 位点转录的 CRISPR RNA（crRNA）及反式激活 crRNA（tracrRNA）——形成复合物，生成位点特异性内切核酸酶，其中切割位点被定位在 crRNA 和靶 DNA 碱基配对区域。这为确定 Cas9 是一种通过单个"引导 RNA"（gRNA，由转录自 CRISPR 位点的 crRNA 和 tracrRNA 组成的嵌合体）进行编辑，可在特定的 DNA 位点进行切割的内切核酸酶打下了基础（Jinek et al.，2012）（图 A-3a）。在 Jinek 等人完成论文手稿的同时（2012），Siksnys 和其同事（Gasiunas et al.，2012）证明了含有 Cas9 和 crRNA 的纯化复合物可以在体外实验中裂解 crRNA 互补位点的双链 DNA。尽管这显然是一个重要的进步，但是要获得 tracrRNA 以形成活性内切核酸酶，必须找到一个关键的缺失元件。

利用一个单体嵌合引导 RNA 能同时发挥 crRNA 和 tracrRNA 的作用（参见图 A-3a），Jinek

A brief synopsis of the key findings is provided below (for a more complete review, see Doudna and Charpentier, 2014).

CRISPR (Clustered Regularly Interspersed Short Palindromic Repeats) loci were first identified based on analyses of bacterial genomes. From these studies, it was inferred that the spacer (i.e., non-repetitive) regions of CRISPR loci were derived from the genomic DNA of bacteriophages (viruses that infect bacteria) leading to the hypothesis that CRISPR provided a defense mechanism against foreign genetic elements (Makarova et al., 2006; Mojica et al., 2005; Pourcel et al., 2005). The key experimental breakthrough came from research showing that CRISPR allowed bacteria to acquire resistance to bacteriophages by integrating segments of the bacteriophage genome into the CRISPR loci, demonstrating that CRISPR was a new form of adaptive immunity (Barrangou et al., 2007). In 2010, type II CRISPR/Cas systems were shown to mediate cleavage of invading bacteriophage DNA (Garneau et al., 2010). And in 2011, an associated RNA, tracrRNA, was identified (Deltcheva et al., 2011) and the CRISPR-associated gene, Cas9, was shown to be the only protein-coding gene in the type II Cas locus required for the defense function (Sapranauskas et al., 2011).

Development of Cas9 as a Programmable Endonuclease

The critical advance in the development of the CRISPR/Cas9-based genome-editing method came in 2012 from the laboratories of Doudna and Charpentier. They established that the CRISPR-associated protein Cas9, in complex with two small RNAs, CRISPR RNA (crRNA) transcribed from the CRISPR locus, and a trans-activating crRNA (tracrRNA), yields a site-specific endonuclease in which the site of cleavage is defined by base pairing of crRNA to the target DNA. This laid the groundwork for establishing that Cas9 is an endonuclease that can be programmed with a single "guide RNA" (gRNA, a chimera of crRNA transcribed from the CRISPR locus and the tracrRNA) to cleave at specific DNA sites (Jinek et al., 2012) (see Figure A-3a). Concurrent with the Jinek et al. (2012) manuscript, Siksnys and coworkers (Gasiunas et al., 2012) demonstrated that purified complexes containing Cas9 and a crRNA could mediate cleavage of double-stranded DNA *in vitro* at sites complementary to the crRNA. While this was clearly an important advance, a key missing element was the identification of the requirement for the tracrRNA to form an active endonuclease.

The Jinek et al. (2012) paper made the key breakthrough of the use of a single chimeric guide

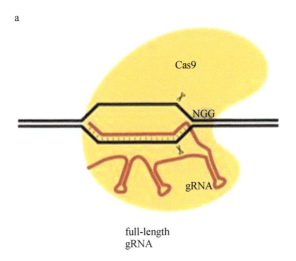

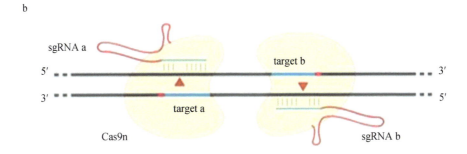

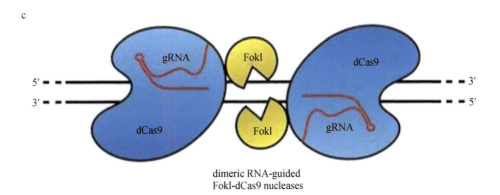

图 A-3　Cas9 及不同 Cas9 变体的示意图。
FIGURE A-3　Schematic of Cas9 and different Cas9 variants.
(a) 未修饰的 Cas9 和引导 RNA(gRNA) 靶向定位的 DNA 序列；DNA 切割位点已标出。(b) 成对的 Cas9 切口酶 (Cas9n)。Cas9 的两个核酸酶结构域之一失去切割活性以产生具有酶活性的切口酶。(c) 二聚 RNA 引导的 FokI-dCas9 核酸酶 (RFNs)。无切割活性的"死" Cas9 (dCas9) 融合到二聚化依赖性 FokI 非特异性核酸酶结构域。一对 FokI-dCas9 单体通过结合前间序列邻近基序高效生产 DSBs。
资料来源：由 Tsai 和 Joung 修订，2016。
(a) Unmodified Cas9 with a guide RNA (gRNA) targeting a specific DNA sequence; DNA cut sites are indicated. (b) Paired Cas9 nickases (Cas9n). One of the two nuclease domains of Cas9 is catalytically inactivated to make an enzymatically active nickase. (c) Dimeric RNA-guided FokI–dCas9 nucleases (RFNs). Catalytically inactivated "dead" Cas9 (dCas9) is fused to the dimerization-dependent FokI non-specific nuclease domain. A pair of FokI–dCas9 monomers oriented in a PAM-out orientation mediates efficient DSBs.
SOURCE: Modified from Tsai and Joung, 2016.

等（2012）的论文作出了关键性的突破。因此，就使用的方便性而言，CRISPR/Cas9系统在机制上优于ZFNs和TALENs；产生位点特异性核酸酶所需要的是设计和合成单链引导RNA（gRNA），以将Cas9核酸酶靶向定位于所需的编辑位点。虽然不是所有预测的引导RNA（gRNA）都能工作，但是它们易于合成，并且能容易且低成本地被合成多个可能的候选gRNA。

可编程核酸酶Cas9的体内应用

在Jinek等（2012）和Siksnys实验室手稿（Sapranauskas et al.，2011；Gasiunas et al.，2012）出版的几个月内，有6个独立的报告使用Cas9/gRNA系统在体内介导可编程基因组编辑。其中有4篇论文报道的Cas9基因编辑是在哺乳动物细胞中进行的（Cho et al.，2013；Cong et al.，2013；Jinek et al.，2013；Mali et al.，2013）；另有一篇在斑马鱼上进行（Hwang et al.，2013）；一篇在细菌中进行（Jiang et al.，2013）。此外，第7篇论文（Qi et al.，2013）表明无催化活性的dCas9可以用于抑制转录。自此以后，Cas9介导的DNA切割得到广泛运用和改进，一些新发现的新型CRISPR系统也被改造用于基因编辑，其中包括一种新的RNA引导的内切核酸酶Cpf1（Zetsche et al.，2015）。最近一些研究正在发现更多具有不同潜能的CRISPR靶向的核酸酶。

基因编辑的准确性

DNA意外改变的潜在影响是对把基因编辑的安全使用作为治疗策略的重要挑战。基因组的意外改变主要是对除靶位点外的其他位点上的DNA被切割造成的。

脱靶毒性的挑战

Cas9切割DNA的位点由DNA靶标与接近前间区序列邻近基序（PAM）序列（如最常用的酿脓链球菌CAS9对应的NGG序列）的引导RNA（通常20个碱基对）的互补性决定。

一般而言，22个碱基序列可以产生足够的多样性，即便是在有30亿碱基对的人类基因组内，切割位点应该也是唯一的。但是事实上，一些碱基

RNA that can fulfill the role of both the crRNA and tracrRNA (see Figure A-3a). Thus, mechanistically, the CRISPR/Cas9 system is superior to both ZFNs and TALENs in terms of ease of use; all that is required to generate a site-specific nuclease is design and synthesis of a single guide RNA (gRNA) to target the Cas9 nuclease to the desired site of editing. Although not all predicted guide RNAs work, their synthesis is easy and multiple possible candidates can be synthesized easily and cheaply.

In Vivo Application of the Cas9 Programmable Nuclease

Within months of the publication of the Jinek et al. (2012) and Siksnys laboratory manuscripts (Sapranauskas et al., 2011; Gasiunas et al., 2012), there were six independent reports using the Cas9–guide RNA system to mediate programmable genome editing *in vivo*.

These included four papers reporting Cas9 editing in mammalian cells (Cho et al., 2013; Cong et al., 2013; Jinek et al., 2013; Mali et al., 2013); one on zebrafish (Hwang et al., 2013); and one on bacteria (Jiang et al., 2013). Additionally, there was a seventh paper (Qi et al., 2013) showing that catalytically dead dCas9 could be used to inhibit transcription. Since that time there has been an explosion in the application and refinement of Cas9-mediated cleavage, as well as the discovery of novel CRISPR systems that can be adapted for genome editing. These include the discovery of a novel RNA-guided endonuclease, Cpf1 (Zetsche et al., 2015), and more recent work has shown that further CRISPR-targetable nucleases with the potential for differing capabilities are being discovered.

ACCURACY OF GENE EDITING

The potential impact of unintended changes to DNA is a key challenge for safe use of genome editing as a therapeutic strategy. Unintended changes to the genome could be caused by cleaving DNA at sites other than those that are being deliberately targeted.

The Challenge of Off-Target Toxicity

The site at which Cas9 cleaves DNA is determined by the complementarity of the DNA target with the RNA guide (typically 20 base pairs) adjacent to a protospacer adjacent motif (PAM) sequence (e.g., NGG for *Streptococcus pyogenes*, the most commonly used species of Cas9). In principle, this 22-base sequence would give enough diversity such that the cut site should be unique within even a 3 billion base-pair human genome. In practice, however, some base mismatches are tolerated leading to significant potential for off-target cutting.

错配能导致显著的脱靶切割风险。这就要求做更多积极的努力，在增强靶向核酸酶的特异性的同时监测脱靶切割的位点。最开始对 Cas9 脱靶切割的研究只是集中在对近同源位点切割的特意搜寻，最近这些研究通过带有更少偏向性的全基因组筛查得到了补充。这些研究可以分为两大类：基于细胞的和无细胞的（体外）。

基于细胞的全基因组分析

表面上，进行单细胞水平的全基因组测序，看起来能够为 Cas9 基因编辑的准确性提供确定性的评估。然而，证明不存在脱靶切割所需的测序深度目前难以在细胞群体中实现。但是，估算该系统检测脱靶编辑的灵敏度还是有可能的。用这种分析方法检测不到基因编辑，说明脱靶编辑速率低于检测水平。

整合酶缺陷型慢病毒载体（IDLV）捕获（图 A-4a）是一种全基因组水平的方法，用于评估基因组编辑核酸酶的特异性，该方法最初应用于设计 ZFNs，后来也应用于 TALENs 和 CRISPR／Cas9（Gabriel et al.，2011；Wang et al.，2015）。这种方法是基于具有线性双链 DNA 基因组的 IDLV 的 NHEJ（非同源末端连接修复）将基因捕获到核酸酶诱导的 DSB（DNA 双链断裂缺口）位点中。虽然 IDLV 捕获法能直接识别发生在活细胞中的 DSB，但它是相对不敏感的，并且背景很高。为了克服这些限制，开发了通过测序（GUIDE-seq）（图 A-4b）实现的 DSBs 的全基因组的无偏差鉴定（Tsai et al.，2015）。GUIDE-seq 利用钝端、末端保护的双链寡脱氧核苷酸（dsODN）标签的有效整合，随后进行标签特异性扩增和高通量测序。GUIDE-seq 可以检测在细胞群体中由 Cas9-sgRNA 诱变的低频率（<0.1%）的脱靶位点，即使在仅有几百万个测序读数的情况下依然能发挥作用。

高通量全基因组移位测序（HTGTS）（图 A-4c）是另一种鉴定活细胞中的 Cas9 脱靶切割的全基因组方法（Chiarle et al.，2011）。HTGTS 方法基于检测核酸酶诱导的"诱饵"DSB 和脱靶"猎物"DSB 之间的易位。HTGTS 的缺陷是，核酸酶诱导的易位是罕见的，因此需要大量的输入基因组用于检测。在固定细胞中检测全基因组核酸酶诱导的 DSBs 的策略称为"BLESS"，利用断裂标记、链霉亲和素富集和下一代测序等方法，通过在固定和透化细胞

This has motivated efforts both to monitor the sites of off-target cutting as well as to enhance the specificity of the targeted nucleases. Initial efforts to define off-target cutting of Cas9 focused on specific searches for cutting at near-cognate sites. More recently these have been complemented by less biased genome-wide efforts. These can be divided into two broad classes: cell-based and cell-free (*in vitro*).

Genome-Wide Cell-Based Assays

Ostensibly, whole-genome sequencing (WGS), when conducted at a single-cell level, would seem to provide a definitive assessment of the accuracy of Cas9 genome editing. However, the depth of sequencing that would be required to certify the absence of off-target cutting is currently difficult to achieve for populations of cells. It should be possible, however, to estimate the sensitivity of the system for detecting off-target editing. Failure to detect editing with the assay would then indicate that the off-target editing rate was below the detection level.

Integrase-defective lentiviral vector (IDLV) capture (see Figure A-4a) is a genome-wide approach used to evaluate the specificity of genome-editing nucleases that was initially applied to engineered ZFNs and then later applied to TALENs and CRISPR/Cas9 (Gabriel et al., 2011; Wang et al., 2015). This method is based on the capture by NHEJ of IDLVs, which have linear double-stranded DNA genomes, into sites of nuclease-induced DSBs. Although the IDLV capture method directly identifies DSBs that occur in living cells, it is relatively insensitive and has a high background. To overcome these limitations, genome-wide unbiased identification of DSBs enabled by sequencing (GUIDE-seq) (see Figure A-4b) was developed (Tsai et al., 2015). GUIDE-seq exploits the efficient integration of a blunt, end-protected, double-stranded oligodeoxynucleotide (dsODN) tag, followed by tag-specific amplification and high-throughput sequencing. GUIDE-seq can detect off-target sites that are mutagenized by Cas9-sgRNAs with low frequencies (<0.1 percent) in a cell population, even with only a few million sequencing reads.

High-throughput genome-wide translocation sequencing (HTGTS) (see Figure A-4c) is another genome-wide method that identifies Cas9 off-target cleavage in live cells (Chiarle et al., 2011). HTGTS is based on the detection of translocations between a nuclease-induced "bait" DSB and off-target "prey" DSBs. A limitation of HTGTS is that nuclease-induced translocations represent rare events and thus require large numbers of input genomes for detection. A strategy for detecting genome-wide nuclease-induced DSBs in fixed cells, termed "BLESS" for breaks labeling, enrichment on

核中生物素化的发夹衔接子的直接原位连接捕获在细胞群中瞬间存在的瞬时 DSBs（图 A-4d）。

体外的全基因组分析

Digenome-seq（图 A-4e）是使用 Cas9 切割的基因组 DNA 的全基因组测序检测基因组 DNA 中核酸酶诱导的 DSBs 的体外方法。基因组 DNA 是通过分离并在体外用高浓度的 Cas9 / 导向 RNA 处理以使脱靶切割达到最大化，并通过 DNA 测序鉴定切割位点。因为该测定在纯化的 DNA 上离体进行，所以其不受基于细胞的因素，如染色质环境、表观遗传因子、亚细胞水平定位或适合效应的限制。Digenome-seq 因此可以在基于细胞的方法中被掩盖的位点处检测潜在的额外脱靶切割，故其用于体内分析会产生高估的脱靶事件。

增强 CRISPR / CAS9 的特异性

鉴于 CRISPR / Cas9 系统的易用性、灵活性和多功能性，CRISPR / Cas9 系统正迅速成为基因编辑的首选工具。然而，对不希望出现的脱靶效应的潜在风险的关注已经主导了最近的许多讨论。大多数检测到显著脱靶的实验都是在癌细胞中进行的（Fu et al., 2013; Hsu et al., 2013），在这些实验中可能存在变异的 DNA 修复途径，从而导致更多脱靶事件的发生。相比之下，在整个生物体，如小鼠（Yang et al., 2013）、灵长类动物（Niu et al., 2014）、斑马鱼（Auer et al., 2014）或秀丽隐杆线虫（Dickinson et al., 2013）中进行的实验所报道的，脱靶频率很低或根本检测不到，这与 CRISPR / Cas9 介导的基因靶向的高特异性一致。有可能在非转化细胞中，细胞通过内源性 DNA 损伤反应有效地逆向选择脱靶切割。人多能干细胞（hPSC）是具有基因完整质量控制机制的原代细胞，并且看起来有可能在 hPSC 或正常体细胞中观察到的脱靶事件相比在癌细胞中的更少。然而，重要的是确定是否存在倾向于脱靶事件积累的特定细胞类型和条件。为了解决对脱靶情况的担忧，目前正在开发用于最小化脱靶风险的各种方法。基于已经取得的进展，预计在不久的将来，许多基因编辑方法中的脱靶风险可以被显著地降低（如果没有消除的话）。以下是三种方法和相关进展。

streptavidin, and next-generation sequencing, captures a snapshot of transient DSBs that are present at a moment in time in a population of cells by direct in situ ligation of biotinylated hairpin adaptors in fixed and permeabilized cell nuclei (see Figure A-4d).

Genome-Wide in *In Vitro* Assays

Digenome-seq (see Figure A-4e) is an *in vitro* method for detection of nuclease-induced DSBs in genomic DNA using whole-genome sequencing of Cas9-cleaved genomic DNA. Genomic DNA is isolated, and treated with high concentrations of Cas9/guide RNA *in vitro* to maximize off-target cleavage and cleavage sites are identified by DNA sequencing. Because this assay is performed *in vitro* on purified DNA, it is not limited by cell-based factors such as chromatin context, epigenetic factors, subnuclear localization, or fitness effects. Digenome-seq thus may detect potential additional off-target cleavage at sites that would otherwise be obscured in cell-based methods. Thus, it could yield an overestimate of off-target events *in vivo*.

ENHANCING SPECIFICITY OF CRISPR/CAS9

Given its ease of use, flexibility, and versatility, the CRISPR/Cas9 system is rapidly becoming the tool of choice for gene editing. However, concerns about the potential risk of unwanted off-target effects have dominated many recent discussions. Most experiments that have detected significant off-targets have been performed in cancer cells (Fu et al., 2013; Hsu et al., 2013), which may have altered DNA repair pathways that could lead to elevated off-target events. In contrast, experiments in whole organisms such as mice (Yang et al., 2013), primates (Niu et al., 2014), zebrafish (Auer et al., 2014), or *Caenorhabditis elegans* (Dickinson et al., 2013) reported off-target frequencies that were low or not detectable, consistent with the high specificity of the CRISPR/Cas9-mediated gene targeting. It is possible that, in non-transformed cells, off-target cleavages are efficiently counter-selected by the endogenous DNA-damage response. Human pluripotent stem cells (hPSCs) are primary cells with genetically intact quality control mechanisms, and it seems possible that off-target events will accumulate less frequently in hPSCs or in normal somatic cells than has been observed in cancer cells. Nevertheless, it will be important to determine whether there are specific cell-types and conditions that predispose for the accumulation of off-target events. To address concerns about off-target events, diverse approaches to minimize mistargeting are being developed. Based on progress already made, it is anticipated that the risk of off-target events may be dramatically reduced, if not eliminated, in the near future for many genome-editing approaches. Below are three approaches and related progress.

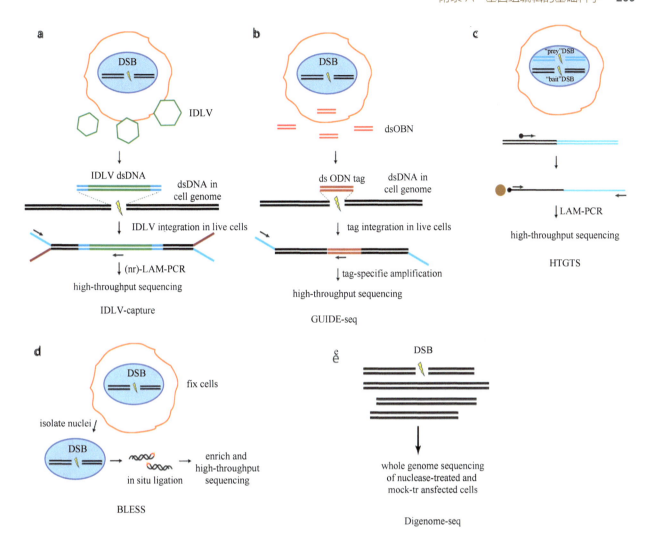

图 A-4　全面检测脱靶切割情况的策略。
FIGURE A-4　Strategies for globally detecting off-target cutting events.
（a）整合酶缺陷型慢病毒载体（IDLV）捕获。IDLV（绿色）与选择性标记整合到活细胞中核酸酶诱导的双链断裂（DSB）的位点中。通过线性扩增介导的 PCR（LAM-PCR）回收整合位点，然后进行高通量测序。（b）通过测序实现的 DSB 的全基因组无偏差鉴定（GUIDE-seq）。末端保护的短双链寡脱氧核苷酸（dsODN）有效地整合入活细胞中核酸酶诱导的 DSB 位点。该短序列用于标签特异性扩增，随后进行高通量测序以鉴定脱靶切割位点。（c）高通量全基因组转位测序（HTGTS）。核酸酶在细胞中表达以产生"猎物"和"诱饵"DSB。使用针对靶诱饵 DSB 连接设计的生物素化引物，通过 LAM-PCR 和基于链霉亲和素的富集来回收猎物和诱饵之间的易位，用于高通量测序。通过分析这些易位连接来鉴定脱靶切割位点（猎物）。（d）断裂标记、链霉亲和素上的富集和下一代测序（BLESS）。核酸酶处理的细胞被固定，完整的核被分离和透化，然后测序衔接子原位连接到瞬时核酸酶诱导的 DSB。衔接子连接的片段被富集并扩增用于高通量测序。（e）消化基因组测序（Digenome-seq）。从细胞中分离基因组 DNA，并在体外用 Cas9 核酸酶处理。将测序衔接子连接，并在标准全基因组测序覆盖率下进行高通量测序。连续序列读数的缺失部分即为切割位点。
来源：Tsai 和 Joung，2016。

(a) Integrase-defective lentiviral vector (IDLV) capture. IDLVs (green) are integrated with a selectable marker into sites of nuclease-induced double-stranded breaks (DSBs) in living cells. Integration sites are recovered by linear amplification-mediated PCR (LAM-PCR), followed by high-throughput sequencing. (b) Genome-wide unbiased identification of DSBs enabled by sequencing (GUIDE-seq). An end-protected, short, double-stranded oligodeoxynucleotide (dsODN) is efficiently integrated into sites of nuclease-induced DSBs in living cells. This short sequence is used for tag-specific amplification followed by high-throughput sequencing to identify off-target cleavage sites. (c) High-throughput genome-wide translocation sequencing (HTGTS). Nuclease is expressed in a cell to generate a "prey" and "bait" DSB. Using a biotinylated primer designed against the targeted bait DSB junction, translocations between prey and bait are recovered by LAM-PCR and streptavidin-based enrichment for high-throughput sequencing. Off-target cleavage sites (prey) are identified by analysis of these translocation junctions. (d) Breaks labeling, enrichment on streptavidin, and next-generation sequencing (BLESS). Nuclease-treated cells are fixed, intact nuclei are isolated and permeabilized, and then sequencing adapters are ligated in situ to transient nuclease-induced DSBs. Adapter-ligated fragments are enriched and amplified for high-throughput sequencing. (e) Digested genome sequencing (Digenome-seq). Genomic DNA is isolated from cells and treated with Cas9 nuclease *in vitro*. Sequencing adapters are ligated and high-throughput sequencing is performed at standard whole-genome sequencing coverage. Absence of continuous sequence reads identifies cleavage sites.
SOURCE: Modified from Tsai and Joung, 2016.

Cas9 结构的修饰

基于蛋白质工程的方法可用于开发更好的 Cas9 变体，这些变体可能更准确和有效。基于 CRISPR / Cas9 结构的研究，最近发表的两篇论文报道了人工修饰 Cas9 蛋白的方法，使得所得 Cas9 蛋白脱靶率显著降低（Kleinstiver et al.，2016；Slaymaker et al.，2016）。虽然这两个研究着眼于 Cas9 的不同 DNA 结合结构域，但是采用了降低其与非特异性 DNA 结合的相对亲和力的共同策略。这显著提高了 Cas9 切割的特异性，而又不明显影响其整体效率。这些尝试是令人振奋的，但通过对 CRISPR / Cas9 结构认识的深入，毫无疑问，其他方法将会很快出现，并进一步加速这一进程（Haurwitz et al.，2012；Jinek et al.，2014；Jore et al.，2011；Staals et al.，2013；Wiedenheft et al.，2009，2011）。

Cas9 与修饰的切割位点的工程组合

这种方法使用两个目标 DNA 切割片段，以确保比单个目标切割更好的保真度。其基本原理是：Cas9 蛋白具有两个活性 DNA 切割位点，包括天冬氨酸 D10 和组氨酸 H840，每个负责切割 DNA 的一条单链，从而产生 DSB（Jinek et al.，2014）。基于这个特征，有两种方法可以使 Cas9 失活：单失活和双失活（Guilinger et al.，2014；Ran et al.，2013；Tsai et al.，2014）。第一种方法使用具有单一失活的 Cas9，也称为 Cas9 切口酶（Cas9n），其原理是其中一个活性位点残基（D10 或 H840）被丙氨酸（A）替换，从而产生能切割双链 DNA 中一条链的 Cas9 蛋白。因此，通过提供两种向导 RNA，在 Cas9n 单切割酶二聚体的介导下于紧密接近的相对链上进行直接剪切，能够高效地获得 DSB，还可以同时激活 NHEJ 和 HDR 机制以产生 DSB（见图 A-3b）。即使间隔远至 100bp，这些切口仍然非常有效。在第二种方法中，Cas9 的两个切割位点都失活，能够产生核酸酶缺陷的"死"Cas9（dCas9），并随后与 FokI 切割结构域融合。与 ZFN 和 TALEN 一样，两种 Cas9 单体为 FokI 二聚化和 DNA 切割提供识别特异性（见图 A-3c）。此外，这两种方法都要求靶点之间有适当长度的间隔序列：如果间隔序列太短，将存在两种竞争性 Cas9 蛋白；如果间隔序列太长，则更难以执行有效的切割。由

Modification of Cas9 Structure

Protein engineering approaches can be deployed to develop better Cas9 variants that may be more accurate and efficient. Based on studies of CRISPR/Cas9 structure, two recent papers report engineered substitutions such that the resulting Cas9 protein has a significantly reduced off-target rate (Kleinstiver et al., 2016; Slaymaker et al., 2016). The two studies focused on different DNA-binding domains of Cas9 but took the common strategy of reducing the relative affinity for nonspecific DNA binding. Remarkably this enhanced the specificity of Cas9 cutting without obviously impairing its overall efficiency. While these attempts are encouraging, other strategies will undoubtedly be forthcoming and should further improve the process, based on knowledge of CRISPR/Cas9 structure (Haurwitz et al., 2012; Jinek et al., 2014; Jore et al., 2011; Staals et al., 2013; Wiedenheft et al., 2009, 2011).

Engineered Combinations of Cas9 with Modified Cleavage Sites

This approach involves the use of two targeted DNA cuts, ensuring better fidelity than a single target cut. The basic rationale is that Cas9 protein has two active DNA-cleaving sites, involving aspartic acid, D10, and histidine, H840, each responsible for cutting a single strand of DNA thus generating the DSB (Jinek et al., 2014). Based on this feature, there are two ways of inactivating Cas9: single inactivation and double inactivation (Guilinger et al., 2014; Ran et al., 2013; Tsai et al., 2014). In one approach, Cas9 with a single inactivation, also named Cas9 nickase (Cas9n), in which only one of the active site residues (D10 or H840) is replaced with alanine (A), yields a Cas9 protein capable of cutting one strand of double-stranded DNA. Consequently, providing two guide RNAs that direct cutting on opposite strands in close proximity mediated by a dimer of Cas9n single cutters leads to an effective DSB and stimulation of both NHEJ and HDR to yield a DSB (see Figure A-3b). Remarkably these nicks can be effective even as far apart as ~100bp. In a second approach, both cleavage sites of Cas9 are inactivated, yielding nuclease-deficient "dead" Cas9 (dCas9), which is then fused to the FokI cleavage domain. As with ZFNs and TALENs, two Cas9 monomers provide recognition specificity for FokI dimerization and DNA cleavage (see Figure A-3c). In addition, these two strategies require appropriate length spacers between the targets: if the spacers are too short, there will be two competing Cas9 proteins, and if the spacers are too long, it is more difficult to execute effective cutting. Because of these stringent requirements, the off-target

于存在这些严格的要求,尽管靶向选择的难度增加,但脱靶率大大降低。已发表的成果显示,这些策略可以显著降低脱靶率(Guilinger et al.,2014;Ran et al.,2013;Tsai et al.,2014)。

Cas9 的碱基编辑:不发生双链 DNA 断裂的基因组编辑

为了提高在基因组 DNA 中进行点突变的效率和精确度,Liu 团队最近开发了名为"碱基编辑"的 Cas9 变体(图 A-5),其由具有胞嘧啶脱氨酶结构域的高度工程化的融合蛋白组成,而且该结构域的胞嘧啶被转化为尿嘧啶,从而在没有 DNA 骨架双链切割的情况下实现不可逆的 C→T 取代(或通过靶向互补链的 G→A 取代)(Komor et al.,2016)。

因为胞嘧啶脱氨酶仅作用于单链 DNA,所以胞嘧啶向尿嘧啶转变的活性将靶向于被取代 DNA 链上的向导 RNA 指定的原型间隔序列 5′ 端附近的大约 5 个核苷酸的"小窗口"。通过避免双链断裂、外源的 DNA 模板入侵和随机的 DNA 修复过程,碱基编辑在未修饰的哺乳动物细胞中引入点突变的效率高达 75%,而点突变校正和插入缺失的比例则

rate is greatly reduced, albeit with increased difficulty in target selection. These strategies have been shown to significantly reduce off-target rates (Guilinger et al., 2014; Ran et al., 2013; Tsai et al., 2014).

Cas9 Base-Editor: Genome Editing Without Double-Stranded DNA Breaks

In an effort to increase the efficiency and precision of making point mutations in genomic DNA, the Liu group recently developed the so called "base editor" variant of Cas9 (see Figure A-5), which consists of a highly engineered fusion protein that recruits a cytosine deaminase domain which converts cytosine to uracil, thereby effecting an irreversible C→T substitution (or G→A by targeting the complementary strand) without double-stranded cleavage of the DNA backbone (Komor et al., 2016).

Because cytosine deaminase acts only on single-stranded DNA, the C to U conversion activity is targeted to a small window of ~5 nucleotides near the 5' end of the guide RNA-specified protospacer sequence on the displaced DNA strand. By avoiding double-stranded breaks, exogenous DNA templates, and stochastic DNA-repair processes, base editing introduces point mutations in unmodified mammalian cells with an efficiency as high as 75 percent and with a ratio of point mutation correction:indels exceeding 20:1. Moving forward it will

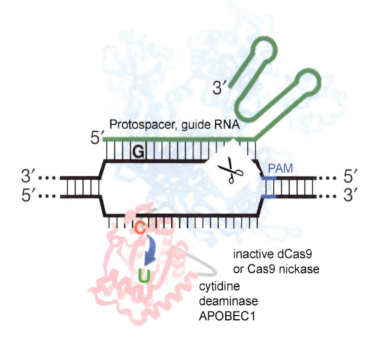

图 A-5 使用 Cas9 核酸酶变体的 CRISPR/Cas9 基因组编辑原理示意图。Cas9 的"碱基编辑者"变体,在改变基因组 DNA 的同时提高了效率和精度。
FIGURE A-5 Schematic of a CRISPR/Cas9 genome-editing system using a variant of the Cas9 nuclease. The "base editor" variant of Cas 9 provides increased efficiency and precision when making changes to genomic DNA.
来源:Komor 等,2016
SOURCE: Komor et al., 2016.

超过20:1。展望未来，开发可替代的碱基编辑器将是非常重要的，它能介导更广泛的遗传变化，我们的理想是它最终能允许在用户自定义的窗口内将任何碱基转化为其他的碱基。此外，有效评估胞嘧啶脱氨基结构域在Cas9介导的RNA靶向的DNA区域之外引入突变的倾向也是很重要的。

另一种可用于改善HDR的替代策略叫"CORRECT"（consecutive re-guide or re-Cas steps to erase CRISPR/Cas-blocked targets），根据观察结果，通过掺入CRISPR/Cas阻断突变，HDR精确度显著增加，并能获得所需的功能突变（Paquet et al., 2016）。这种方案能防止可能破坏的HDR产物的重新破裂和潜在的NHEJ修复。作者表示，通过控制所引入的Cas9介导的双链断裂的位置相关的点突变，可以改变突变的效率，从而在人诱导多能性干细胞（human induced pluripotent stem cell，hiPSC）中产生杂合或纯合的改变。

基因编辑的质量控制和质量保证

基因组编辑的特异性在临床应用中比在实验室研究中更重要。作为一种医疗产品或医疗手段，基因组编辑必须是安全的、有效的和划算的。开发基因组编辑的监管框架需要解决与这些需求相关的各种问题。如上所述，尽管技术进步可能使脱靶事件成为可控的问题，但是这种考虑仍然是必须由质量控制（quality contorl，QC）和质量保证（quality assurance，QA）程序来解决的问题。在体细胞基因组编辑中，可以相对容易地建立用于解决脱靶改变的方法或程序，但是这样的比例可能在细胞类型之间变化，因此需要检测靶向的每种细胞类型中的脱靶事件。虽然具体细节可能因情况而异，但也必须遵循检测确保安全性和有效性的一般原则。相比之下，如果人们要对胚胎进行基因编辑，那么这种监测将是相当困难的。需要开发功能等同性测定方法，并将其作为质量控制的有效措施。同时也可以考虑替代方案，比如对精子祖细胞进行基因编辑。

使用DCAS9调节转录或进行表观遗传修饰

基因组编辑的替代策略包括使用Cas9的催化

be important to develop alternate base-editors capable of mediating a wider range of genetic changes ideally ultimately allowing conversion of any base to any other base within a user-defined window. Additionally, it will be important to robustly evaluate the propensity of the cytosine deaminase domain to introduce mutations outside of the DNA region targeted by the Cas9 guide RNA.

An alternative strategy for improving HDR, termed "CORRECT," (consecutive re- guide or re -Cas steps to erase CRISPR/Cas-blocked targets), exploits the observation that HDR accuracy is increased dramatically by incorporating silent CRISPR/Cas-blocking mutations along with the desired functional mutations (Paquet et al., 2016). This prevents recleavage and potential NHEJ repair that might disrupt successful HDR products. The authors show that by controlling the location of the introduced point mutation relative to the Cas9-mediated double -stranded break they can alter the efficiency of mutagenesis, generating either heterozygous or homozygous alterations in human-induced pluripotent stem cells (hiPSCs).

QUALITY CONTROL AND QUALITY ASSURANCE FOR GENE EDITING

The specificity of genome editing is even more important in clinical applications than in laboratory research. As a medical product or medical practice, genome editing must be safe, efficacious, and cost-effective. Development of a regulatory framework for genome editing will need to address various issues associated with these requirements. As discussed above, although technical advances will likely make off-target events a manageable issue, this consideration will remain a concern that must be addressed by quality control (QC) and quality assurance (QA) procedures. In somatic genome editing, it may be relatively easy to set up assays or procedures to address off-target changes but it is probable that such rates will vary among cell types, necessitating measurement of off-target events in each cell type targeted. Although the specifics may vary from case to case, the general principle of monitoring to ensure safety and efficacy should be implemented. In contrast, it would be quite difficult to monitor embryos if they were to undergo editing. Functional equivalence assays need to be developed and agreed upon to serve as quality control measures. Alternatives may be considered, such as editing performed on sperm progenitor cells.

USE OF DCAS9 TO REGULATE TRANSCRIPTION OR TO MAKE EPIGENETIC MODIFICATIONS

An alternate strategy for genome editing involves the use of catalytically dead variants of Cas9 (dCas9). This yields

死亡变体（dCas9）。它能产生可编程的 DNA 结合蛋白，而不产生单链或双链断裂，因此通常不导致基因组 DNA 序列的任何改变。然而，通过将不同的效应结构域融合到dCas9，可用于打开（CRISPRa）（Gilbert et al.，2014；Konermann et al.，2015；Perez-Pinera et al.，2013）或关闭（CRISPRi）转录（Gilbert et al.，2013；Qi et al.，2013），使表观遗传标记（调节基因表达的染色质的修饰）的特异性位点发生改变。也许，大多数（如果不是全部）这样的表观遗传改变不能传递给后代，从而减轻了围绕种系编辑方法的一些担忧。出于同样的原因，这些变化的瞬时性质限制了它们用于校正由基因突变引起的疾病的效用。然而，这种瞬时的种系编辑工程的可能用途包括：扩增生殖细胞，或在体外产生所需的干细胞，或终末分化的细胞。此外，这种变化的瞬时性质可以扩大安全靶向的基因的数量。例如，CCR5 HIV 共同受体的瞬时下调可以防止 HIV 的垂直传递，而且这种策略可以应用到其他的病毒受体。此外，可以想象，基因表达的瞬时改变可以导致胚胎发育的永久改变，这可以改善引起疾病的遗传突变的影响。因为没有对个体进行永久性、可遗传的改变，使用 dCas9 改变基因表达确实可减轻一些伦理问题。然而，目前 dCas9 在胚胎的潜在用途似乎相对于涉及种系编辑的方法相当受限，而 dCas9 更直接的医疗应用可能涉及体细胞基因表达的改变。

转基因动物的基因组靶向

基因突变可能导致发育异常和疾病。在过去几十年中，科学家在研究基因突变所造成的后果方面有了重大进展，也就是成功开发了将实验上设计的靶向突变引进如小鼠、果蝇和斑马鱼等生物体的技术，从而提供了理解胚胎发育和疾病遗传学基础的重要工具。这些方法也可以用于纠正缺陷基因。在描述当前精确的基因组编辑方法的应用之前，我们将总结最初开发的用于遗传修饰动物的主要步骤。

外源 DNA 的随机插入

动物的遗传操作已然是理解胚胎发育和人类疾病等大量研究的基础。一种有效的技术是基于在体

a programmable DNA-binding protein that is incapable of generating either single- or double-stranded breaks and thus does not typically lead to any changes in the DNA sequence of the genome. However, by fusing different effector domains to dCas9, it can be used either to turn on (CRISPRa) (Gilbert et al., 2014; Konermann et al., 2015; Perez-Pinera et al., 2013) or turn off (CRISPRi) transcription (Gilbert et al., 2013; Qi et al., 2013) or to make locus-specific changes to the epigenetic marks (modifications of the chromatin that regulate gene expression). It is likely that most if not all such epigenetic changes would fail to be passed to subsequent generations thus alleviating some of the concerns surrounding germline-editing approaches. By the same token, the transient nature of these changes limits their utility for correcting diseases caused by genetic mutations. Possible uses for such transient germline engineering, however, include the ability to expand germ cells, or the *in vitro* generation of desired stem cells or terminally differentiated cells. Additionally, the transient nature of the changes could expand the number of genes that could be safely targeted. For example, transient down-regulation of the HIV CCR5 coreceptor could protect against vertical passage of HIV, and this strategy could be expanded to other viral receptors. Additionally, it is possible to imagine that transient alterations to gene expression could lead to permanent developmental changes in an embryo, which could ameliorate the effects of disease-causing inherited mutations. Because no permanent, heritable changes are made to the individual, the use of dCas9 to alter gene expression does alleviate some ethical concerns. Nonetheless, at present the potential uses of dCas9 on embryos seem rather limited compared to approaches involving germline editing, and the more immediate therapeutic applications of dCas9 likely involve somatic alteration in gene expression.

GENOME TARGETING IN TRANSGENIC ANIMALS

Gene mutations can lead to abnormal development and to disease. Over the past several decades, a major advance in studying the consequence of mutations has been the development of the ability to experimentally introduce designed, targeted mutations into genes of organisms such as mice, fruit flies, and zebrafish, thus providing an important tool to understand the molecular-genetic basis of embryonic development and of disease. These methods can also be used to correct defective genes. Before describing the application of current precise genome-editing methods, we will summarize the major steps that were initially developed to genetically modify animals.

Random Insertion of Foreign DNA

The genetic manipulation of animals has been the basis for much of the research aimed at understanding embryonic development and human diseases. One powerful

外操作小鼠胚胎,并将胚胎转移到寄养母亲体内,以产生遗传背景发生改变的动物(图 A-6)。使用这些方法,最初采用 SV40 DNA(Jaenisch and Minz,1974)和随后的逆转录病毒(Janeisch,1976)将外源基因转入早期小鼠胚胎,遵循孟德尔期望理论,会产生将外源 DNA 转移到下一代的第一代转基因小鼠。最广泛使用的产生遗传修饰动物的方法是将 DNA 显微注射到受精的小鼠或果蝇的卵中,从而产生大量在其生殖细胞携带外源 DNA 的转基因小鼠或果蝇(Brinster et al.,1981;Costantini and Lacy,1981;Gordon and Ruddle,1981;Rubin and Spradling,1982)。

外源 DNA 随机整合到动物基因组中,可导致内源基因的破坏,从而失活。在这种"插入突变"的方法中,整合到基因组中的 DNA 可用来作为分离和鉴定突变基因的分子标签。胶原 I 基因是通过逆转录病毒插入突变失活的第一个基因,导致突变小鼠产生类似于脆性骨病的表型(Schnieke et al.,1983),这是由胶原基因突变引起的骨骼系统的主要疾病。类似的方法还有,将 DNA 注射到合子原核,通过插入突变的方法,产生突变小鼠(Mahon et al.,1988)。除了插入突变外,基因整合到基因组中还可以引起邻近基因的转录激活,其已经被广泛应用于研究异位表达转基因的结果(Hammer et al.,1984)。

尽管整合到基因组中导致插入突变或异位表达转基因在产生转基因动物中是有效的,但是该方法具有不可预测性,因为 DNA 插入基因组是随

technology is based on manipulating mouse embryos *in vitro* and transferring the embryos to a foster mother to produce genetically altered animals (see Figure A-6).

Using these techniques, initially SV40 DNA (Jaenisch and Mintz, 1974) and later retroviruses (Jaenisch, 1976) were introduced into early mouse embryos leading to the generation of the first transgenic mice that transmitted the foreign DNA to the next generation according to Mendelian expectations. The most widely used method to generate genetically modified animals was microinjection of DNA into fertilized mouse or *Drosophila* eggs leading to the production of a large number of transgenic mice or flies that carried foreign DNA in their germline (Brinster et al., 1981; Costantini and Lacy, 1981; Gordon and Ruddle, 1981; Rubin and Spradling, 1982).

The random integration of foreign DNA into the genome of an animal can cause disruption of an endogenous gene leading to its inactivation. In this "insertional mutagenesis" approach the integrated DNA is used as a molecular tag for the isolation and identification of the mutated gene. The collagen I gene was the first endogenous gene inactivated by retroviral insertional mutagenesis resulting in mutant mice whose phenotype resembled brittle bone disease (Schnieke et al., 1983), a major disease of the skeletal system caused by mutations in a collagen gene. Similarly, injection of DNA into the zygote pronucleus produced mutant mice by insertional mutagenesis (Mahon et al., 1988). In addition to insertional mutagenesis, the integration of a gene into the genome can result in transcriptional activation of nearby genes, which has been widely used to study the consequences of ectopic transgene expression (Hammer et al., 1984).

While integration into the genome leading to insertional mutagenesis or ectopic transgene expression is efficient in generating transgenic animals, the approach suffers

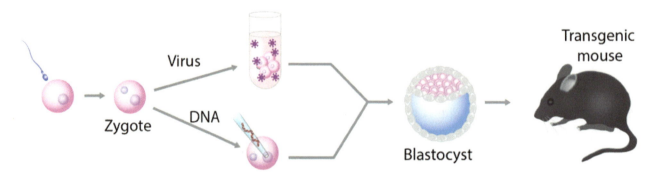

图 A-6 随机插入转基因的转基因小鼠的产生。4~8 细胞期的胚胎用逆转录病毒感染或将 DNA 注射到受精卵的雄原核中。孵育胚胎直到它们达到囊胚期(约 100 个细胞),并移植到代孕母鼠体内。外源 DNA 序列被随机整合到所得的转基因小鼠的基因组中。

FIGURE A-6 The generation of transgenic mice with randomly inserted transgenes. Embryos at the 4 to 8 cell stage are infected with retroviruses or DNA is injected into the male pronucleus of the zygote. The embryos are incubated until they reach the blastocyst stage (~100 cells) and are transplanted into a foster mother. The exogenous DNA sequences are randomly integrated into the genome of the resulting transgenic mice.

机的，无法做到靶向基因或实现可预测的转基因表达。

基因靶向胚胎干细胞

最初从小鼠囊胚分离出来的胚胎干细胞（embtyonic stem cell，ES）可以分化为成体的各种细胞类型（Evans and Kaufman，1981；Martin，1981）。有趣的是，当注射到小鼠囊胚中，ES 细胞可以整合到发育的胚胎中，用于产生所有体细胞组织，并产生"嵌合小鼠"。特别重要的是，ES 细胞能够分化为生殖细胞，从而实现从培养的细胞中衍生出转基因动物。因此，该方法作为 ES 细胞体外操作的结果，可产生内源基因改变的小鼠。第一个来源于 ES 细胞的小鼠品系携带有 HPRT 失活基因，该基因突变在人类中会导致严重的精神障碍，即 Lesch-Nyhan 综合征。通过使用 HAT 培养基可以杀死正常细胞，筛选出 HPRT 突变的 ES 细胞，从而实现 HPRT 突变的 ES 细胞的分离（Kuehn et al.，1987）。这种筛选方法虽然对 HPRT 成功，但不能用于编辑其他基因。

同源重组

同源重组的发现代表一个重大突破，因为它允许编辑任何基因（Doetschman et al.，1987；Thomas and Capecchi，1987）。基因的靶向编辑需要制备靶向载体，该载体中需含有与所需修饰的内源基因侧翼序列同源的 DNA 片段（见图 A-1）。将载体转染到 ES 细胞中，并挑选正确靶向修饰的克隆（图 A-7a）。将携带目的修饰的细胞注射到小鼠囊胚中从而产生嵌合小鼠（图 A-7b），将嵌合小鼠与正常小鼠交配从而获得携带突变等位基因的后代（图 A-7c）。同源重组与 ES 细胞结合，允许科学家有效地获得向下一代传递特定基因突变的小鼠。在最初的携带 β_2 微球蛋白和 c-abl 基因的靶向突变的小鼠产生后（Schwartzberg et al.，1989；Zijlstrastra et al.，1989），ES 细胞的同源重组已经成为哺乳动物发育和人类遗传疾病的动物模型研究中广泛使用的手段（Solter，2006）。因为具有嵌合体能力的 ES 细胞仅在小鼠体系中可用，通过同源重组的基因编辑仅限于小鼠，并无现成方法用于其他物种。

from unpredictability because the insertion of DNA into the genome is random and does not allow the targeting of predetermined genes or predictable transgene expression.

Gene Targeting in Embryonic Stem Cells

Embryonic stem (ES) cells, initially isolated from mouse blastocysts, are able to differentiate into all cell types of the body (Evans and Kaufman, 1981; Martin, 1981). Of great interest was that ES cells, when injected into a mouse carrier blastocyst, could integrate into the developing embryo and contribute to all somatic tissues and generate "chimeric mice." Of particular importance was the fact that the cells were able to contribute to the germline, thus allowing the derivation of animals from the cultured cells. Thus, the approach allowed the generation of mice carrying the alteration of an endogenous gene as a result of *in vitro* manipulation of the ES cells.

The first mouse strain derived from an ES cell carrying a mutation in a predetermined gene was a strain with inactivation of the HPRT gene, which is mutated in human patients with the severe mental disorder Lesch-Nyhan syndrome. The isolation of HPRT mutant ES cells was straightforward using a culture medium (HAT) that kills normal cells and selects for cells carrying an inactivated HPRT gene (Kuehn et al., 1987). This selective approach, while successful for HPRT, cannot be used for editing other genes.

Homologous Recombination

The discovery of homologous recombination represented a major breakthrough as it allowed the editing of any gene (Doetschman et al., 1987; Thomas and Capecchi, 1987). Targeting of genes requires the generation of a targeting vector containing DNA segments homologous to sequences of the endogenous gene flanking the desired modification (see Figure A-1). The vector is transfected into ES cells, and correctly targeted clones are selected (see Figure A-7a). Cells carrying the desired modification are injected into mouse blastocysts to generate chimeric mice (see Figure A-7b), which are bred with normal mice to obtain offspring carrying the mutant allele (see Figure A-7c). Homologous recombination in combination with ES cells has allowed scientists to efficiently create mice transmitting specific gene mutations to the next generation. Following the initial generation of mice carrying targeted mutations of the 2-microglubulin and the c-abl gene (Schwartzberg et al., 1989; Zijlstra et al., 1989), homologous recombination in ES cells has become a widely used tool for the study of mammalian development and the generation of animal models of human genetic diseases (Solter, 2006). Because chimera–competent ES cells were only available in the murine system, gene editing by homologous recombination was restricted to mice and could not readily be used in other species.

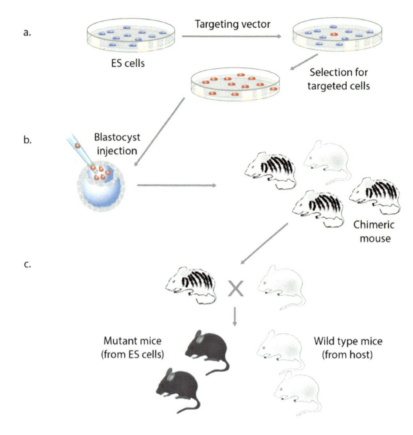

图 A-7 通过同源重组产生的确定的遗传改变的小鼠（三步法）。
FIGURE A-7 The generation of defined genetically altered mice by homologous recombination (three-step process).
(a) 在第一步中，将与靶基因具有同源性的靶向载体引进 ES 细胞中，并在培养过程中选择目的克隆。(b) 将正确靶向修饰的 ES 细胞注射到毛发白色的小鼠囊胚中，将其移植到代孕母鼠体内产生嵌合小鼠，其中供体 ES 细胞对动物组织器官形成有作用（主要从 ES 细胞得到的小鼠的毛发颜色来判断这种作用）。(c) 通过将嵌合小鼠和毛发白化的品系交配来验证 ES 细胞的生殖传递能力。毛发有颜色的后代来源于供体 ES 细胞。
(a) In the first step a targeting vector with homology to the target gene is introduced into ES cells and correctly targeted clones are selected in culture. (b) The targeted ES cells are injected into an albino host blastocyst, which is transplanted into a foster mother to produce chimeric mice where the donor ES cells contribute to the tissues of the animal (as seen by coat color contribution derived from the ES cells). (c) Germline transmission of the ES cell clone is verified by mating of the chimeric mouse with the albino host strain. Pigmented offspring are derived from the donor ES cells.

细胞核克隆和变异动物的产生

体细胞核移植到无卵核中，重置细胞核的表观遗传状态为胚胎状态，并允许产生动物，如"多利"羊（Wakayama et al.，1998；Wilmut et al.，1997）。通过细胞核克隆从体细胞产生动物的方法，允许在 ES 难以获得的物种中产生变异动物。第一例成功应用细胞核克隆与同源重组结合产生基因变异的研究中使用的农场动物细胞类型是绵羊成纤维细胞。将人 α1 抗胰酶基因插入到 COL1A1 基因的 3′UTR 中，其中 COL1A1 基因是方便的"安全港"基因座，可以使转基因得到可预测的表达。转基因绵羊通过细胞核克隆从靶向修饰的成纤维细胞衍生而来，产生表达治疗人 α1- 抗胰蛋白的动物（MaCreath et al.，2000）。

如上所述，获得携带工程改造的基因改变的动

Nuclear Cloning and the Generation of Mutant Animals

The transfer of a somatic nucleus into an enucleated egg resets the epigenetic state of the nucleus to an embryonic state and allows the generation of animals such as Dolly, the first cloned mammal (Wakayama et al., 1998; Wilmut et al., 1997). The production of animals from somatic cells by nuclear cloning allowed the generation of mutant animals in species where no ES cells were available. The first successful application of nuclear cloning in combination with homologous recombination to produce gene- altered farm animals used sheep fibroblasts. The human α1-antitrypsin gene was inserted into the 3′ UTR of the COL1A1 gene, a convenient "safe harbor" locus giving predictable expression of the transgene. Transgenic sheep were derived from the targeted fibroblasts by nuclear-cloning, generating animals that expressed the therapeutically important human α1-antitrypsin protein (McCreath et al., 2000).

The strategies to produce animals carrying engineered

物的策略依赖于对ES细胞的操作或者细胞核克隆。然而，这两种方法都是劳动密集型的，同时需要特殊技能。当有了ZFN、TALEN和CRISPR/Cas9等新型基因组编辑方法后（Doudna and Charpentier, 2014），这种情况发生了巨大的变化。上述这些方法彻底改变了研究人员在任何物种中编辑基因并获得遗传改变的动物的能力，相比基于对ES细胞操作或者核转移等策略，上述这些方法所需的努力和时间更短，而且产生基因编辑动物所需的实验技能更加简单。

胚胎的基因组编辑

常规的基因靶向中的同源重组效率低下，并且需要在细胞培养过程中筛选正确靶向的细胞克隆。在第二步中，将正确靶向的ES细胞克隆注射到宿主囊胚中以产生嵌合动物，其在第三步中通过交配获得所需的突变动物，该过程可能需要1~2年（比较图A-7）。相比之下，通过TALENs或CRISPR/Cas9的基因靶向效率很高，因为不需要筛选正确靶向的ES细胞克隆（Sakuma and Woltjen, 2014），使得可能通过直接对受精卵进行遗传操作，从而在一个步骤中获得有遗传修饰的动物（图A-8）。

CRISPR/Cas9介导的受精卵基因编辑

将向导RNA和Cas9 RNA一起注入受精卵（合子）中，可用于产生携带几个基因突变的小鼠。Cas9介导的DNA双链断裂效率很高，可以导致80%的幼崽携带两个不同基因的两个等位基因发生突变（Wang et al., 2013）。当向导RNA和Cas9 RNA与携带点突变的寡核苷酸共同注射时，

gene alterations as summarized above relied on the manipulation of ES cells or on nuclear cloning, both of which are labor-intensive and require special skills. This changed dramatically when the new genome-editing methods based on ZFNs, TALENs, and CRISPR/Cas9 became available (Doudna and Charpentier, 2014). These approaches, described above, revolutionized the ability of researchers to edit genes in any species and to produce genetically altered animals with a fraction of the effort and time and much less sophistication and experimental skills needed than were required for the generation of gene-edited animals by strategies based on ES cells or nuclear transfer.

GENOME EDITING IN EMBRYOS

Homologous recombination in conventional gene targeting is an inefficient process and requires the selection of correctly targeted cell clones in cell culture. In a second step the targeted ES cell clone is injected into a host blastocyst to create a chimeric animal, which, in a third step, is mated to produce the desired mutant animal, a process that may take as much as one or two years (compare Figure A-7). In contrast, gene targeting by TALEN or by CRISPR/Cas9 is so efficient that no selection for correct targeting is required (Sakuma and Woltjen, 2014), making it possible to derive genetically modified animals in one step by direct genetic manipulation of the fertilized egg (see Figure A-8).

CRISPR/Cas9-Mediated Gene Editing in the Zygote

The injection of guide RNAs together with Cas9 RNA into the fertilized egg (zygote) was used to generate mice carrying mutations in several genes. The efficiency of Cas9-mediated DNA double-strand breaks was high and resulted in 80 percent of pups carrying mutations in both alleles of two different genes (Wang et al., 2013). When the guide and Cas9 RNAs were co-injected with an oligonucleotide carrying a point mutation, the mutation

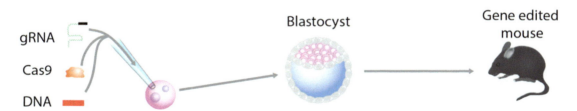

图A-8　CRISPR/Cas9靶向的受精卵基因编辑小鼠的获得（一步法）。将靶向目标基因的Cas9 RNA和向导RNA（gRNA）注射到受精卵的细胞质中。携带靶基因的两个等位基因突变的小鼠将在一个步骤中获得，且效率非常高。如果与DNA载体共同注射，外源序列将在造成双链断裂的情况下以某种概率插入到基因组中。
FIGURE A-8　The generation of gene-edited mice by CRISPR/Cas9 targeting in the zygote (one-step process). Cas9 RNA and guide RNA (gRNA) targeting a gene of interest are injected into the cytoplasm of the zygote. Mice carrying a mutation in both alleles of the targeted gene are derived in one step at a high frequency. If co-injected with a DNA vector, exogenous sequences will be inserted at some frequency at the double-strand break.

在 60% 的幼崽中将产生两个等位基因的突变。此外，需要在相同等位基因中插入两个 LoxP 位点的条件突变体也能有效地在受精卵中获得。因此，NHEJ 介导的突变及在双链断裂位点插入 DNA 是非常有效的，允许携带复杂突变的小鼠在 3 周（小鼠的妊娠时间）内即可获得，而不是像用 ES 细胞介导的基因靶向所需的 1~2 年时间。CRISPR/Cas9 介导的基因编辑方法不仅在小鼠体系中可以有效工作，在其他物种中也同样适用，如大鼠（Li et al., 2013）、斑马鱼（Hwang et al., 2013）、线虫（Friedland et al., 2013）和果蝇（Zeng et al., 2015）。重要的是，该方法使携带特定基因突变的灵长类动物的产生得以实现（Niu et al., 2014）。最近，有两个报告显示，在有缺陷的植入前人类胚胎中使用基因组编辑，能够造成无法怀孕的结果（Kang et al., 2016; Liang et al., 2015）。

上文总结的证据表明，可以通过操作受精卵在一个步骤中获得携带多个基因特定突变的动物。然而，如果用于基因治疗（校正突变基因），则需要考虑可能出现的几种重要的并发症。这些并发症包括被修饰胚胎的频繁嵌合、预期突变等位基因数量不一致，以及无法对单细胞胚胎进行基因分型。

镶嵌现象

靶基因的切割和 DNA 在双链断裂点的插入可以发生在比合子较晚的阶段，比如二细胞期。在比单细胞合子期晚的阶段发生整合的结果是半数胚胎细胞（或者更少，取决于 DNA 插入的时间）会携带改变的基因，而其他细胞则不携带。仅在一部分细胞中具有遗传改变的动物被称为"嵌合体"。已有的证据表明，嵌合体的发生率可高达 50% 或更高（Wang et al., 2013）。嵌合体的高发生率具有重要的实际应用：植入前遗传学诊断（PGD）——胚胎单个细胞的活体检测不能用于确定基因靶向是否导致期望的突变，因为单个细胞不能反映胚胎其他细胞的基因型。

Cas9 介导的切割导致的野生型等位基因突变

Cas9 切割比通过同源重组在切割位点插入供体 DNA 更加有效。如果胚胎中基因编辑的目的是校正突变的等位基因，则会造成并发症。为了校正

was introduced into two target genes in 60 percent of the pups. In addition, the generation of conditional mutants requiring the insertion of two LoxP sites into the same allele was shown to be effective in the zygote. Thus, NHEJ-mediated mutation as well as insertion of DNA at the double-strand break site is extremely efficient, allowing the generation of mice carrying complex mutations within 3 weeks (the gestation time of the mouse) instead of 1 to 2 years when using ES cell–mediated gene targeting. The CRISPR/Cas9 gene-editing method was shown not only to work efficiently in mice but also in other species including rats (Li et al., 2013), zebrafish (Hwang et al., 2013), *C. elegans* (Friedland et al., 2013), and *Drosophila* (Zeng et al., 2015). Importantly, the approach allowed the generation of primates carrying mutations in specific genes (Niu et al., 2014). Most recently, two reports appeared that used genome editing in preimplantation human embryos that were defective and therefore could not be used to generate pregnancy (Kang et al., 2016; Liang et al., 2015).

The evidence summarized above indicates that animals carrying defined mutations in multiple genes can be generated in one step by manipulation of the fertilized egg. However, if intended for gene therapy (correction of mutant genes), several significant complications need to be considered. These include frequent mosaicism of manipulated embryos, the mutation of both alleles when the goal is to correct one mutant allele and the impossibility of genotyping the one-cell embryo.

Mosaicism

The cleavage of the target gene and the insertion of DNA at the double-strand breakpoint may occur at a later stage than the zygote—such as the two-cell stage. The consequence of integration at a stage later than the one-cell zygote stage is that half (or less, depending on the time of DNA insertion) of the embryo's cells will carry the altered gene whereas the others will not. Animals with genetic alterations in only a subset of the cells are designated as "mosaics." The available evidence indicates that the incidence of mosaicism may be as high as 50 percent or higher (Wang et al., 2013). The high incidence of mosaicism has an important practical consequence: Preimplantation genetic diagnosis (PGD)—the biopsy of a single cell of the manipulated embryo—cannot be used to ascertain whether gene targeting resulted in the desired mutation because a single cell may not reflect the genotype of the other cells of the embryo.

Mutation of the Wild-Type Allele by Cas9-Mediated Cleavage

Cleavage by Cas9 is significantly more efficient than insertion of a donor DNA at the cleavage site by homologous recombination. This poses a complication if

给定的突变，将指导 RNA 和 DNA 靶向构建的载体共同注射到胚胎中。虽然 DNA 将在双链断裂时整合到突变体等位基因中并校正突变，但是另一个等位基因通常也会被切割，由 NHEJ 产生新的突变。考虑到目前的技术，这对基因治疗构成严重的问题，因为校正突变的等位基因，同时又产生了新的等位基因突变。使用小分子通过末端连接进行修复的抑制有助于改善这个问题，已有证据表明，相比于 NHEJ，通过 HDR 更有利于 DNA 插入（Maruyama et al., 2015）。然而，正常等位基因的非预期突变目前仍然是 CRISPR/Cas9 介导的基因校正的并发症。

单细胞胚胎的基因分型

如果胚胎的基因组编辑是校正突变等位基因，那么还有一个问题：如何区分野生型胚胎和突变体胚胎。如果一个亲本携带显性突变基因，那么 50% 的胚胎受到影响，50% 为野生型；如果两个亲本都携带隐形突变，则会有 75% 的胚胎为野生型，25% 受到影响。因为目前任何分子测试都不能被用来区分合子阶段的突变胚胎和正常胚胎，很多基因编辑都尝试靶向（和修改）大部分正常胚胎的基因。在可预见的未来，即使技术进步了，也可能无法解决这个问题。

基因驱动：通过性繁殖群体传播遗传改变的机制

已经有人提出天然存在的"归巢内切酶"能够导致等位基因的突变在有性繁殖的群体中快速扩散（Burt, 2003, 2014），这种现象被命名为"基因驱动"。这种突变的等位基因在群体的扩散不需要突变基因载体的选择性优势，而是像"自私基因"一样繁殖（Esvelt et al., 2014；Oye et al., 2014）。

最近，同时编码 Cas9 和向导 RNA 的一个载体被引入到控制角质层颜色的果蝇基因组位点，并产生了一个敲除突变。在生殖系发育过程中，Cas9 所介导的切割刺激了野生型位点的插入型拷贝，使得所有的雌配子中携带了这种插入（图 A-9）。重要的是，当这些卵受精后，所得到的杂合动物中这种突变在减数分裂过程中由 Cas9 所介导的靶基因切割，将会使得野生型等位基因转变为突变的等位基因，

the goal of gene editing in embryos is the correction of a mutant allele. To correct a given mutation, a guide RNA and a DNA target construct are injected into the embryo. While the DNA will integrate into the mutant allele at the double-strand break and correct the mutation, the other allele will often be cleaved, creating a new mutation by NHEJ. Given present technology, this poses a possibly serious problem for gene therapy as the mutant allele is corrected, but a new mutant allele is created. The inhibition of repair by end joining using a small molecule may help to mitigate this problem as this has been shown to favor the insertion of DNA by HDR over that by NHEJ (Maruyama et al., 2015). However, the unwanted mutation of the normal allele currently remains a complication of CRISPR/Cas9-mediated gene correction.

Genotyping of the One-Cell Embryo

Genome editing in embryos with the goal being to correct a mutant allele faces another problem: how to distinguish a wild type from a mutant embryo. If one parent carries a dominant mutant gene, 50 percent of the embryos will be affected and 50 percent will be wild type, and if both parents carry a recessive mutation, 75 percent of the embryos will be normal and 25 percent will be affected. Because it is not possible to use any current molecular test to distinguish mutant from normal embryos at the zygote stage, any gene-editing attempt will target (and modify) a large fraction of normal embryos. It is unlikely that a technological advance could resolve this dilemma in the foreseeable future.

GENE DRIVE: A MECHANISM TO SPREAD GENETIC ALTERATIONS THROUGH SEXUALLY REPRODUCING POPULATIONS

It has been proposed that naturally occurring "homing endonucleases" could cause mutant alleles to spread rapidly through sexually reproducing populations by a process designated as "gene drive" (Burt, 2003, 2014). The spreading of such a mutant allele through a population does not require a selective advantage for the carrier of the mutant gene but would rather propagate like a "selfish gene" (Esvelt et al., 2014; Oye et al., 2014).

Recently, a vector encoding both Cas9 and a guide RNA was introduced into a *Drosophila* genomic locus that governs cuticle color, creating a knock-out mutation. Cas9-mediated cleavage during development of the germline stimulated the copying of the insertion into the wild-type locus, such that all female gametes carried the insertion (see Figure A-9). Importantly, when these eggs were fertilized the mutation in the resulting heterozygous animals converted the wild-type allele into a mutant allele during meiosis by Cas9-mediated target gene

随后通过同源介导的双链 DNA 修复（HDR）产生纯核突变。因此，当将同时携带 Cas9 和向导 RNA 的载体整合到一个等位基因中时，在减数分裂期间的其他等位基因有 98% 的转换效率，造成突变的等位基因在群体中快速扩散（Gantz and Bier，2015）。这种自动催化过程被称为"诱变链反应"。

除了 Cas9 和向导 RNA 被插入到一个等位基因外，如果一个载体携带能够纠正给定突变的序列，则这段序列在减数分裂期间可以用作模版，并通过同源重组转变其他等位基因，产生两个修复的等位基因（图 A-9a）。同样的，除了 Cas9 和向导 RNA

cleavage, followed by HDR resulting in a homozygous mutation. Thus, the vector carrying the Cas9 gene and a guide RNA, when integrated into one allele, led to the conversion of the other allele during meiosis with an efficiency of 98 percent, causing the rapid spread of the mutant allele through the population (Gantz and Bier, 2015). This autocatalytic process was dubbed as a "mutagenic chain reaction."

If a vector carrying sequences correcting a given mutation in addition to the Cas9 and the guide RNA was inserted into one allele, these sequences could serve as a template during meiosis and convert the other allele by homologous recombination resulting in two repaired alleles (see Figure A-9a). Similarly, if a vector carrying

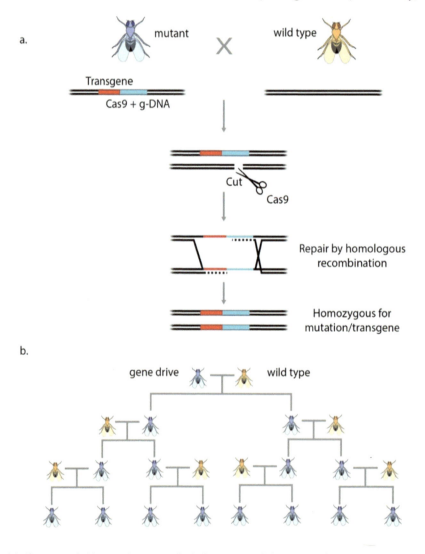

图 A-9 基因驱动。(a) 将 Cas9、向导 RNA（g-DNA，靶向基因）和"货物"DNA 序列插入到目的基因中。当在减数分裂过程中表达时，向导 RNA/Cas9 复合物将在其他等位基因上引入双链断裂，随后通过同源重组驱动修复，其将 Cas9/gRNA/转基因序列插入到野生型等位基因上。(b) 基因修饰将通过性繁殖群体迅速传播。

FIGURE A-9　Gene drive.(a) Insertion of Cas9, guide RNA (g-DNA, targeting the gene) and "cargo" DNA sequences are inserted into a gene of interest. When expressed during meiosis, the guide RNA/Cas9 complex will introduce a double-strand break onto the other allele followed by homologous recombination driven repair, which will insert the Cas9/gRNA/transgene sequences into the wild-type allele. (b) The gene modification will rapidly spread through a sexually reproducing population

SOURCE: Modified from Esvelt et al., 2014.

被插入到一个等位基因外，如果一个载体携带能够编码另一个基因的序列，Cas9 在减数分裂过程中介导的野生型等位基因的切割就能够将外源"载体"基因转移到其他等位基因上，导致产生纯合转基因动物（Esvelt et al., 2014）。在动物繁殖过程中，通过基因驱动机制，突变等位基因和新插入的基因就能在动物种群中有效地传播扩散（图 A-9b）。

到目前为止，突变载体已经被证明可以通过基因驱动在昆虫种群扩散，并且已经有人提出可将此运用于蚊子的群体控制。只是，基因驱动尚未在哺乳动物中显示出相似的表现。然而，考虑到在哺乳动物胚胎中基因组编辑的效率，这也是完全可能发生的，即使不是一模一样，潜在的基因驱动机制在哺乳动物中也是有效的，并且原则上可以产生能通过群体传播的基因修饰。然而，考虑到人类的世代时间和繁殖模式，任何这样的基因驱动扩散到人类中将需要很多很多年，似乎难以想象。

可遗传性的生殖系编辑的替代方法

CRISPR/Cas9 基因编辑技术的效率和精度提高了在细胞中进行精确基因组编辑的潜力，有助于人类生殖系的研究。对于传递到下一代的遗传改变，其必须在能够产生配子（卵子和精子）的祖细胞、卵子和精子本身，或处于受精阶段的合子或早期胚胎中进行，而且所有这些细胞需要仍处于可以对未来的生殖系统有贡献的时期。

正如上述所讨论的那样，用于生殖系编辑的方法在小鼠中被高度开发，并且应用于许多其他哺乳动物种中，特别体现在农业相关的需要或者针对用于人类遗传疾病的临床前研究的疾病动物模型的产生。直到最近出现了先进的基因组编辑工具，小鼠的生殖系基因改变主要是通过非定向转基因 DNA 导入受精卵的基因组或通过 ES 细胞中的定向诱变实现的。后面一种方法包括将靶向载体同源重组到宿主基因组中，选择正确靶向的克隆，产生传递种系的嵌合动物（细节请参见上述关于种系遗传改变的部分）。尽管这种方法已经被证实在制造敲除型小鼠、条件性突变、报道系统和多种人类疾病模型中非常强大，但它仍然相对低效，并且不容易应用于受精卵中来直接靶向基因组改造。因此，潜在目的基因改变或校正人类胚胎的想法本没有予以

sequences coding for another gene in addition to the Cas9 and the guide RNA were integrated into one allele, Cas9-mediated cleavage of the wild-type allele during meiosis would transfer the exogenous "cargo" gene into the other allele, leading to homozygous transgenic animals (Esvelt et al., 2014). Thus, gene-drive mechanisms can effectively propagate and spread mutant alleles or newly inserted genes through animal populations (see Figure A-9b).

So far gene-drive constructs have been demonstrated to spread through insect populations and have been proposed for use in mosquito population control. Gene drives have not as yet been shown to behave similarly in mammalian species. However, given the efficiency of genome editing in mammalian embryos it is possible, if not likely, that the underlying gene-drive mechanisms would also be effective in mammals and could in principle create gene modifications that could spread through the population. However, given the generation time and breeding patterns of humans, any such gene-drive application in the human species would require an inordinate numbers of years and seems inconceivable.

ALTERNATIVE ROUTES TO HERITABLE GERMLINE EDITING

The efficiency and precision of CRISPR/Cas9 gene-editing approaches have raised the potential that precise genome editing might be possible in cells that could contribute to the germline of the human species. For a genetic alteration to be passed on to the next generation, it has to be made in (1) *progenitor cells* that can give rise to the *gametes (eggs and sperm)*, (2) the eggs and sperm themselves, or (3) in the *fertilized zygote or early embryo*, when all cells can still contribute to the future germline.

As discussed in the preceding section, methods for germline editing have been most highly developed in the mouse and applied to a number of other mammalian species, particularly in relation to agricultural needs or the generation of preclinical disease models for human genetic diseases. Until the recent advent of advanced genome-editing tools, germline gene alterations in mice were primarily achieved via non-targeted introduction of transgenic DNA into the genome of the zygote or by targeted mutagenesis in ES cells. The latter approach involved the homologous recombination of targeting vectors into the host genome, the selection of the correctly targeted clones, and the generation of germline-transmitting chimeric animals (see details in the above section on germline genetic alteration). While this approach has proven to be extraordinarily powerful in producing knock-out mice, conditional mutations, reporter lines, and a variety of human disease models, it is still relatively inefficient and has not been readily applicable

考虑。

然而，近期CRISPR/Cas已经在靶向特定位点的核酸酶强化的基因组编辑效率上展现出了新的应用前景，包括可能的人类种系编辑。在证明CRISPR/Cas可以非常有效地靶向识别哺乳动物细胞基因组中的特定位点后不久，实验显示该方法可以直接应用于小鼠的受精卵，而不需要中间的胚胎干细胞的步骤（见上面章节和图A-8）。因此，直接在人类胚胎中进行基因编辑将成为可能。在目前唯一一篇有关在人类受精卵中测试CRISPR的文章，证明人类受精卵中的靶向突变是可以做到的，但所得到的突变通常非常复杂，且仅有一部分胚胎或胚胎细胞能够携带目标突变（Kang et al.，2016；Liang et al.，2015）。如上所述，CRISPR编辑的这些内在问题使得利用编辑受精卵基因组来纠正人类遗传疾病的概念非常具有挑战性。

然而，编辑胚胎基因组不是实现基因组在种系中修饰的唯一可能方式。在受精之前直接修饰配子——卵子和精子——的基因组的方法将克服嵌合问题，并有允许在体外受精之前预选适当靶向的配子的潜力。

配子基因编辑的现状

目前，配子基因编辑有多种方法，其中一些已经在小鼠中使用，另外一些仍有待完全开发。

直接导入编辑因子到卵母细胞

由于Cas9核酸酶方便使用，在小鼠中，母系遗传的Cas9提供了一个在目标受精卵中进行靶向突变的高效方法（Sakurai et al.，2016）。当然，尽管这种方法不适用于人类，但它的确证明在体外受精之前用编辑因子预处理排卵后的卵母细胞，可能是避免嵌合编辑并提高效率的手段。实际上，这是否可以促进卵母细胞基因组而不是受精后的基因靶向修改仍有待研究。突变或校正的卵母细胞的预选仍然是一项挑战，但嵌合现象减少了，可考虑使用种植前基因诊断（preimplantation genetic diagnosis，PGD）以鉴定正确靶向的胚胎。

精子的体外基因编辑

从鱼到猪的许多物种中，精子介导的基因

to direct targeted genome alterations in zygotes. Thus, the idea of potential targeted gene alterations or corrections in human embryos was not under consideration.

However, more recently, the efficiency of CRISPR/Cas in targeting nuclease-enhanced editing to specific sites in the genome has raised new vistas including possible human germline editing. Soon after the demonstration that CRISPR/Cas could very efficiently target specific sites in the genome of mammalian cells, it was shown that this approach could be applied directly to the mouse zygote without the need for the intermediate ES cell step (see above section and Figure A-8). Thus, it became possible to consider genome editing directly in human embryos. The only publications to date on testing CRISPR on human zygotes demonstrated that targeted mutations could be generated, but also demonstrated that the resulting mutations were often complex and that only some embryos or embryonic cells carried the targeted event (Kang et al., 2016; Liang et al., 2015). As explained above, these intrinsic issues of CRISPR editing make the concept of using zygote genome editing to correct human genetic disease very challenging.

Editing the embryo genome, however, is not the only potential way to achieve modification of the genome in the germline. Approaches that directly modify the genome of the gametes—the eggs and sperm—before fertilization would overcome problems of mosaicism and would potentially allow preselection of the appropriately targeted gamete before *in vitro* fertilization.

Gamete Gene Editing: Current Status

There are a number of potential routes to gamete gene editing, some of which are already in use in the mouse and some of which remain to be fully developed.

Direct Introduction of Editing Factors into Oocytes

In the mouse it has been shown that maternally inherited Cas9 nuclease provides a very efficient means of generating targeted alterations in the resulting zygotes (Sakurai et al., 2016), presumably because of the immediate availability of the enzyme. While such an approach is, of course, not applicable to humans, it does suggest that preloading ovulated oocytes with the editing factors prior to *in vitro* fertilization might be a means of avoiding mosaic editing and enhancing efficiency. Whether this could actually promote gene targeting in the oocyte genome rather than after fertilization remains to be reported. Preselection of mutated or corrected oocytes would still be a challenge, but the reduction in mosaicism would allow PGD to be contemplated for identifying correctly targeted embryos.

Gene Editing in Sperm In Vitro

Sperm-mediated gene transfer is a fairly well established, although inefficient, route to transgenesis in a

转移是一种相当完善的转基因路径，虽效率低下（Lavitrano et al., 2013）。因此，应可将基因编辑系统的组件导入精子并使其携带到受精卵中以促进基因编辑。在精子核中直接进行基因组编辑则是另一项有趣的尝试。鉴于精子是非分裂细胞，目前只有非同源末端连接（non-homologous end joining，NHEJ）介导的基因编辑可实现，尽管修复机制可能不同于体细胞（Ahmed et al., 2010）。迄今为止，同源重组介导的基因修正或改变只有在分裂的细胞中是可行的。然而，有研究表明可冲破这个瓶颈（Orthwein et al., 2015）。Izpisua Belmonte 的研究小组最近开发了一种基于 NHEJ 的基因敲入法——不依赖于同源性的靶向基因整合（homology-independent targeted gene integration，HITI）。HITI 能够在有丝分裂后的细胞如神经元细胞的特定基因组位点直接敲入 DNA 序列（Suzuki et al., 2016b）。HITI 可为精子（甚至卵母细胞）的基因编辑开辟新的途径。插入瞬时荧光报告基因以识别携带编辑因子的精子，可帮助收集潜在的经过基因编辑的精子。在体外受精或卵胞浆内精子注射（intracytoplasmic sperm injection，ICSI）之后的胚胎是否正确编辑，将需要种植前基因诊断进行最终确认（preimplantation genetic diagnosis，PGD）。

生殖干细胞的基因编辑

可在体外无限增殖的干细胞系产生的配子具有相当大的生物学和临床效益。可从小鼠睾丸分离出来精原干细胞（SSCs），移植到生殖细胞缺失的成体睾丸时，具有再生成有受精功能精子的能力（Kanatsu-Shinohara and Shinohara, 2013）。精原干细胞的基因编辑能够预先选择具有正确靶向突变的无性系，以及在产生配子之前预选出脱靶效应、其他不需要的基因组或表观基因组的修改。这种方法的原理证明已经发表（Wu et al., 2015），其中，作者通过 CRISPR / Cas9 编辑 SSCs 修正了一个可导致小鼠白内障的基因突变，将 SSCs 转移回睾丸，并收集圆形精子细胞用于 ICSI。后代得以正确编辑的效率达到了 100%。

将这项工作转化到人类身上面临许多挑战。虽然类似的精原干细胞已可以从人类睾丸中分离出来（Wu et al., 2015），但稳定的、具有自我更新能力的细胞系还无法建立。即便冲破了这项瓶颈，从精原干细胞中产生可以用于细胞质内精子注射的有效

number of species, from fish to pigs (Lavitrano et al., 2013). Thus, it should be possible to introduce the components of genome-editing systems into sperm and have them carried into the zygote to promote genome editing there. More interesting is the possibility of direct genome editing in the sperm nuclei. Given that sperm are non-dividing cells, currently only NHEJ-mediated gene editing would be possible, although the repair mechanism is presumably different from that in somatic cells (Ahmed et al., 2010). Homologous recombination-mediated gene correction or alteration has to date only been possible in dividing cells. However, there are indications that this block can be overcome (Orthwein et al., 2015).

Izpisua Belmonte's group has recently developed an NHEJ-based gene knock-in method— homology-independent targeted gene integration (HITI). HITI allows the direct knock-in of DNA sequences at specific genomic loci in post-mitotic cells, for example, neurons (Suzuki et al., 2016b). HITI may open up new avenues to sperm (and even oocyte) gene editing. Insertion of a transient fluorescent reporter to identify sperm carrying the editing factors could help enrich for potentially gene-edited sperm. Final confirmation of the appropriately edited embryos after fertilization *in vitro* or intracytoplasmic sperm injection (ICSI) would require PGD.

Gene Editing in Germline Stem Cells

There is considerable biological and clinical interest in generating gametes from stem cell lines that can be propagated indefinitely *in vitro*. Spermatogonial stem cells (SSCs) have been isolated from mouse testes and have the capacity to regenerate fertilization-competent sperm when retransplanted to the germ cell–depleted adult testis (Kanatsu-Shinohara and Shinohara, 2013). Gene editing in SSCs would allow for the preselection of clonal lines with appropriate targeted mutations and the potential to prescreen for off-target effects or other unwanted genomic or epigenomic alterations before generating gametes. Proof of principle for such an approach has been published (Wu et al., 2015), in which the authors corrected a gene mutation that causes cataracts in mice by CRISPR/Cas9 editing in SSCs. SSCs were transferred back to the testis and round spermatids collected for ICSI. Offspring were correctly edited at 100 percent efficiency.

Translating this work into humans has many challen-ges. While SSC-like cells have been isolated from human testes (Wu et al., 2015), stable, self-renewing cell lines have not yet been achieved. If this challenge is overcome, there still remains the challenge of generating ICSI-competent gametes from the SSCs. In the mouse, this is achieved by transfer into the

配子方面仍然存在挑战。这在小鼠中是通过转移到生殖细胞缺失的睾丸中实现的，但是在人类中却并不能轻易实现。替代方法包括用混合的精原干细胞和睾丸支持细胞产生"重组睾丸"，并将其移植到睾丸囊下。这种方法在人类中也面临伦理上的挑战。如种间重组和移植到免疫缺陷小鼠中也可带来其自身的科学和伦理挑战。最佳解决方案是在体外明确定义的培养系统中促进精原干细胞向成熟单倍体配子的分化，然而这是一个在任何体系中都仍未实现的目标。

尽管在雌性生殖细胞系中应用类似方法很具有吸引力，但是雌性干细胞存在的证据仍然存有争议（Johnson et al., 2004）。大多数证据表明，卵母细胞在成年哺乳动物卵巢中是一种有限的资源（Eggan et al., 2006），并且没有任何内源干细胞存在的证据。

多能性干细胞中的基因编辑以及随后的生殖细胞分化

多能性胚胎干细胞或诱导多能性干细胞可以从雄性和雌性产生，它易于进行CRISPR编辑，并且可以分化成具有有丝分裂能力的生殖细胞。在小鼠中，产生胚胎干细胞来源的生殖细胞最可信的报道是关于模拟早期胚胎中由多能性外胚层细胞诱导产生原始生殖细胞的已知途径。使用这种方法，Saitou的实验室已经生成了来自雄性和雌性胚胎干细胞的原始生殖细胞样细胞（primordial germ cell–like cell，PGC-LC）。当PGC-LC分别和来自睾丸或卵巢的支持细胞重建并移植回睾丸或卵巢环境的时候，能够恢复产生精子或卵母细胞。而当与正常卵子和精子结合时，就可产生可存活的后代（Hayashi et al., 2011, 2012）。最近，周琪实验室进一步利用这种方法，把PGC-LC与睾丸细胞在培养基中共培养2周后可以分离精子细胞样细胞，并能够在细胞质内精子注射之后使卵母细胞受精，产生可存活的后代（Zhou et al., 2016）。虽然有一些关于表观遗传重编程是否会在这个培养系统中起作用的顾虑，但总体结果是相当惊人的。另一个有趣的研究进展是来自Izpisua、Belmonte、Okuda和Matsui研究组，他们的研究显示在小鼠胚胎干细胞中敲降或敲除*Max*基因会强烈激活生殖细胞相关基因的表达，并导致产生类似于经历减数分裂的重大细胞机制变化（Maeda et al., 2013；Suzuki et al., 2016a）。使

germ cell–depleted testis—not an easy solution in humans. Alternate approaches include generating a "reconstituted testis" with mixed SSCs and supporting cells of the testis and transplanting this under the testis capsule. This approach would also be ethically challenging in humans. The possible use of interspecies reconstitutions and transplants into immune-deficient mice would bring its own scientific and ethical challenges. The best solution would be to promote differentiation of the SSCs to mature haploid gametes in a fully defined culture system *in vitro*— a challenge not yet achieved in any system.

While the possibility of applying similar approaches to the female germline is attractive, the evidence for the existence of oogonial stem cells is controversial (Johnson et al., 2004). Most evidence suggests that there is a limited resource of oocytes in the adult mammalian ovary (Eggan et al., 2006) and no evidence for any endogenous stem cells.

Gene Editing in Pluripotent Stem Cells Followed by Germ Cell Differentiation

Pluripotent embryonic stem cells or induced pluripotent stem cells can be generated from both males and females, are readily amenable to CRISPR editing, and can be differentiated down the pathway toward meiotically competent germ cells. In the mouse, the most reliable reports of germ cell generation from ES cells have come from mimicking the known pathways that induce primordial germ cells from the pluripotent epiblast in the early embryo. Using this approach, Saitou's lab has generated primordial germ cell–like cells (PGC-LCs) from both male and female ES cells. When PGC-LCs were reconstituted with support cells from the testis or ovary respectively and transplanted back to the testis or ovary environment, investigators were able to recover spermatids or oocytes that could be used to generate viable offspring when combined with normal eggs and sperm (Hayashi et al., 2011, 2012). Most recently the Zhou lab has taken this approach one step further and reports that coculture of the PGC-LCs with testis cells in culture over a 2-week period allowed the isolation of spermatid-like cells that were capable of fertilizing oocytes after ICSI and generating viable offspring (Zhou et al., 2016). There are some concerns about whether epigenetic reprogramming would be complete in this culture system, but the overall result is quite remarkable. Another interesting development is from the Izpisua Belmonte, Okuda, and Matsui groups, who have shown that knockdown or knock-out of *Max* in mouse ESCs strongly activates expression of germ cell–related genes and results in profound cytological changes to resemble cells undergoing meiotic division (Maeda et al., 2013; Suzuki et al., 2016a). Whether functional

用此方法是否可以产生有功能的单倍体细胞仍有待研究。

在鼠科系统中的这些研究成果，提升了从人多能性细胞中产生单倍体配子的预期，这为探明配子发生的原理、不孕不育的原因和治疗不育夫妇开辟了新的道路。它也将打开干细胞基因修饰的大门，旨在修复由已知遗传原因导致的不育或修复显性基因突变。然而，迄今为止，尽管最近两篇研究都报道了从人胚胎干细胞可以产生早期的原始PGC-LC（Irie et al.，2015；Sasaki et al.，2015），但人类配子仍尚未能从多能性干细胞中成功产生。这些研究揭示了人类与小鼠生殖细胞分化途径的相似性和差异。这意味着需要更多与小鼠胚胎相对的人类或非人灵长类胚胎的知识来推动这项研究的进步。

单倍体胚胎干细胞中的基因编辑

大多数动物是二倍体，天然单倍体细胞通常仅限于成熟的生殖细胞。最近在小鼠和大鼠中已经获取了孤雄生殖（雄性）和孤雌生殖（雌性）的单倍体胚胎干细胞（haploid ES cell，haESC）（Leeb and Wutz，2011；Li et al.，2012，2014；Yang et al.，2012）。单倍体胚胎干细胞只含有二倍体细胞等位基因的一个拷贝，并且可以用传统的基因靶向方法和新的基于核酸酶基因编辑的策略进行基因修饰（Li et al.，2012，2014）。更有趣的是，含有Y染色体而不是X染色体的孤雄生殖的单倍体胚胎干细胞可以通过在胞质内注射入成熟卵母细胞后产生能存活能生育的后代（Li et al.，2012，2014）。把单倍体孤雌生殖的小鼠的haESC注射入卵母细胞替代母系基因组后，也能够产生可育的小鼠（Wan et al.，2013）。这两种策略都可以用于向后代引入遗传修饰。最近，有报道指出孤雌生殖的人类单倍体胚胎干细胞也可在体外生成（Sagi et al.，2016）。目前仍未有关于人类孤雄生殖单倍体胚胎干细胞的报道。

单倍体胚胎干细胞也有一些局限性。首先，在培养中发现单倍体的表型不稳定。单倍体胚胎干细胞经历了自发性自体二倍体化，并且需要通过流式分选细胞进行几轮单倍体纯化，才能在体外培养中变得稳定。此外，尚未获得含有Y染色体的孤雄生殖的单倍体胚胎干细胞（Li et al.，2012）。这是由于含有YY染色体孤雄生殖的胚胎发育潜能很

haploid cells can be generated using this approach remains to be seen.

These results in murine systems raise expectations that human haploid gametes could be generated from human pluripotent cells, with implications for understanding gametogenesis, causes of infertility and potentially offering new avenues for reproduction in infertile couples. It also would open up genetic modification of the stem cells to repair known genetic causes of infertility or to repair dominant gene mutations. However, to date, human gametes have not been generated successfully from pluripotent stem cells, although two recent papers report the generation of early PGC-LCs from human ES cells (Irie et al., 2015; Sasaki et al., 2015). Those studies revealed similarities and differences from the mouse germ-cell differentiation pathway. This suggests that more knowledge of how germ cells actually develop in the human, or perhaps the nonhuman primate, embryo versus the mouse embryo is needed to move this research forward.

Gene Editing in Haploid ES Cells

Most animals are diploids, and natural haploid cells are typically limited to mature germ cells. Recently both androgenetic (male) and parthenogenetic (female) haploid ES cells (haESCs) have been derived in mice and rats (Leeb and Wutz, 2011; Li et al., 2012, 2014; Yang et al., 2012). haESCs contain only one copy of allelic genes of diploid cells and are amenable to genetic modification with traditional gene-targeting approaches and with new nuclease-based genome-editing strategies (Li et al., 2012, 2014). More interestingly, androgenetic haESCs, which contain a Y rather than an X chromosome, can produce viable and fertile offspring after intracytoplasmic injection into mature oocytes (Li et al., 2012, 2014). Haploid parthenogenetic mouse haESCs were also shown to be able to produce fertile mice when injected into oocytes in place of the maternal genome (Wan et al., 2013). Both strategies are possible to be used for introduction of genetic modifications to progeny. Most recently parthenogenetic human haESCs have also been successfully generated (Sagi et al., 2016). Human androgenetic haESCs have not yet been reported.

There are several limitations of haESCs. First, the haploid phenotype has been found to be unstable in culture. haESCs undergo spontaneous auto-diploidization and need several rounds of haploid purification by flow-activated cell sorting before becoming stable in culture. Also, there is a lack of androgenetic haESCs containing the Y chromosome (Li et al., 2012). This is due to the poor developmental potential of androgenetic embryos with YY chromosomes (Latham et al., 2000; Tarkowki, 1977). Therefore only female animals can currently be created. With further

差（Latham et al.，2000；Tarkowki，1977）。因此，目前只能获得雌性动物。随着育种技术的进一步发展，雄性动物也将能获得。另一个主要的缺点是孤雄生殖的单倍体胚胎干细胞使卵子受精的效率非常低（小鼠中低于5%，大鼠中低于2%）。

总之，尽管目前无法实现从干细胞系产生人类"人工"配子，但是小鼠的研究进展表明了未来的可能性。体外重建睾丸或卵巢环境可以通过从人胚胎干细胞的体外分化获得的生殖细胞和支持细胞（例如，塞尔托利氏细胞和卵丘细胞）来实现。对促进生殖细胞发育和减数分裂成熟作用的内源信号通路的进一步理解将有助于未来体外生成人类配子。该类细胞将迅速帮助理解配子生成的原理和剖析生育力问题，但是在它们能够用于经过或不经过基因编辑的人类生育治疗之前，都需要克服某些安全性顾虑。通常认为生殖细胞系与体细胞不同，在一定程度上被保护免受遗传损伤，并且在配子发生完成之前还经历了广泛的表观遗传重塑。这两个方面都需要在体外产生的人工配子中再现。

编辑线粒体的基因组

线粒体疾病是由线粒体DNA（mtDNA）突变引起的线粒体功能障碍导致的一些疾病。线粒体疾病与具有高能量需求的组织和器官的退化相关，包括肌肉、心脏和脑。病理包括肌病、心肌病、神经病、脑病、乳酸性酸中毒、中风样综合征和失明（Taylor and Turnbull，2005）。突变的线粒体DNA分子的百分比通常决定患者是否有症状。目前，线粒体疾病仍无法治愈。对于仍可健康生育的患者，遗传咨询和种植前基因诊断是预防疾病传播的最佳选择。然而，由于线粒体DNA的非孟德尔遗传和不同卵裂球之间潜在的不同异质水平，种植前基因诊断只能减少而不能消除疾病传播的风险。最近开发的线粒体置换技术涉及在患者和供体卵母细胞核基因组之间的的一系列复杂的技术操作，产生携带来自三个不同来源的遗传物质的胚胎（Paull et al.，2012；Tachibana et al.，2012）。由于这些原因，线粒体置换技术引发了生物、医学和伦理问题（Hayden，2013；Reinhardt et al.，2013）。线粒体置换技术成功率低，并且已经报道在低等生物中核酸和线粒体

breeding, males can then be obtained. Another major drawback is that the efficiency for androgenetic haESCs to fertilize an egg is very low (less than 5 percent in mice and less than 2 percent in rats).

In summary, although the generation of human "artificial" gametes from stem cell lines is not currently achievable, work in the mouse suggests that this will likely be possible. Reconstituting the testis or ovary environment *in vitro* may be achieved by deriving both germ cells and supporting cells, such as Sertoli cells, and cumulus cells from *in vitro* differentiation of human ES cells. Further understanding of the endogenous signaling pathways that promote germ-cell development and meiotic maturation will aid in the future derivation of human gametes *in vitro*. Such cells will be immediately useful for understanding gametogenesis and dissecting fertility problems, but safety concerns will need to be overcome before they could be used for human reproduction, with or without genome editing. The germline is generally considered to be somewhat protected from genetic damage, unlike that of somatic cells, and also undergoes extensive epigenetic remodeling before completion of gametogenesis. Both aspects would need to be replicated in the artificial gametes generated *in vitro*.

EDITING THE MITOCHONDRIAL GENOME

Mitochondrial diseases are a group of maladies caused by the dysfunction of mitochondria due to mutations in mitochondrial DNA (mtDNA). Mitochondrial diseases are associated with the degeneration of tissues and organs that have high energetic demands—including muscle, heart, and brain—that lead, among other pathologies, to myopathies, cardiomyopathies, neuropathies, encephalopathies, lactic acidosis, stroke- like syndrome, and blindness (Taylor and Turnbull, 2005). The percentage of mtDNA molecules that is mutated generally determines whether or not a patient is symptomatic. Currently, there are no cures for mitochondrial diseases, and for patients healthy enough to have children, genetic counseling and PGD represent the best options for preventing disease transmission. However, due to the non-Mendelian inheritance of mtDNA and the potentially different heteroplasmy levels among different blastomeres, PGD can only reduce, not eliminate, the risk of transmitting the disease. Recently developed mitochondrial replacement techniques involve a series of complex technical manipulations of nuclear genome between patient and donor oocytes that results in the generation of embryos carrying genetic material from three different origins (Paull et al., 2012; Tachibana et al., 2012). For these reasons, mitochondrial replacement techniques have raised biological, medical, and ethical concerns (Hayden,

DNA 之间进行线粒体置换时，因为不相容性而引起了潜在问题（Reinhardt et al.，2013）。

最近开发的一种新的替代治疗方法可以消除生殖细胞中突变的线粒体 DNA。使用线粒体靶向的内切核酸酶，成功地阻止了小鼠生殖细胞系中的目标线粒体 DNA 传递到下一代（Reddy et al.，2015）。由于限制性内切核酸酶只能靶向有限数量的线粒体 DNA 突变，研究者们努力通过使用线粒体靶向的转录激活因子样效应物核酸酶（mito-TALENs）和锌指核酸酶（ZFNs）来处理大多数线粒体 DNA 突变（Bacman et al.，2013；Gammage et al.，2014）。mito-TALENs 能够特异性消除小鼠生殖细胞中的靶向线粒体 DNA（Reddy et al.，2015）。重要的是，将核酸酶（如 mito-TALENs）注射到卵母细胞或早期胚胎中的技术仅涉及了一个编码核酸酶的信使 RNA 的简单显微注射。此外，线粒体定位信号（如 Cox8 和 ATP5b）的使用将其易位限制在线粒体中。这项技术的使用注意事项是，在卵母细胞中消除高水平的突变线粒体 DNA 将导致只拥有低数目正常线粒体 DNA 的胚胎产生，如果植入后细胞不能复制，可能导致流产。种植前基因诊断可用于选择和转移含有更高水平的正常线粒体 DNA 的胚胎。关键点在于，不同于核编辑，线粒体 DNA 的编辑并非旨在校正突变，而是要消除突变的 DNA。因在卵母细胞中存在多个拷贝的线粒体 DNA，所以达到这项效果是有可能的。此外，由于线粒体中修复机制的活性非常低，靶向线粒体 DNA 的重新连接和引入新突变的频率也将非常低。另外，类似的线粒体编辑工具在未来也可以用于清除干细胞来源的配子中突变的线粒体 DNA。最后，线粒体基因编辑工具联合线粒体置换技术可成为未来的备选，以便能够防止线粒体 DNA 中的突变在生殖细胞系间传递。这不仅可以用于治疗线粒体特异性疾病，而且还可以用于线粒体功能病变，有助于防治诸如癌症、糖尿病和衰老相关疾病。

传递技术难题

尽管已有一些技术上的进步，体内使用基因编辑系统的一个关键难题是如何有效传递。表 A-1 精选了一些正在开发的基因组编辑传递组件的策略，并讨论了它们的优缺点。

2013; Reinhardt et al., 2013). Mitochondrial replacement techniques have low rates of success, and studies in lower organisms have reported potential issues arising from incompatibility between nuclear and mtDNA upon mitochondrial replacement (Reinhardt et al., 2013).

A novel alternative therapeutic approach was recently developed to eliminate the mutated mtDNA in the germline. Using mitochondria-targeted endonuclease, the targeted mtDNA in the mouse germline was successfully prevented from transmission to the next generation (Reddy et al., 2015). Due to the limited number of mtDNA mutations that can be targeted by restriction endonucleases, efforts have been made to target most of the mtDNA mutations using mitochondria-targeted transcription activator-like effector nucleases (mito-TALENs) and ZFNs (Bacman et al., 2013; Gammage et al., 2014). The mito-TALENs were able to specifically eliminate the targeted mtDNA in the mouse germline (Reddy et al., 2015). Importantly, the technique of injection of nucleases (e.g., mito-TALENs) into oocytes or early embryos involves a simple microinjection of mRNA that encodes the nucleases. Moreover, the use of mitochondrial localization signal (e.g., Cox8 and ATP5b) restricts the translocation to mitochondria alone. A caveat of this technology is that elimination of high levels of mutated mtDNA in oocytes will lead to the generation of embryos with a low number of normal mtDNA that, if failing to replicate after implantation, could lead to pregnancy loss. PGD could be used for the selection and transfer of embryos containing higher levels of normal mtDNA. Importantly, unlike nuclear editing, mtDNA editing is not aimed to correct the mutations, but to eliminate mutated DNA, which is possible due to the presence of multiple copies of mtDNA in the oocytes. Moreover, due to the very low activity of repair mechanisms in mitochondria, the frequency of re-ligation of target mtDNA and introduction of new mutations would be very rare. In addition, similar mitochondrial editing tools in the future could also be used to eliminate mutated mtDNA in gametes derived from stem cells. Finally, a combination of mitochondrial gene-editing tools with mitochondrial replacement techniques may represent an alternative option, in the future, to prevent the germline transmission of mutations in the mtDNA responsible not only for mitochondrial specific diseases but also for situations where alterations in mitochondrial function contribute to pathologies such as cancer, diabetes, and aging-associated diseases.

THE CHALLENGE OF DELIVERY

In addition to the technical advances being made to genome-editing systems, an important challenge for *in vivo* use is effective delivery. Table A-1 highlights a number of strategies being explored for the delivery of genome-editing components, including a discussion of their advantages and disadvantages.

表 A-1　传递基因组编辑组件的通用方法

方法	传送的组件	说明	优点	缺点	首选应用
非病毒					
转染	核酸酶，如质粒 DNA、RNA 或者蛋白质 引导 RNA 以质粒 DNA 或者寡核苷酸的形式与核酸酶 RNA 混合（与核酸酶组装成核糖核蛋白 RNP）	所有组件用糖/脂聚合物载体组装以利于进入细胞；在体外将复合物导入细胞或在体内注射到组织或血液中	相对简单；可以一起传递所有编辑工具；瞬时表达编辑工具的细胞毒性和免疫原性有限	细胞摄取和入核的速率受限；在体内载体会引起细胞毒性和炎症；体内递送需要对 RNA 进行修饰以提高稳定性；不能针对特定组织；制剂通常是专利产品	细胞系和一些离体的原代细胞
纳米颗粒	质粒 DNA 或寡核苷酸作为模板（可以被预组装成 RNP）	如上所述把复合物与纳米颗粒（比如金、葡聚糖）偶联以增强细胞摄取、递送及体内分布	可以增强细胞摄取；可以靶向特定组织；瞬时表达编辑工具的细胞毒性和潜在免疫原性有限	少	组织或全身
电穿孔	（一些方案中通过非病毒途径，以信使 RNA、蛋白质或 RNP 等形式递送核酸酶，最常见的是电穿孔，模板 DNA 在传递核酸酶之前或之后由病毒载体单独递送）	利用短暂的电脉冲穿过含有核酸酶编辑试剂的细胞溶液，里面可有模板，亦可没有	方法非常有效，可以传递到体外的多种细胞；瞬时表达编辑工具的细胞毒性和潜在的免疫原性有限	会引起细胞毒性（从蛋白质到信使 RNA 到 DNA 毒性依次增加；使用修饰的核酸会减轻毒性），但比转染具有更好的耐受性；体内传递仍然是个挑战	细胞系和一些离体的原代细胞
挤压穿孔		细胞群穿过一个小于细胞直径的小通道，在膜上产生小孔以允许核酸酶和供体试剂进入细胞	新方法尚未得到广泛验证	无法实施体内传递	
病毒					
慢病毒载体	核酸酶和引导 RNA 作为基因表达盒由载体基因组传送（可以由单独的和组织特异性启动子驱动）	复制缺陷型病毒，可以包装 8kb 的模板核酸，具有进入几乎所有细胞类型的能力；载体随机地整合到细胞中允许其稳定表达并可以传递到后代细胞；用突变的整合酶缺陷型慢病毒载体（IDLV）制备的修改版，不能并入基因组并可在增殖细胞中快速丢失	是目前最常见的体外基因传递工具，体内应用也正在探索开发中；慢病毒是稳定表达的，整合缺陷的慢病毒是瞬时表达的，它们可以是组织特异性调节或条件性调节；转导人类细胞具有良好的耐受性；IDLV 使核酸酶瞬时表达，从而限制细胞毒性和免疫原性，并且非常适合递送模板	慢病毒稳定表达核酸酶可能导致细胞毒性和免疫原性；慢病毒的插入诱变有潜在的风险；相比非病毒平台，载体制作更复杂和更昂贵	细胞系和一些离体的原代细胞
腺相关病毒载体（AAV）	模板 DNA（可用于单独传送模板，或与电穿孔并用传送核酸酶 RNA、蛋白质或 RNP，或在同一个或分开的载体上组合递送模板和引导 RNA 和/或核酸酶）	复制缺陷型病毒具有大量的变体（天然和改造血清型），可以包装 4.7kb 的模板 DNA，可以靶向不同类型的细胞	是目前最常见的体内基因转移工具；大部分在靶细胞核内成为游离基因；在各种类型的细胞和组织中具有高效的传送能力和强大的表达能力；转导人类细胞具有良好的耐受性；在分裂细胞中瞬时表达，很适合递送核酸酶和模板	在非增殖细胞和组织中能力有限，载体可以持续存在数年，导致核酸酶的长期表达和潜在的细胞毒性及免疫原性；某些人群拥有 AAV 的预存免疫性，这会抑制体内的基因转移或者清除转导的细胞	离体原代细胞和组织或体内全身
腺病毒载体		复制缺陷型病毒可以包装大于 20kb 的模板 DNA，能够在体外和体内转导多种类型的细胞	能够包装大片段的 DNA；可用于体内和离体递送；在分裂的细胞和组织中瞬时表达	在一些临床试验中显示出严重的急性毒性；许多人都拥有预存免疫性；不再常用的治疗载体	离体原代细胞

注：IDLV 指整合酶缺陷型慢病毒载体；RNP 指核糖核蛋白复合物；sgRNA 指单链的引导 RNA。

TABLE A-1 General Approaches to Delivering Genome-Editing Components

Method	Delivered Component	Explanation	Advantages	Disadvantages	Preferred Applications
Non-viral Transfection	Nuclease(s) as plasmid DNA, RNA, or protein guide RNA as plasmid DNA, or oligonucleotide is mixed with the nuclease RNA (can be complexed with nuclease as Ribonucleoprotein, RNP)	All components are assembled with a glyco/lipo-polymer vehicle that favors cell entry; the complex is applied to cells *ex vivo* or injected into a tissue or blood *in vivo*.	relatively simple; can deliver all editing machinery together; transient expression limiting cytotoxicity and immunogenicity of editing machinery.	Cellular uptake and nuclear access can be rate-limiting; vehicle can cause cytotoxicity and inflammation *in vivo*; for *in vivo* delivery RNA needs to be modified for improved stability; cannot be targe-ted to specific tissues; formulation often proprietary	Cell lines and some primary cells *ex vivo*
Nano-particle	template as plasmid DNA or oligonucleotide (can be pre-complexed with RNP)	The complex, as above, is coupled to a nanopar-ticle (i.e., by gold, dex-tran) to enhance cellular uptake and delivery and biodistribution *in vivo*.	Cellular uptake can be enhanced; can be targeted to specific tissues; transient expression limiting cyto-xicity and potential immu-nogenicity of editing ma-chinery.	Few	Tissues or systemic
Electro-poration	(in some protocols nuclease(s) are delivered as mRNA, protein, or RNP by a non-viral method, most often electroporation, and template DNA is delivered separately by a viral vector, before or after delivery of the nuclease)	A brief electric pulse is passed across a population of cells in a solution that contains the editing nuclease reagents with or without template.	Very effective method of delivery to a wide variety of cell types *ex vivo*; transient expression limiting cytotoxicity and potential immunogenicity of editing machinery.	Can cause cytotoxicity (increasingly from protein to mRNA to DNA; alleviated by using modified nucleic acids) but better tolerated than transfection; challen-ging to get delivery *in vivo*.	Cell lines and primary cells *ex vivo*
Squeeze Poration (No-Viral)		A population of cells is passed through a small channel that is smaller than the diameter of the cells, creating small pores in the membrane to allow the nuclease and donor reagents to enter the cell.	New strategy that has not been widely validated.	Would not work for in vivo delivery.	
Viral					
Lenti-viral Vector (LV)	Nuclease(s) and guideRNA as gene expression cassettes delivered by the vector genome (can be driven by separate and tissue specific promoters) template DNA (can be used for the delivery of template alone, in combination with electroporation of nuclease RNA, protein or RNP, or for the combined delivery of template and guideRNA and/or nuclease, on the same or separate vectors)	Replication-defective virus that can package ~8 kilobases of nucleic acid with an ability to enter almost all cell types; vector integrates quasi-randomly into cell genome, allowing for stable expression and transmission to the cell progeny; a modified version made with mutant integrase-defective lentiviral vector (IDLV), fails to integrate and is rapidly lost in proliferating cells.	Currently the most common tool for *ex vivo* gene transfer, also being explored for *in vivo* use; expression is stable for LV and transient for IDLV and can be made tissue-specific, regulated or conditional; transduction of human cells well tolerated; IDLV provides for transient nuclease expression, thus limiting cytotoxicity and immunogenicity, and is well suited for template delivery.	Stable expression of nuclease by LV potentially leading to cytotoxicity and immunogenicity; potential risk of insertional mutagenesis by LV; vector manufacturing more complex and expensive than non-viral platform	Cell lines and primary cells *ex vivo*
Adeno-Asso-ciated Virus (AAV) Vector		Replication-defective virus that can package ~4.7 kilobases of DNA with a broad number of variants (both natural and engineered serotypes) that can target different cell types.	Currently the most common tool for gene transfer *in vivo*; remains mostly episomal in the target cell nucleus; efficient delivery and robust expression in wide variety of cell types and tissues; transduction of human cells well tolerated; transient expression in dividing cells, well suitable for nuclease and template delivery.	Limited capacity In non-proliferating cells and tissues, the vector can persist for years, leading to long-term expression of nuclease and potential cytotoxicity and immuno-genicity; some people have preexisting immunity to AAV, which can inhibit *in vivo* gene transfer or clear transduced cells.	Primary cells *ex vivo* and tissues or syste-mic *in vivo*
Adeno-viral Vector		Replication-defective virus that can package >20 kilobases of DNA able to transduce a wide variety of cell types both *ex vivo* and *in vivo*.	Able to package large fragments of DNA; can be used for *in vivo* and *ex vivo* delivery; transient expression in dividing cells and tissues.	Has shown severe acute toxicity in some clinical trials; many people have pre-existing immunity; no longer commonly used gene therapy vector.	Primary cells *ex vivo*

NOTE: IDLV = integrase-defective lentiviral vector; RNP = ribonuclear protein complex; sgRNA = single guide RNA.

参 考 文 献

Ahmed, E. A., P. de Boer, M. E. P. Philippens, H. B. Kal, and D. G. de Rooij. 2010. Parp1-XRCC1 and the repair of DNA double strand breaks in mouse round spermatids. *Mutation Research/Fundamental and Molecular Mechanisms of Mutagenesis* 683(1-2): 84-90.

Auer, T. O., K. Duroure, A. De Cian, J. P. Concordet, and F. Del Bene. 2014. Highly efficient CRISPR/Cas9-mediated knock-in in zebrafish by homology-independent DNA repair. *Genome Research* 24(1): 142-153.

Bacman, S. R., S. L. Williams, M. Pinto, S. Peralta, and C. T. Moraes. 2013. Specific elimination of mutant mitochondrial genomes in patient-derived cells by mitoTALENs. *Nature Medicine* 19(9): 1111-1113.

Barrangou, R., C. Fremaux, H. Deveau, M. Richards, P. Boyaval, S. Moineau, D. A. Romero, and P. Horvath. 2007. CRISPR provides acquired resistance against viruses in prokaryotes. *Science* 315(5819): 1709-1712.

Bibikova, M., D. Carroll, D. Segal, J. K. Trautman, J. Smith, Y. G. Kim, and S. Chandrasegaran. 2001. Stimulation of homologous recombination through targeted cleavage by chimeric nucleases. *Molecular and Cellular Biology* 21(1): 289-297.

Bibikova, M., M. Golic, M., K. G. Golic, and D. Carroll. 2002. Targeted chromosomal cleavage and mutagenesis in *Drosophila* using zinc-finger nucleases. *Genetics* 161(3): 1169-1175.

Bibikova, M., K. Beumer, J. K. Trautman, and D. Carroll. 2003. Enhancing gene targeting with designed zinc finger nucleases. *Science* 300(5620): 764.

Brinster, R. L., H. Y. Chen, M. Trumbauer, A. W. Senear, R. Warren, and R. D. Palmiter. 1981. Somatic expression of herpes thymidine kinase in mice following injection of a fusion gene into eggs. *Cell* 27(1 Pt. 2): 223-231.

Burt, A. 2003. Site-specific selfish genes as tools for the control and genetic engineering of natural populations. *Proceedings of The Royal Society B: Biological Sciences* 270(1518): 921-928.

Burt, A. 2014. Heritable strategies for controlling insect vectors of disease. *Philosophical Transactions of the Royal Society B: Biological Sciences* 369(1645): 20130432.

Carroll, D. 2014. Genome engineering with targetable nucleases. *Annual Review of Biochemistry* 83: 409-439.

Chevalier, B. S., T. Kortemme, M. S. Chadsey, D. Baker, R. J. Monnat, and B. L. Stoddard. 2002. Design, activity, and structure of a highly specific artificial endonuclease. Molecular Cell 10(4): 895-905.

Chiarle, R., Y. Zhang, R. L. Frock, S. M. Lewis, B. Molinie, Y. J. Ho, D. R. Myers, V. W. Choi, M. Compagno, D. J. Malkin, D. Neuberg, S. Monti, C. C. Giallourakis, M. Gostissa, and F. W. Alt. 2011. Genome-wide translocation sequencing reveals mechanisms of chromosome breaks and rearrangements in B cells. *Cell* 147(1): 107-119.

Cho, S. W., S. Kim, J. M. Kim, and J. S. Kim. 2013. Targeted genome engineering in human cells with the Cas9 RNA-guided endonuclease. *Nature Biotechnology* 31(3): 230-232.

Cong, L., F. A. Ran, D. Cox, S. Lin, R. Barretto, N. Habib, P. D. Hsu, X. Wu, W. Jiang, L. A. Marraffini, and F. Zhang. 2013. Multiplex genome engineering using CRISPR/Cas systems. *Science* 339(6121): 819-823.

Costantini, F., and E. Lacy. 1981. Introduction of a rabbit beta-globin gene into the mouse germ line. *Nature* 294(5836): 92-94.

Cox, D. B. T., R. J. Platt, and F. Zhang. 2015. Therapeutic genome editing: Prospects and challenges. *Nature Medicine* 21(2): 121-131.

Deltcheva, E., K. Chylinski, C. M. Sharma, K. Gonzales, Y. Chao, Z. A. Pirzada, M. R. Eckert, J. Vogel, and E. Charpentier. 2011. CRISPR RNA maturation by trans-encoded small RNA and host factor RNase III. *Nature* 471(7340): 602-607.

Desjarlais, J. R., and J. M. Berg. 1992. Toward rules relating zinc finger protein sequences and DNA binding site preferences. *Proceedings of the National Academy of Sciences of the United States of America* 89(16): 7345-7349.

Dickinson, D. J., J. D. Ward, D. J. Reiner, and B. Goldstein. 2013. Engineering the *Caenorhabditis elegans* genome using Cas9-triggered homologous recombination. *Nature Methods* 10(10): 1028-1034.

Doetschman, T., R. G. Gregg, N. Maeda, M. L. Hooper, D. W. Melton, S. Thompson, and O. Smithies. 1987. Targeted correction of a mutant HPRT gene in mouse embryonic stem cells. *Nature* 330: 576-578.

Doudna, J. A., and E. Charpentier. 2014. Genome editing: The new frontier of genome engineering with CRISPR-Cas9. *Science* 346(6213): 1258096.

Eggan, K., S. Jurga, R. Gosden, I. M. Min, and A. J. Wagers. 2006. Ovulated oocytes in adult mice derive from non-circulating germ cells. *Nature* 441(7097): 1109-1114.

Esvelt, K. M., A. L. Smidler, F. Catteruccia, and G. M. Church. 2014. Concerning RNA-guided gene drives for the alteration of wild populations. *eLife* e03401.

Evans, M. J., and M. H. Kaufman. 1981. Establishment in culture of pluripotential cells from mouse embryos. *Nature* 292(5819):

154-156.

Friedland, A. E., Y. B. Tzur, K. M. Esvelt, M. P. Colaiacovo, G. M. Church, and J. Calarco. 2013. Heritable genome editing in *C. elegans* via a CRISPR-Cas9 system. *Nature Methods* 10(8): 741-743.

Fu, Y., J. A. Foden, C. Khayter, M. L. Maeder, D. Reyon, J. K. Joung, and J. D. Sander. 2013. High-frequency off-target mutagenesis induced by CRISPR-Cas nucleases in human cells. *Nature Biotechnology* 31(9): 822-826.

Gabriel, R., A. Lombardo, A. Arenas, J. C. Miller, P. Genovese, C. Kaeppel, A. Nowrouzi, C. C. Bartholomae, J. Wang, G. Friedman, M. C. Holmes, P. D. Gregory, H. Glimm, M. Schmidt, L. Naldini, and C. von Kalle. 2011. An unbiased genome-wide analysis of zinc-finger nuclease specificity. *Nature Biotechnology* 29(9): 816-823.

Gammage, P. A., J. Rorbach, A. I. Vincent, E. J. Rebar, and M. Minczuk. 2014. Mitochondrially targeted ZFNs for selective degradation of pathogenic mitochondrial genomes bearing large-scale deletions or point mutations. *EMBO Molecular Medicine* 6(4): 458-466.

Gantz, V., and E. Bier. 2015. The mutagenic chain reaction: A method for converting heterozygous to homozygous mutations. *Science* 348(6233): 442-444.

Garneau, J. E., M. E. Dupuis, M. Villion, D. A. Romero, R. Barrangou, P. Boyaval, C. Fremaux, P. Horvath, A. H. Magadan, and S. Moineau. 2010. The CRISPR/Cas bacterial immune system cleaves bacteriophage and plasmid DNA. *Nature* 468(7320): 67-71.

Gasiunas, G., R. Barrangou, P. Horvath, and V. Siksnys. 2012. Cas9-crRNA ribonucleoprotein complex mediates specific DNA cleavage for adaptive immunity in bacteria. *Proceedings of the National Academy of Sciences of the United States of America* 109(39): E2579-E2586.

Gilbert, L. A., M. H. Larson, L. Morsut, Z. Liu, G. A Brar, S. E. Torres, N. Stern-Ginossar, O. Brandman, E. H. Whitehead, J. A. Doudna, W. A. Lim, J. S. Weissman, and L. S. Qi. 2013. CRISPR-mediated modular RNA-guided regulation of transcription in eukaryotes. *Cell* 154(2): 442-451.

Gilbert, L. A., M. A. Horlbeck, B. Adamson, J. E. Villalta, Y. Chen, E. H. Whitehead, C. Guimaraes, B. Panning, H. L. Ploegh, M. C. Bassik, L.S. Qi, M. Kampmann, and J. S. Weissman. 2014. Genome-scale CRISPR-mediated control of gene repression and activation. *Cell* 159(3): 647-661.

Gordon, J. W., and F. H. Ruddle. 1981. Integration and stable germ line transmission of genes injected into mouse pronuclei. *Science* 214(4526): 1244-1246.

Guilinger, J. P., D. B. Thompson, and D. R. Liu. 2014. Fusion of catalytically inactive Cas9 to FokI nuclease improves the specificity of genome modification. *Nature Biotechnology* 32(6): 577-582.

Hammer, R. E., R. D. Palmiter, and R. L. Brinster. 1984. Partial correction of murine hereditary growth disorder by germ-line incorporation of a new gene. *Nature* 311(5981): 65-67.

Haurwitz, R. E., S. H. Sternberg, and J. A. Doudna. 2012. Csy4 relies on an unusual catalytic dyad to position and cleave CRISPR RNA. *The EMBO Journal* 31(12): 2824-2832.

Hayashi, K., H. Ohta, K. Kurimoto, S. Aramaki, and M. Saitou. 2011. Reconstitution of the mouse germ cell specification pathway in culture by pluripotent stem cells. *Cell* 146(4): 519-532.

Hayashi, K., S. Ogushi, K. Kurimoto, S. Shimamoto, H. Ohta, and M. Saitou. 2012. Offspring from oocytes derived from *in vitro* primordial germ cell-like cells in mice. *Science* 338(6109): 971-975.

Hayden, E. C. 2013. Regulators weigh benefits of "three-parent" fertilization. *Nature* 502(7471): 284-285.

Hsu, P. D., D. A. Scott, J. A. Weinstein, F. A. Ran, S. Konermann, V. Agarwala, Y. Li, E. J. Fine, X. Wu, O. Shalem, T. J. Cradick, L. A. Marraffini, G. Bao, and F. Zhang. 2013. DNA targeting specificity of RNA-guided Cas9 nucleases. *Nature Biotechnology* 31(9): 827-832.

Hwang, W. Y., Y. Fu, D. Reyon, M. L. Maeder, S. Q. Tsai, J. D. Sander, R. T. Peterson, J. R. Yeh, and J. K. Joung. 2013. Efficient genome editing in zebrafish using a CRISPR-Cas system. *Nature Biotechnology* 31(3): 227-229.

Irie, N., L. Weinberger, W. W. Tang, T. Kobayashi, S. Viukov, Y. S. Manor, S. Dietmann, J. H. Hanna, and M. A. Surani. 2015. SOX17 is a critical specifier of human primordial germ cell fate. *Cell* 160(1-2): 253-268.

Jaenisch, R. 1976. Germ line integration and Mendelian transmission of the exogenous Moloney leukemia virus. *Proceedings of the National Academy of Sciences of the United States of America* 73(4): 1260-1264.

Jaenisch, R., and B. Mintz. 1974. Simian virus 40 DNA sequences in DNA of healthy adult mice derived from preimplantation blastocysts injected with viral DNA. *Proceedings of the National Academy of Sciences of the United States of America* 71(4): 1250-1254.

Jamieson, A. C., S. H. Kim, and J. A. Wells. 1994. *In vitro* selection of zinc fingers with altered DNA-binding specificity. *Biochemistry* 33(19): 5689-5695.

Jiang, W., D. Bikard, D. Cox, F. Zhang, and L. A. Marraffini. 2013. RNA-guided editing of bacterial genomes using CRISPR-Cas systems. *Nature Biotechnology* 31(3): 233-239.

Jinek, M., K. Chylinski, I. Fonfara, M. Hauer, J.A. Doudna, and E. Charpentier. 2012. A programmable dual-RNA-guided DNA endonuclease in adaptive bacterial immunity. *Science* 337(6096): 816-821.

Jinek, M., A. East, A. Cheng, S. Lin, E. Ma, and J. Doudna. 2013. RNA-programmed genome editing in human cells. *eLife* 2: e00471.

Jinek, M., F. Jiang, D. W. Taylor, S. H. Sternberg, E. Kaya, E. Ma, C. Anders, M. Hauer, K. Zhou, S. Lin, M. Kaplan, A. T. Iavarone, E. Charpentier, E. Nogales, and J. A. Doudna. 2014. Structures of Cas9 endonucleases reveal RNA-mediated conformational activation. *Science* 343(6176): 1247997.

Johnson, J., J. Canning, T. Kaneko, J. K. Pru, and J. L. Tilly. 2004. Germline stem cells and follicular renewal in the postnatal mammalian ovary. *Nature* 428(6979): 145-150.

Jore, M. M., M. Lundgren, E. van Duijn, J. B. Bultema, E. R. Westra, S. P. Waghmare, B. Wiedenheft, Ü. Pul, R. Wurm, R. Wagner, M. R. Beijer, A. Barendregt, K. Zhou, A. P. L. Snijders, M. J. Dickman, J. A. Doudna, E. J. Boekema, A. J. R. Heck, J. van der Oost, and S. J. J. Brouns. 2011. Structural basis for CRISPR RNA-guided DNA recognition by Cascade. *Nature Structural & Molecular Biology* 18(5): 529-536.

Joung, J. K., and J. D. Sander. 2013. TALENs: A widely applicable technology for targeted genome editing. *Nature Reviews Molecular Cell Biology* 14(1): 49-55.

Kanatsu-Shinohara, M., and T. Shinohara. 2013. Spermatogonial stem cell self-renewal and development. *Annual Reviews Cell and Developmental Biology* 29: 163-187.

Kang, X., W. He, Y. Huang, Q. Yu, Y. Chen, X. Gao, X. Sun, and Y. Fan. 2016. Introducing precise genetic modifications into human 3PN embryos by CRISPR/Cas-mediated genome editing. *Journal of Assisted Reproduction and Genetics* 33(5): 581-588.

Kim, Y. G., J. Cha, and S. Chandrasegaran. 1996. Hybrid restriction enzymes: Zinc finger fusions to Fok I cleavage domain. *Proceedings of the National Academy of Sciences of the United States of America* 93(3): 1156-1160.

Kleinstiver, B. P., V. Pattanayak, M. S. Prew, S. Q. Tsai, N. T. Nguyen, Z. Zheng, and J. K. Joung. 2016. High-fidelity CRISPR-Cas9 nucleases with no detectable genome-wide off-target effects. *Nature* 529(7587): 490-495.

Komor, A. C., Y. B. Kim, M. S. Packer, J. A. Zuris, and D. R. Liu. 2016. Programmable editing of a target base in genomic DNA without double-stranded DNA cleavage. *Nature* 533(7603): 420-424.

Konermann, S., M. D. Brigham, A. E. Trevino, J. Joung, O. O. Abudayyeh, C. Barcena, P. D. Hsu, N. Habib, J. S. Gootenberg, H. Nishimasu, O. Nureki, and F. Zhang. 2015. Genome-scale transcriptional activation by an engineered CRISPR-Cas9 complex. *Nature* 517(7536): 583-588.

Kuehn, M., A. Bradley, E. Robertson, and M. Evans. 1987. A potential animal model for Lesch-Nyhan syndrome through introduction of HPRT mutations into mice. *Nature* 326(6110): 295-298.

Latham, K. E., B. Patel, F. D. Bautista, and S. M. Hawes. 2000. Effects of X chromosome number and parental origin on X-linked gene expression in preimplantation mouse embryos. *Biology of Reproduction* 63(1): 64-73.

Lavitrano, M., R. Giovannoni, and M. G. Cerrito. 2013. Methods for sperm-mediated gene transfer. In *Spermatogenesis: Methods and protocols*, Vol. 927, edited by D. Carrell, and K. I. Aston. New York: Humana Press. Pp. 519-529.

Leeb, M., and A. Wutz. 2011. Derivation of haploid embryonic stem cells from mouse embryos. *Nature* 479(7371): 131-134.

Le Page, M. 2015. Gene editing saves girl dying from leukemia in world first. *New Scientist*, November 5. https://www.newscientist.com/article/dn28454-gene-editing-saves-life-of-girl-dying-from-leukaemia-in-world-first (accessed October 31, 2016).

Li, W., L. Shuai, H. Wan, M. Dong, M. Wang, L. Sang, C. Feng, G.Z. Luo, T. Li, X. Li, L. Wang, Q. Y. Zheng, C. Sheng, H. J. Wu, Z. Liu, L. Liu, L. Wang, X. J. Wang, X. Y. Zhao, and Q. Zhou. 2012. Androgenetic haploid embryonic stem cells produce live transgenic mice. *Nature* 490(7420): 407-411.

Li, W., F. Teng, T. Li, and Q. Zhou. 2013. Simultaneous generation and germline transmission of multiple gene mutations in rat using CRISPR-Cas systems. *Nature Biotechnology* 31(8): 684-686.

Li, W., X. Li, T. Li, M. G. Jiang, H. Wan, G. Z. Luo, C. Feng, X. Cui, F. Teng, Y. Yuan, Q. Zhou, Q. Gu, L. Shuai, J. Sha, Y. Xiao, L. Wang, Z. Liu, X. J. Wang, X. Y. Zhao, and Q. Zhou. 2014. Genetic modification and screening in rat using haploid embryonic stem cells. *Cell Stem Cell* 14(3): 404-414.

Liang, P., Y. Xu, X. Zhang, C. Ding, R. Huang, Z. Zhang, J. Lv, X. Xie, Y. Chen, Y. Li, Y. Sun, Y. Bai, Z. Songyang, W. Ma, C. Zhou, and J. Huang. 2015. CRISPR/Cas9-mediated gene editing in human tripronuclear zygotes. *Protein & Cell* 6(5): 363-372.

Maeda, I., D. Okamura, Y. Tokitake, M. Ikeda, H. Kawaguchi, N. Mise, K. Abe, T. Noce, A. Okuda, and Y. Matsui. 2013. Max is a repressor of germ cell-related gene expression in mouse embryonic stem cells. *Nature Communications* 4: 1754.

Mahon, K. A., P. A. Overbeek, and H. Westphal. 1988. Prenatal lethality in a transgenic mouse line is the result of a chromosomal translocation. *Proceedings of the National Academy of Sciences of the United States of America* 85(4): 1165-1168.

Makarova, K. S., N. V Grishin, S. A Shabalina, Y. I. Wolf, and E. V. Koonin. 2006. A putative RNA-interference-based immune system in prokaryotes: Computational analysis of the predicted enzymatic machinery, functional analogies with eukaryotic RNAi, and hypothetical mechanisms of action. *Biology Direct* 1(1): 7.

Mali, P., L. Yang, K. M. Esvelt, J. Aach, M. Guell, J. E. DiCarlo, J. E. Norville, and G. M. Church. 2013. RNA-guided human genome engineering via Cas9. *Science* 339(6121): 823-826.

Martin, G. R. 1981. Isolation of a pluripotent cell line from early mouse embryos cultured in medium conditioned by teratocarcinoma stem cells. *Proceedings of the National Academy of Sciences of the United States of America* 78(12): 7634-7638.

Maruyama, T., S. K. Dougan, M. C. Truttmann, A. M. Bilate, J. R. Ingram, and H. L. Ploegh. 2015. Increasing the efficiency of precise genome editing with CRISPR-Cas9 by inhibition of nonhomologous end joining. *Nature Biotechnology* 33(5): 538-542.

McCreath, K. J., J. Howcroft, K. H. Campbell, A. Colman, A. E. Schnieke, and A. J. Kind. 2000. Production of gene-targeted sheep by nuclear transfer from cultured somatic cells. *Nature* 405(6790): 1066-1069.

Mojica, F. J. M., C. Diez-Villasenor, J. Garcia-Martinez, and E. Soria. 2005. Intervening sequences of regularly spaced prokaryotic repeats derive from foreign genetic elements. *Journal of Molecular Evolution* 60(2): 174-182.

Naldini, L. 2015. Gene therapy returns to centre stage. *Nature* 526(7573): 351-360.

Niu, Y., B. Shen, Y. Cui, Y. Chen, J. Wang, L. Wang, Y. Kang, X. Zhao, W. Si, W. Li, A. P. Xiang, J. Zhou, X. Guo, Y. Bi, C. Si, B. Hu, G. Dong, H. Wang, Z. Zhou, T. Li, T. Tan, X. Pu, F. Wang, S. Ji, Q. Zhou, X. Huang, W. Ji, and J. Sha. 2014. Generation of gene-modified cynomolgus monkey via Cas9/RNA-mediated gene targeting in one-cell embryos. *Cell* 156(4): 836-843.

Orthwein, A., S. M. Noordermeer, M. D. Wilson, S. Landry, R. I. Enchev, A. Sherker, M. Munro, J. Pinder, J. Salsman, G. Dellaire, B. Xia, M. Peter, and D. Durocher. 2015. A mechanism for the suppression of homologous recombination in G1 cells. *Nature* 528(7582): 422-426.

Oye, K. A., K. Esvelt, E. Appleton, F. Catteruccia, G. Church, T. Kuiken, S. B. Lightfoot, J. McNamara, A. Smidler, and J. P. Collins. 2014. Biotechnology: Regulating gene drives. *Science* 345(6197): 626-628.

Paquet, D., D. Kwart, A. Chen, A. Sproul, S. Jacob, S. Teo, K. M. Olsen, A. Gregg, S. Noggle, and M. Tessier-Lavigne. 2016. Efficient introduction of specific homozygous and heterozygous mutations using CRISPR/Cas9. *Nature* 533(7601): 125-129.

Paull, D., V. Emmanuele, K.A. Weiss, N. Treff, L. Stewart, H. Hua, M. Zimmer, D. J. Kahler, R. S. Goland, S. A. Noggle, R. Prosser, M. Hirano, M. V. Sauer, and D. Egli. 2012. Nuclear genome transfer in human oocytes eliminates mitochondrial DNA variants. *Nature* 493(7434): 632-637.

Perez-Pinera, P., D. D. Kocak, C. M. Vockley, A. F. Adler, A. M. Kabadi, L. R. Polstein, P. T. Thakore, K. A. Glass, D. G. Ousterout, K. W. Leong, F. Guilak, G. E. Crawford, T. E. Reddy, and C. A. Gersbach. 2013. RNA-guided gene activation by CRISPR-Cas9-based transcription factors. *Nature Methods* 10(10): 973-976.

Pourcel, C., G. Salvignol, and G. Vergnaud. 2005. CRISPR elements in *Yersinia pestis* acquire new repeats by preferential uptake of bacteriophage DNA, and provide additional tools for evolutionary studies. *Microbiology* 151(Pt. 3): 653-663.

Qasim, W., H. Zhan, S. Samarasinghe, S. Adams, P. Amrolia, S. Stafford, K. Butler, C. Rivat, G. Wright, and K. Somana. 2017. Molecular remission of infant B-ALL after infusion of universal TALEN gene-edited CAR T cells. *Science Translational Medicine* 9(374): eaaj2013

Qi, L. S., M. H. Larson, L. A. Gilbert, J. A. Doudna, J. S. Weissman, A. P. Arkin, and W. A. Lim. 2013. Repurposing CRISPR as an RNA-guided platform for sequence-specific control of gene expression. *Cell* 152(5): 1173-1183.

Ran, F. A., P. D. Hsu, C. Y. Lin, J. S. Gootenberg, S. Konermann, A. E. Trevino, D. A. Scott, A. Inoue, S. Matoba, Y. Zhang, and F. Zhang. 2013. Double nicking by RNA-guided CRISPR Cas9 for enhanced genome editing specificity. *Cell* 154(6): 1380-1389.

Rebar, E. J., and C. O. Pabo. 1994. Zinc finger phage: Affinity selection of fingers with new DNA-binding specificities. *Science* 263(5147): 671-673.

Reddy, P., A. Ocampo, K. Suzuki, J. Luo, S. R. Bacman, S. L. Williams, A. Sugawara, D. Okamura, Y. Tsunekawa, J. Wu, D. Lam, X. Xiong, N. Montserrat, C. R. Esteban, G. H. Liu, I. Sancho-Martinez, D. Manau, S. Civico, F. Cardellach, M. del Mar O'Callaghan, J. Campistol, H. Zhao, J. M. Campistol, C. T. Moraes, and J. C. I. Belmonte. 2015. Selective elimination of mitochondrial mutations in the germline by genome editing. *Cell* 161(3): 459-469.

Reinhardt, K., D. K. Dowling, and E. H. Morrow. 2013. Medicine: Mitochondrial replacement, evolution, and the clinic. *Science* 341(6152): 1345-1346.

Rouet, P., F. Smih, and M. Jasin. 1994. Introduction of double-strand breaks into the genome of mouse cells by expression of a rare-cutting endonuclease. *Molecular and Cellular Biology* 14(12): 8096-8106.

Rubin, G. M., and A. C. Spradling. 1982. Genetic transformation of *Drosophila* with transposable element vectors. *Science* 218(4570): 348-353.

Sagi, I., G. Chia, T. Golan-Lev, M. Peretz, U. Weissbein, L. Sui, M. V. Sauer, O. Yanuka, D. Egli, and N. Benvenisty. 2016. Derivation and differentiation of haploid human embryonic stem cells. *Nature* 532(7597): 107-111.

Sakuma, T., and K. Woltjen. 2014. Nuclease-mediated genome editing: At the front-line of functional genomics technology. *Development, Growth & Differentiation* 56(1): 2-13.

Sakurai, T., A. Kamiyoshi, H. Kawate, C. Mori, S. Watanabe, M. Tanaka, R. Uetake, M. Sato, and T. Shindo. 2016. A non-inheritable maternal Cas9-based multiple-gene editing system in mice. *Scientific Reports* 6: 20011.

Sapranauskas, R., G. Gasiunas, C. Fremaux, R. Barrangou, P. Horvath, and V. Siksnys. 2011. The *Streptococcus thermophilus* CRISPR/Cas system provides immunity in *Escherichia coli*. *Nucleic Acids Research* 39(21): 9275-9282.

Sander, J. D., and J. K. Joung. 2014. CRISPR-Cas systems for editing, regulating and targeting genomes. *Nature Biotechnology* 32(4): 347-350.

Sasaki, K., S. Yokobayashi, T. Nakamura, I. Okamoto, Y. Yabuta, K. Kurimoto, H. Ohta, Y. Moritoki, C. Iwatani, H. Tsuchiya, S. Nakamura, K. Sekiguchi, T. Sakuma, T. Yamamoto, T. Mori, K. Woltjen, M. Nakagawa, T. Yamamoto, K. Takahashi, S. Yamanaka, and M. Saitou. 2015. Robust *in vitro* induction of human germ cell fate from pluripotent stem cells. *Cell Stem Cell* 17(2): 178-194.

Schnieke, A., K. Harbers, and R. Jaenisch. 1983. Embryonic lethal mutation in mice induced by retrovirus insertion into the a1(I) collagen gene. *Nature* 304(5924): 315-320.

Schwartzberg, P., S. Goff, and E. Robertson. 1989. Germ-line transmission of a c-abl mutation produced by targeted gene disruption in ES cells. *Science* 246(4931): 799-803.

Silva, G., L. Poirot, R. Galetto, J. Smith, G. Montoya, P. Duchateau, and F. Paques. 2011. Meganucleases and other tools for targeted genome engineering: Perspectives and challenges for gene therapy. *Current Gene Therapy* 11(1): 11-27.

Slaymaker, I. M., L. Gao, B. Zetsche, D. A. Scott, W. X. Yan, and F. Zhang. 2016. Rationally engineered Cas9 nucleases with improved specificity. *Science* 351(6268): 84-88.

Solter, D. 2006. From teratocarcinomas to embryonic stem cells and beyond: A history of embryonic stem cell research. *Nature Reviews Genetics* 7: 319-327.

Staals, R. H., J. Y. Agari, S. Maki-Yonekura, Y. Zhu, D. W. Taylor, E. van Duijn, A. Barendregt, M. Vlot, J. J. Koehorst, K. Sakamoto, A. Masuda, N. Dohmae, P. J. Schaap, J. A. Doudna, A. J. R. Heck, K. Yonekura, J. van der Oost, and A. Shinkai. 2013. Structure and activity of the RNA-targeting Type III-B CRISPR-Cas complex of *Thermus thermophilus*. *Molecular Cell* 52(1): 135-145.

Suzuki, A., M. Hirasaki, T. Hishida, J. Wu, D. Okamura, A. Ueda, M. Nishimoto, Y. Nakachi, Y. Mizuno, Y. Okazaki, Y. Matsui, J. C. I. Belmonte, and A. Okuda. 2016a. Loss of MAX results in meiotic entry in mouse embryonic and germline stem cells. *Nature Communications* 7: 11056.

Suzuki, K., Y. Tsunekawa, R. Hernandez-Benitez, J. Wu, J. Zhu, E. J. Kim, F. Hatanaka, M. Yamamoto, T. Araoka, Z. Li, M. Kurita, T Hishida, M. Li, E. Aizawa, S. Guo, S. Chen, A. Goebl, R. D. Soligalla, J. Qu, T. Jiang, X. Fu, M. Jafari, C. R. Esteban, W. T. Berggren, J. Lajara, E. Nuñez-Delicado, P. Guillen, J. M. Campistol, F. Matsuzaki, G. H. Liu, P. Magistretti, K. Zhang, E. M. Callaway, K. Zhang, and J. C. Belmonte. 2016b. *In vivo* genome editing via CRISPR/Cas9 mediated homology-independent targeted integration. *Nature* 540: 144-149.

Tachibana, M., P. Amato, M. Sparman, J. Woodward, D. M. Sanchis, H. Ma, N. M. Gutierrez, R. Tippner-Hedges, E. Kang, H. S. Lee, C. Ramsey, K. Masterson, D. Battaglia, D. Lee, D. Wu, J. Jensen, P. Patton, S. Gokhale, R. Stouffer, and S. Mitalipov. 2012. Towards germline gene therapy of inherited mitochondrial diseases. *Nature* 493(7434): 627-631.

Tarkowki, A. K. 1977. *In vitro* development of haploid mouse embryos produced by bisection of one-cell fertilized eggs. *Journal of Embryology and Experimental Morphology* 38: 187-202.

Taylor, R. W., and D. M. Turnbull. 2005. Mitochondrial DNA mutations in human disease. *Nature Reviews Genetics* 6(5): 389-402.

Thomas, K., and M. Capecchi. 1987. Site directed mutagenesis by gene targeting in mouse embryo-derived stem cells. *Cell* 51(3): 503-512.

Tsai, S. Q., and J. K. Joung. 2016. Defining and Improving the genome-wide specificities of CRISPR-Cas9 nuclease. *Nature Reviews Genetics* 17(5): 300-312.

Tsai, S. Q., N. Wyvekens, C. Khayter, J.A. Foden, V. Thapar, D. Reyon, M. J. Goodwin, M. J. Aryee, and J. K. Joung. 2014. Dimeric CRISPR RNA-guided FokI nucleases for highly specific genome editing. *Nature Biotechnology* 32(6): 569-576.

Tsai, S. Q., Z. Zheng, N. T. Nguyen, M. Liebers, V. V. Topkar, V. Thapar, N. Wyvekens, C. Khayter, A. J. Iafrate, L. P. Le, M. J. Aryee, and J. K. Joung. 2015. GUIDE-seq enables genome-wide profiling of off-target cleavage by CRISPR-Cas nucleases. *Nature Biotechnology* 33(2): 187-197.

Wakayama, T., A. C. Perry, M. Zuccotti, K. R. Johnson, and R. Yanagimachi. 1998. Full-term development of mice from enucleated oocytes injected with cumulus cell nuclei. *Nature* 394(6691): 369-374.

Wan, H., Z. He, M. Dong, T. Gu, G. Z. Luo, F. Teng, B. Xia, W. Li, C. Feng, X. Li, T. Li, L. Shuai, R. Fu, L. Wang, X.J. Wang, X.Y. Zhao, and Q. Zhou. 2013. Parthenogenetic haploid embryonic stem cells produce fertile mice. *Cell Research* 23(11): 1330-1333.

Wang, H., H. Yang, C.S. Shivalila, M. M. Dawlaty, A. W. Cheng, F. Zhang, and R. Jaenisch. 2013. One-step generation of mice carrying mutations in multiple genes by CRISPR/Cas-mediated genome engineering. *Cell* 153(4): 910-918.

Wang, X., Y. Wang, X. Wu, J. Wang, Y. Wang, Z. Qiu, T. Chang, H. Huang, R. J. Lin, and J. K. Yee. 2015. Unbiased detection of off-target cleavage by CRISPR-Cas9 and TALENs using integrase-defective lentiviral vectors. *Nature Biotechnology* 33(2): 175-178.

Wiedenheft, B., K. Zhou, M. Jinek, S.M. Coyle, W. Ma, and J. A. Doudna. 2009. Structural basis for DNase activity of a conserved protein implicated in CRISPR-mediated genome defense. *Structure* 17(6): 904-912.

Wiedenheft, B., G. C. Lander, K. Zhou, M. M. Jore, S. J. J. Brouns, J.van der Oost, J. A. Doudna, and E. Nogales. 2011. Structures of the RNA-guided surveillance complex from a bacterial immune system. *Nature* 477(7365): 486-489.

Wilmut, I., A. E. Schnieke, J. McWhir, A. J. Kind, and K. H. Campbell. 1997. Viable offspring derived from fetal and adult mammalian cells. *Nature* 385(6619): 810-813.

Wu, Y., H. Zhou, X. Fan, Y. Zhang, M. Zhang, Y. Wang, Z. Xie, M. Bai, Q. Yin, D. Liang, W. Tang, J. Liao, C. Zhou, W. Liu, P. Zhu, H. Guo, H. Pan, C. Wu, H. Shi, L.Wu, F. Tang, and J. Li. 2015. Correction of a genetic disease by CRISPR-Cas9-mediated gene editing in mouse spermatogonial stem cells. *Cell Research* 25(1): 67-79.

Yang, H., L. Shi, B. A. Wang, D. Liang, C. Zhong, W. Liu, Y. Nie, J. Liu, J. Zhao, X. Gao, D. Li, G. L. Xu, and J. Li. 2012. Generation of genetically modified mice by oocyte injection of androgenetic haploid embryonic stem cells. *Cell* 149(3): 605-617.

Yang, H., H. Wang, C. S. Shivalila, A. W. Cheng, L. Shi, and R. Jaenisch. 2013. One-step generation of mice carrying reporter and conditional alleles by CRISPR/Cas-mediated genome engineering. *Cell* 154(6): 1370-1379.

Zeng, H., S. Wen, W. Xu, Z. He, G. Zhai, Y. Liu, Z. Deng, and Y. Sun. 2015. Highly efficient editing of the actinorhodin polyketide chain length factor gene in Streptomyces coelicolor M145 using CRISPR/Cas9-CodA(sm) combined system. *Applied Microbiology and Biotechnology* 99(24): 10575-10585.

Zetsche, B., J. S. Gootenberg, O. O. Abudayyeh, I. M. Slaymaker, K. S. Makarova, P. Essletzbichler, S. E. Volz, J. Joung, J. Van der Oost, A. Regev, E. V. Koonin, and F. Zhang. 2015. Cpf1 is a single RNA-guided endonuclease of a class 2 CRISPR-Cas system. *Cell* 163(3): 759-771.

Zhou, Q., M. Wang, Y. Yuan, X. Wang, R. Fu, H. Wan, M. Xie, M. Liu, X. Guo, Y. Zheng, G. Feng, Q. Shi, X.Y. Zhao, J. Sha, and Q. Zhou. 2016. Complete meiosis from embryonic stem cell-derived germ cells *in vitro*. *Cell Stem Cell* 18(3): 330-340.

Zijlstra, J., E. Li, F. Sajjadi, S. Subramani, and R. Jaenisch. 1989. Germ line transmission of a disrupted b2-microglobulin gene produced by homologous recombination in embryonic stem cells. *Nature* 342(6248): 435-438.

附录 B 国际研究监管法规
Appendix B International Research Oversight and Regulations

人类基因组编辑的研究和临床试验的管理应当借鉴适应用于其他临床研究和开发领域的国际和国家法规、政策及指南,包括其他类型的遗传技术、干细胞、生殖医学和涉及人类胚胎的研究。本附录提供了有关这些系统的更多信息,虽然并不全面,仅仅是提供一些关于在美国以外的国家是如何处理问题的观点。

公众咨询可以构成管理策略的一部分

世界上有许多关于使用公共咨询来制定广泛的生物医学和环境政策的例子(详细的两个实例参见延伸内容 B-1)。在美国,《国家环境政策法》不是寻常的环境法案,因为它不是直接管控,而是规定当政府作出重大决定时,必须受到高于正常程度的公众监督。通过纳入公众意见,这种公众监督产生的政治压力,可以以某种方式推动决策,并且使得政府专家/权威人士与公众咨询之间存在一些相互影响。加拿大在审查许多不同形式的协助生殖案例时,针对全国流行的新的生殖技术组成了皇家委员会,举行公开听证会。在欧盟(European Union,EU),转基因食品特别受关注,欧盟指令要求当产品可能影响生物多样性或其他环境因素时应当信息公开。公众咨询被认为是指导性集中治理形式的替代方案,公众可以通过自己的权力下放的过程,对政府或公司施加压力,改变生物技术创新

The governance of research and clinical trials using human genome editing is expected to draw on the foundation of international and national regulations, policies, and guidance that apply to other areas of clinical research and development, including other types of genetic technologies, stem cells, reproductive medicine, and research involving human embryos. This appendix provides further information on some of these systems. It is not meant to be comprehensive, but rather to provide perspectives on how issues are addressed in countries other than the United States.

PUBLIC CONSULTATION CAN FORM PART OF A GOVERNANCE STRATEGY

There are numerous examples around the world of the use of public consultation on a wide range of biomedical and environmental policies (see Box B-1 for two examples described in greater detail). In the United States, the National Environmental Policy Act is unusual among environmental laws, because rather than directly regulating action, it simply provides that when the government makes a major decision, it must be subjected to a higher than usual degree of public scrutiny. By incorporating public comment, such public scrutiny creates political pressure that can drive decisions in one way or another, and it allows for some interplay between government expertise/authority and public consultation. Canada, when it looked at assisted reproduction across many different forms, formed a royal commission on new reproductive technologies that traveled the country from east to west, holding public hearings on the topic. In the European Union (EU), genetically engineered foods are of special interest, and an EU directive requires

的方向或速度（Charo，2016b）。

遵循准则的自愿性规制是管理的另一个组成部分

超越协商的是自愿自律和非约来性协议。这些都是自我强加的规则，严格地限制捐赠组织、招募捐赠者，以及引起关注的实验（如嵌合体的使用）。例如，国际干细胞研究学会所采用的指南，

public access to information whenever a product potentially affects biodiversity or other environmental elements. Public consultation is considered an alternative to a directive centralized form of governance, in which the public can, through its own decentralized processes, exert pressure on government or on industry and alter the direction or the speed of biotechnology innovation (Charo, 2016b).

VOLUNTARY REGULATION THROUGH GUIDELINES IS ANOTHER COMPONENT OF GOVERNANCE

Beyond consultation are voluntary self-regulation and

延伸内容 B-1　公众咨询的两个实例

在这里讨论了法国和英国就应用到人类的新兴技术进行公众咨询的例子。咨询的方法大同小异，可能提供一些关于其他国家考虑如何处理与人类基因组编辑相关的科学和伦理问题的信息。

法国在2009年修订《法国生物伦理法》之前组织了一次公众讨论（"Les états généraux de la bioéthique"，生命伦理的现状）。除了机构的报告和一个互动网站外，咨询还包括了在生命伦理问题上的三个"共识会议"，包含胚胎研究及获取新的生殖技术。通过投票选择了25个公民代表参加了每个会议。这些代表参加了周末研讨会，并由一个多学科专家小组在这些问题上对他们进行了指导。公民也被邀请参加公开辩论，并由专家回答他们的问题。在该过程结束时，每个公民群体起草的建议被综合在咨询会最后的报告中。咨询结束后对在随后的法律修订中没有纳入公民的建议，以及法律修订没有解决公民小组提出的一些社会问题而提出了批评。另一方面，咨询的过程提供了一个扩大参与修订生物伦理学法律群体的机会。该法律于1994年首次实施，主要是基于国家生物伦理委员会的建议，该委员会包括医生、生物医学研究人员、哲学家，以及宗教派别的代表。咨询公民提出了一些社会价值观，例如，希望儿童知道他们的来历或不论性取向为一对忠诚的夫妇获得生殖技术，这些社会价值观应当有机会在媒体覆盖的官方论坛上得以表达，并为随后的公开讨论做出贡献。

英国在2009年修订了《人类受精和胚胎法》（Human Fertilisation and Embryology Act，HFEA）后，最近就线粒体替代治疗进行了磋商。公共研讨会和辩论及互动网站聚焦于线粒体技术的使用可能引发的伦理问题，以及这种技术是否应被允许在英国的临床实践中使用。政府平衡各方意见后在2013年报告称，应该允许这种治疗技术的使用，但是应谨慎控制。2014年进一步磋商了条例草案。为了确保有大量的参与者，咨询了一系列组织，包括患者团体、专业团体、研究机构、遗传利益团体、教徒、社区组织及个人；收到了有关各方及一些个人对规章草案提出意见的若干答复（1857份）。与法国的法律制定过程相比，英国的《人类受精和胚胎法》制定过程包括了多场所的公共论坛和更多的公民参与，以及就单一问题磋商而提出的更集中的建议。另一方面，也收到一些批评，包括法律制定的过程和形式是以规则为导向的，而不是基于对话和协商，并且在咨询和最终的立法结果之间缺乏明确的联系。

Box B-1　Two Examples of Public Consultation

This box discusses examples from France and the United Kingdom in which public consultations have been undertaken on human application of emerging technologies. The consultative approaches have similarities and differences that may be informative as other countries consider how to address scientific and ethical issues associated with human genome editing.

In France, a General Public Discussion ("Les états généraux de la bioéthique") was organized prior to the 2009 revision of French Bioethics Laws. In addition to institutional reports and an interactive website, the consultation included three "Consensus Conferences" on bioethical issues, including embryo research and access to new reproductive technologies. A representative sample of 25 citizens was chosen using an opinion-poll method to participate in each conference. These participants attended weekend seminars and received instruction on the issues by a multidisciplinary team of experts. The citizens were also invited to a public debate where topical experts answered their questions. At the end of the process, each group of citizens drafted recommendations that were synthesized in the consultation's final report. Criticisms raised after the consultation included an impression that few citizen recommendations were incorporated in subsequent legal revisions and that the revisions did not address some of the social concerns raised by citizen panels. On the other hand, the consultative process provided an opportunity to broaden participation in the revision of the bioethics law, which was first implemented in 1994, based largely on recommendations of a national bioethics committee that consisted of experts such as doctors, biomedical researchers, philosophers, and representatives of religious denominations. Social values raised by consulted citizens, such as the desire for children to know their history or for a committed couple to access reproductive technologies regardless of sexual orientation, also had the opportunity to be expressed in an official forum, covered by the media, and to contribute to subsequent public discussions.

More recently, the United Kingdom undertook consultations on mitochondrial replacement therapy, following a 2009 revision of the Human Fertilisation and Embryology Act (HFEA). Public workshops and debates and interactive websites focused on ethical issues the use of mitochondrial techniques might raise and whether such techniques should be permitted for use in clinical practice in the United Kingdom. The balance of views from this exercise, reported in 2013, was that such treatment techniques should be allowed but that their use should be carefully controlled. Further consultation was undertaken in 2014 on draft regulations. To ensure a large audience, a range of organizations was solicited, including patient groups, professional bodies, research bodies, genetic interest groups, and faith and community organizations, as well as individuals. A number of responses (1,857) were received from interested parties as well as from a number of individuals giving their views on the draft regulations. In comparison to the French process, the HFEA process included multisite public forums and a larger number of citizens providing input, as well as more focused recommendations due to the single-issue consultation. On the other hand, criticisms levied included that the processes and modalities were rule-guided rather than based on dialogue and deliberation, and that there was a lack of identifiable links between the consultation and the ultimate legislative outcome.

该指南已经过修订，涵盖了从基础科学到干细胞临床实验的所有形式的胚胎研究（ISSCR，2016）。也有的指南是以有说服力但不强制执行的国际文书的形式出现，如国际医学科学组织理事会（CIOMS）发布的《关于人类受试者研究的全球标准》（Gallagher et al.，2000）。

更进一步的管理条例当然还有规章和立法。特别是在基因治疗和种系操作方面存在一些具有不同可执行程度的国际文书。例如，在欧洲委员会制定的《奥维耶多公约》中，预测性遗传检测只应用于医疗目的，规定还特别提出要求禁止使用种系的遗传工程或改变后代的基因组成。《奥维耶多公约》建立在早期的《欧洲公约》基础之上，但与许多国际文书一样，它并没有得到所有成员国的批准，即使获得批准，也不一定能够通过国内立法予以执行。虽然它具有很大的规范价值，但并不具备相同等级的执行能力。

规范方法各国不一

根据一份近期来自监管和其他来源的转基因实验信息的综述，截至2012年6月，已有超过1800件转基因实验在31个国家获得批准、启动或完成（Ginn et al.，2013；IOM，2014）。到2016年中期，这一数字已经超过2400件。这些实验出现在以美洲和欧洲为主的每个有人口居住的洲，并基本保持逐年增长的态势[70]。2013年度的报告表示这些试验有65.1%发生在美国，28.3%发生在欧洲，3.4%发生在亚洲。2015~2016年的数据也与此类似。因为其中过半的试验（63.7%，1174件）与美国的研究员或机构有关（Ginn et al.，2013）。那些管理由美国国家卫生研究院（NIH）资助或由于美国国内保护而受制于NIH规章的研究的法规将对这项工作如何在美国境外进行产生一定影响。美国食品药品监督管理局（FDA）的规定也将适用于需要FDA批准的产品，以便其可在美国进行销售，无论它们的资金来源或试验地点是否在美国国内。

各国选择了不同的方法来构架其监管途径。日本采用了预估试验风险等级来进行相应等级的管理。美国在其医疗器械的监管中遵循了类似的过程。

non-binding agreements. These are self-imposed rules that are seriously constraining with respect to donation of tissues, recruitment of donors, and experimentation that raises concerns, such as the use of chimeras. Examples include the guidelines adopted by the International Society for Stem Cell Research, which have been amended to cover all forms of embryo research, from basic science to clinical trials with stem cells (ISSCR, 2016). Guidance also comes in the form of persuasive, albeit unenforceable, international instruments, such as those issued by the Council for International Organizations of Medical Sciences (CIOMS) for global standards for human subjects research (Gallagher et al., 2000).

At the far end of the spectrum, of course, there is regulation and legislation. Specifically with respect to gene therapy and germline manipulation, there are a number of international instruments of varying degrees of enforceability. For example, the Council of Europe's Oviedo Convention says that predictive genetic tests should be used only for medical purposes. It specifically calls for a prohibition on the use of genetic engineering of the germline or changing the makeup of the following generations. It builds on earlier European conventions, but like many international instruments, it is not ratified by every member country and, even when ratified, does not necessarily get implemented with domestic legislation. It has great normative value, but its enforcement-level value is uneven.

REGULATORY APPROACHES VARY BY COUNTRY

According to one recent review of gene-transfer trial information from regulatory and other sources, as of June 2012 more than 1,800 trials have been approved, initiated, or completed in 31 countries (Ginn et al., 2013; IOM, 2014). By mid-2016 that number had grown to more than 2,400, with trials primarily located in the Americas and Europe but nonetheless ongoing on every populated continent, and the number of studies generally growing every year.[70] The 2013 review reported that 65.1 percent of the trials were based in the Americas, 28.3 percent in Europe, and 3.4 percent in Asia. Data from 2015 and 2016 show a similar pattern. Because more than half of all trials (63.7 percent, or 1,174) are associated with U.S. investigators or institutions (Ginn et al., 2013), U.S. regulations that govern research funded by the National Institutes of Health (NIH) or subject to the NIH rules due to a federal-wide assurance will have some effect on how the work proceeds outside the United States. The U.S. Food and Drug Administration (FDA) rules will also apply for products for which FDA approval is sought so that sale can proceed in the United States, regardless of funding source or whether the trial site is in the United States or abroad.

Countries approach the structure of their regulatory pathways in different ways. Japan has a regulatory pathway

---

[70] Gene Therapy Clinical Trials Worldwide, provided by the Journal of Gene Medicine. http://www.abedia.com/wiley/years.php (accessed January 30, 2017).

但对于药物，美国选择从一开始就视全部药物为同等危险，并用相同的规则测试每一种拟议药物的安全性和有效性。与此相反，日本选择对拟议药物中可能存在的风险等级和其严谨性进行预估。因此，为了使用此监管途径，日本还为再生医学和基因治疗产品增加了条件性批准途径，但是想要评价这些新医疗依旧为时过早。

新加坡同样拥有一个与日本类似的风险等级评估途径，并在评估细胞治疗时选取了一些变量，包括治疗中操作量的多少、使用的是否为同源基因，以及治疗方案是否包含使用其他药物、装置或其他生物制剂。

作为一个累积规范管理的例子，巴西曾批准过有关转基因食品、干细胞研究和细胞治疗专门相关的法律，包括禁止出售任何种类的人体组织的宪法禁令和1996年关于人类生物材料专利的法律。但那些法律只是属于早些时候、较一般的基本法规，因而造成了局面的混乱。结果是这些法律相互作用、同时管理时，出现了某种程度的瘫痪。

一般来说，拉丁美洲有关人类体细胞基因编辑的话题已经在遗传修饰的植物和动物、生物剽窃、生物安全性，以及干细胞在临床护理中的应用方面引起关注。墨西哥在其基本卫生法和研究规范中的转基因生物及生物安全部分提到了基因工程[71]。巴西则将相关条目写入生物安全法，默许进行少量人体细胞的基因编辑，虽然其重点显然是转基因生物[72]。同样，厄瓜多尔的宪法也在转基因生物和生物剽窃方面对基因组遗传做出了规定，并在保障个人完整性方面，禁止使用遗传材料进行侵犯人权的科学研究[73]。

一部分拉丁美洲国家对体细胞基因组编辑加强了限制，以禁止使用那些被认为对可能是"加强"而不是治疗或预防疾病与损伤的基因编辑。智利在一项涉及知识产权、歧视和保护遗传特性，以及禁止"优生做法"（遗传咨询除外）的涉及面广泛的法律中，表示仅允许将体细胞基因编辑用于

that tries to identify prospectively those things that are going to be high, medium, or low risk, and regulate them accordingly. The United States follows a similar process in its regulation of medical devices. But for drug regulation, the United States treats everything ab initio as equally dangerous and runs every proposed drug through the same rules for testing safety and efficacy. By contrast, in Japan there is an initial determination about the level of risk that is likely to be present for each proposed drug and the degree of stringency that the regulatory process must apply as a result. Japan has also added a conditional approval pathway specifically for regenerative medicine and gene therapy products, but it is too new for evaluation (Charo, 2016b).

Singapore also has a risk-based approach similar to Japan's, and for cell therapy it uses variables that include whether the manipulation is substantial or minimal, whether or not the intended use is homologous or non-homologous, and whether or not this is going to be combined with some drug, device, or other biologic.

Brazil provides an example of regulation and governance by accretion. It has approved laws related specifically to genetically engineered foods and stem cell research and cell therapy, but they are layered on top of earlier, more general rules, including constitutional prohibitions on the sale of any kind of human tissue and 1996 laws on the patenting of human biological materials, creating a situation of confusion. The result has been a degree of paralysis while the interplay among the laws is being managed.

More generally, discourse in Latin America on human somatic cell genome editing has been informed by concerns about genetically modified plants and animals, biopiracy, biosecurity, and use of stem cells for clinical care. Mexico addresses genetic engineering in the context of GMOs and biosecurity in its general health law and in its research regulations.[71] Brazil addresses gene editing in its Biosafety Law, implicitly permitting at least some somatic gene-editing research in humans, although its primary focus is clearly on GMOs.[72] Similarly, Ecuador's Constitution has provisions addressing genomic heritage in the setting of GMOs and biopiracy, and in its guarantee of personal integrity it prohibits the use of genetic material for scientific research in violation of human rights.[73]

A few jurisdictions in Latin America have explicitly addressed somatic genome editing, typically imposing restrictions aimed at prohibiting uses that might be perceived as "enhancement" rather than treatment or prevention of disease and injury. Chile states that gene editing "in somatic cells will

---

[71] Ley General de Salud, Titulo Decimo Segundo. Capitulo XII, Artículo 282. And Regulamento de la Ley General de Salud en Materia de Investigacíon para la Salud, Titulo Cuarto, Capitulo II, Articulos 85-88 (recombinant DNA research).
[72] Public Law No. 11.105, Chapter 1, Article 6, as translated by WIPO.
[73] Constitución de la Republica del Ecuador 2008, Titulo II, Articulo 66, 1.3(d). Interestingly, Ecuador promulgated an extensive set of regulations governing the use of biological samples and genomic data, specifically citing, inter alia, the biopiracy of DNA from an indigenous population in that country. Ministero de Salud Pública (MSP), Reglamento para uso del material genético humano en Ecuador. Ministereo de Salud Pública, Dirección Nacional de Noamrtización y Programa Nacional de Genética, 2013.

疾病治疗和防范[74]。在巴拿马和墨西哥城，进行用于治疗严重缺陷或疾病以外的遗传操作可处以 2~6 年徒刑[75]。哥伦比亚的刑法同样允许遗传修饰用于治疗、诊断，以及减轻痛苦或改善人类健康方面的研究，同时对用于其他用途判的人处以 1~5 年徒刑[76]。

在欧盟，评定与监管人类及兽用药以保障公众和动物的健康是欧洲药品管理局（EMA）的职责（EMA，2013）。EMA 于 2007 年成立了高端治疗委员会，以评估那些被称为"高端治疗"的药物——基因与细胞药物的质量、安全性和有效性。该委员会为面向欧盟市场销售的药物提供了一个统一的审批程序，包括用于基因和细胞治疗的产品及许多其他产品，例如，用于治疗 HIV / AIDS 和癌症的药物在内的生物制剂必须通过此项审批（Cichutek，2008）。

然而，EMA 并没有权力审批包括转基因研究在内的临床研究方案（Pignatti，2013）。这份权力是属于国家监管机构的，但每个欧盟国家都通过了欧盟的临床试验指南（Kong，2004）。这要求成员国采用一种遵循国际公认的良好临床标准的系统来评判临床与研究的一致性，用于审查那些符合伦理和科学有效性的设计、行为和实验报告（Kong，2004）。参与制定此项标准的国际进程的 FDA 也承认并将这些标准作为指导文件发布（FDA，2012）[77]。

与欧洲一样，中国也已经有面对人类医疗产品的开发和使用的规章制度。虽然尚未针对基因组编辑进行修订，但管理基因和细胞治疗的制度已经实施，国家食品药品监督管理局（现任中国食品药品监督管理局的前身）也已批准了面向市场的基因治疗产品。同时，中国的相关机构也已发布了人类胚胎研究和体外受精（IVF）实践的监管指南（China Ministry of Health，2001，2003）。在目前的规章制度内，人类体细胞基因组编辑可以被认为是第三类治疗技术而不是药物。若是这样，它将

be authorized only for the treatment of diseases or to prevent their occurrence" in a far reaching law that also addresses intellectual property, discrimination, and protection of genetic identity, as well as prohibiting "eugenic practices" (with an exception for genetic counseling).[74] In Panama and Mexico City, use of genetic manipulation except for the elimination or treatment of a serious defect or disease is punishable by a prison sentence of 2 to 6 years.[75] Colombia's penal code similarly permits genetic modification for treatment, diagnosis, and research to alleviate suffering or improve human health, while imposing a prison sentence of 1 to 5 years for other uses.[76]

In the European Union, the European Medicines Agency (EMA) has the responsibility to evaluate and supervise human and veterinary medicines to protect public and animal health (EMA, 2013). In 2007, the EMA established the Committee for Advanced Therapies as the unit responsible for assessing the quality, safety, and efficacy of medicines made from genes and cells—medicines that are termed "advanced therapy medicinal products." This committee provides a centralized procedure for the assessment and approval of medicines for marketing in the European Union. The process is mandatory for biologics, including gene and cell therapy products, and a number of other product categories, including medicines for the treatment of HIV/AIDS and cancer (Cichutek, 2008).

EMA, however, does not have authority to review and approve protocols for clinical research, including gene-transfer research (Pignatti, 2013). That authority resides with national regulatory agencies, but every EU state has adopted the EU Directive on Clinical Trials (Kong, 2004). It requires member states to adopt a system for the review of clinical research consistent with internationally recognized standards for good clinical practice for the ethical and scientifically valid design, conduct, and report of trials (Kong, 2004). FDA, which participated in the international process for developing these standards, also recognizes these standards and publishes them as guidance documents (FDA, 2012).[77]

As with Europe, China has a regulatory framework for the development and use of human medical products. Although it is not yet amended to address genome editing specifically, frameworks governing gene and cell therapies have been implemented and the State Food and Drug Administration (the predecessor of the current China Food and Drug Administration [CFDA]) approved a gene-therapy product for marketing. In addition, regulatory guidelines for human embryo research and *in vitro* fertilization (IVF) practices have been published

---

[74]Public Law No. 20.120 On the Scientific Investigation of the Human Genome, Its Genoma, & Prohibition of Human Cloning, Articles 1, 3, 4, 7, 8, 12, and 13, 2006 (English translation).
[75]Ley Penal en General. Capítulo II, Artículo 145. (2010); Código Penal para el Distrito Federal. Capítulo II, Artículo 154 (also bars employment and other benefits this period).
[76]Código Penal Colombiano, Capítulo Octavo, Artículo 132 (2015).
[77]Information on the national regulatory frameworks likely to apply to somatic and germline human genome editing in a number of European countries is described in a background document produced for a 2016 workshop on human genome editing in the EU. http://acmedsci.ac.uk/file-download/41517-573f212e2b52a.pdf (accessed January 30, 2017).

属于 CFDA 的监管范畴，监管程序将包括临床前测试，以及临床试验的安全性和有效性的评估，类似于 FDA 和 EMA 使用的过程。除 CFDA 之外，监管 IVF 诊所的卫生和计划生育委员会（HFPC），将同样可能参与人类基因组编辑的监督。包括科学技术部、中国科学院、中国医学科学院和中国工程院等在内的机构应会对将此项写入法规进行磋商。

可遗传的基因修饰带来了更多问题

在对人类胚胎的使用和人类胚系的遗传的潜在改变中可能出现很多特殊的监管和治理问题。关于这些问题的讨论已经分为诸如胚胎干细胞、克隆辅助生殖技术和生命起源几个方面。

正如上面提到的，《保护生命和医学应用人权和人格尊严公约：人权与生物医学公约》（《奥维耶多公约》）制定了有关若干会引起生命伦理问题的医学课题的原则。对于已经签署和批准公约的国家，应保护个人的遗传构成，以免遭到旨在修改种系的非法干预。一些国家还颁布了限制人类种系修改的国家法律、法规或准则。

尽管正式的协调管理似乎缺乏可行性，在管理人类基因组编辑上的国际合作仍不可缺少

鉴于科学和医学的全球性发展，以及管理人类遗传技术的方法的多样性和复杂性，人们一直呼吁对此开展国际协作、合作，甚至统一建立有关人类基因组编辑的法规。可以提出支持或反对国际上统一的基因组技术的国家监督的论点（Breggin et al.，2009；Marchant et al.，2012）。同时，另一个支持创建统一或一致的基因编辑规则的有力论据是这样可以避免"监管避风港"——提供者或消费者为了进行某些被限制的操作而前往管理更宽松或不存在法规的管辖区（Charo，2016a）。这种潜在的利润丰厚的医疗旅游可能会形成一个"竞争的底线"，鼓励那些在寻求医疗旅游收入的国家（Abbott et al.，2010）。一个统一的标准还可以促使每个国家公民得到平等的健康保护，并为该领域的公司和科学家提供一致的要求，降低交易成本并扩大经济规模（Vogel，1998）。一个统一的标准还为监管者提供规

by Chinese agencies (China Ministry of Health, 2001, 2003). Within the current regulatory framework, human somatic cell genome editing may be considered a third category therapeutic technology rather than a drug. If so, it would be regulated by CFDA and procedures would include evaluations for safety and efficacy through preclinical testing and clinical trials, similar to processes used by the FDA and the EMA. In addition to CFDA, the Health and Family Planning Commission (HFPC), which regulates IVF clinics, is likely to be involved in oversight of human genome editing. Consultations would likely occur with agencies such as the Ministry of Science and Technology, Chinese Academy of Sciences, Chinese Academy of Medical Sciences, and Chinese Academy of Engineering to enable their positions to be incorporated into regulations.

HERITABLE GENETIC MODIFICATIONS RAISE ADDITIONAL ISSUES

A number of special regulatory and governance issues may arise around use of human embryos and the potential for genetic changes to be made to the human germline. Discussions of such issues have been informed by debates about topics such as embryonic stem cells, cloning assisted reproductive technologies, and the beginning of life.

As noted above, the Convention for the protection of Human Rights and Dignity of the Human Being with regard to the Application of Biology and Medicine: Convention on Human Rights and Biomedicine ("Oviedo Convention") develops principles concerning a number of medical topics that raise bioethics issues. For countries that have signed and ratified the treaty, the genetic constitution of the individual is to be protected against unlawful interventions seeking to modify the germline. A number of countries have also enacted national laws, regulations, or guidelines that restrict human germline modifications (see Figure B-1).

INTERNATIONAL COOPERATION ON THE GOVERNANCE OF HUMAN GENOME EDITING IS DESIRABLE ALTHOUGH FORMAL REGULATORY HARMONIZATION SEEMS INFEASIBLE

Given the global nature of scientific and medical advances and the diversity and complexity of approaches to regulating human genetic technologies, there have been calls for international collaboration, cooperation, and even harmonization of regulations governing human genome editing. Arguments can be made both for and against an international convergence of national regulation of genomic technologies (Breggin et al., 2009; Marchant et al., 2012). A compelling argument for creating uniform or consistent regulations of gene editing is to avoid "regulatory havens" that circumvent restrictions if providers or consumers

模经济，降低通过和管理国家法律的行政成本，并增加共享监管资源和共同承担工作的机会。最后，协调进程可以促进良性的习惯做法以交流和建立监管能力（OECD，2013）。

另一方面，各国具有不同的历史、经济、社会和文化系统与价值观，这可能导致其对强大的技术，如人类基因组编辑的管理方法产生差异。统一的国际标准还可能使每个国家仅能受到最低的共同标准的监管，这都将取决于这种统一在现实中将要如何被实现。实际上，试图在 100 个或更多的国家就任何技术的监管达成共识都是一项耗费精力和资源的工作，而且可能最终无法成功。然而，监管方法的多样性也可以被看成是一个用来评估不同规章制度的效果的实验，提供一个"国家实验室"，以"迅速了解创新与预防措施之间达成的不同平衡对促进创新的影响"（Evans，2015）。然而，要实现这一益处还需要促进国家间的学习和信息交流。

如表 B-1 所示，串联起国家的监管需求通常需要一系列措施（Breggin et al.，2009）。这种统一协调通常涉及执行参与国的相同或等效的国家法律。通常是通过一项国际条约或其他具有约束力的正式法律文书，通过修订使其符合条约要求以实施（OECD，2013）。关于国际条约和其他正式协定的谈判十分困难且耗时，并常常存在执法困难的问题[78]。在这些障碍面前，各国在相关技术和产品的监管方面已经表现出了偏离国际统一协调管理的倾向（Falkner，2013；Susskind，2008）。

国际协调和合作的非正式机制并不指定可具体实施的法律条款，而是以不具约束力的指导、建议、协商、原则声明或自愿标准的形式在政府之间提供一般性协议。这种规范性准则可以在独立谈判中商定，但通常由适当的国际组织在商议中决定（Abbott，2014）。国际合作和协调也可以在非政府组织中决定，例如，提出胚胎干细胞研究指南的国际干细胞研究协会（ISSCR）这样的科学家学会。

travel to jurisdictions with more lenient or non-existent regulations in order to undertake the restricted procedures (Charo, 2016a). The potential for lucrative medical tourism may create a "race to the bottom" that encourages laxer standards in nations seeking revenues from medical tourism (Abbott et al., 2010). Consistent standards may also promote equal health protection for citizens of all nations and provide consistent requirements for companies and scientists in the field, reducing transaction costs and increasing economies of scale (Vogel, 1998). Harmonized standards also provide economies of scale for regulators, reduce administrative costs in adopting and administering national laws, and increase opportunities to share regulatory resources and workload. Finally, the process of harmonization can promote the exchange of good practices and build regulatory capacity (OECD, 2013).

On the other hand, nations have different historical, economic, social, and cultural systems and values, which may translate into different approaches to the regulation of a powerful technology such as human genome editing. Uniform national regulations may also subject every nation to the lowest common regulatory denominator, depending on how harmonization is achieved. As a practical matter, reaching consensus among 100 or more nations on regulatory requirements for any technology is a laborious and resource-intensive undertaking that in the end may not be successful. Diversity in regulatory approaches also provides a natural experiment to evaluate the effects of different regulatory frameworks, providing a "laboratory of nations" that "fosters innovation and rapid learning about the impact of striking different balances between innovation and precaution" (Evans, 2015). However, realizing this benefit requires procedures that facilitate the exchange of information and promote learning.

As summarized in Table B-1, there is a continuum of approaches to aligning national regulatory requirements (Breggin et al., 2009). Harmonization usually involves the enforcement of identical or equivalent regulatory requirements under the national laws of participating countries. Harmonization is usually accomplished through an international treaty or other formal and binding legal instrument, implemented through the amendment of national laws to conform to treaty requirements (OECD, 2013). International treaties and other formal agreements

表 B-1 统一国际监管的三个通常方法

统一手段	定义	举例
跨国监管对话与监管网	监管者之间的非正式沟通和政策学习	关于纳米科技的国际对话
国际协调合作	不具约束力的国际文书，如准则、原则和标准	ISSCR 胚胎干细胞研究指南
基于条约的协调	具有约束力条约的正式谈判	联合国克隆公约（失败）

来源：改编自 Breggin 等（2009）。我们在此感谢英国皇家国际事务研究所允许使用此文中的图表：Securing the Promise of Nanotechnologies: Towards Transatlantic Regulatory Cooperation, 作者为 Linda Breggin, Robert Falkner, Nico Jaspers, John Pendergrass 和 Read Porter，2009.

TABLE B-1 Three General Approaches to International Regulatory Convergence

Convergence Process	Definition	Example
Transnational Regulatory Dialogue and Networking	Informal process of communication and policy learning between regulators	International Dialogue on Responsible Nanotechnology
International Coordination/Cooperation	Non-binding international instruments such as guidelines, principles, and standards	ISSCR Guidelines for Embryonic Stem Cell Research
Treaty-Based Harmonization	Formal negotiation of binding treaties	United Nations Convention on Cloning (failed)

SOURCE: Adapted from Breggin et al. (2009). We are grateful to Chatham House, the Royal Institute of International Affairs, for permission to reproduce a figure from the work titled: Securing the Promise of Nanotechnologies: Towards Transatlantic Regulatory Cooperation, authored by Linda Breggin, Robert Falkner, Nico Jaspers, John Pendergrass and Read Porter, 2009.

国际趋同化最正式的机制是通过跨国监管对话和网络政策传播。这种方法通常不涉及为各国提出实质或程序性建议的文书的制定，而是被用于为来自不同国家的监管者提供一个分享信息、方法、挑战和想法的论坛。其中一个例子就是关于纳米科技的国际对话，它是来自25个国家政府的专家每2年召开一次的一系列会议，内容为各方监管活动和面临挑战的汇报（Meridian Institute，2004）。参与国际协调活动的监管机构指出，这种面对面的接触与沟通是促进国际合作和理解的最有效机制之一（Saner and Marchant，2015）。

人类基因组编辑监管方法的统一确实将会带来上述的一些好处。但一些国家已经通过了与人类基因组编辑相关的各种法律，因此，正式或完全的协调似乎并不可行——甚至在这个时间点上也并不是很理想的选择。此外，国家对人类基因组编辑的反应表明了其独特的历史、文化、经济和社会因素，尽管这些重要的差异阻碍了统一的国际标准，但为不同国家的监管者提供强有力的沟通和协调仍是十分有益的，并为具体实质或技术方面发现共同点，以及产生学习效益方面提供了潜在的机会（Zhai et al.，2016）。

are difficult and time consuming to negotiate, and often present difficult enforcement issues.[78] Given these obstacles, there has been a trend away from treaties in the international governance of technologies and products in favor of mechanisms of international cooperation and coordination (Falkner, 2013; Susskind, 2008).

Informal mechanisms of international coordination and cooperation do not create legal requirements to implement specific provisions, but rather provide general agreement between governments in the form of non-binding guidelines, recommendations, consensus documents, statements of principles, or voluntary standards. Such normative guidelines may be agreed upon in free-standing negotiations, but are often negotiated within an appropriate international organization (Abbott, 2014). International cooperation and coordination approaches can also be achieved through nongovernmental organizations, such as scientific societies, for example through the International Society for Stem Cell Research (ISSCR) Guidelines for Embryonic Stem Cell Research (Daley et al., 2007).

The least formal mechanism of international convergence is policy diffusion through transnational regulatory dialogue and networking. This approach usually does not involve the creation of instruments that set forth specific substantive or procedural recommendations for nations to follow. Rather, it provides a forum for regulators from different nations to share information, approaches, challenges, and ideas. An example is the International Dialogue on Responsible Nanotechnology, in which experts from 25 national governments convened in a series of meetings every 2 years to report on their regulatory activities and challenges (Meridian Institute, 2004). Regulators who have been involved in international coordination activities state that such person-to-person contacts and communications provide one of the most effective mechanisms for promoting international cooperation and understanding (Saner and Marchant, 2015).

A convergence of regulatory approaches for human genome editing would have some beneficial effects as described above, but countries have already adopted diverse laws relevant to human genome editing and a formal or complete harmonization does not seem feasible—and may

---

[78] An example of the challenges associated with negotiating and enforcing treaties was the unsuccessful attempt through the United Nations system in the early 2000s to create a binding international treaty banning human cloning (Cameron and Henderson, 2008).

not even be entirely desirable—at this time. Moreover, national responses to human genome editing reflect unique historical, cultural, economic, and social factors. Notwithstanding these important differences that prevent uniform international standards, there are important benefits for providing for robust communication and coordination between regulators in different countries, and potential opportunities for identifying common ground on specific substantive or technical aspects as well as opportunities to produce learning benefits (e.g., Zhai et al., 2016).

参 考 文 献

Abbott, K. W. A. 2014. International organisations and international regulatory co-operation: Exploring the links. In *International regulatory co-operation and international organisations: The Cases of the OECD and the IMO.* Paris: OECD Publishing. Pp. 17-44.

Abbott, K. W. A., D. J. Sylvester, and G. E. Marchant. 2010. Transnational regulation: Reality or romanticism? In *International handbook on regulating nanotechnologies,* edited by G. Hodge, D. Bowman, and A. Maynard. Cheltenham, UK: Edward Elgar Publishing. Pp. 525-544.

Araki, M. and T. Ishii. 2014. International regulatory landscape and integration of corrective genome editing into *in vitro* fertilization. *Reproductive Biology and Endocrinology* 12: 108. http: //www.rbej.com/content/12/1/108 (accessed January 25, 2017).

Breggin, L., R. Falkner, N. Jaspers, J. Pendergrass, and R. Porter. 2009. *Securing the promise of nanotechnologies: Towards transatlantic regulatory cooperation.* London, UK: Chatham House. https: //www.chathamhouse.org/sites/files/chathamhouse/public/Research/Energy,% 20Environment%20and%20Development/r0909_nanotechnologies.pdf (accessed November 7, 2016).

Cameron, N. M. de S., and A.V. Henderson. 2008. Brave new world at the General Assembly: The United Nations Declaration on Human Cloning. *Minnesota Journal of Law, Science and Technology* 9(1): 145-238.

Charo, R. A. 2016a. On the road (to a cure?): Stem-cell tourism and lessons for gene editing. *New England Journal of Medicine* 374(10): 901-903.

Charo, R. A. 2016b. The legal and regulatory context for human gene editing. *Issues in Science and Technology* 32(3). http: //issues.org/32-3/the-legal-and-regulatory-context-for-human-gene-editing (accessed November 7, 2016).

China Ministry of Health (People's Republic of China Ministry of Health). 2001. *Guidelines on human assisted reproductive technologies* [in Chinese]. http: //go.nature.com/1ztc8qb (accessed November 7, 2016).

China Ministry of Health (People's Republic of China Ministry of Health). 2003. *Guidelines on human embryonic stem cell research.* http: //www.cncbd.org.cn/News/Detail/3376 (accessed November 7, 2016).

Cichutek, K. 2008. Gene and cell therapy in Germany and the EU. *Journal fur Verbraucherschutz und Lebensmittelsicherheit* 3(Suppl. 1): 73-76.

Daley, G. Q., L. Ahrlund-Richter, J. M. Auerbach, N. Benvenisty, R. A. Charo, G. Chen, H. K. Deng, L. S. Goldstein, K. L. Hudson, I. Hyun, S. C. Junn, J. Love, E. H. Lee, A. McLaren, C. L. Mummery, N. Nakatsuji, C. Racowsky, H. Rooke, J. Rossant, H. R. Scholer, J. H. Solbakk, P. Taylor, A. O. Trounson, I. L. Weissman, I. Wilmut, J. Wu, and L. Zoloth. 2007. The ISSCR guidelines for human embryonic stem cell research. *Science* 315: 603-604.

EMA (European Medicines Agency). 2013. *Legal foundation.* http: //www.ema.europa.eu/ema/index.jsp?curl=pages/about_us/general/general_content_000127.j sp&mid=WC0b01ac0580029320 (accessed January 25, 2017).

Evans, B. J. 2015. *Panel: Governance at the institutional and national levels: National regulatory frameworks.* Presentation at International Summit on Gene Editing, Washington, DC, December 2.

Falkner, R. 2013. The crisis of environmental multilateralism: A liberal response. In *The green book: New Directions for liberals in government*, edited by D. Brack, P. Burall, N. Stockley, and M. Tuffrey. London, UK: Biteback Publishing. Pp. 347-358.

FDA (U.S. Food and Drug Administration). 2012. *Vaccine, blood, and biologics: SOPP 8101.1: Scheduling and conduct of regulatory review meetings with sponsors and applicants.* Rockville, MD: FDA. http: //www.fda.gov/BiologicsBloodVaccines/GuidanceComplianceRegulatoryInformation/Proced uresSOPPs/ucm079448.htm (accessed January 25, 2017).

Gallagher, J., S. Gorovitz, and R. J. Levine. 2000. *Biomedical research ethics: Updating international guidelines: A consultation.* Geneva, Switzerland: Council for International Organizations of Medical Sciences (CIOMS). http: //www.cioms.ch/index.

php/publications/available-publications/540/view_bl/61/bioethics-and-health-policy/3/biomedical-research-ethics-updating-international- guidelines-a-consultation?tab=getmybooksTab&is_show_data=1 (accessed November 7, 2016).

Ginn, S. L., I. E. Alexander, M. L. Edelstein, M. R. Abedi, and J. Wixon. 2013. Gene therapy clinical trials worldwide to 2012—An update. *Journal of Gene Medicine* 15(2): 65-77.

IOM (Institute of Medicine). 2014. *Oversight and review of clinical gene transfer protocols: Assessing the role of the Recombinant DNA Advisory Committee.* Washington, DC: The National Academies Press.

Ishii, T. 2015. Germline genome-editing research and its socioethical implications. *Trends in Molecular Medicine* 21(8): 473-481.

ISSCR (International Society for Stem Cell Research). 2016. *Guidelines for stem cell research and clinical translation.* http://www.isscr.org/docs/default-source/guidelines/isscr-guidelines-for-stem-cell-research-and-clinical-translation.pdf?sfvrsn=2 (accessed November 7, 2016).

Kong, W. M. 2004. The regulation of gene therapy research in competent adult patients, today and tomorrow: Implications of EU directive 2001/20/EC. *Medical Law Review* 12(2): 164-180.

Marchant, G. E., K. W. Abbott, D. J. Sylvester, and L. M. Gaudet. 2012. Transnational new governance and the international coordination of nanotechnology oversight. In *The nanotechnology challenge: Creating law and legal institutions for uncertain risks,* edited by D. A. Dana. Cambridge, UK: Cambridge University Press. Pp. 179-202.

Meridian Institute. 2004. *International dialogue on responsible research and development of nanotechnology.* Washington, DC: Meridian Institute. http://www.temas.ch/nano/nano_homepage.nsf/vwRes/SafetyAlexandria/$FILE/Final_Report_Re sponsible_Nanotech_RD_040812.pdf (accessed November 7, 2016).

OECD (Organisation for Economic Co-operation and Development). 2013. *International regulatory co-operation: Addressing global challenges.* Paris: OECD Publishing.

Pignatti, F. 2013. *Harmonizing across regions.* Presentation at Implementing a National Cancer Clinical Trials System for the 21st Century, Washington, DC, February 12.

Saner, M. A., and G. E. Marchant. 2015. Proactive international regulatory cooperation for governance of emerging technologies. *Jurimetrics* 55(2): 147-178.

Susskind, L. 2008. Strengthening the global environmental treaty system. *Issues in Science and Technology* 25(1): 60-68.

Vogel, D. 1998. Globalization of pharmaceutical regulation. *Governance* 11(1): 1-22.

Zhai, X., V. Ng, and R. Lie. 2016. No ethical divide between China and the West in human embryo research. *Developing World Bioethics* 16(2): 116-120.

附录 C 资料来源及方法
Appendix C Data Sources and Methods

美国国家科学院和医学院人类基因编辑委员会的任务是研究人类基因编辑技术（包括人类胚系编辑）的科学基础，考量其科技、医学和伦理问题，及其使用所带来的临床、伦理、法律和社会影响。该委员会还探寻了在人类基因编辑方面全球通用的基本原则和指导方针。为了全面应对此项任务，委员会查阅了各种来源的数据，包括现有文献、公开会议和电话会议、公开证言和意见，以及其他可用的公共资源。

委员会构成

美国国家科学院、工程院和医学院（美国国家科学院）任命了一个由 22 名专家组成的委员会来承担此项任务。该委员会由基础科学、临床研究与医学、法律法规、伦理宗教、患者权益、科学传播、公众参与和生物医药产业中各方面的专家组成。各委员会成员的履历信息详见附录 D。

会议及信息收集活动

委员会于 2015 年 12 月至 2017 年 1 月进行了研讨，并开始评估和收集与其任务相关的信息和资料，其中包括审查现有文献、邀请利益相关者在公开会议上分享观点，以及在线上和线下征求公众意见。

The National Academy of Sciences and the National Academy of Medicine Committee on Human Gene Editing: Scientific, Medical, and Ethical Considerations was tasked with studying the scientific underpinnings of human gene-editing technologies—including human germline editing—and the clinical, ethical, legal, and social implications of their use. The committee also explored fundamental, underlying principles that could be adapted by any nation considering the development of guidelines for human gene editing. To respond comprehensively to its charge, the committee examined data from a variety of sources, including a review of the literature, open-session meetings and conference calls, public testimony and input, and other publicly available resources.

COMMITTEE COMPOSITION

The National Academies of Sciences, Engineering, and Medicine (the National Academies) appointed a committee of 22 experts to undertake the statement of task. The committee was composed of members with expertise in basic science, clinical research and medicine, law and regulation, ethics and religion, patient advocacy, science communication, public engagement, and the biomedical industry. Appendix E provides the biographical information for each committee member.

MEETINGS AND INFORMATION-GATHERING ACTIVITIES

The committee deliberated from December 2015 to January 2017 to conduct its assessment, and gathered information and data relevant to its statement of task by conducting a review of available literature, inviting

文献评估

委员会使用了几种策略来识别任务相关的文献。首先，工作人员在包括 PubMed、Scopus、Web of Science、ProQuest Research Library、Medline、Embase 和 LexisNexis 在内的文献数据库中查阅同行审评期刊上的相关文章，例如，讨论与人类基因编辑相关的基础研究、临床应用、患者安全、科学标准、伦理、监督和社会问题的文章。同时，工作人员也阅读了近期的新闻和文学作品，以寻找与委员会任务相关的文章，并以此组成引文数据库。此外，委员会成员、演讲者、赞助者和其他对此感兴趣的组织也提交了一些相关主题的文章、报告和政策声明。现在，委员会的数据库已包含数百篇相关文章和报告，这个数字在研究过程中不断更新。

公开会议

委员会共举行了五次会议，其中四次涉及从利益相关者及公众中收集意见的调研环节。四次会议中的三次在美国华盛顿特区举行（2015年12月、2016年2月和2016年7月），剩下一次则由法国国家医学科学院在法国巴黎主办（2016年4月）。

委员会在2015年12月首次举行的会议是由美国国家科学院、美国国家医学院、中国科学院和英国皇家学会共同主办的为期3天的"人类基因编辑国际峰会：全球论坛"。虽然是一个由特设委员会筹划的峰会，它仍为研究委员会提供了一个重要的收集信息的机会。峰会召集了来自世界各地的专家，讨论与人类基因编辑研究相关的科学、伦理和监管问题[79]。

委员会在2016年2月举行的第二次会议中引入了潜在的利益相关方组织，如患者组织和开发基因编辑疗法的公司代表的观点。这次会议还特别展示了来自公众参与模型和联邦及机构的监督机关的专家的演讲。

stakeholders to share perspectives at public meetings, and soliciting public comments both online and in person.

Literature Review

Several strategies were used to identify literature relevant to the committee's charge. A search of bibliographic databases, including PubMed, Scopus, Web of Science, ProQuest Research Library, Medline, Embase, and LexisNexis, was conducted to obtain articles from peer-reviewed journals that discussed basic research, clinical applications, patient safety, scientific standards, ethics, oversight, and social issues associated with human gene editing. Staff reviewed recent news and literature to identify articles relevant to the committee's charge and created a database of references. In addition, committee members, speakers, sponsors, and other interested parties submitted articles, reports, and policy statements on these topics. The committee's database included several hundred relevant articles and reports, and was updated continuously throughout the study process.

Public Meetings

Four of the five meetings held over the course of the study included sessions in which committee members obtained input from a range of stakeholders and members of the public. Three meetings were held in Washington, DC (December 2015, February 2016, and July 2016), and one meeting was held in Paris, France, hosted by the French National Academy of Medicine (April 2016).

The committee's first meeting in December 2015 was held in association with the 3-day International Summit on Human Gene Editing: A Global Discussion, co-hosted by the U.S. National Academy of Sciences, U.S. National Academy of Medicine, Chinese Academy of Sciences, and The Royal Society of the United Kingdom. Although a separate ad hoc committee planned the summit, it provided a critical opportunity for the study committee to gather information. The summit convened experts from around the world to discuss scientific, ethical, and governance issues associated with human gene-editing research.[79]

The committee's second meeting in February 2016 included perspectives from potentially affected stakeholder communities such as patient groups and representatives from companies developing gene editing–based therapeutics. It also featured presentations from

---

[79] The list of speakers at the International Summit included Peter Braude, Annelien Bredenoord, Philip Campbell, Alta Charo, George Church, Ralph Cicerone, Chad Cowan, George Daley, Marcy Darnovsky, Victor Dzau, Fola Esan, Barbara Evans, William Foster, Bärbel Friedrich, Hille Haker, John Harris, John Holdren, Rudolf Jaenisch, Weizhi Ji, Pierre Jouannet, J. Keith Joung, Daniel Kevles, Jonathan Kimmelman, Eric Lander, Ephrat Levy-Lahad, Jinsong Li, Robin Lovell-Badge, Gary Marchant, Jennifer Merchant, Keymanthri Moodley, Indira Nath, Staffan Normark, Kyle Orwig, Pilar Ossorio, Duanquing Pei, Matthew Porteus, K. Vijay Raghavan, Klaus Rajewsky, Thomas Reiss, Janet Rossant, Ismail Serageldin, Bill Skarnes, John Skehel, Azim Surani, Sharon Terry, Adrian Thrasher, Fyodor Urnov, Marco Weinberg, Ernst-Ludwig Winnacker, Zhihong Xu, and Feng Zhang. Presentations and other materials from the Summit are available at http://nationalacademies.org/gene-editing/Gene-Edit-Summit/index.htm (accessed January 7, 2017).

在由法国国家医学科学院在法国巴黎举办主办的第三次会议（2016 年 4 月）中，发言者们探讨了基因编辑管理的基本原则，并提出了具有国际视角的宽容、中立、审慎的观点和防范性的管理方法。会议同时讨论了人类胚系基因编辑的潜在临床应用。这次会议是在欧洲医学科学院院联合会（FEAM / UK）召开的基因编辑研讨会之后举行的，这使得一部分委员会成员有机会参加之前 FEAM 组织的研讨会。这为进一步了解欧洲共同体基因编辑规范管理问题和战略提供了一个重要的机会。

最后，委员会在 2016 年 7 月举办的第四次会议上提出了关于人类基因编辑的几个社会问题的意见，其中包括美国历史上的种族、遗传学和道德观念与公共政策的交叉领域。在往届会议中的发言者名单详见下文。

公众意见

委员会的数据收集会议还为委员会提供了亲自参与并与各种利益相关者互动的机会。每次公开会议都包括一个公众评议期，在此期间委员会邀请任何有意者发表意见。委员会还致力于使其活动尽可能透明和便利，以适应有特殊需要的人或无法亲自到场的人。

委员会的研究网站保持定期更新，以展示委员会近期的活动和计划。研究推广还专设了一个用于接收评论和问题，以及来自社交媒体的反馈和标签的专用电子邮件地址。电子邮箱信息订阅服务可用于分享更多信息，并获得更多对委员会的反馈和意见。

在研究全程中提供的带有隐藏字幕的视频直播和在线公开留言工具的链接，使得无法亲自参加会议的人也能有机会参与提供反馈意见。所有在线评论和呈件均已归类为该项研究的公开文件。任何外部来源或通过在线公开留言工具向委员会提供的信息都可通过国家科学院公开档案室查询。

演讲者名单

以下为委员会在资料收集会议中所邀请的演讲者：

experts on models for public engagement as well as federal and institutional oversight bodies.

The study's third meeting in April 2016, hosted by the French National Academy of Medicine in Paris, France, explored the principles underlying governance of gene editing. Speakers during the meeting provided international perspectives spanning permissive, neutral, precautionary, and preventive governance approaches. Meeting discussions also addressed potential therapeutic clinical applications for human germline gene editing. This meeting was held the day following a gene-editing workshop convened by the Federation of European Academies of Medicine (FEAM/UK). A number of committee members were able to participate in the preceding FEAM workshop, which provided an important opportunity to learn further about gene-editing regulatory and governance concerns and strategies across the European community.

Finally, the study's fourth meeting in July 2016 provided input on several of the social issues associated with human gene editing, including race and genetics in U.S. history and the intersection of moral views and public policy. The list of speakers who provided input to the committee in these meeting sessions is below.

Public Comments

The committee's data-gathering meetings also provided opportunities for the committee to engage and interact with a variety of stakeholders. Each public meeting included a public comment period, in which the committee invited input from any interested party. The committee also worked to make its activities as transparent and accessible as possible and to accommodate those with special needs or those who may not have been able to attend in person.

The study website was updated regularly to reflect the recent and planned activities of the committee. Study outreach also included a study-specific email address for comments and questions and social media feeds and tags. A subscription to regular email updates was available to share further information and solicit additional comments and input to the committee.

Live video streams with closed captioning and links to an online public comment tool were provided throughout the course of the study to allow the opportunity for input from those unable to attend meetings in person. All online comments and submissions were catalogued in the study's public access file. Any information provided to the committee from outside sources or through the online comment tool is available by request through the National Academies' Public Access Records Office.

Speakers

The following individuals were invited speakers at data-gathering sessions of the committee:

Roberto Andorno
苏黎世大学
University of Zurich

Mónica López Barahona
西班牙生物卫生研究中心
Centro de Estudios Biosanitarios

Pierre Bégué
法国国家医院
Académie Nationale de Médecine, France

Nick Bostrom
牛津大学
University of Oxford

Abby Bronson
肌肉萎缩症母体计划
Parent Project Muscular Dystrophy

Dominique Brossard
威斯康星大学麦迪逊分校
University of Wisconsin–Madison

Jacqueline Chin
新加坡国立大学
National University of Singapore

Hans Clevers
豪博瑞特研究所
Hubrecht Institute

Ronald Cole-Turner
匹兹堡神学院
Pittsburgh Theological Seminary

Francis Collins
美国国立卫生研究院
National Institutes of Health

George William Foster
美国国会议员，IL-11
Congressman, IL-11

Søren Holm
曼彻斯特大学
University of Manchester

Rahman Jamal
马来西亚国立大学
National University of Malaysia

Bartha Knoppers
麦吉尔大学
McGill University

Fredrik Lanner
卡罗林斯卡学院
Karolinska Institute

James Lawford-Davies
英国 Hempsons 律师事务所
Hempsons Law Firm, United Kingdom

John Leonard
英特利亚治疗
Intellia Therapeutics

Bruce Lewenstein
康奈尔大学
Cornell University

Andrew May
Caribou Biosciences 公司
Vic Myer Editas

Editas Medicine
爱迪塔斯医药公司

Alondra Nelson
哥伦比亚大学
Columbia University

Erik Parens
黑斯廷斯中心
Hastings Center

Guido Pennings
比利时根特大学
Ghent University, Belgium

Pearl O'Rourke
Partners HealthCare 公司

Jackie Leach Scully
纽卡斯尔大学
Newcastle University

Oliver Semler
科隆大学
University of Cologne

Trevor Thompson
田纳西州镰状细胞基金会
Sickle Cell Foundation of Tennessee

Anna Veiga
西班牙巴塞罗那再生医学中心
Center of Regenerative Medicine, Barcelona, Spain

Thomas Voit
伦敦大学学院
University College London

Elizabeth Vroom
肌肉萎缩症联合母体计划
United Parent Projects Muscular Dystrophy

Keith Wailoo
普林斯顿大学
Princeton University

Michael Werner
再生医学联盟
Alliance for Regenerative Medicine

Nancy Wexler
遗传性疾病基金会
Hereditary Disease Foundation

Bethan Wolfenden
Bento Bioworks 公司

Carrie Wolinetz
美国国立卫生研究院
National Institutes of Healt

Philip Yeske
联合线粒体疾病基金会
United Mitochondrial Dis

翟晓梅
中国北京协和医学院
Peking Union Medical C

附录 D 组委会成员简介
Appendix D　Committee Member Biographies

R. Alta Charo，法律博士（联合主席） 美国医学院院士，威斯康星大学麦迪逊分校法律与生物伦理学教授，法律特聘教授。她在麦迪逊分校的法学院和医学院教授公共卫生法、生物技术法规和生物伦理学。她毕业于哈佛大学生物学专业并获得学士学位，后在哥伦比亚大学获得法学博士学位。Charo 教授曾是奥巴马总统过渡小组成员，专注于科技政策方面。2009~2011 年，她担任了美国食品药品监督管理局专员办公室的新兴技术问题高级政策顾问。Charo 教授的其他政府任职包括国会技术评估办公室（the Congressional Office of Technology Assessment）成员、美国国际发展署（the U.S. Agency for International Development）成员、国立卫生研究院人类胚胎研究小组成员，以及克林顿总统时期美国国家伦理顾问委员会委员。在美国国家科学院、工程院和医学院（the National Academies of Sciences, Engineering, and Medicine），她曾与 Richard Hynes 联合主持胚胎干细胞研究指导委员会。她也曾是生命科学委员会、人口健康与公共卫生实践委员会、科技与法律委员会、健康科学委员会等组织的委员。

Richard O. Hynes，博士（联合主席） 麻省理工学院（MIT）癌症研究教授，霍华德休斯医学研究所（Howard Hughes Medical Institute）研究员。他本科毕业于英国剑桥大学生物化学专业，在 1971 年获得 MIT 生物学博士学位。在伦敦帝国癌症研究基金会（the Imperial Cancer Research Fund Imperial

R. Alta Charo, J.D. (*Co-Chair*), is a member of the National Academy of Medicine and is the Warren P. Knowles Professor of Law and Bioethics, and the Sheldon B. Lubar Distinguished Research Chair in Law at the University of Wisconsin-Madison, where she is on the faculties of the law and medical schools and teaches public health law, biotechnology regulation, and bioethics. She received her B.A. in biology from Harvard and J.D. from Columbia. Professor Charo was a member of President Obama's transition team, focusing on science policy, and from 2009 to 2011 was a senior policy advisor on emerging technology issues in the Office of the Commissioner at the U.S. Food and Drug Administration. Her other federal service includes the congressional Office of Technology Assessment, the U.S. Agency for International Development, the National Institutes of Health Human Embryo Research Panel and President Clinton's National Bioethics Advisory Commission. At the National Academies of Sciences, Engineering, and Medicine she co-chaired (with Richard Hynes) the Committee on Guidelines for Embryonic Stem Cell Research and has been a member of the Board on Life Sciences; Board on Population Health and Public Health Practice; Committee on Science, Technology and Law; and Board on Health Sciences Policy.

Richard O. Hynes, Ph.D. (*Co-Chair*), is Daniel K. Ludwig Professor for Cancer Research at the Massachusetts Institute of Technology (MIT) and an Investigator of the Howard Hughes Medical Institute. He received his B.A. in biochemistry from the University of Cambridge, United Kingdom, and his Ph.D. in biology from MIT in 1971. After postdoctoral work at the Imperial Cancer Research Fund in London, where he initiated his

Cancer Research Fund）获得了细胞黏着方面的博士后之后，Hynes 博士回到了 MIT 任职。Hynes 博士是伦敦皇家学会（the Royal Society of London）、美国科学促进会（the American Academy for the Advancement of Science）会士，也是美国科学院、美国医学院、美国艺术与科学院（the American Academy of Arts and Sciences）院士。他曾获盖尔德纳基金会（the Gairdner Foundation）国际医学成就奖及 Pasarow 心血管研究奖。他曾是 MIT 癌症研究中心的副主任，后晋升主任，在任十年。从 2007 年到 2016 年，他担任英国惠康基金会（Wellcome Trust）总裁。他在美国科学院和美国医学院曾与 Jonathan Moreno 和 Alta Charo 共同主持人类胚胎干细胞研究指导委员会。

David W. Beier，法律博士 Bay City Capital 公司的董事总经理，自 2013 年以来一直在该公司工作。他是医疗保健政策、定价、知识产权、政府事务、法规事务、卫生保健经济学和产品商业化等方面世界公认的领军人物。他曾任职于世界最大的两个生物技术公司——安进公司（Amgen）和基因技术公司（Genentech）的高层管理团队二十年。期间他为生物技术企业的战略、对潜在收购者的需求、全球医疗保健行业等贡献良多。克林顿执政期间，Beier 先生曾在白宫担任副总统戈尔的首席国内政策顾问。他曾被克林顿总统任命为贸易政策和谈判顾问委员会委员，并加入美国医学院未来医疗卫生与公共事业专家组（the Institute of Medicine panel on the Future of Health and Human Services），并作为总统科技顾问委员会的顾问。Beier 先生也曾是国际律师事务所 Hogan and Hartson 的合伙人，曾任美国众议院司法委员会顾问。他曾在国会和联邦贸易委员会法庭作证，撰写了大量法学综述和技术性法律论文，定期被邀作为公共卫生专栏作者，撰写了关于知识产权、贸易、隐私和司法问题等的著作。他目前被加利福尼亚州州长杰里·布朗任命到加州政府组织和经济委员会，同时也是全球企业中心（Center for Global Enterprise）的研究员，并在加州大学伯克利分校哈斯商学院任职客座讲师。Beier 先生从联合大学奥尔巴尼法学院（Albany Law School at Union University）获得法学博士学位，在科尔盖特大学（Colgate University）获得本科学位。他拥有在纽约和哥伦比亚特区的律师执照

work on cell adhesion, he returned to MIT as a faculty member. Dr. Hynes is a fellow of The Royal Society of London, the American Academy of Arts and Sciences, and the American Association for the Advancement of Science, and is a member of the National Academy of Sciences and the National Academy of Medicine. He has received the Gairdner Foundation International Award for achievement in medical science and the Pasarow Award for Cardiovascular Research. He was previously associate head and then head of the Biology Department and served for 10 years as director of the MIT Center for Cancer Research. He was a Governor of the Wellcome Trust U.K. from 2007-2016. At the National Academies he has previously co-chaired (with Jonathan Moreno and Alta Charo) committees on guidelines for human embryonic stem cell research.

David W. Beier, J.D., is a managing director of Bay City Capital and has been with the firm since 2013. He is a globally recognized leader in health care policy, pricing, intellectual property, government affairs, regulatory affairs, health care economics, and product commercialization. In addition, having spent two decades as part of the senior management teams for Amgen and Genentech, the two largest biotechnology companies in the world, he contributes invaluable perspective regarding strategy for entrepreneurial biotechs, the needs of potential acquirers, and the global health care industry in general. Mr. Beier served in the White House as the Chief Domestic Policy Advisor to Vice President Al Gore during the Clinton Administration. He has served as an appointee of President Clinton on his Advisory Committee for Trade Policy and Negotiations, on the Institute of Medicine panel on the Future of Health and Human Services, and as an advisor to the President's Council of Advisors on Science and Technology. Mr. Beier was also formerly a partner in the international law firm Hogan and Hartson and was formerly Counsel to the U.S. House of Representatives Committee on the Judiciary. He has testified before Congress and the Federal Trade Commission, has written numerous law review articles and technical legal works, is regularly invited to author expert op-eds on health care, and has contributed to books on topics ranging from intellectual property, trade, privacy and justice issues. He currently serves as an appointee of Governor Brown on the California State Government Organization and the Economy Commission, as a fellow of the Center for Global Enterprise, and teaches as an adjunct lecturer at the Haas School of Business at the University of California, Berkeley. Mr. Beier received his J.D. from Albany Law School at Union University and his undergraduate degree at Colgate University. He is admitted to practice law in New York and the District of Columbia.

Juan Carlos Izpisua Belmonte, Ph.D., is a professor

许可。

Juan Carlos Izpisua Belmonte，博士 从 1993 年开始在加利福尼亚州拉荷亚的索尔克生物研究学院（the Salk Institute for Biological Studies in La Jolla）的基因表达实验室任职教授。2005 年至 2013 年，他担任了巴塞罗那再生医学中心的主任。Izpisua Belmonte 博士毕业于西班牙巴伦西亚大学，获得药学与科学学士学位及药理学硕士学位。1987 年，他获得意大利博洛尼亚大学和西班牙巴伦西亚大学的博士学位。他在德国海德堡的欧洲分子生物学实验所（EMBL）和美国加州大学洛杉矶分校完成了博士后工作。Izpisua Belmonte 博士的研究方向为干细胞生物学、器官发生、再生和衰老，在国际同行评审的期刊上发表了 400 多篇相关文章和专著章节。他的终级研究目标是促进细胞和基因疗法研究，以及开发用于治疗人类疾病的新分子。

Ellen Wright Clayton，医学博士，法律博士 在法学和遗传学领域是一位国际知名的领军人物，现在美国范德堡大学的法学院和医学院任职，在此她创立并领导了生物医学伦理和社会研究中心。她出版了两本书，并且在医学期刊、跨学科期刊和法学期刊上发表了关于法律、医学和公共卫生交叉学科的 150 多篇学术论文和专著章节。此外，她还与范德堡大学，以及全美及世界各地的许多机构的学生与研究者合作开展跨学科研究项目，并帮助了许多国家和国际组织的政策声明的制定。她目前在主持国际公共人口基因组学计划（the International Public Population Program in Genomics）的儿科平台。她是政策辩论的积极参与者，已为国立卫生研究院，以及其他美国联邦和国际机构就儿童健康与涉及人类研究的伦理行为等一系列议题提供咨询意见。Clayton 教授曾为美国医学院的多个项目工作，她是美国国家顾问理事会执行委员，美国国家科学院、工程院和医学院（the National Academies of Sciences, Engineering, and Medicine）的人口健康和公共健康实践委员会主席，也是肌痛性脑脊髓炎/慢性疲劳综合征释义委员会的主席。她是科学院联盟报告审查委员会的成员，以及美国科学促进学会（the American Academy for the Advancement of Science）的当选会士。

in the Gene Expression Laboratories at the Salk Institute for Biological Studies in La Jolla, California since 1993. From 2005 to 2013 he was also director of the Center for Regenerative Medicine in Barcelona. Dr. Izpisua Belmonte graduated from the University of Valencia, Spain with a Bachelor's degree in Pharmacy and Science and a Master's degree in Pharmacology and received his Ph.D. from the University of Bologna, Italy and the University of Valencia, Spain, in 1987. He completed his postdoctoral work at the EMBL in Heidelberg, Germany and UCLA, Los Angeles, USA. Dr. Izpisua Belmonte's research interests are focused on the understanding of stem cell biology, organogenesis, regeneration and ageing. He has published more than 400 articles in internationally peer-reviewed journals and book chapters in these areas. The ultimate goal of his research is to translate it towards the development of cell and gene based therapies as well as new molecules for the treatment of human disease. Dr. Izpisua Belmonte graduated from the University of Valencia, Spain with a bachelor's degree in pharmacy and science and a master's degree in pharmacology, and he received his Ph.D. from the University of Bologna, Italy, and the University of Valencia, Spain, in 1987. He completed his postdoctoral work at the European Molecular Biology Lab in Heidelberg, Germany, and at the University of California, Los Angeles.

Ellen Wright Clayton, M.D., J.D., is an internationally respected leader in the field of law and genetics who holds appointments in both the law and medical schools at Vanderbilt, where she also co-founded and directed the Center for Biomedical Ethics and Society. She has published two books and more than 150 scholarly articles and chapters in medical journals, interdisciplinary journals, and law journals on the intersection of law, medicine, and public health. In addition, she has collaborated with faculty and students throughout Vanderbilt and in many institutions around the country and the world on interdisciplinary research projects, and helped to develop policy statements for numerous national and international organizations. She currently chairs the Paediatric Platform of the international Public Population Program in Genomics. An active participant in policy debates, she has advised the National Institutes of Health as well as other federal and international bodies on an array of topics ranging from children's health to the ethical conduct of research involving human subjects. Professor Clayton has worked on a number of projects for the National Academy of Medicine, of which she is a member of the Executive Committee of its National Advisory Council, chair of the National Academies of Sciences, Engineering, and Medicine's Board on Population Health and Public Health Practice, and was chair of its committee to define myalgic encephalomyelitis/chronic fatigue syndrome. She is also a member of the National Academies' Report Review Committee. She is an elected fellow of the American

Barry S. Coller，医学博士 主任医师，医学事务部（Medical Affairs）副总裁，洛克菲勒大学医学教授，血液和血管生物学实验室负责人。他在血小板生理学、血管生物学和镰状细胞病的黏附现象方面的研究水平世界领先。他发明了抑制血小板聚集和预防镰状红细胞黏附到血管壁的单克隆抗体，以及世界最早的血小板单克隆抗体之一，使其发展成用于预防冠状动脉血管成形术和支架置入后血栓症的主要作用药物。他还验证了导致人类出血性疾病的基因突变。Coller 博士于 2005 年获得 Pasarow 奖，于 2001 年获得 Warren Alpert 基金奖，并于 1998 年获得美国心脏协会的国家研究成就奖。他是美国医学院、美国科学院、美国艺术与科学院（American Academy of Arts and Sciences）院士。Coller 博士还是美国血液学学会的前主席，并且是临床和转化科学学会的创办主席。他目前在美国国家科学院、工程院和医学院（the National Academies of Sciences，Engineering，and Medicine）的健康科学政策委员会任职。Coller 博士于 1966 年在哥伦比亚大学本科毕业，并于 1970 年在纽约大学医学院获得博士学位。他在纽约市 Bellevue 医院完成了内科住院医生实习，并在国立卫生研究院接受了血液学和临床病理学的高级培训。他于 1976 年至 1993 年在石溪大学（Stony Brook University）任职，1993 年至 2001 年在西奈山医学院担任医学教授和医系主任。Coller 博士于 2001 年来到洛克菲勒大学，目前担任该大学临床和转化科学奖资助的研究员及临床和转化科学中心主任。

John H. Evans，博士 加州大学圣地亚哥分校社会学教授。Evans 博士从马卡莱斯特学院（Macalester College）取得学士学位，在普林斯顿大学取得博士学位。他曾是耶鲁大学博士后研究员、新泽西州普林斯顿高等研究院（the Institute for Advanced Study in Princeton）的访问学者，并参与了爱丁堡大学和明斯特大学（University of Muenster）的访问学者项目。他的研究领域涉及宗教、文化、政治和科学，曾出版一本关于 20 世纪下半叶人类遗传工程伦理辩论的著作。另一著作研究了美国宗教人士对生殖遗传技术的观点。最近出版的一本书探讨了如何将社会舆论纳入关于遗传修饰等问题的公共生物伦理辩论中。准备出版的一本书剖析美国人认为的人的本质是什么，以及我们应该如何对待

Academy for the Advancement of Science.

Barry S. Coller, M.D., is the Physician in Chief, Vice President for Medical Affairs, David Rockefeller Professor of Medicine, and Head, Allen and Frances Adler Laboratory of Blood and Vascular Biology at Rockefeller University. He is a leader in investigating platelet physiology, vascular biology, and adhesion phenomena in sickle cell disease. He produced monoclonal antibodies that inhibit platelet aggregation and adhesion of sickle red blood cells to the blood vessel walls. He produced one of the earliest monoclonal antibodies to platelets and played a leading role in its development into a drug used to prevent thrombosis after coronary artery angioplasty and stent placement in humans. He has also identified mutations in genes that cause human bleeding disorders. Dr. Coller received the Pasarow Award in 2005, the Warren Alpert Foundation Award in 2001, and a National Research Achievement Award from the American Heart Association in 1998. He is a member of the National Academy of Medicine, the National Academy of Sciences, and the American Academy of Arts and Sciences. Dr. Coller is a past president of the American Society of Hematology and was founding president of the Society for Clinical and Translational Science. He currently serves on the National Academies of Sciences, Engineering, and Medicine's Board on Health Science Policy. Dr. Coller received his B.A. from Columbia University in 1966 and his M.D. from the New York University School of Medicine in 1970. He completed his residency in internal medicine at Bellevue Hospital in New York City and received advanced training in hematology and clinical pathology at the National Institutes of Health. He was at Stony Brook University from 1976 to 1993, and from 1993 to 2001 he served as a professor of medicine and chairman of the department of medicine at Mount Sinai School of Medicine. Dr. Coller came to Rockefeller in 2001 and currently serves as principal investigator of the university's Clinical and Translational Science Award and director of The Rockefeller University Center for Clinical and Translational Science.

John H. Evans, Ph.D., is a Professor of Sociology at the University of California, San Diego. Dr. Evans earned his B.A. from Macalester College and his Ph.D. from Princeton. He has been a postdoctoral fellow at Yale, a visiting member of the Institute for Advanced Study in Princeton, New Jersey, and has held visiting professorial fellowships at the University of Edinburgh and the University of Muenster. His research concerns religion, culture, politics and science. He has published a book about the ethical debates about human genetic engineering in the second half of the 20th century. Another book examines what religious people in the United States think about reproductive genetic technologies. A recent book concerns how societal views can be included in public bioethical debates about issues like genetic modification. A forthcoming book examines

彼此。他正在写一本关于美国公民宗教和科学之间关系的书。除了这些著作，埃文斯博士还写了40多篇关于宗教、文化、政治和科学的论文及专著章节。

Rudolf Jaenisch，医学博士 麻省理工学院怀特海生物医学研究所（Whitehead Institute for Biomedical Research）的创始人之一，生物学教授。Jaenisch博士研究基因表达的表观遗传调控，其目标是将分化的细胞有效地转变为另一种类型。在哺乳动物胚胎干细胞分化，以及将成体细胞重编程为胚胎干细胞样［称为诱导性多能干（iPS）细胞］等方面取得重大突破。Jaenisch博士推广了iPS细胞方法学，并已证明iPS细胞在镰状细胞性贫血和帕金森病模型中的治疗潜力。鉴于他的成果，Jaenisch博士荣获了第一届Peter Gruber基因遗传学奖、Brupracher基金会癌症奖、美国国家科学院学报（PNAS）的Cozzarelli奖、Robert Koch科学成就卓越奖、Meira and Shaul G. Massry奖、Ernst Schering奖、Vilcek奖、Wolf医学奖等，并获得美国国家科学奖章。Jaenisch博士是美国科学院、美国医学院院士，同时也是美国艺术与科学院（the American Academy of Arts and Sciences）的研究学者。

Jeffrey Kahn，博士，公共卫生硕士 约翰·霍普金斯大学伯曼生物伦理学院（Berman Institute of Bioethics）院长。他还是生物伦理学和公共政策、彭博公共卫生学院（Bloomberg School of Public Health）卫生政策与管理系教授。他的研究兴趣包括科研伦理、伦理与公共卫生、伦理与新兴生物医学技术等。他的言论遍及美国内外，至今出版了4本著作，并在生物伦理学和医学文献上发表了超过125篇论文。他是美国医学院的当选院士、黑斯廷斯中心（Hastings Center）的研究学者，并曾担任国立卫生研究院、疾病控制和预防中心、美国医学院的委员会或专家小组的主席或委员。他目前是美国国家科学院、工程院和医学院（the National Academies of Sciences, Engineering, and Medicine）健康科学政策委员会主席。他的教育背景包括加利福尼亚大学洛杉矶分校学士学位、乔治敦大学（Georgetown University）博士学位、约翰霍普金斯大学彭博公共卫生学院的公共卫生硕士学位。

what Americans think a human is, and how that relates to how we should treat each other. He is writing a book about what the relationship is between religion and science for American citizens. In addition to these books, Dr. Evans has written more than 40 articles and book chapters on topics in religion, culture, politics and science.

Rudolf Jaenisch, M.D., is a Founding Member of Whitehead Institute for Biomedical Research and Professor of Biology at the Massachusetts Institute of Technology. Dr. Jaenisch studies the epigenetic regulation of gene expression with the goal of efficiently changing one differentiated cell type into another. This has led to groundbreaking work with mammalian embryonic stem cells and adult cells that have been reprogrammed to an embryonic stem cell like-state, called induced pluripotent stem (iPS) cells. Dr. Jaenisch continues to push iPS cell methodology forward and has demonstrated iPS cells' therapeutic potential in models of Sickle Cell anemia and Parkinson's disease. For his work, Dr. Jaenisch has been honored with the first Peter Gruber Foundation Award in Genetics, the Brupracher Foundation Cancer Award, Cozzarelli Prize from the Proceeding of the National Academy of Sciences (PNAS), Robert Koch Prize for Excellence in Scientific Achievement, Meira and Shaul G. Massry Prize, Ernst Schering Prize, Vilcek Prize, the Wolf Prize in Medicine and is a recipient of the United States National Medal of Science. Dr. Jaenisch is a Member of the National Academy of Sciences, a Member of the Institute of Medicine, and a Fellow of American Academy of Arts and Sciences.

Jeffrey Kahn, Ph.D., M.P.H., is the Andreas C. Dracopoulos Director of the Johns Hopkins Berman Institute of Bioethics. He is also Robert Henry Levi and Ryda Hecht Levi Professor of Bioethics and Public Policy, and Professor in the Department of Health Policy and Management in the Johns Hopkins University Bloomberg School of Public Health. His research interests include the ethics of research, ethics and public health, and ethics and emerging biomedical technologies. He speaks widely both in the United States and abroad, and has published four books and more than 125 articles in the bioethics and medical literature. He is an elected member of the National Academy of Medicine and a Fellow of the Hastings Center, and has chaired or served on committees and panels for the National Institutes of Health, the Centers for Disease Control and Prevention, and the Institute of Medicine/National Academy of Medicine. He is currently chair of the National Academies of Sciences, Engineering, and Medicine's Board on Health Sciences Policy. His education includes a B.A. from the University of California, Los Angeles, a Ph.D. from Georgetown University, and an M.P.H. from the Johns Hopkins Bloomberg School of Public Health.

Ephrat Levy-Lahad, M.D., is a professor of internal

Ephrat Levy-Lahad,医学博士 耶路撒冷希伯来大学内科医学和医学遗传学教授,耶路撒冷 Shaare Zedek 医学中心医学遗传学研究所所长。她在希伯来大学-哈达萨医学院(Hadassah Medical School)获得医科学位,在以色列取得内科医师执照,并且是临床遗传学和临床分子遗传学(以色列和美国)执业医师。Levy-Lahad 博士的临床实验室涉及癌症遗传学诊断和植入前诊断服务,专攻乳腺癌遗传学,特别是 BRCA1 和 BRCA2 基因,以及影响与这些突变相关风险的遗传和环境因素。她还研究了遗传检测在人群筛选和大规模预防工作中的应用。她的研究的另一个方向是探明罕见疾病的遗传基础,包括发现神经病学表型和卵巢发育缺陷的相关基因。Levy-Lahad 博士积极参与遗传学相关的生物伦理学研究,目前是以色列国家生物伦理委员会的联合主席。她是以色列国家妇科、围产期医学、遗传学学会成员,也是数字健康创新学会(National Council for Digital Health Innovation)成员。在国际上,她是联合国教科文组织国际生物伦理委员会委员(2006—2009)和国际干细胞研究学会干细胞临床转化工作组成员。

Robin Lovell-Badge,博士 弗朗西斯·克里克研究所(the Francis Crick Institute)的高级团队领导。Lovell-Badge 博士对干细胞生物学有着长期的兴趣,尤其是胚胎发育背景下基因的作用及细胞命运决定。他当前的研究工作主要包括性别决定、神经系统和垂体的发育,以及早期胚胎中的干细胞生物学。他非常积极参与公共舆论和政策工作,特别是关于干细胞、遗传学、人类胚胎、动物研究,以及科学的监管传播方式等方面。他是人类受孕及胚胎学管理局(HFEA)的科学和临床进展顾问委员会的增选委员,并且是探讨预防线粒体疾病的理论和安全方式专家组成员。同时他是英国医学科学院"种间人类胚胎"委员会、"含人类物质的动物"委员会和"增强人类与未来展望"("Human Enhancement and the future of work")联合学术委员会委员。他也是 Hinxton Group 指导委员会、英国皇家学会公众参与委员会,以及英国科学传媒中心顾问委员会委员。他于 1993 年当选为欧洲分子生物学组织(EMBO)成员,1999 年成为美国医学院院士,并于 2001 年成为英国皇家学会院士。他于 1995 年获得 Louis Jeantet 医学奖,1996 年获得了美国艺术与

medicine and medical genetics at the Hebrew University of Jerusalem and director of the Medical Genetics Institute at Shaare Zedek Medical Center in Jerusalem, Israel. She received her medical degree from the Hebrew University-Hadassah Medical School in Jerusalem, and is board-certified in Internal Medicine (Israel) and in Clinical Genetics and Clinical Molecular Genetics (Israel and the United States). Dr. Levy-Lahad's clinical laboratory includes cancer genetics diagnostics and a large preimplantation diagnosis service. Her research laboratory focuses on the genetics of breast cancer, in particular the BRCA1 and BRCA2 genes, and on genetic and environmental factors that affect the risk associated with these mutations. She also studies the application of genetic testing to population screening and large-scale prevention efforts. Another focus of her research is elucidating the genetic basis of rare diseases, including discoveries of novel genes for neurological phenotypes and for defects in ovarian development. Dr. Levy-Lahad is highly involved in bioethical aspects of genetic research, and is currently co-chair of the Israel National Bioethics Council. She is a member of Israel's National Council for Gynecology, Perinatal Medicine, and Genetics and the National Council for Digital Health Innovation. Internationally, she was a member of the United Nations Educational, Scientific and Cultural Organization's International Bioethics Committee (2006-2009) and the International Society for Stem Cell Research's Task Force on the Clinical Translation of Stem Cells.

Robin Lovell-Badge, Ph.D., is a Senior Group Leader at The Francis Crick Institute. Dr. Lovell-Badge has had long-standing interests in the biology of stem cells, in how genes work in the context of embryo development, and how decisions of cell fate are made. Major themes of his current work include sex determination, development of the nervous system and pituitary, and the biology of stem cells within the early embryo. He is also very active in both public engagement and policy work, notably around stem cells, genetics, human embryo and animal research, and in ways science is regulated and disseminated. He is a co-opted member of the HFEA's Scientific and Clinical Advances Advisory Committee and a member of their panel looking at the science and safety of ways to avoid mitochondrial diseases. He was a member of the UK Academy of Medical Science's committees on "Interspecies human embryos", "Animals Containing Human Material" and a Joint Academies committee on "Human Enhancement and the future of work". He is also a member of the steering committee of the Hinxton Group, of the Royal Society's Public Engagement Committee and of the UK Science Media Centre's Advisory Board. He was elected a member of EMBO in 1993, a fellow of the Academy of Medical Sciences in 1999, and a fellow of The Royal Society in 2001. He has received the Louis Jeantet Prize for Medicine in 1995, the Amory Prize, awarded by the American Academy of Arts and Sciences, in 1996, the

科学院（the American Academy of Arts and Sciences）Amory 奖，2008 年获 Feldberg 基金奖，2010 年获英国发育生物学学会的 Waddington 奖章。他是香港大学（2009—2015）的特聘客座教授及动物技术研究所所长。Lovell-Badge 博士 1975 年在伦敦大学学院获得动物学学士学位，1978 年在伦敦大学学院取得胚胎学博士学位。

Gary Marchant，法学博士，博士 美国亚利桑那州立大学（ASU）Sandra Day O'Connor 法学院新兴技术、法律与伦理学教授。他也是 ASU 生命科学教授，以及 ASU 法律、科学和技术研究中心的执行主任。Marchant 教授拥有英属哥伦比亚大学的遗传学博士学位、肯尼迪政治学院的公共政策硕士学位及哈佛大学的法学学位。在 1999 年加入 ASU 之前，他是华盛顿特区律师事务所的合伙人，执业领域是环境和行政法律。Marchant 教授在环境法、风险评估与风险管理、遗传学与法律、生物技术法规、食品和药物法规、纳米技术法律问题，以及法律与科技等领域从事教职与研究。

Jennifer Merchant，博士 巴黎第二大学（Pan-théon-Assas）英美法律政治研究院（Anglo-American legal and political institution）的教授。她是比较性公共政策中的生物伦理问题的领军研究学者，专攻北美和欧洲的比较政策、政治，以及涉及人类生殖系统的医疗技术监管问题。她还是法国胚胎研究和辅助生殖技术方面的法律和政策专家。她的学术兴趣包括比较公共政策、生殖、生物伦理、公民社会、科学和政府等。Merchant 博士是法国国立大学研究院（French National University Institute）、行政和政治科学研究中心（the Centre for the Study and Research of Administrative and Political Sciences）、法国国家健康与医学研究院（the French National Institute of Health and Medical Research，INSERM）伦理委员会和国际女权视角生物伦理协会（the International Network on Feminist Approaches to Bioethics，FAB）成员、FAB 法国国家代表。她自 2001 年起担任期刊 *Revue Tocqueville/Tocqueville Review* 的主编，并自 2005 年起成为全球伦理观察站和联合国教科文组织的成员。

Luigi Naldini，医学博士，博士 意大利米兰 San Raffaele 大学医学院细胞与组织生物学及基因 /

Feldberg Foundation Prize in 2008, and the Waddington Medal of the British Society for Developmental Biology in 2010. He is also a distinguished visiting professor at the University of Hong Kong (2009-2015) and the president of the Institute of Animal Technologists. Dr. Lovell-Badge obtained his B B.Sc. in zoology at the University College London in 1975 and his Ph.D. in embryology at the University College London in 1978.

Gary Marchant, J.D., Ph.D., is Regents' Professor and Lincoln Professor of Emerging Technologies, Law and Ethics at the Sandra Day O'Connor College of Law at Arizona State University (ASU). He is also a professor of life sciences at ASU and executive director of the ASU Center for the Study of Law, Science and Technology. Professor Marchant has a Ph.D. in genetics from the University of British Columbia, a Masters of Public Policy degree from the Kennedy School of Government, and a law degree from Harvard. Prior to joining the ASU faculty in 1999, he was a partner in a Washington, DC, law firm where his practice focused on environmental and administrative law. Professor Marchant teaches and researches in the subject areas of environmental law, risk assessment and risk management, genetics and the law, biotechnology law, food and drug law, legal aspects of nanotechnology, and law, science, and technology.

Jennifer Merchant, Ph.D., is a professor of Anglo-American legal and political institutions at the Université de Paris II (Panthéon-Assas). She is a leading researcher in bioethical issues of comparative public policy with expertise in comparative North American and European policy, politics, and regulation of medical technologies involving human reproduction. She is also an expert in French law and politics on embryo research and assisted reproductive technology. Her academic interests include comparative public policy, reproduction, bioethics, civil society, science, and government. Dr. Merchant is a member of the French National University Institute, the Centre for the Study and Research of Administrative and Political Sciences, the Ethics Committee of the French National Institute of Health and Medical Research (INSERM), and the International Network on Feminist Approaches to Bioethics (FAB) Association as well as FAB Country Representative for France. She has been the co-editor-in-chief of the *Revue Tocqueville/Tocqueville Review* since 2001 and a member of the Global Ethics Observatory and the United Nations Educational, Scientific and Cultural Organization since 2005.

Luigi Naldini, M.D., Ph.D., is professor of cell and tissue biology and of gene and cell therapy at the San Raffaele University School of Medicine and scientific director of the San Raffaele Telethon Institute for Gene Therapy, Milan, Italy. Dr. Naldini has pioneered the development and

细胞疗法教授，San Raffaele Telethon 基因治疗研究所的研发总监。Naldini 博士率先开发和应用基因转移慢病毒载体，已成为生物医学研究中最广泛使用的工具之一。慢病毒载体技术最近进入了临床测试，有望开启几种绝症的治愈契机。他一直在研究新策略克服困难，使基因转移更加安全高效，并可转化为遗传疾病和癌症的新型治疗方案，试图探明造血干细胞功能、免疫耐受诱导和肿瘤血管生成等机制。他近期的工作促进了工程核酸酶在细胞和基因疗法中的靶向基因编辑的使用。Naldini 博士是欧洲分子生物学组织（EMBO）的成员，也是欧洲基因和细胞治疗学会（ESGCT）的主席。他于 2009 年获得欧洲研究委员会高级研究员奖，2014 年和 2015 年分别获美国基因和细胞治疗学会和 ESGCT 的杰出成就奖。2015 年，布鲁塞尔 Vrije 大学授予他荣誉博士学位，2016 年授予 Jimenez Diaz 奖。他在意大利都灵大学获医科博士学位，后获罗马第一大学（University of Rome La Sapienza）的细胞与组织生物学博士学位。

裴端卿，博士 中国科学院广州生物医药与健康研究院（GIBH）教授。裴博士于 2002 年加入清华大学医学院，并于 2004 年加入新成立的 GIBH。在此之前，他曾在明尼苏达大学医学院任职。裴博士的博士论文研究乙型肝炎病毒（HBV）的转录调节，博士后及教职工作主要围绕细胞外基质重塑。回国后，他首先投入干细胞多能性研究，随后专攻细胞重编程。在清华大学，裴端卿实验室发表了关于 Oct4、Sox2、FoxD3、Essrb 和 Nanog 的结构和功能，以及它们对多能性的相互依赖关系的论文。裴端卿实验室使用非选择性系统创建了小鼠诱导性多能干（iPS）细胞，再系统地改进 iPS 过程，属国内首次。课题组随后通过提供资源和开展培训班，在中国传播 iPS 技术。裴端卿实验室最近的研究成果包括发现维生素 C 作为生成 iPSC 的强力促进剂，而间充质向上皮细胞的转分化启动了小鼠成纤维细胞的重编程过程。裴端卿实验室继续探索新方法来改进 iPS 技术，剖析由 Oct4 / Sox2 / Klf4 或更少因子驱动的重编程机制，开发替代的重编程方案，在体外使用 iPSC 模拟人类疾病，在再生医学方面利用基因编辑工具来校正干细胞突变。裴博士于 1991 年在宾夕法尼亚大学获博士学位，曾在密歇根大学攻读博士后。

applications of lentiviral vectors for gene transfer, which have become one of the most widely used tool in biomedical research and, upon recently entering clinical testing, are providing a long sought hope of cure for several currently untreatable and otherwise deadly human diseases. Since then he has continued to investigate new strategies to overcome the major hurdles to safe and effective gene transfer, translate then into new therapeutic strategies for genetic disease and cancer, and allowed novel insights into hematopoietic stem cell function, induction of immunological tolerance, and tumor angiogenesis. His recent work also contributed to advance the use of engineered nucleases for targeted genome editing in cell and gene therapy. Dr. Naldini is member of the European Molecular Biology Organization (EMBO), has been president of the European Society of Gene and Cell Therapy (ESGCT), and was awarded an European Research Council Advanced Investigator Grant in 2009, the Outstanding Achievement Award from the American Society of Gene and Cell Therapy in 2014 and from ESGCT in 2015, an Honorary doctorate from the Vrije University, Brussel, in 2015, and the Jimenez Diaz Prize in 2016. He received his M.D. from the University of Torino and his Ph.D. in cell and tissue biology from the University of Rome La Sapienza.

Duanqing Pei, Ph.D., is professor and director general of Guangzhou Institutes of Biomedicine and Health, Chinese Academy of Sciences. Dr. Pei joined the medical faculty at Tsinghua University in Beijing, China, in 2002 and moved to the newly formed Guangzhou Institutes of Biomedicine and Health in 2004. Prior to this appointment, he served as a faculty member at the University of Minnesota School of Medicine. Dr. Pei studied the transcription regulation of hepatitis B virus (HBV) for his Ph.D. thesis and worked on extracellular matrix remodeling as a postdoctoral fellow and faculty member. Upon returning to China, he first started to work on stem cell pluripotency and then on reprogramming. The Pei lab in Tsinghua has published on the structure and function of Oct4, Sox2, FoxD3, Essrb, and Nanog, and their interdependent relationship toward pluripotency. The Pei lab was the first in China to create mouse induced pluripotent stem (iPS) cells using a non-selective system, and then improved the iPS process systematically. The Pei lab subsequently disseminated the iPS technology in China by providing resources and training workshops. Recent publications from the Pei lab includes the discovery of vitamin C as a potent booster for iPSC generation and that a mesenchymal to epithelial transition initiates the reprogramming process of mouse fibroblasts. Dr. Pei's lab continues to explore new ways to improve iPS technology, dissect the reprogramming mechanisms driven by Oct4/Sox2/Klf4 or fewer factors, develop alternative reprogramming methods, employ iPSCs to model human diseases *in vitro*, and use gene editing tools to correct mutations in stem cells for regenerative medicine. Dr. Pei obtained his Ph.D. from

Matthew Porteus 医学博士，博士 斯坦福大学医学院的儿科、干细胞移植和再生医学科、血液学/肿瘤学及人类基因疗法的副教授。他在斯坦福大学医学院获得他的医科博士和理学博士学位，学位论文研究哺乳动物前脑发育的分子基础，论文题目为"TES-1／DLX-2 的提取和表征：在哺乳动物前脑发育期间表达的新型同源盒基因"（"Isolation and Characterization of TES-1/DLX-2: A Novel Homeobox Gene Expressed During Mammalian Forebrain Development."）。完成他的双学位课程后，他在波士顿儿童医院进行儿科实习，随后他在波士顿儿童医院/达纳法伯癌症研究所（Dana Farber Cancer Institute）联合访学项目进修儿科血液学/肿瘤学。他的访学项目和博士后研究是与 David Baltimore 博士合作的。在麻省理工学院和加州理工学院，他开始研发同源重组作为纠正干细胞致病突变的终极手段，针对患有血液遗传疾病，特别是镰状细胞病的儿童。在完成了 Baltimore 博士的培训后，他在得克萨斯大学西南医学院（University of Texas Southwestern）的儿科和生物化学科获得独立教职位，在2010年再次回到斯坦福大学担任副教授。在这段时间里，他的工作包括首次证明基因修正可在人类细胞中达到足够的频率用以治愈患者。他因此被认为是基因组编辑研究领域的先驱者和创始人之一，该领域现吸引了全世界数千个实验室和数个创新企业。他在主要工程核酸酶平台上拥有丰富经验，包括锌指核酸酶、模拟转录激活子的效应核酸酶及 CRISPR／Cas9 核酸酶。他在各种不同的干细胞类型中应用基因编辑手段，包括造血干细胞和祖细胞、神经干细胞、精原干细胞、人胚胎干细胞和诱导性多能干细胞等。他的研究课题继续致力于通过同源重组开发基因组编辑技术，作为遗传性和非遗传性疾病患者的治疗方案。在临床方面，Porteus 博士在 Lucille Packard 儿童医院出诊，照料造血干细胞移植后的儿科患者。在行政管理上，Porteus 博士是斯坦福医学科学家培训计划的副主任，负责监督斯坦福大学双博士学位计划的录取和学业进展。

Janet Rossant 博士 加拿大病童医院（the Hospital for Sick Children）发育生物学与干细胞生物学计划的高级科学家、分子遗传科的教授，也是多伦多大学妇产科教授。她的研究专注于利用细胞学和

the University of Pennsylvania in 1991 and trained as a postdoctoral fellow at the University of Michigan.

Matthew Porteus, M.D., Ph.D., is associate professor of Pediatrics, Divisions of Stem Cell Transplantation and Regenerative Medicine, Hematology/Oncology, and Human Gene Therapy at Stanford School of Medicine. He completed his combined M.D./Ph.D. at Stanford Medical School, with his Ph.D. focusing on understanding the molecular basis of mammalian forebrain development with his Ph.D. thesis titled "Isolation and Characterization of TES-1/DLX-2: A Novel Homeobox Gene Expressed During Mammalian Forebrain Development." After completion of his dual-degree program, he was an intern and resident in pediatrics at Boston Children's Hospital and then completed his pediatric hematology/oncology fellowship in the combined Boston Children's Hospital/Dana Farber Cancer Institute program. For his fellowship and post-doctoral research he worked with Dr. David Baltimore at the Massachusetts Institute of Technology and the California Institute of Technology where he began his studies in developing homologous recombination as a strategy to correct disease causing mutations in stem cells as definitive and curative therapy for children with genetic diseases of the blood, particularly sickle cell disease. Following his training with Dr. Baltimore, he took an independent faculty position at University of Texas Southwestern in the Departments of Pediatrics and Biochemistry before again returning to Stanford University in 2010 as an associate professor. During this time his work has been the first to demonstrate that gene correction could be achieved in human cells at frequencies that were high enough to potentially cure patients and is considered one of the pioneers and founders of the field of genome editing—a field that now encompasses thousands of labs and several new companies throughout the world. He has extensive experience with the major engineered nuclease platforms including zinc finger nucleases, transcription activator-like effector nucleases, and CRISPR/Cas9 nucleases. He has used genome-editing strategies in a variety of different stem cells including hematopoietic stem and progenitor cells, neural stem cells, spermatogonial stem cells, human embryonic stem cells, and induced pluripotent cells. His research program continues to focus on developing genome editing by homologous recombination as curative therapy for patients with both genetic and non-genetic diseases. Clinically, Dr. Porteus attends at the Lucille Packard Children's Hospital where he takes care of pediatric patients undergoing hematopoietic stem cell transplantation. Administratively, Dr. Porteus is the associate director of the Stanford Medical Scientist-Training Program where he oversees the admission and progress of students obtaining both M.D. and Ph.D. degrees at Stanford.

Janet Rossant, Ph.D., is a senior scientist in the Developmental and Stem Cell Biology program at The Hospital for Sick Children and is a professor in the

遗传学操控手段探明小鼠胚胎早期的正常和异常发育中的遗传调控。她对早期胚胎的研究兴趣，使其发现一种新型胎盘干细胞类型——滋养层干细胞。Rossant 博士是 Gairdner 基金会的总裁和研发总监。她积极参加国际发育生物学/干细胞生物学社团活动，并参与讨论与干细胞研究相关的公众议题的科学和伦理问题。她主持了加拿大卫生研究院（the Canadian Institutes of Health Research，CIHR）干细胞研究工作组，编撰了 CIHR 对该领域研究课题的资助指南。Rossan 博士曾在英国牛津大学和剑桥大学受训，自 1977 年以来一直在加拿大，先在布鲁克大学（Brock University）任职，在 1985 年至 2005 年间在多伦多西奈山医院的 Samuel Lunenfeld 研究所工作。

Dietram A. Scheufele，博士 威斯康星大学麦迪逊分校的科学传播学教授、Vilas 杰出成就教授、莫格里奇研究所（Morgridge Institute for Research）教授，是德国德累斯顿大学通信系荣誉教授。Scheufele 博士联合主持美国国家科学院、工程院和医学院（the National Academies of Sciences，Engineering，and Medicine）的生命科学公众交谈圆桌会议（Roundtable on Public Interfaces of the Life Sciences）、全国律师和科学家会议、美国科学促进会和美国律师协会的联合委员会等。他是美国科学促进会、国际通信协会、威斯康星科学艺术和文学学会的会士，以及德国国家科学与工程院（German National Academy of Science and Engineering）院士。他目前在美国国家科学院、工程院和医学院联盟地球与生命研究部（DELS）咨询委员会任职。Scheufele 博士曾是康奈尔大学的终身教员、哈佛大学 Shorenstein 研究学者和宾夕法尼亚大学 Annenberg 公共政策中心的访问学者。他曾为美国公共广播系统、世界卫生组织和世界银行担任顾问。

Ismail Serageldin，博士 于 2002 年成立的埃及新亚历山大图书馆（BA）的创办馆长，也是隶属于 BA 的研究所和博物馆的董事会主席。他是埃及总理在文化、科学和博物馆事务的顾问。他担任了数个学术、研究、科学和国际机构及民间团体顾问委员会的主席或委员，包括 2013 年世界社会科学报告顾问委员会、联合国教育科学及文化组织的"世界用水计划"（World Water Scenario，

Department of Molecular Genetics, and the Department of Obstetrics and Gynaecology at the University of Toronto. Her research focuses on understanding the genetic control of normal and abnormal development in the early mouse embryo using both cellular and genetic manipulation techniques. Her interests in the early embryo have led to the discovery of a novel placental stem cell type, the trophoblast stem cell. Dr. Rossant is also the president and scientific director of the Gairdner Foundation. She is actively involved in the international developmental and stem cell biology communities and has contributed to the scientific and ethical discussion on public issues related to stem cell research. She chaired the working group of the Canadian Institutes of Health Research (CIHR) on Stem Cell Research, which came up with guidelines for CIHR-funded research in this area. Dr. Rossant trained at Oxford University and Cambridge University in the United Kingdom and has been in Canada since 1977, first at Brock University and then at the Samuel Lunenfeld Research Institute, Mount Sinai Hospital, Toronto, from 1985 to 2005.

Dietram A. Scheufele, Ph.D., is the John E. Ross Professor in Science Communication and Vilas Distinguished Achievement Professor at the University of Wisconsin–Madison and in the Morgridge Institute for Research. He is also an Honorary Professor of Communication at the Technische Universität Dresden, Germany. Dr. Scheufele has co-chaired the of Sciences, Engineering, and Medicine's Roundtable on Public Interfaces of the Life Sciences and the National Conference of Lawyers and Scientists, a joint committee of the American Association for the Advancement of Science and the American Bar Association. He is a fellow of the American Association for the Advancement of Science, the International Communication Association, and the Wisconsin Academy of Sciences, Arts, and Letters, and a member of the German National Academy of Science and Engineering. He currently serves on the National Academies of Sciences, Engineering, and Medicine's Division on Earth and Life Studies (DELS) Advisory Committee.

In the past, Dr. Scheufele has been a tenured faculty member at Cornell University, a Shorenstein fellow at Harvard University, and a visiting scholar at the Annenberg Public Policy Center of the University of Pennsylvania. His consulting experience includes work for the Public Broadcasting System, the World Health Organization, and the World Bank.

Ismail Serageldin, Ph.D., is the founding director of the Bibliotheca Alexandrina (BA), the new Library of Alexandria, inaugurated in 2002. He also chairs the Boards of Directors for each of the BA's affiliated research institutes and museums. He is advisor to the Egyptian Prime Minister in matters concerning culture, science and museums. He serves as chair or member of a number of advisory committees for academic, research, scientific

2013年)、世界数字图书馆执行委员会(2010年)、"生命百科全书"执行委员会(2010年)、展望互联网未来-互联网名称和数字地址分配公司专家组(the Internet Corporation for Assigned Names and Numbers Panel for the Review of the Internet Future,2013年)等。他还担任非洲联盟生物技术高水平专家组(2006年)和科学、技术与创新高水平专家组(2012—2013)的联合主席。他曾担任过世界银行副行长(1992—2000)和国际农业研究咨询小组主席(1994—2000)、全球用水合伙人计划(Global Water Partnership)的创始人和前主席(1996—2000)、帮助贫困者小额信贷计划(Assist the Poorest,a microfinance program)顾问团创始人与前主席(1995—2000)、巴黎法兰西学院(Collège de France,Paris)国际"知识对抗贫穷"(Savoirs Contre Pauvreté)教授、荷兰瓦赫宁根大学(Wageningen University)的特聘教授。他是众多科学院与学会的院士/会士,包括美国科学院(公共福利奖章获得者)、美国哲学学会、美国艺术与科学院(the American Academy of Arts and Sciences)、世界科学院、世界艺术与科学院、欧洲艺术与科学院(the European Academy of Sciences and Arts)、非洲科学院、埃及科学院、比利时皇家科学院、孟加拉国科学院、印度国家农业科学院等。Serageldin博士发表了超过100本书籍和专著,以及500多篇关于各种主题的论文,包括生物技术、农村发展、可持续性和科学对社会的价值等。他在埃及的电视台主办了一个文化节目(超过130集),开启了阿拉伯语和英语的双语科普连续剧。他拥有开罗大学工程学士学位、哈佛大学的硕士和博士学位。此外拥有34个荣誉博士学位。

Sharon Terry,文学硕士 Genetic Alliance总裁兼首席执行官,该联盟覆盖超过一万个组织,其中1200个是疾病志愿宣传团体。Genetic Alliance团结了个人、家庭和社区关心健康问题。Terry女士也是PXE International的创办CEO,PXE International是弹性假黄瘤(PXE)形成条件的科研倡导组织,Terry女士的两个成年子女也是罹患此病。她是PXE相关基因的发现者之一,拥有ABCC6的专利,作为其管事,她将专利权转让给基金会。她开发了相关的诊断测试并进行临床试验,发表了140篇同行评审论文,其中30篇是关

and international institutions and civil society efforts, including the Advisory Committee of the World Social Science Report for 2013, as well as the United Nations Educational, Scientific and Cultural Organization–supported World Water Scenarios (2013) and Chairs the Executive Council of the World Digital Library (2010) and the executive council of the Encyclopedia of Life (2010) and the Internet Corporation for Assigned Names and Numbers Panel for the Review of the Internet Future (2013). He also co-chaired the African Union's High-Level Panel for Biotechnology (2006) and again for Science, Technology and Innovation in 2012-2013. He has previously held positions including Vice President of the World Bank (1992-2000), and Chairman Consultative Group on International Agricultural Research (1994-2000), founder and former Chairman of the Global Water Partnership (1996-2000) and the Consultative Group to Assist the Poorest, a microfinance program (1995-2000) and was professor of the International Savoirs Contre Pauvreté (Knowledge Against Poverty), at Collège de France, Paris, and distinguished professor at Wageningen University in the Netherlands. He is a member of many academies, including the U.S. National Academy of Sciences (Public Welfare Medalist), the American Philosophical Society, the American Academy of Arts and Sciences, the World Academy of Sciences, the World Academy of Arts and Sciences, the European Academy of Sciences and Arts, the African Academy of Sciences, Institut d'Egypte (Egyptian Academy of Science), the Royal Belgian Academy, the Bangladesh Academy of Sciences, the Indian National Academy of Agricultural Sciences. Dr. Serageldin has published more than 100 books and monographs and more than 500 papers on a variety of topics, including biotechnology, rural development, sustainability, and the value of science to society. He has hosted a cultural program on television in Egypt (more than 130 episodes) and developed a TV Science Series in Arabic and English. He holds a Bachelor of Science degree in Engineering from Cairo University and a master's degree and a Ph.D. from Harvard University and has received 34 honorary doctorates.

Sharon Terry, M.A., is president and CEO of Genetic Alliance, a network of more than 10,000 organizations, of which 1,200 are disease advocacy organizations. Genetic Alliance engages individuals, families, and communities to transform health. Ms. Terry is also the founding CEO of PXE International, a research advocacy organization for the genetic condition pseudoxanthoma elasticum (PXE), which affects Ms. Terry's two adult children. As co-discoverer of the gene associated with PXE, she holds the patent for ABCC6 to act as its steward and has assigned her rights to the foundation. She developed a diagnostic test and conducts clinical trials. She is the

于 PXE 临床研究。Terry 女士也是 Genetic Alliance 注册处和生物样本库的联合创始人。她关注着消费者参与遗传学研究、服务和政策的最前沿信息，领导了许多美国内外许多组织，包括："医药加速合伙人"（Accelerating Medicines Partnership），美国国家科学院、工程院和医学院政策委员会（the National Academies of Sciences, Engineering, and Medicine's Science and Policy Board），科学院联盟基因组研究与健康圆桌会议（Roundtable on Translating Genomic-Based Research for Health），PubMed Central 国家顾问委员会，PhenX 科学顾问委员会，全球基因组与健康联盟，国际稀有疾病研究联盟执行委员会，瑞士日内瓦 EspeRare 基金会创办主席，等等。她是数本期刊的编委会成员，也是 Genome 期刊的编辑。她在《遗传信息非歧视法案》通过过程中发挥了重要作用。鉴于她在众多团体组织中的贡献，她于 2006 年获得了艾奥那学院（Iona College）的荣誉博士学位；2007 年获北卡罗来纳大学药物基因组学和个体化治疗研究所的第一个患者服务奖；2009 年 Research!America 杰出组织倡导奖；2011 年获得临床研究论坛与基金会公共倡导领袖年度奖。2012 年，她成为位于中国唐山的河北联合大学荣誉教授，并获得了"勇于直面癌症风险"（Facing Our Risk of Cancer Empowered，FORCE）精神赋予倡导奖（Spirit of Empowerment Advocacy Award）。她于 2013 年被美国食品药品监督管理局列入"纪念罕见病用药法案三十周年的 30 位英雄"（"30 Heroes for the Thirtieth Anniversary of the Orphan Drug Act"）。在 2012 年和 2013 年，Terry 女士参加了"人人尽责参与平台"（Platform for Engaging Everyone Respon-sibly，PEER）的三个大型比赛，获得一等奖并赢取了 40 万美元的奖金。

Jonathan Weissman，博士 加利福尼亚大学旧金山分校细胞和分子药理学教授，霍华德休斯医学研究所（Howard Hughes Medical Institute）的研究员。他的研究关注细胞如何确保蛋白质折叠成正确形状，以及蛋白质错误折叠在疾病和正常生理中的影响。他还设计实验和分析方法，探索生物系统的组织原理，并通过核糖体图谱整体监测蛋白质翻译。他的最终目标是结合大规模手段和深入的机理研究，以揭示基因组密码信息。

Keith Yamamoto，博士 加利福尼亚大学旧

author of 140 peer-reviewed papers, of which 30 are PXE clinical studies. Ms. Terry is also a co-founder of the Genetic Alliance Registry and Biobank. In her focus at the forefront of consumer participation in genetics research, services and policy, she serves in a leadership role on many of the major international and national organizations, including the Accelerating Medicines Partnership, the National Academies of Sciences, Engineering, and Medicine's Science and Policy Board, the National Academies' Roundtable on Translating Genomic-Based Research for Health, the PubMed Central National Advisory Committee, the PhenX scientific advisory board, the Global Alliance for Genomics and Health, the International Rare Disease Research Consortium Executive Committee and as Founding President of EspeRare Foundation of Geneva, Switzerland. She is on the editorial boards of several journals and is an editor of *Genome*. She led the coalition that was instrumental in the passage of the Genetic Information Nondiscrimination Act. She received an honorary doctorate from Iona College for her work in community engagement in 2006; the first Patient Service Award from the University of North Carolina Institute for Pharmacogenomics and Individualized Therapy in 2007; the Research!America Distinguished Organization Advocacy Award in 2009; and the Clinical Research Forum and Foundation's Annual Award for Leadership in Public Advocacy in 2011. In 2012, she became an honorary professor of Hebei United University in Tangshan, China, and also received the Facing Our Risk of Cancer Empowered (FORCE) Spirit of Empowerment Advocacy Award. She was named one of the U.S. Food and Drug Administration's "30 Heroes for the Thirtieth Anniversary of the Orphan Drug Act" in 2013. In 2012 and 2013, Ms. Terry won $400,000 in first prizes in three large competitions for the Platform for Engaging Everyone Responsibly (PEER). PEER was awarded a $1M contract from the Patient-Centered Outcomes Research Institute in 2014.

Jonathan Weissman, Ph.D., is a professor of cellular and molecular pharmacology at the University of California, San Francisco, and a Howard Hughes Medical Institute investigator. His research explores how cells ensure that proteins fold into their correct shape, as well as the role of protein misfolding in disease and normal physiology. He is also developing experimental and analytical approaches for exploring the organizational principles of biological systems and globally monitoring protein translation through ribosome profiling. A broad goal of his work is to bridge large-scale approaches and in-depth mechanistic investigations to reveal the information encoded within genomes.

Keith Yamamoto, Ph.D., is University of California, San Francisco vice chancellor of Science

金山分校科技政策与战略事务副校长，同时兼任医学院科研事务副院长、细胞与分子药理学的教授。Yamamoto 博士的研究兴趣是核受体的信号转导和转录调节，从而介导关键激素和细胞信号的活动。他运用了系统的方法和机理研究，从纯化后的分子、细胞和生物体整体水平进行探索。Yamamoto 博士曾领导和服务于公共科学政策、公众对生物学研究的理解与支持、科学教育等方面的多个国家委员会。他主持了生命科学联盟（Coalition for the Life Sciences），也在美国医学院理事会和美国国家科学院、工程院和医学院（the National Academies of Sciences, Engineering, and Medicine's Science）地球与生命研究部顾问委员会任职。此外，他也曾在美国国立卫生研究院主持和服务过众多委员会，负责监管生物医药人员、培训、研究经费、同行评议过程和相关政策等。此外，他还是劳伦斯伯克利国家实验室（Lawrence Berkeley National Laboratory）顾问委员会和 Research!America 董事会成员。他是美国国家科学院、美国医学院、美国艺术与科学院（the American Academy of Arts and Sciences）院士，美国微生物学会和美国科学促进会（American Association for the Advancement of Science）会士。

Policy and Strategy. He also serves as vice dean for Research for the School of Medicine, and Professor of Cellular and Molecular Pharmacology. Dr. Yamamoto's research focuses on signaling and transcriptional regulation by nuclear receptors, which mediate the actions of essential hormones and cellular signals; he uses mechanistic and systems approaches to pursue these problems in pure molecules, cells and whole organisms. He has led or served on numerous national committees focused on public and scientific policy, public understanding and support of biological research, and science education; he chairs the Coalition for the Life Sciences, and sits on the National Academy of Medicine Council and the National Academies of Sciences, Engineering, and Medicine's Division on Earth and Life Studies Advisory Committee. He has chaired or served on many committees that oversee training and the biomedical workforce, research funding, and the process of peer review and the policies that govern it at the National Institutes of Health. He is a member of the advisory board for Lawrence Berkeley National Laboratory and the board of directors of Research!America. He was elected to the National Academy of Sciences, the National Academy of Medicine, the American Academy of Arts and Sciences, and the American Academy of Microbiology, and is a fellow of the American Association for the Advancement of Science.

附录 E 术语表[80]
Appendix E Glossary[80]

成体干细胞——一种存在于生物体分化组织中的未分化细胞。可自我更新，并分化生成其所在组织的特化细胞。

等位基因——染色体上特定基因座的不同的基因变体。不同的等位基因产生不同的遗传性状。

非整倍体——细胞中存在异常数量的染色体。

辅助生殖技术 (ART)——一种涉及体外配子（卵子和精子）或胚胎操作的生育治疗或程序。其实例包括体外受精（IVF）和胞质内精子注射（ICSI）（NAS，2002）。

自体移植——受体的移植组织来源于自身。自体移植有助于避免移植所带来的免疫排斥。

囊胚——哺乳动物胚胎着床前的特定阶段（在人类中约为受精后 5 天），包含 50~150 个细胞。囊胚为由外部细胞层（滋养层）、液体填充的腔（囊胚腔）和内细胞团组成的球体。来自内细胞团的细胞经体外培养可生成胚胎干细胞系。

CRISPR 关联蛋白 9(Cas9)——一种可切割 DNA 序列的特殊核酸酶。Cas9 是 CRISPR-Cas9 基因编辑体系的重要"工具包"。

嵌合体——由源自至少两个不同遗传背景的个体的细胞组成的生物体

绒毛膜癌——一种源于胎盘前体——滋养层的肿瘤，可入侵子宫壁。

Adult stem cell An undifferentiated cell found in a differentiated tissue in an adult organism that can renew itself and can differentiate to yield specialized cell types of the tissue in which it is found.

Allele A variant form of a gene at a particular locus on a chromosome. Different alleles produce variations in inherited characteristics.

Aneuploidy The presence of an abnormal number of chromosomes in a cell.

Assisted reproductive technology (ART) A fertility treatment or procedure that involves laboratory handling of gametes (eggs and sperm) or embryos. Examples of ART include *in vitro* fertilization (IVF) and intracytoplasmic sperm injection (ICSI) (NAS, 2002).

Autologous transplant Transplanted tissue derived from the intended recipient of the transplant. Such a transplant helps to avoid complications of immune rejection.

Blastocyst A preimplantation embryo in placental mammals (about 5 days after fertilization in humans) of 50-150 cells. The blastocyst consists of a sphere made up of an outer layer of cells (the trophectoderm), a fluid-filled cavity (the blastocoel or blastocyst cavity), and a cluster of cells in the interior (the inner cell mass). Cells from the inner cell mass, if grown in culture, can give rise to embryonic stem cell lines.

Cas9 (CRISPR Associated Protein 9) A specialized enzyme known as a nuclease that has the ability to cut DNA sequences. Cas9 makes up part of the "toolkit" for the CRISPR/Cas9 method of genome editing.

Chimera An organism composed of cells derived from at least two genetically different individuals.

Choriocarcinoma A type of tumor that originates

---

[80]Definitions for a number of terms draw from the reports NAS (2002), NASEM (2016b), and NRC and IOM (2005).

染色质——形成染色体的 DNA 和蛋白质的复合物。其中一些蛋白质是结构性的,帮助组织和保护 DNA;另一些蛋白质是调节性的,负责调控基因活性,促进 DNA 复制或修复。

染色体——包含单一长度 DNA 的线状结构分子,通常携带数以百计的基因。通常与蛋白质组装形成染色质。一个细胞内完整的染色体组(人类为 23 对)包含了两个基因组的拷贝,来自两个亲本。染色体通常驻留在细胞核中,只有当细胞进行分裂时,核膜结构解体,染色体浓缩形成可视的、分立的实体。

卵裂——胚胎发育早期形成囊胚以前所进行的细胞分裂活动。

切割——用于描述断裂或切割 DNA 的过程。

临床应用——使用生物医学试剂、操作或设备来处理临床状况的过程。

临床试验——对患者实施新开发的临床应用的实验过程,实验在监管下进行,以确保风险最小化和疗效最优化。新型的治疗策略在正式运用前均需进行临床试验。

成簇规律间隔短回文重复序列 (CRISPR)——在细菌中发现的一种为抵御外界病毒而保留一些外来 DNA 片段的天然免疫机制。有时,该系统也被称为 CRISPR / Cas9 以代指整个基因编辑平台。其中与靶向 DNA 同源的 RNA 与 Cas9(CRISPR 关联蛋白9)组合形成用于此体系的"工具包",而 Cas9 是 CRISPR / Cas9 基因编辑体系中重要的 DNA 切割酶(核酸酶)。

CRISPR 激活 (CRISPRa)——使用向导 RNA 和核酸酶活性衰弱或被抑制的、连接到一个或多个激活域的 Cas9(dCas9)以增加目标基因的转录。

CRISPR 抑制或 CRISPR 干扰 (CRISPRi/ CRISPRr)——使用向导 RNA 和 dCas9 或 dCas9 抑制子以减少目标基因的转录。

培养细胞——在培养基中生存并继续可增殖的细胞。

核酸酶活性衰弱或被抑制的 Cas9 蛋白变体 (dCas9)——一种能与向导 RNA 一同结合 DNA,但不能切割 DNA 的 Cas9 突变体。通常设计将其与转录因子、染色质修饰酶或荧光蛋白相连以改变目的基因的表达或标记特定位点。

from the trophoblast, the precursor of the placenta, and invades the uterine wall.

Chromatin The complex of DNA and proteins that forms chromosomes. Some of the proteins are structural, helping to organize and protect the DNA, while others are regulatory, acting to control whether genes are active or not, and to promote DNA replication or repair.

Chromosome A thread-like structure that contains a single length of DNA, usually carrying many hundreds of genes. This is packaged with proteins to form chromatin. The DNA within the complete cellular set of chromosomes (23 pairs in humans) comprises two copies of the genome, one from each parent. The chromosomes usually reside in the nucleus of a cell, except during cell division when the nuclear membrane beaks down and the chromosomes become condensed and can be visualized as discrete entities.

Cleavage The process of cell division in the very early embryo before it becomes a blastocyst. Also used to describe breaking or cutting DNA

Clinical application The use of a biomedical reagent, procedure, or device to treat a clinical condition.

Clinical trial A supervised and monitored experimental test in patients of a newly developed clinical application to ensure minimization of risk and optimization of efficacy. Clinical trials are required before a treatment is approved for general use.

CRISPR (Clustered Regularly-Interspaced Short Palindromic R epeats) A naturally occurring mechanism found in bacteria that involves the retention of fragments of foreign DNA, providing the bacteria with some immunity to viruses. The system is sometimes referred to as CRISPR/Cas9 to denote the entire gene-editing platform in which RNA homologous with the targeted gene is combined with Cas9 (C RISPR Associated Protein 9), which is a DNA-cutting enzyme (nuclease) to form the "toolkit" for the CRISPR/Cas9 method of genome editing.

CRISPRa CRISPR activation, using a guide RNA and nuclease-deficient or nuclease-dead Cas9 (dCas9) linked to one or more activation domains to increase transcription of a target gene.

CRISPRr/ CRISPRi CRISPR repression, or CRISPR interference, using a dCas9 or dCas9-repressor with a guide RNA to decrease transcription of a target gene.

Cultured cell A cell maintained in a tissue culture allowing expansion of its numbers.

dCas9 (Nuclease-deficient Cas9 or nuclease-dead Cas9)—This can still bind DNA, together with a guide RNA, but not cut it. It is often linked to a transcription factor, chromatin-modifying enzyme, or fluorescent protein to mediate alterations to gene expression or to mark specific sites.

伦理道德理论——关于某项选择在道德层面上是否被需要、允许或禁止的规范性理论。

脱氧核糖核酸 (DNA)——在所有已知生物体中携带有关指导发育、运转和繁殖的遗传指令的双链双螺旋分子。

分化——早期胚胎细胞由非特化状态表现出特化性状的过程，如心脏、肝脏或肌肉细胞等。

二倍体——包含来自每个亲本的全套 DNA 的细胞。人类二倍体细胞含有 46 条染色体（23 对）。

进化趋异——在进化过程中基因序列会产生多样性，那些能在自然选择中带来优势的序列改变更容易被保留和扩散。不同的选择压力会选择不同的改变，因此不同的群体所选择的序列改变也不同。

显性——基因或性状的遗传模式，生物体中特定等位基因（基因变体）的单拷贝赋予与其二倍体细胞中基因的另一拷贝无关的功能。

双链断裂 (DSB)——DNA 的双螺旋均断裂，与单链断裂或"缺刻"不同。

外胚层——胚胎的三个原始胚层的最外层，它将分化为皮肤、神经和大脑。

异位——某物在一个不寻常的位置被发现，如宫外孕。

胚胎——动物生长发育的早期阶段。其特征为卵裂（受精卵的细胞分裂）、基本细胞和组织的分化，以及原始器官和系统的形成。对人类来说是从受精卵着床开始到受孕后第 8 周结束。之后它被称为胎儿。

胚胎生殖 (EG) 细胞——在发育早期迁移至未来将形成性腺的区域，以形成卵或精细胞的祖细胞的多功能干细胞。胚胎生殖细胞的性质与胚胎干细胞的性质相似，但在一些印证区域的 DNA 甲基化状态可能不同。

胚胎干 (ES) 细胞——来自胚胎的原始（未分化）细胞，其具有分化成多种特化细胞（即多能性）的潜力。来自囊胚的内细胞团。胚胎干细胞不等于胚胎，它无法自行分化为必要类型的细胞，如滋养外胚层细胞，更无法产生完整的生物体（NAS, 2002）。胚胎干细胞可以在培养物中维持其多能性，并可被诱导分化为许多不同类型的细胞。

内胚层——胚胎的三个原始胚层的最内层，它将分化为肺、肝脏和消化器官。

Deontology ethics A normative theory regarding which choices are morally required, forbidden, or permitted.

Deoxyribonucleic acid (DNA) A two-stranded molecule, arranged as a double helix, that contains the genetic instructions used in the development, functioning, and reproduction of all known living organisms.

Differentiation The process whereby an unspecialized early embryonic cell acquires the features of a specialized cell, such as a heart, liver, or muscle cell.

Diploid Cells that contain a full set of DNA—half from each parent. In humans, diploid cells contain 46 chromosomes (in 23 pairs).

Divergence (evolutionary) During evolution, variations occur in the sequences of genes; if these variations confer some advantage natural selection increases their prevalence. Different selective pressures select for different variations so that the prevalence of different gene variants diverges in different populations.

Dominant A pattern of inheritance of a gene or trait in which a single copy of a particular allele (gene variant) confers a function independent of the nature of the second copy of the gene in a diploid cell of an organism.

Double-strand break (DSB) A break in the DNA double helix in which both strands are cut, as distinct from a single-strand break or "nick."

Ectoderm The outermost of the three primitive germ layers of the embryo; it gives rise to skin, nerves, and brain.

Ectopic Found in an unusual location, such as an ectopic pregnancy outside the uterus.

Embryo An animal in the early stages of growth and differentiation that are characterized by cleavage (cell division of the fertilized egg), differentiation of fundamental cell types and tissues, and the formation of primitive organs and organ systems; the developing human individual from the time of implantation to the end of the eighth week after conception, after which stage it becomes known as a fetus.

Embryonic germ (EG) cell A pluripotent stem cell that migrates during early development to the future gonads to form the progenitors of egg or sperm cells. The properties of EG cells are similar to those of embryonic stem cells, but may differ in the DNA methylation of some imprinted regions.

Embryonic stem (ES) cell A primitive (undifferentiated) cell from the embryo that has the potential to become a wide variety of specialized cell types (that is, is pluripotent). It is derived from the inner cell mass of the blastocyst. An embryonic stem cell is not an embryo; by itself, it cannot produce the necessary cell types, such as trophectoderm cells, so as to give rise to a complete organism (NAS, 2002). Embryonic stem cells can be maintained as pluripotent cells in culture and induced to differentiate into many different cell types.

Endoderm Innermost of the three primitive germ layers of the embryo; it later gives rise to the lungs, liver,

内生——起源于细胞或生物体内。

子宫内膜——子宫内壁上层，胚胎着床的地方。

内切核酸酶——通过切断核苷酸链内部的磷酸二酯键，将其分解成两个或多个较短链核酸分子的酶。

增强——改变某个状态或性状使其超出典型或正常水平。

去核细胞——去掉细胞核的细胞。

去核——去除细胞核，仅留下细胞质的过程。当应用于卵子时，去核可以应用于去除不被核膜包围的母性遗传物质。

酶——作为生物催化剂，加速化学反应的蛋白质。

上胚层——在脊椎动物胚胎早期产生的除卵黄囊和胎盘之外的整个胚胎的特定细胞层。上胚层细胞是多能性的，并且可以产生胚胎干细胞。

表观遗传效应——在不改变基因的 DNA 序列的情况下发生的基因表达的改变。例如，在称为基因组印记的表观遗传效应中，DNA 发生甲基化并改变其表达状态。

表观基因组——一系列可影响基因是否表达及如何表达的在基因组 DNA 和 DNA 结合蛋白上的化学修饰。

离体 (*ex vivo*)——拉丁语意为"生物体外"。

外源性——引入或源自细胞或生物体外部的。

受精作用——男性和女性配子（精子和卵子）结合的过程。

FokI——一种核酸酶，其具有切割功能的结构域已被用于偶联锌指（ZF）或转录激活子样效应物（TALE）等 DNA 结合结构域。FokI 的切割结构域仅切割 DNA 的一条链（缺刻），因此需要一对 ZFN 或 TALEN 以产生双链断裂。FokI 切割结构域还可连接到 dCas9，并且该融合物也必须二聚化才可以切割 DNA。

功能获得——一种在突变后可带来全新功能或表达状态的基因突变形式。

配子——生殖细胞（卵子或精子）。配子是单倍体（仅含有体细胞中染色体数目的一半，对人类来说是 23 条），因此当两个配子在受精时结合，所得的单细胞胚胎（合子）具有全部数目的染色体（人类为 46 条）。

原肠胚形成——动物胚胎发育早期形成三个

and digestive organs.

Endogenous Originating from within a cell or an organism.

Endometrium The inner epithelial lining of the uterus into which embryos implant.

Endonuclease An enzyme that breaks down a nucleotide chain into two or more shorter chains by cleaving at internal phosphodiester bonds.

Enhancement—improving a condition or trait beyond a typical or normal level.

Enucleated cell A cell whose nucleus has been removed.

Enucleation A process whereby the nuclear material of a cell is removed, leaving only the cytoplasm. When applied to an egg, can be applied to the removal of the maternal chromosomes, when they are not surrounded by a nuclear membrane.

Enzyme A protein that acts as a biological catalyst, speeding up chemical reactions.

Epiblast A specific layer of cells in an early vertebrate embryo that gives rise to the entire embryo other than yolk sac and placenta. Epiblast cells are pluripotent and can give rise to embryonic stem cells.

Epigenetic effects Changes in gene expression that occur without changing the DNA sequence of a gene; for example, in the epigenetic effect called genomic imprinting, chemical molecules called methyl groups attach to DNA and alter the gene's expression.

Epigenome A set of chemical modifications to the DNA of the genome and to proteins that bind to DNA in the chromosomes to affect whether and how genes are expressed.

Ex vivo—Latin: "out of the living"; outside an organism.

Exogenous—Introduced or originating from outside a cell or an organism.

Fertilization—The process whereby male and female gametes (sperm and egg) unite.

FokI— The nuclease from which the cleavage domain has been abstracted and joined to zinc finger (ZF) or transcription activator-like effector (TALE) DNA-binding domains. The FokI cleavage domain cuts only one strand of the DNA (a nick), so a pair of ZFNs or TALENs is required to create double-strand breaks. The FokI cleavage domain has also been linked to nuclease-deficient Cas9 (dCas9), and this fusion must also dimerize to cut DNA.

Gain of function—A type of mutation that results in an altered gene product that possesses a new molecular function or a new pattern of gene expression.

Gamete—A reproductive cell (egg or sperm). Gametes are haploid (having only half the number of chromosomes found in somatic cells—23 in humans), so that when two gametes unite at fertilization, the resulting one-cell embryo (zygote) has the full number of chromosomes (46 in humans).

Gastrulation—The procedure by which an animal embryo at an early stage of development produces the three

胚层（外胚层、中胚层和内胚层）的过程。

基因——遗传的功能单元，对应于染色体上特定位点的DNA区段。基因通常指导某个蛋白或RNA分子的形成。

基因驱动——一种基因偏好性的遗传系统技术，特定遗传序列通过有性生殖从亲本传递到其后代致其能力得到增强。基因驱动技术会主动将一条染色体上的序列复制到同源染色体上，使得生物体携带两个拷贝的修饰基因。这一过程确保所有的后代和之后的世代将继承编辑后的基因和其相关的性状。因此，基因驱动的结果就是特定基因型的有偏好性地在传代中增加，并有在人群中扩散的可能性。

基因编辑——一种允许研究人员改变细胞或生物体的DNA，包括插入、删除或修改基因或基因序列，以实现基因的沉默、增强或其他改变其特征的技术。

基因表达——由基因编码合成RNA和蛋白质的过程。基因表达由与基因组或其RNA拷贝结合的蛋白质和RNA分子控制，调节其生成水平和其产物水平。基因表达的改变会改变细胞、组织、器官或整个生物的功能，并可能带来与特定基因相关的可见特征的改变。

基因靶向——用于改变特定基因的操作。

基因治疗——以改善病症为目的，将外源基因引入细胞。

基因转移——任何用于描述将基因转移到细胞中的过程，如在基因治疗中使用的。

遗传因子——基因组中具有的由其序列赋予的一些特性的DNA区段，如编码蛋白质或RNA的基因。更常用于除此之外的可以控制基因表达或基因组组织的序列。

基因组——构成生物体的完整DNA组。人类基因组由23对同源染色体组成。

基因组编辑——通过引入DNA断裂或其他DNA修饰来改变基因组序列的过程。

基因型——个体的遗传构成。

生殖细胞——在细胞谱系中可产生精子或卵子的细胞。有性生殖中卵子和精子融合形成胚胎，种系得以繁衍下一代。

胚层——在发育早期，胚胎分化成三个不同的胚层（外胚层、内胚层和中胚层），每个胚层各

primary germ layers—ectoderm, mesoderm, and endoderm.

Gene—A functional unit of heredity that is a segment of DNA in a specific site on a chromosome. A gene typically directs the formation of a protein or RNA molecule.

Gene drive— A system of biased inheritance in which the ability of a particular genetic sequence to pass from a parent to its offspring through sexual reproduction is enhanced. Gene drive technology actively copies a sequence on one chromosome to its partner chromosome, so that the organism carries two copies of the intentionally modified gene. This process ensures that all of an organism's offspring and subsequent generations will inherit the edited genome and related trait(s). Thus, the result of a gene drive is the preferential increase of a specific genotype from one generation to the next, and potentially throughout a population.

Gene editing—A technique that allows researchers to alter the DNA of cells or organisms to insert, delete, or modify a gene or gene sequences to silence, enhance, or otherwise change the gene's characteristics.

Gene expression—The process by which RNA and proteins are made from the instructions encoded in genes. Gene expression is controlled by proteins and RNA molecules that bind to the genome or to the RNA copy and regulate their levels of production and those of their products. Alterations in gene expression change the functions of cells, tissues, organs, or whole organisms and sometimes result in observable characteristics associated with a particular gene.

Gene targeting - A procedure used to produce an alteration in a specific gene.

Gene therapy—Introduction of exogenous genes into cells with the goal of ameliorating a disease condition.

Gene transfer—Any process often used to describe the transfer of genes into cells—as used in gene therapy.

Genetic element—A segment of the DNA in a genome that has some particular property conferred by its sequence, such as a gene encoding a protein or RNA—more often used to refer to sequences that are not such genes but may control gene expression or genome organization.

Genome The complete set of DNA that makes up an organism. In humans, the genome is organized into 23 pairs of homologous chromosomes.

Genome editing The process by which the genome sequence is changed through the intervention of a DNA break or other DNA modification. .

Genotype—Genetic constitution of an individual.

Germ cell (or germline cell) A cell at any point in the lineage of cells that will give rise to sperm or eggs. The germline is this lineage of cells. Eggs and sperm fuse during sexual reproduction to create an embryo. In so doing, the germline continues into the next generation.

Germ layer In early development, the embryo differentiates into three distinct germ layers (ectoderm, endoderm, and mesoderm), each of which gives rise to

自分化为生物体的不同部分。

妊娠——从卵子受精到出生的新生命体的孕育期。

管制——通过传统（实践标准）或规章行使监督的过程，个人和委员会由此承担责任。管制通常涉及诸如专业实践标准和行为守则等政策工具，正式准则、协定和条约，以及立法或其他政府法规。

向导分子——将基因组编辑中负责将编辑酶引导至靶序列的蛋白质或 RNA 短片段。

向导 RNA(gRNA)——用于将 DNA 切割酶导向基因组中靶序列的 RNA 短片段。gRNA 片段含有与靶序列（通常 20 个碱基）同源的区域，以及与核酸酶（如 Cas9）相互作用的序列。在基因组编辑中使用的 gRNA 是人工合成的，在自然中不存在。

单倍体——指仅具有一组染色体（人类中为 23 条）的细胞（通常是配子或其直接前体）。与此相反，生物体细胞（体细胞）是二倍体，具有两组染色体（人类中为 46 条）。

遗传改变——对可以通过世代传递的基因的修饰。

杂合——细胞或生物体的两条同源染色体上的特定基因（等位基因）上包含两种不同变体。

同源重组——两个相似的 DNA 分子发生重组，包括通过基因靶向产生特定基因改变的过程。

同源指导修复——一种依赖同源序列模板进行断裂 DNA 修复的自然过程。通常发生在 DNA 合成期间或之后，以彼此提供模板。

借助同源指导修复进行基因组编辑时，通常采用人工合成或重组 DNA 技术来制备 DNA 模板，且通常使模板的两个末端包含与靶基因座精确同源性的序列，需改变的 DNA 序列则位于其中间。

纯合子——细胞或生物体的两条同源染色体上的特定基因（等位基因）上含有相同的变体。

人类受精和胚胎学管理局 (HFEA)——英国负责监督生殖细胞和胚胎在生育治疗和研究中的使用的独立监管机构。它还代表"人类受精和胚胎学法"，管理局运作和维护的法律。

植入 (着床)——胚胎附着到子宫内膜的过程（人类为 7~14 天）。

宫内 (In utero)——拉丁语意为"在子宫里"。

different parts of the developing organism.

Gestation The period of development of an organism from fertilization of the egg until birth.

Governance—The process of exercising oversight through traditions (standards of practice) or regulations by which individuals and communities are held accountable. Governance often involves such policy tools as professional standards of practice and codes of conduct; formal guidelines, agreements, and treaties; and legislation or other governmental regulation.

Guide molecule—A protein or short section of RNA used to guide the genome editing machinery to the desired location in the DNA sequence.

guide RNA (gRNA) Short segments of RNA used to direct the DNA-cutting enzyme to the target location in the genome. gRNA segments contain the region of homology to the target sequence (usually 20 bases), and a sequence that interacts with the nuclease (e.g., Cas9). gRNAs used in genome editing are synthetic and do not occur in nature.

Haploid—Refers to a cell (usually a gamete or its immediate precursor) having only one set of chromosomes (23 in humans). In contrast, body cells (somatic cells) are diploid, having two sets of chromosomes (46 in humans).

Heritable genetic change—Modifications to genes that could be passed down through generations.

Heterozygous—Having two different variants (alleles) of a specific gene on the two homologous chromosomes of a cell or an organism.

Homologous recombination—Recombining of two like DNA molecules, including a process by which gene targeting produces an alteration in a specific gene.

Homology-directed repair (HDR)—A natural repair process used to repair broken DNA, which relies on a DNA "template" with homology to the broken stretch of DNA. This usually occurs during or after DNA synthesis, which provides this template.

In genome editing via HDR, the DNA template is synthesised or made by recombinant DNA techniques, and usually contains regions of exact homology to the target locus at each end, with the desired alteration contained within the middle.

Homozygous—Having the same variant (allele) of a specific gene on both homologous chromosomes of a cell or an organism.

Human Fertilisation and Embryology Authority (HFEA)—The United Kingdom's independent regulator overseeing the use of germ cells and embryos in fertility treatment and research. It also stands for the Human Fertilisation and Embryology Act, the law under which the Authority operates and which it upholds.

Implantation—The process by which an embryo becomes attached to the inside of the uterus (7-14 days in humans).

In utero—Latin: "in the uterus."

体外 (*In vitro*)——拉丁语意为"在玻璃器皿中";指在实验室器皿或试管中,或在人工环境中。

体外受精 (IVF)——辅助生殖技术,其中受精的过程在体外完成。

体内 (*In vivo*)——拉丁语意为"在活体中";指在自然环境中,通常在受试者的身体内。该术语通常也用于指培养中"活"细胞中。

插入缺失突变 (Indel)——DNA 序列的插入或缺失。小的插入缺失突变(1~4 个碱基)通常伴随非同源末端连接发生。通常源于移动开放阅读框和(或)产生提前终止密码子而导致基因破坏。

诱导多能干细胞 (iPS)——含有被导入或激活的可赋予细胞多能性和干细胞样特征的基因的细胞。由此,可以将已分化的细胞(如皮肤细胞)诱导成具有多能性的细胞。这在再生医学中是有应用价值的,因为将 iPS 细胞移植回供体后产生免疫排斥反应的风险更小。

插入诱变——在基因原序列中插入新序列的突变形式,如插入病毒序列。

机构审查委员会 (IRB)——在研究机构(大学或医院)中,为保障参加其有关人类的研究活动的人的基本权益和福利而设立的行政机构。IRB 有权根据联邦法规和地方机构政策的规定批准,要求修改或否决其管辖范围内的研究活动。

慢病毒——属于逆转录病毒的亚纲。其基因组为 RNA,可被逆转录成 DNA 并被整合到细胞的 DNA 基因组中。其通常作为基因载体(质粒)把目的基因导入到细胞中。

连接酶——一种催化两条 DNA 片段黏合反应的酶。

功能丧失——由突变引起基因产物的改变,导致野生型基因的分子功能缺失。

巨大核酸酶——一种特殊的天然酶,可结合到特异性 DNA 序列并剪切 DNA,通常在基因组的少数位点上起作用。这种酶(及其合成衍生物)可通过 DNA 裂解促使 DNA 的重排。巨大核酸酶可促进非同源性末端连接,以及通过同源性定向修复介导的 DNA 修改。相关研究首次发现了 DNA 裂解的基本原理和 DNA 修复过程——基因组编辑的基础。

中胚层——胚胎的中间层,由囊胚的内细胞团的一部分细胞发育而成。中胚层形成在原肠胚

In vitro—Latin: "in glass"; in a laboratory dish or test tube; in an artificial environment.

In vitro fertilization (IVF)—An assisted reproduction technique in which fertilization is accomplished outside the body.

In vivo—Latin: "in the living"; in a natural environment, usually in the body of the subject. This term is often also used to refer to events in "living" cells in culture.

Indel—An insertion or deletion of DNA sequence. Small indels (e.g., one to four base pairs) are often associated with non-homologous end joining. These often result in the disruption of a gene by shifting the open reading frame and/or creating premature stop codons.

Induced pluripotent stem (iPS) cell—A cell induced by the introduction or activation of genes conferring pluripotency and stem cell-like properties. Thus, cells already committed to a particular fate (e.g., skin) can be induced to become pluripotent. This is useful in regenerative medicine because the iPS cells can be introduced back into the donor of the original cells with much less risk of transplant rejection.

Insertional mutagenesis—The alteration of the sequence of a gene by the insertion of exogenous sequence such as by integration of viral sequences.

Institutional review board (IRB)—An administrative body in an institution (such as a hospital or a university) established to protect the rights and welfare of human research subjects recruited to participate in research activities conducted under the auspices of that institution. The IRB has the authority to approve, require modifications in, or disapprove research activities in its jurisdiction, as specified by both federal regulations and local institutional policy.

Lentivirus—A subclass of retroviruses, viruses whose genome are made of RNA but during viral replication becomes copied into a DNA form that can integrate into the DNA genome of a cell. Often used as carriers of genes (vectors) to introduce genes into cells.

Ligase—An enzyme that catalyzes joining of two pieces of DNA.

Loss of function—A type of mutation in which the altered gene product lacks the molecular function of the wild-type gene.

Meganuclease—A special type of enzyme that binds to and cuts DNA at specific DNA sequences of a length that occurs at few sites in the genome. These are natural enzymes (and their synthetic derivatives) that catalyze DNA rearrangement events via DNA cleavage. They can be used in genome editing for both non-homologous end joining and homology directed repair–mediated alterations. It was the study of these that first revealed the basic mechanisms of DNA cleavage and the DNA repair processes on which genome editing depends.

Mesoderm—The middle layer of the embryo, which

形成阶段，它是血液、骨骼、肌肉和结缔组织的前体。

线粒体移植（或者线粒体置换）——为了预防母系传递的线粒体 DNA（mtDNA）疾病所设计的新技术。

线粒体——细胞质中的一种细胞结构，它可以为细胞提供能量。每一个细胞都含有许多的线粒体。对于人类而言，细胞核中的 DNA 含有 35 000 个基因，而每一个线粒体中，一个环状线粒体 DNA 包含 37 个基因。

镶嵌现象——部分细胞的变异，使所有细胞不完全相同。例如，一个仅部分细胞被编辑过的人工胚胎。

亚多能干细胞——一种来源于胚胎、胎儿或者成体的干细胞。可分化成多种，如某组织、器官或生理系统的所需细胞类型，但不一定能分化为所有类型。

鼠源——来自小鼠的。

突变——DNA 序列的改变。可在细胞分裂期间自行发生，或受环境压力诱发，如紫外线、辐射和化学品。

切口酶——一种核酸酶，可以切割 DNA 双螺旋中的其中一条链。

非同源性末端连接 (NHEJ)——一种自然修复程序，用于把两个断裂 DNA 的末端连接起来。易发生错误，导致短片段（通常是 2~4 个 DNA 碱基对）的插入或缺失突变。

规范理论——关于人们应该怎样做出决定，而不是怎样做或将怎样做出决定的理论。

核酸酶——一种可以切断 DNA 或 RNA 链的酶。

脱靶效应——对生物体进行干预所产生的偏离预期的、直接或间接的、短期或长期的结果。

脱靶事件（或脱靶切割）——指基因编辑的核酸酶在非目标位点对 DNA 进行切割。通常是由于脱靶序列与预期的目标序列非常相似而引起的。

卵母细胞——发育中的卵子，通常是一个静止的较大的细胞。

表型——生物体的可观测特性，它是由基因型和环境共同决定的。

质粒——具有自我复制功能的环形 DNA。可通过设计质粒使其将目标基因运载到目标细胞中并表达。

consists of a group of cells derived from the inner cell mass of the blastocyst; it is formed at gastrulation and is the precursor to blood, bone, muscle, and connective tissue.

Mitochondrial transfer (or mitochondrial replacement)—Novel procedures designed to prevent the maternal transmission of mitochondrial DNA (mtDNA) diseases.

Mitochondrion (plural, Mitochondria)—A cellular structure in the cytoplasm that provides energy to the cell. Each cell contains many mitochondria. In humans, a single mitochondrion contains 37 genes on a circular mitochondrial DNA, compared with about 35,000 genes contained in the nuclear DNA.

Mosaicism—Variation among cells, such that the cells are not all the same—for example, in an embryo when not all the cells are edited.

Multipotent stem cells—Stem cells from the embryo, fetus, or adult, whose progeny are of multiple differentiated cell types and usually, but not necessarily, all of a particular tissue, organ, or physiological system.

Murine—Derived from mice.

Mutation—A change in a DNA sequence. Mutations can occur spontaneously during cell division or can be triggered by environmental stresses, such as sunlight, radiation, and chemicals.

Nickase—A nuclease that cuts only one strand of the DNA double helix.

Non-homologous end joining (NHEJ)— A natural repair process used to join the two ends of a broken DNA strand back together. This is prone to errors where short indels (usually of two to four base pairs of DNA) are introduced.

Normative theory—A theory of how people should make decisions, as opposed to how they actually do or will make decisions.

Nuclease—An enzyme that can cut through DNA or RNA strands.

Off-target effect—A direct or indirect, unintended, short- or long-term consequence of an intervention on an organism other than the intended effect on that organism.

Off-target event (or off-target cleavage)—when a genome-editing nuclease cuts DNA at a location other than the one for which it was targeted. This can occur because the off-target sequence is similar to but not identical with the intended target sequence.

Oocyte—Developing egg; usually a large and immobile cell.

Phenotype—Observable properties of an organism that are influenced by both its genotype and its environment.

Plasmid—A self-replicating circular DNA molecule. A plasmid can be engineered to carry and express genes of interest in target cells.

多能干细胞 (PSC)—— 一种可以分化成植入后期胚胎、胎儿或发育个体中所有细胞类型的干细胞。

种群—— 在一定的生态空间内某个物种的所有个体。

临床前研究—— 对潜在的临床应用的研究，如对分子、细胞、组织或者动物。但不包括以人类为实验对象的研究。

前体细胞/祖细胞—— 在胎儿或成人的组织中，已部分定向但尚未彻底分化的细胞。它们可以分裂并产生分化的细胞。

胚胎植入前的基因诊断 (PGD)—— 体外受精胚胎可在植入女性子宫，可前筛查已知的会导致基因疾病的特殊基因突变或染色体异常。通过从植入前的胚胎中抽离出一个或多个细胞进行检测，可确认被植入的胚胎中没有出现这些遗传异常。

产前诊断—— 指当胎儿在子宫发育的时候，对胎儿是否异常或患病等方面进行检测。许多产前诊断技术都需要提供胎盘组织，或者对羊水、母胎循环系统中的胎儿细胞或组织进行取样，如绒毛膜绒毛取样和羊膜穿刺术。另有如超声波扫描术，则不需要细胞或组织取样便能完成产前诊断。

原条—— 细胞组成的细长条带。早在原肠胚形成时，原条沿着胚胎的枢椎生成，起始时侧面的细胞向枢椎移动，逐渐沿着中线发育成凹槽，同时细胞向胚胎内部移动形成中胚层。

前核—— 指精子或者卵母细胞的单倍体细胞核，也指受精前后，精子和卵子的细胞核未融合成一个双倍体细胞核时的单倍体细胞核。

蛋白质—— 由一条或多条氨基酸链所构成的复杂大分子。蛋白质在细胞行使着各种职责。

隐性性状—— 隐性等位基因的表达效果被双倍体细胞或生物体中的第二对等位基因的表达效果所掩盖，以至于子代中隐性等位基因表达的性状未能展现。而这个第二对等位基因就称为显性基因。

重组 DNA—— 重组 DNA 分子是由经过人工修饰或接合的 DNA 序列所组成的，因此重组的基因序列与天然存在的遗传物质有差异。

重组 DNA 顾问委员会 (RAC)—— 监管和评审美国国立卫生研究院（NIH）及其资助的研究所中涉及 DNA 的重组和合成课题，如基因疗法。

重组—— 天然或人工设计的使两段 DNA 经历了断裂和重接过程，从而产生新的重组 DNA 片段。

Pluripotent stem cell (PSC)—A stem cell that includes in its progeny all cell types that can be found in a postimplantation embryo, fetus, or developed organism.

Population—All of the individuals of a given species within a defined ecological area.

Preclinical research—Research conducted to investigate potential clinical applications but not involving humans. For example, research on molecules, cells, tissues, or animals.

Precursor cell or Progenitor cell—In fetal or adult tissues, it is a partially committed but not fully differentiated cell that divides and gives rise to differentiated cells.

Preimplantation genetic diagnosis (PGD)—Before an *in vitro*–fertilized embryo is implanted in a woman's uterus, it can be screened for specific genetic mutations that are known to cause particular genetic diseases or for chromosomal abnormalities. One or more cells are removed from the preimplantation embryo for testing and the surviving embryo is implanted is not carrying the genetic abnormality.

Prenatal diagnosis—Detection of abnormalities and disease conditions while a fetus is developing in the uterus. Many techniques for prenatal diagnosis, such as chorionic villus sampling and amniocentesis, require sampling placental tissue or fetal cells found in the amniotic fluid or fetomaternal circulation. Others, such as ultrasonography, can be performed without cell or tissue samples.

Primitive streak—An elongated band of cells that forms along the axis of an embryo early in gastrulation by the movement of lateral cells toward the axis and that develops a groove along its midline through which cells move to the interior of the embryo to form the mesoderm.

Pronucleus—The haploid nucleus of an oocyte or sperm, either prior to fertilization or immediately after fertilization, before the sperm and egg nuclei have fused into a single diploid nucleus.

Protein—A large complex molecule made up of one or more chains of amino acids. Proteins perform a wide variety of activities in the cell.

Recessive—A recessive allele of a gene is one whose effects are masked by the second allele present in a diploid cell or organism, which is referred to as dominant.

Recombinant DNA—An recombinant DNA molecule is made up of DNA sequences that have been artificially modified or joined together so that the new genetic sequence differs from naturally occurring genetic material.

Recombinant DNA Advisory Committee (RAC)—Oversees and reviews proposals for research funded by the National Institutes of Health (NIH) or similar projects conducted at institutions funded by NIH that involve recombinant or synthetic DNA, such as gene therapy.

Recombination—The process, natural or engineered, in which two pieces of DNA undergo breakage and reunion to generate a new combination of DNA segments.

再生医学——通过细胞或组织工程及人工植入的方法来寻找替代有缺陷的、损伤或缺失的组织的医疗手段，通常与干细胞相关。

限制性内切核酸酶——一种来自细菌的酶，在特定的序列上对 DNA 进行切割。常被用于 DNA 分析和连接不同 DNA 片段之前对 DNA 进行切割。

逆转录病毒——一种遗传物质由 RNA 组成的病毒，但在病毒复制的过程中，病毒的 RNA 会逆转录成 DNA 从而结合到细胞的 DNA 基因组里面。通常，逆转录病毒被用作基因载体，把基因导入到受体细胞中。慢病毒是其中的一种。

风险——由于一个或多个应激源所带来的一个效应影响一个或多个结果的可能性，这个效应可好可坏。

风险评估——通过收集、评估和解读一切可用的关于结果概率的根据来预计结果的总概率的过程。

核糖核酸 (RNA)——一个结构与 DNA 相似的化学分子。它的一个主要功能就是把 DNA 的基因编码翻译成结构蛋白。

核糖核蛋白复合物 (RNP)——通常指细胞中存在的多种类型的核糖核蛋白复合物。在基因编辑的语境中，核糖核蛋白复合物通常指一个向导 RNA 分子与 DNA 切割酶的结合体，如 Cas9。

选择优势——通过自然选择可以筛选出由基因的某些变异所引起的具生存或繁殖优势的性状，因此导致这种带有繁殖优势的基因在某个种群内盛行。

单链向导 RNA(sgRNA)——能与核酸酶结合的 RNA 短片段，如 Cas9。单链向导 RNA 可与特定 DNA 序列结合，引导核酸酶到基因组中的一个特定位置。这个术语在大多数用法上与上文中的向导 RNA 同义。

体细胞——植物或动物体中除生殖细胞或者生殖前体细胞之外的细胞。

体细胞核移植 (SCNT)——把体细胞的细胞核转移到去除了细胞核的卵子（卵母细胞）中。

精原干细胞——精子的可自我复制的前体细胞。

干细胞——一种未分化的细胞。它可以在培养基中无限分裂并分化成多种更成熟和分化细胞。

干细胞疗法——运用干细胞在再生医学中替代有缺陷的、损伤及缺失的组织。

合胞体滋养层细胞——来源于早期哺乳动物

Regenerative medicine Medical treatments that seek to replace defective, damaged, or missing tissue by engineered cells, tissues, or implants, often involving stem cells.

Restriction enzyme—An enzyme from bacteria that is used to cut DNA at defined sequences, used in DNA analysis and in joining DNA fragments through the cut ends.

Retrovirus—A virus whose genome is made of RNA but during viral replication becomes copied into a DNA form that can integrate into the DNA genome of a cell. Often used as carriers of genes (vectors) to introduce genes into cells. A subset of retroviruses is called lentiviruses.

Risk—The probability of an effect on a specific endpoint or a set of endpoints due to a specific set of a stressor or stressors. An effect can be beneficial or harmful.

Risk assessment—The process by which all available evidence on the probability of effects is collected, evaluated, and interpreted to estimate the probability of the sum total of effects.

RNA (ribonucleic acid)— A chemical that is similar in structure to DNA. One of its main functions is to translate the genetic code of DNA into structural proteins.

RNP (ribonuclear protein complex)—Many types exist within cells and this is a general term encompassing all of these, but in the context of genome editing it is often used to refer to a guide RNA molecule combined with a DNA-cutting enzyme such as Cas9.

Selective advantage—Some variants of genes provide a trait that confers a survival or a reproductive advantage that can be selected by natural selection and therefore increases in prevalence in a population.

sgRNA (single guide RNA)—A short piece of RNA that binds to a nuclease such as Cas9 and also to a specific DNA sequence to guide the nuclease to a specific location in the genome. This term is synonymous with guide RNA (vide infra) in most usages.

Somatic cell—Any cell of a plant or animal other than a reproductive cell or reproductive cell precursor. Latin: soma = body.

Somatic cell nuclear transfer (SCNT)—The transfer of a cell nucleus from a somatic cell into an egg (oocyte) whose nucleus has been removed.

Spermatogonial stem cells The self-replicating precursors of sperm cells.

Stem cell—A nonspecialized cell that has the capacity to divide indefinitely in culture and to differentiate into more mature cells with specialized functions.

Stem cell therapy—The use of stem cells in regenerative medicine to replace defective, damaged or missing tissue.

Syncytiotrophoblast cell—A cell derived from

胚胎中滋养外胚层细胞。它们相互融合（成多核合胞体）并有助于形成胎盘的结构和功能。

合成生物学——将独立的遗传部件组建合成活细胞，以及使用工程原理将所期望的功能装载进活生物体内的学科。

合成 DNA——通过化学或其他手段合成或扩增的 DNA 分子。虽然可能经过化学或其他方法修饰，合成 DNA 可以与天然 DNA 分子进行碱基配对或重组。

T 细胞——属于白细胞的一类，是免疫系统非常重要的组成。它们会配合其他免疫细胞去杀死被感染或癌变的细胞。然而当它们被激活时，亦可对抗生物体自身的细胞和组织，引起炎症或自身免疫疾病。

靶序列——基因组里特定的 DNA 碱基序列，是基因编辑工具的目标。例如，对 CRISPR/Cas9 方法而言，靶序列是一个 20 个核苷酸长的、设计为可被 gRNA 识别的序列（例如，gRNA 会包含与靶序列等长的互补序列）。

疗法 (或干预治疗)——疾病或残疾的治疗或预防。

组织培养——在体外，把细胞或组织碎片作为研究对象放到人工培养基中培养。

全能性细胞——一种具有无限增殖能力的干细胞。早期胚胎（在囊胚阶段之前）的全能性细胞具有分化成胚外组织、膜、胚胎和所有胚后期组织和器官的能力。

转录——从基因或其他 DNA 序列中制造出 RNA 模板。转录是基因表达的第一步。

转录激活因子样效应核酸酶 (TALEN)——工程限制酶的一类。它是由转录激活因子样效应子的 DNA 结合区域（结合到特定的 DNA 序列）与用作基因组编辑工具的 DNA 切割结构域（核酸酶）融合所产生的。转录激活因子样效应核酸酶、锌指核酸酶及最新的 CRISPR/Cas9 都是基因编辑工具。

转录因子——一种结合到基因调控区（增强子和启动子）去激活或抑制特定基因转录（或表达）的蛋白质。

转染——把实验 DNA 导入到细胞中的方法。

转基因——指被导入到细胞或生物体中的基因或遗传物质。转基因可以通过同源重组或使用同源定向修复的基因编辑方式，随意或定点整合到

trophectodermal cells from the early mammalian embryo that fuse (into multinucleate syncytia) and contribute to the structure and function of the placenta

Synthetic biology—The development of living cells from separate genetic components, using engineering principles to build desired functions into living organisms.

Synthetic DNA—DNA molecules that are chemically or by other means synthesized or amplified; they may be chemically or otherwise modified but can base pair or be recombined with, with naturally occurring DNA molecules.

T cells—Types of white blood cells that are of crucial importance in the immune system. They cooperate with other immune cells in killing infected or cancerous cells but can also participate in inflammation or in autoimmunity when they become activated against an organism's own cells or tissues.

Target sequence—Specific sequence of DNA bases within the genome that is the target of genome editing tools. For CRISPR/Cas9 methods this will be a 20 nucleotide sequence that the gRNAs are designed to recognize (i.e., they will contain a complementary sequence of the same length).

Therapy (or therapeutic intervention)—The treatment or prevention of disease or disability.

Tissue culture—The growth of cells or tissue segments *in vitro* in an artificial medium for experimental research.

Totipotent cell—A stem cell that has unlimited developmental capability. The totipotent cells of the very early embryo (an embryo prior to the blastocyst stage) have the capacity to differentiate into extraembryonic tissues, membranes, the embryo, and all postembryonic tissues and organs.

Transcription—Making an RNA copy from a gene or other DNA sequence. Transcription is the first step in gene expression.

Transcription Activator-Like Effector Nuclease (TALEN)—A class of engineered restriction enzymes generated by the fusion of a transcription activator-like effector DNA-binding domain (that binds to a specific DNA sequence) to a DNA-cleavage domain (nuclease) to be used as a genome editing tool. TALENs followed zinc finger nucleases and preceded CRISPR/Cas9 as genome-editing tools.

Transcription factor—A protein that binds to control regions (enhancers and promoters) of genes to activate or repress their transcription (or expression).

Transfection—A method by which experimental DNA may be introduced into a cell.

Transgene—A gene or genetic material that has been introduced into a cell or organism. Transgenes can be integrated at random, or targeted to a specific site by homologous recombination or by genome editing using

胞或生物体中。

转基因生物——带有一个或多个从其他物种中转移或人工导入的基因的生物体。

超人类主义——一种人生哲学。致力于通过科技寻找能够超越目前人类形态和人类极限的方法，从而延续和加速智慧生命的进化。它遵从于提升生活质量的理论和价值观（More，1990）。

翻译——基于信使 RNA（mRNA）的信息形成蛋白质分子的过程。这是在基因表达过程中转录（从 DNA 中复制出 RNA）后的一步。

滋养外胚层——发育中的囊胚的最外层。其最终会形成胎盘的胚胎侧。

未分化——还没发育形成特化的细胞或组织。

单能干细胞——可以分裂并分化成单一类型成熟细胞的干细胞，如只能分化成精子的精原干细胞。单能性干细胞也可称为祖细胞。

实用主义——产生最多"好处"的行动就是道德层面上最正确的行动。

变种——种群中具有许多变体的基因。基因变种会导致在功能上某种程度的差异。有一些是有利的，但是也有一些是有害的或是使其丧失功能的。

野生型——生物体或基因的"通常"型。

X 染色体失活——是指雌性哺乳类细胞中两条 X 染色体的其中之一失去活性的现象，因此，只有一条 X 染色体中的基因被表达。

载体——把目的基因转移到新位点的工具（类似于转移病毒或寄生虫到新的动物宿主中）。载体通常被用在分子细胞生物学和基因工程中，包括运用质粒和设计改造的病毒，把目的基因转运和表达到目标细胞中。临床上基因转移常用的病毒载体包括逆转录病毒载体、慢病毒载体、腺病毒载体和腺相关病毒载体。

美德伦理——指注重道德品格多于注重职责（义务论）或结果（效果论）。

锌指——与自然存在的转录因子共同作用的小型蛋白质，与特定的 DNA 序列结合后，可控制周围基因的转录和表达。锌指可用在基因工程中，被定制设计为与 DNA 序列的特定部位结合。

锌指核酸酶（ZFN）——锌指 DNA 结合结构域与 DNA 切割酶（通常是 FokI）结合所产生的一类工程酶。锌指核酸酶可被用作基因编辑的工具，这是最早的可靠的基因编辑的方法之一。

methods of homology-directed repair.

Transgenic organism—An organism into which one or more genes from another species (transgenes) have been transferred or otherwise artificially introduced.

Transhumanism—A class of philosophies of life that seek the continuation and acceleration of the evolution of intelligent life beyond its currently human form and human limitations by means of science and technology, guided by life-promoting principles and values (More, 1990).

Translation—The process of forming a protein molecule from information contained in a messenger RNA—a step in gene expression following transcription (copying of RNA from DNA).

Trophectoderm—The outer layer of the developing blastocyst that will ultimately form the embryonic side of the placenta.

Undifferentiated—Not having developed into a specialized cell or tissue type.

Unipotent stem cell—A stem cell that both divides and gives rise to a single mature cell type, such as a spermatogenic stem cell, which only gives rise to sperm. Alternatively called a progenitor.

Utilitarianism—The morally right action as the action that produces the most "good."

Variant—Genes have many variants in a population that can differ somewhat in function, some being advantageous and some being deleterious or nonfunctional.

Wild type (noun); Wild-type (adjective)—The "normal" type of an organism or a gene.

X-inactivation—The process in which one X chromosome of the two present in a female mammalian cell is inactivated so that only the genes of one X chromosome are expressed.

Vector—A vehicle that transfers a gene into a new site (analogous to insect vectors that transfer a virus or parasite into a new animal host). Vectors used in molecular cell biology and genetic engineering include plasmids and modified viruses engineered to carry and express genes of interest in target cells. The most clinically relevant viral vectors for gene transfer include retroviral, lentiviral, adenoviral, and adeno-associated viral vectors.

Virtue ethics—A focus on moral character as opposed to duties (deontology) or consequences (consequentialism).

Zinc finger—A small protein structure based on naturally occurring transcription factors that bind to defined DNA sequences to control the activity of nearby genes. Zinc fingers can be custom engineered to target a specific section of the DNA sequence for use in genome engineering.

Zinc finger nuclease (ZFN)—A class of engineered enzymes generated by the fusion of zinc finger DNA-binding domains to a DNA-cleavage enzyme (usually FokI) that can be used as a genome editing tool. One of the first and a reliable method of genome editing.

受精卵——通过精子和卵子在受精时融合所形成的单细胞胚胎。

Zygote—The one-cell embryo formed by the union of sperm and egg at fertilization.

参 考 文 献

Grossinger, R. 2000. *Embryogenesis: Species, gender, and identity*. Berkeley, CA: North Atlantic Books.

HFEA (Human Fertilisation and Embryology Authority). 2013. *What we do*. http://www.hfea. gov.uk/133.html (accessed May 10, 2017).

IOM (Institute of Medicine). 2005. *Guidelines for human embryonic stem cell research*, Vol. 23. Washington, DC: The National Academies Press.

IOM. 2014. *Oversight and review of clinical gene transfer protocols: Assessing the role of the recombinant DNA advisory committee*. Washington, DC: The National Academies Press.

MGI (Mouse Genome Informatics). 2017. *MGI glossary*, s.v. "loss of function." http://www. informatics.jax.org/glossary/loss-of-function (accessed May 10, 2017).

More, M. 1990. "Religion, Eupraxophy, and Transhumanism"in *Transhumanism: Towards a futurist philosophy*. https://www.scribd.com/doc/257580713/Transhumanism-Towarda-Futurist-Philosophy (accessed January 25, 2017).

NASEM (National Academies of Sciences, Engineering, and Medicine). 2016a. *Gene drives on the horizon: Advancing science, navigating uncertainty, and aligning research with public values*. Washington, DC: The National Academies Press.

NASEM. 2016b. *Mitochondrial replacement techniques: Ethical, social, and policy considerations*. Washington, DC: The National Academies Press.

NRC (National Research Council). 2002. *Scientific and medical aspects of human reproductive cloning*. Washington, DC: National Academy Press.

NRC and IOM (Institute of Medicine). 2015. *Potential risks and benefits of gain-of-function research: Summary of a workshop*. Washington, DC: The National Academies Press.